studies in physical and theoretical chemistry 14

RADIATION CHEMISTRY
OF
HYDROCARBONS

studies in physical and theoretical chemistry

Other titles in this series

studies in physical and theoretical chemistry 14

RADIATION CHEMISTRY OF HYDROCARBONS

Edited by

G. Földiák

Institute of Isotopes
of the Hungarian Academy of Sciences
Budapest, Hungary

ELSEVIER SCIENTIFIC PUBLISHING COMPANY
Amsterdam—Oxford—New York   1981

Translated by

Zoltán Paál (Chapters 1, 2 and 5), István György (Chapters 3 and 6),
László Guczi (Chapter 4)

*The distribution of this book is being handled by the following publishers
for the U.S.A. and Canada*

Elsevier/North-Holland, Inc.
52 Vanderbilt Avenue
New York, New York 10017, U.S.A.

*for the East European countries, China, Korean People's Republic,
Cuba, People's Republic of Vietnam and Mongolia*

Akadémiai Kiadó, The Publishing House of the
Hungarian Academy of Sciences, Budapest

for all remaining areas

Elsevier Scientific Publishing Company
1 Molenwerf
P. O. Box 211, 1014 AG Amsterdam, The Netherlands

Library of Congress Cataloging in Publication Data
Main entry under title:

Radiation chemistry of hydrocarbons.

(Studies in physical and theoretical chemistry; 14)
Bibliography: p.
Includes index.
9. Hydrocarbons.   2. Radiation chemistry.
I. Földiák, G., 1929.   II. Series.
QD651.H93R33   547′.010458   80-27145

ISBN 0-444-99746-6 (Vol. 14)
ISBN 0-444-41699-4 (Series)

Joint edition published by

Elsevier Scientific Publishing Company, Amsterdam, The Netherlands and
Akadémiai Kiadó, The Publishing House of the Hungarian Academy of
Sciences, Budapest, Hungary

Printed in Hungary

LIST OF CONTRIBUTORS TO THIS VOLUME

GY. CSERÉP

Institute of Isotopes of the Hungarian Academy of Sciences
Budapest, Hungary

I. GYÖRGY

Institute of Isotopes of the Hungarian Academy of Sciences
Budapest, Hungary

M. RODER

Institute of Isotopes of the Hungarian Academy of Sciences
Budapest, Hungary

L. WOJNÁROVITS

Institute of Isotopes of the Hungarian Academy of Sciences
Budapest, Hungary

CONTENTS

PREFACE

Ever since the pioneering studies of S. C. Lind on the α-particle radiolysis of gaseous hydrocarbons in the mid 1920's, the radiation chemistry of hydrocarbons has been studied in many laboratories throughout the world. This research became the subject of considerable interest during the early 1940's when the development of reactor technology required basic information on the stability of lubricants and other reactor components in the intense radiation fields to which they were expected to be subjected. At that time the sole published report on the radiolysis of liquid hydrocarbons, the early work (1931) of Schoepfle and Fellows on gas evolution under electron bombardment, became of considerable importance. The status of knowledge in the field as of the end of World War II was quite well summarized at a "Symposium on Radiation Chemistry and Photochemistry" chaired by Lind and held at Notre Dame in the summer of 1947. The reader is referred to *The Journal of Physical and Colloid Chemistry* **52**, 437–611 (1948) for a published account of this symposium.

Interest increased rapidly during the 1950's as the result of a variety of factors. Of prime importance was, of course, the development of electron accelerators and γ-ray sources capable of producing experimental radiation fields of sufficient penetration and intensity to enable many different types of studies to be carried out readily. During this period interest in the United States was substantially catalyzed by the Gordon Conference on Radiation Chemistry which provided an excellent forum for informal discussions among the various scientists involved. The first of these conferences was organized in 1953 by Professor M. Burton and held at the Gordon Conference site at New Hampton, New Hampshire. Except for 1973 and 1977 these conferences have been held annually and have ensured excellent communications between investigators. Attendance has been international with most of the major radiation chemical laboratories of the world having been included in the lists of speakers and discussants. Since 1959 a similar function has been served in Europe by the biennial Miller Conferences, named in honor of Nicholas Miller, an important contributor in the 1940's and 1950's and much esteemed by those that knew him, and also by the four Tihany meetings held in Hungary. The radiation chemist has, indeed, been very fortunate in having these excellent communication channels open to him and for the general spirit of collegiality purveying the relationships which he has been able to establish with his fellow scientists, a spirit which owes much to the informal atmosphere in which these conferences have been conducted. As far as hydrocarbons are concerned, being such a central subject, their radiation chemistry has, of course, been touched on in essentially all of these conferences.

The 1950's saw exploration by the reactor industry into possible development of organic-moderated reactor systems and also a substantial but rather misguided interest of the petroleum industry in the application of radiation processing of basic petroleum products. Various industrial laboratories established significant research efforts on radiation chemical studies on hydrocarbons during this period. These efforts have now been all but abandoned and industrial interest has been largely transferred to the far more reasonable studies of radiation effects on polymers.

The late 1950's and early 1960's saw two very important experimental breakthroughs: the development of electron spin resonance and optical spectroscopic methods whereby one could directly observe radical intermediates even in liquids, and the development of very sensitive vapor phase chromatography, which allowed accurate analysis of a wide variety of products even at low dose levels. Prior researches were limited for the most part to the studies of radicals by scavenger methods, or by the analytical necessity to focus measurements on readily measurable gaseous products such as hydrogen or low molecular weight hydrocarbons. The introduction of the spectroscopic methods has provided completely conclusive evidence that much of the effect of ionizing radiation on hydrocarbons takes place *via* radical intermediates and has given us rather detailed knowledge on the structure and reaction kinetics of at least the major radicals involved in the radiolysis of the simpler hydrocarbons. Information from product analysis approaches has evolved more slowly even though gas chromatography has given us a powerful tool by which to examine overall effects. During recent years the authors of the present volume have been the principal investigators carrying out studies along this latter line. High resolution liquid chromatography, which offers considerable potential for determining the yields of oxidation and high molecular weight products otherwise difficult to examine by gas chromatography, has yet to be applied to studies of hydrocarbons.

In 1954 Hamill and Williams attributed the production of methyl radicals in the radiolysis of cyclohexane solutions of methyl iodide to dissociative electron capture by the methyl iodide. However, distinguishing between the reactions of electrons and hydrogen atoms was difficult and conclusive proof that solutes could react with electrons and interfere with the ion recombination process in hydrocarbons awaited the 1964 study of Scholes and Simic who employed nitrous oxide as an electron scavenger. This study, together with the investigations of Williams in 1964 and of Scala, Lias and Ausloos in 1966 where it was shown, respectively, that ammonia and cyclopropane reacted with the positive ions produced in hydrocarbon radiolysis, catalyzed very intensive investigations of ion scavenging processes. The development of appropriate kinetic models to describe the competition between ion scavenging and ion recombination ensued. This work carried through the early 1970's and today we know quite well, at least in many hydrocarbons, how to treat the kinetic aspects of the radiation chemistry of solutions containing ion scavengers. Although ion recombination occurs on a very short time scale, it has become possible in recent years to make direct time-resolved measurements on ionic intermediates and these direct measurements have for the most part confirmed conclusions from the scavenging studies. The various physical measurements which have been carried out on the yields and mobilities of ions and the results of these studies have importance not only to radiation chemistry but also to discussions of many physical properties of dielectrics.

During the last five years or so, investigations on the radiation chemistry of hydrocarbons have been relatively dormant. It is not that we have anywhere near complete knowledge, but rather that many of the basic experiments which can produce information relatively easy to interpret have already been carried out. We are proceeding into an era where one must develop new and more sophisticated experiments and interpretive approaches designed to provide understanding of the details of fundamental processes involved. Studies, for example, of structural effects on the radiation chemical processes, such as those being carried out by Földiák's team in Hungary, focus on localization of energy in the individual bonds as manifest in the distribution of products. Experimental results are, however, admittedly not always easy or straight to interpret.

Theoretical investigations relevant to hydrocarbon radiolysis carried out to date largely involve considerations on the nature and mobility of the electron and ions produced by radiation. Yet we still know relatively little even on these subjects. For example, the kinetic models for scavenging electrons have been developed largely from a viewpoint which treats the electron classically as a reacting particle. Whether or not this is justified and to what extent a more proper approach requires quantum mechanical treatment has only been discussed in the most qualitative terms. Recent observations that the rate constant for ion scavenging can decrease with temperature imply inadequacies to a completely classical view. Although we have known since the studies of Manion and Burton in the early 1950's that hydrocarbon mixtures do not necessarily behave additively, we know very little from a theoretical viewpoint about the way in which is localized in a particular molecule of a mixture. In comparing theory and experiment one must, of course, extract from the latter effects that are intrinsic and not merely secondary complications, and this is not always easy. Even within a single molecule, although we know that certain types of bonds are selectively ruptured, we do not know, why. It is clear that bond rupture is not simply a matter of energetics, e.g. in a linear hydrocarbon scission of the carbon-carbon backbone is far less frequent than that of the energetically stronger C–H bonds, but the principal features controlling the reactions are still largely a matter of speculation. It should be pointed out that in liquid hydrocarbons relatively simple processes seem to predominate so that theoretical work aimed at describing the very complex fragmentation patterns usually observed in mass spectrometry is certainly not directly applicable. All in all, from the viewpoint of theory based on first principles we can only regard our current understanding of hydrocarbon radiolysis as being very primitive.

I might comment here that while some aspects of hydrocarbon radiolysis are reasonably well defined, at least experimentally, there are still vast areas that have been touched on only sporadically and superficially. For example while attention has, to date, been focused on processes involving free radicals as intermediates, we know relatively little about the importance of non-radical processes. I expect that, among other approaches, experiments using lasers either by themselves or in conjunction with electron beam irradiation may well be applied to correct this situation in the future. Another area that has received only limited attention is that involving radiolysis with heavy particles, even though it is known that with heavy particles qualitatively different effects are superimposed on those found with low LET radiations. As far as transfer of information from one system to the next is concerned we are only now coming to realize, particularly as a result of detailed studies of the ionic processes, that rather subtle differences in structure or form can often result in rather profound differences in radiation chemical behavior. Understanding this in detail will require a substantial experimental effort along with appropriate theoretical backup. These are, of course, only a few of the areas that need to be opened up better.

How does the present volume fit into all of this? Basically it provides us with a comprehensive summary of the literature on the radiolysis of hydrocarbons and with an extensive abstract of data contained in that literature in the form of numerous tables (147) and figures (67). Citations to the literature exceed 1000 and are well referenced in the text so that this summary makes the literature pertinent to a particular topic readily available. The inclusion of the large number of tables and figures is helpful in many cases in providing a qualitative idea on the type of information available and on the level to which a particular subject has been developed. The material is organized very much along the lines that the subjects have been developed experimentally. After introductory material (Chapter 1), saturated hydrocarbons (Chapters 2 and 3), then olefinic (Chapters 4 and 5) and finally aromatic systems (Chapter 6) are treated. Because of the smaller number of bond types, radiation chemists have devoted considerable attention to studies of the radiolysis of cycloalkanes and the authors have treated this subject accordingly in Chapter 3.

This summary of the radiation chemistry of hydrocarbons is the first extensive treatment on the subject to appear since the book by Topchiev in the early 1960's. Clearly much has gone on in the meantime. I know it is the hope of the authors that by providing more ready access to the literature and by focusing attention on certain aspects of what has been accomplished and where future problems may be that they can catalyze further activity in this area of research. This writer, while not necessarily agreeing with all of the statements or interpretations in the text, applauds the rather considerable effort to which the authors have gone to summarize quite well the present status of the subject and concurs with them in their hope that this volume will help in revitalizing the interest with which future researches will be undertaken. All in all, there is still much to be done and the young and able investigator is encouraged to participate.

Robert H. Schuler
Notre Dame, Indiana, U.S.A.

INTRODUCTION

This book is intended to give a comprehensive account of the radiation chemistry of hydrocarbons. The basic principle of treatment will be to seek relations between molecular structure and radiation chemical reactivity.

Although we intend to emphasize first of all intrinsic connections between individual observations, the most important experimental results will also be given in tabulated form. In our efforts to achieve completeness of coverage, results that may not be entirely convincing have been included, with due criticism.

Although this book has been written for those who are actually active in the field of radiation chemistry; it should provide, however, useful information and data for anybody interested in hydrocarbon or petroleum chemistry, especially physicochemical or thermochemical aspects. It should be pointed out that high-temperature pyrolysis, being of great theoretical and practical interest, has many features in common with radiolysis.

Polymerization of hydrocarbons and decomposition of hydrocarbon polymers do not come within the scope of this volume. These topics are only briefly mentioned. The existing excellent monographs (such as those listed in Chapter 4, refs [6–10]) justify this approach.

Most of the conclusions have been drawn on the basis of room temperature data obtained from product yields extrapolated to zero dose or, where zero dose yields were not available, from yields obtained at low doses, usually a few times 10^{19} eV g^{-1}. Any variation in these standard conditions will be noted.

This book is based mainly on the literature published between 1965 and 1977; in addition, work published after the manuscript had been completed but before proofs were corrected has wherever possible, been included. As a result, there are slight inconsistencies in the numbering of references.

The reader holds in his/her hands the result of a collective effort. Every chapter has its own author in the narrower sense of the word: Chapter 1-I. György and L. Wojnárovits; Chapter 2-I. György; Chapter 3-L. Wojnárovits; Chapters 4 and 5- Gy. Cserép; Chapter 6- M. Roder; nevertheless, not only the Editor, but also every author, did his/her best to contribute to the work as a whole.

The authors wish to express their gratitude to all those who kindly permitted their figures or tables to be reproduced here.

1. BASIC PROCESSES IN HYDROCARBON RADIOLYSIS, AND METHODS FOR INVESTIGATING THEM

This chapter gives a brief summary of the most important physical, physicochemical and chemical processes of radiolysis. In Section 1.5 some experimental methods of radiation chemistry and of related fields are reviewed, taking into account the interests of readers who are not familiar with the subject. The selection of the material might appear somewhat arbitrary to experienced radiation chemists. However, it reflects the aim of the authors to concentrate their efforts on the interpretation of the results of radiation chemistry for the benefit of workers in other fields of chemistry who may be interested in gleaning new information on the correlations between molecular structure and reactivity.

1.1. ABSORPTION OF RADIATION ENERGY

When high-energy (as a rule, 10^3–10^7 eV) radiation passes through matter energy is transferred to the molecules resulting in the generation of electronically excited and ionized species. 'Ionizing radiation' is the common name for all such radiation, including both corpuscular (accelerated heavy ions, α-particles, electrons) and electromagnetic (X-rays, γ-rays) radiations. Deeply penetrating electron and electromagnetic radiations are generally most suitable and therefore most frequently used.

Electrons may interact either with the Coulomb field of the nucleus of an atom or with the electron shell in elastic or non-elastic interactions. When the energetic electrons interact with the electrons of the absorbing material in non-elastic interaction they are slowed down, and ionization or excitation of the absorbing material results. In the course of elastic scattering the direction of motion changes without conversion of kinetic energy into any other form of energy. The relative frequencies of the different types of interaction are determined by the energy of the electrons and the nature of the absorbing material.

Electromagnetic radiation may bring about ionization by the following three types of interaction: the photoelectric effect, ionization by the Compton effect and electron–positron pair formation, which generates high-energy particles that may cause ionization.

The photoelectric effect is characteristic of photons of lower energy; it involves the transfer of the total energy of radiation to one of the electrons of the electron shell. Compton scattering involves the transfer of only a part of the energy of the electromagnetic radiation to the electron; the latter leaves the parent atom and, as a so-called δ- or secondary electron, brings about further ionization and excitation in a cascade process. Pair formation requires 1.02 MeV as a minimum energy; a photon possessing such energy and passing through the force field of a nucleus can be transformed into an electron and a positron, the annihilation of the latter producing a further photon.

The most usual type of interaction observed with ^{60}Co γ-radiation (1.12 and 1.33 MeV), which is the source most frequently used in radiation chemistry, is Compton scattering [1–5].

The primary species formed in radiolysis are excited neutral molecules and ions formed by direct ionization. The literature on radiation chemistry suggests that several types of excited species are produced by the interaction of radiation with matter, including optically allowed as well as forbidden singlet and triplet states.

In 1962 Platzman pointed out that the greater part of the oscillator strength of simple molecules lies in a region above the lowest ionization potential [6]. Within this region electronically excited molecules are formed, the energy content of which is higher than the energy corresponding to the ionization threshold. For these states Platzman introduced the name 'superexcited states', as suggested to him by Hurst. The role played in decomposition processes by these states has been verified by many authors [7–14].

Makarov and Polak proposed the following possibilities for the primary events of hydrocarbon radiolysis in the gas phase [15]:

$$RH \xrightarrow{\wedge\wedge} RH^{**} \qquad \text{superexcitation} \qquad (10^{-16}\,s)$$

$$RH \xrightarrow{\wedge\wedge} RH^{*} \qquad \text{excitation} \qquad (10^{-16}\,s)$$

$$RH \xrightarrow{\wedge\wedge} RH^{+} + e^{-} \qquad \text{direct ionization} \qquad (10^{-16}\,s)$$

$$RH^{**} \xrightarrow{} RH^{+} + e^{-} \qquad \text{autoionization} \qquad (\geqslant 10^{-14}\text{--}10^{-12}\,s)$$

Besides autoionization, superexcited molecules may also undergo dissociation or internal conversion into lower excited states. It is known from photolytic studies [16] in the vacuum ultraviolet region that the quantum yield of ionization rises above the ionization potential and at about 16 eV approaches unity for alkanes, indicating that the majority of superexcited states undergo autoionization in the gas phase.

Table 1.1 shows mean values for the energy (W) consumed in the formation of a positive ion-electron pair in gas-phase irradiation of various hydrocarbons, and their first photoionization energies (E_i). The value of W is between 22 and 27 eV, that of E_i between 8 and 13 eV; both values decrease somewhat as the number of carbon atoms in the molecule increases. Ionization potentials of aromatics, unsaturated compounds and branched alkanes are lower than those of n-alkanes [19–25].

The photoionization potentials are often referred as 'adiabatic' ionization potentials, while the other most commonly used ionization potential values determined by the electron impact method are often called 'vertical' ionization potentials.

Table 1.1

Energy values of ion-pair formation (W)
and photoionization (E_i). Data taken from refs [17–22]

Hydrocarbon	W, eV	E_i, eV
Methane	26.92	12.8–12.99
Ethane	24.18	11.49, 11.65
Propane	23.69	11.07
n-Butane	23.16	10.50, 10.63
n-Pentane	22.80	10.33, 10.35
n-Hexane	22.61	10.17, 10.18
n-Heptane	22.45	10.06–10.20
2-Methylpropane	23.50	10.55–10.78
2-Methylbutane	23.07	10.30, 10.32
2-Methylpentane	22.78	10.09
3-Methylpentane	22.67	10.06
2,3-Dimethylbutane	22.80	10.00, 10.02
2,2-Dimethylpropane	23.40	10.35, 10.37
2,2-Dimethylbutane	22.88	10.04
Cyclopropane	23.7	10.06
Cyclobutane		10.56
Cyclopentane		10.53
Cyclohexane	25.05	9.88
Ethylene	25.2, 24.6	10.52
Propene		9.73
But-1-ene		9.61, 9.58
Pent-1-ene		9.50
Hex-1-ene		9.46
But-2-ene		9.13
Buta-1,3-diene		9.07
cis-Penta-1,3-diene		8.59
Cyclopentene		9.01
Cyclohexene		8.95
Benzene	21.4, 20.9	9.25
Toluene		8.81

During the 'vertical' ionization the transition follows the Franck—Condon rule, which requires that the internuclear distances do not change. In contrast the 'adiabatic' ionization potential is the energy required to remove an electron from a molecule in its lowest vibrational level, i.e. transition from the $v = 0$ ground state of the neutral molecule to the $v = 0$ ground state of the positive ion (Fig. 1.1). The electron impact values are in most cases about 0.2 to 0.3 eV greater than the corresponding photoionization values [21]. Recently, numerous papers have published or interpreted ionization potential values obtained by means of the so-called photoelectron [24, 25] or electron energy loss [14, 26] spectroscopic methods. These methods provide principally 'vertical' values.

The difference of 12–14 eV between the values of W and E_i is due to the fact that a considerable part of the energy absorbed during radiolysis is used for excitation.

There are significant differences between the early stages of radiation action on gaseous and on condensed molecular media [13]. Experimental

and theoretical information on characteristic energy losses of fast electrons
in solids and liquids indicates that, to a great extent, the primary excitations involve plasmon states, i.e. collective oscillations of valence electrons
in dense material. The plasmon states decay very rapidly (10^{-16} s) forming
so-called exciton states with high excitation energies (10–30 eV). The upper
exciton states decay with an estimated rate of the order of $10^{14}\,\mathrm{s}^{-1}$ by internal conversion, autoionization or fission in biexciton states. In 'exciton

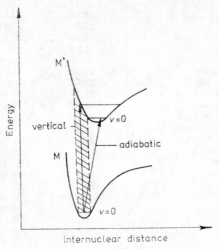

Fig. 1.1. Potential energy diagram of a diatomic molecule

fission' one high singlet excitation energy quantum is converted into a pair
of lower electronic energy quanta located in two different molecules, the
excitation state of which can be either triplet or singlet [13, 27–29].

Several suggestions have been put forward that ion yields are independent
of the state of aggregation, i.e. more or less the same in the condensed phases
as in the gas phase. Other results, mainly the most recent ones, indicate that
ion yields in the liquid phase differ considerably from those observed in the
gas phase [30–34]. Whereas in the gas phase the G-value of ion pairs [G(ion
pair) = number of produced ion pairs/100 eV = $100/W$] is about 4, in the
liquid phase it seems to be closer to 5 than to 4. It should be noted, however,
that these high ion yields were established by means of the scavenger method and, as discussed in Section 1.4.1, there are many possible reactions
between the intermediates of radiolysis and the scavenger; the side-reactions
can interfere with the charge-scavenging effect and contribute to the high
G-values obtained. The charge scavengers can also interact with radicals
and neutral excited molecules.

According to Onsager [35], the probability [$\Phi(r)$] of the electron escaping
from the positive ion is determined by the dielectric constant of the medium:

$$\Phi(r) = \exp\left(-e^2/\varepsilon\, rkT\right)$$

where e is the elementary charge, ε the dielectric constant, and r the distance between ions and electrons. The critical distance, $r_c = 2\,e^2/\varepsilon\,kT$, where the energy of coulombic interaction between electron and positive ion is equal to kT, is **25–32 nm** in hydrocarbons.

If $r > r_c$, the fate of the electron will be determined mainly by thermal diffusion; whereas if $r < r_c$, electric interactions are predominant.

The electrons and their geminate partners that escape from each others' coulombic force field and become homogeneously distributed throughout the solution are often called free electrons and free ions, respectively, while the electrons and positive ions that are unable to separate and consequently undergo a rapid neutralization are usually referred as geminate electrons and ions, respectively.

Several authors have dealt with the interpretation of the free ion yield and with its semiempirical determination. The models differ from each other mainly in the formulation of ion-electron separation. Thus, e.g. Freeman et al. [36–39] and Abell and Funabashi [40] derived various distribution functions for initial internal separation. If these functions are known, the free ion yield can be derived from the following equations:

$$G_{fi} = G_{tot} \int_0^\infty F(r)\,\Phi(r)\,dr$$

Assuming a simple Gaussian distribution:

$$F(r) = (4r^2/\pi^{1/2}b_g^3)\exp(-r^2/b_g^2)$$

Exponential [40] and combined Gaussian and power functions [38, 39] have also been proposed for describing the separation of ion-electron pairs:

$$F(r) = (1/b)\exp(-r/b)$$

or

$$F(r) = (4r^2/\pi^{1/2}b_{gp}^3)\exp(-r^2/b_{gp}^2) + 0.5(b_{gp}^2/r^3)$$

if

$$r > 2.4\,b_{gp}$$

where the constants b_g, b and b_{gp} show the most probable range of charge separation corresponding to the given distribution functions. Thus, for example the Gaussian function proved to be a good approximation for several alkanes if $G_{fi} > 0.2$ [37].

Table 1.2 contains values of free ion yields in the liquid phase, determined by a special pulse radiolysis method called 'clearing field' [41]. The value is relatively low for unsaturated and aromatic hydrocarbons ($G_{fi} = 0.05$–0.15 ion/100 eV) and somewhat higher for branched chain alkanes [41–43] (see also Section 2.1). The table also contains data on values of the dielectric constant (ε), the critical distance (r_c) and the most probable range corresponding to the Gaussian distribution (b_g), too. These latter have a dependence on chemical structure similar to the free ion yields. The values corrected by the density (ϱ) change little from one compound to another; $b_g\varrho$ values for a given hydrocarbon are nearly constant, independent of the density.

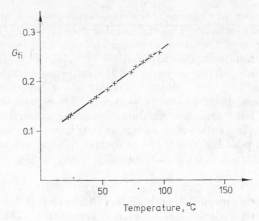

Fig. 1.2. Free ion yield from n-hexane as a function of temperature. After Schmidt and Allen [42]

Table 1.2

Free ion yields from pure liquids at 25°C
After Schmidt and Allen [41]

Compound	G_{fi}, ion pair/ 100 eV	ε	r_c, nm	b_g, nm	ϱ g cm^3	$b_g \varrho$ g cm$^{-2} \cdot 10^8$
n-Pentane	0.145	1.842	30.6	7.2	0.623	44.6
Neopentane	0.857	1.777	31.8	17.8	0.588	104.9
Cyclopentane	0.155	1.960	28.8	6.9	0.742	51.1
Hex-1-ene	0.062	2.046	27.6	4.9	0.670	32.9
Cyclohexane	0.148	2.022	27.9	6.6	0.776	51.3
Cyclohexene	0.150	2.222	25.4	6.0	0.806	48.5
Benzene	0.053	2.278	24.8	4.2	0.876	36.9
2,2,4-Trimethylpentane	0.332	1.936	29.1	9.5	0.689	65.4

It follows from the above-mentioned facts that free ion yields decrease with increasing pressure in gases [44–46], and increase with increasing temperature in gases and liquids (Fig. 1.2) [37–43, 47–50]. Thus, free ion yields are considerably influenced by the experimental conditions.

Based on photolytically induced energy transfer studies it was deduced that cyclohexane molecules have a rather large (1–2 nm) interaction radius [51–53]. Such an excited molecule, in which the electron is visualized as being smeared out over several molecules, may not be essentially different from the so-called geminate ion pair in liquid alkanes, as the energy levels of excited molecules are probably blurred and overlap with the beginning of ionized states [52].

On the other hand, though the most probable thermalization distance in liquids is about 5–15 nm, there is some degree of probability that electrons thermalize at an interpair distance of 1–2 nm. Geminate ions with so short internal separation may show similar behaviour (for instance towards additives) to that of the excited molecules.

1.2. THE CHEMICAL STEPS OF RADIOLYSIS

1.2.1. Reactions of primary species

As pointed out in the previous section, several types of excited and ionized molecules are produced during energy absorption.Their decay involves both competing and consecutive processes. From the radiation chemical point of view, one of the most important characteristics of different types of radiation is the linear energy transfer (LET) value, which is equal to the mean rate of energy loss along the path (generally keV μm^{-1}) of the ionizing particle (see Section 1.3.2).

Along the track of the primary ionizing species, in the so-called 'spur', the concentration of very reactive transient species is higher by about 4–8 orders of magnitude than in the bulk (about 1 against 10^{-8}–10^{-4} M, respectively) [36]. The primary activated species usually decay very rapidly. However, it is sometimes assumed (see refs [54] and [55] for reviews) that the yields of reactions taking place between primary activated molecules might be significant, if their concentrations are great enough, i.e. if radiation with high LET value is used. Examples of such reactions are

$$^3RH + {}^3RH \longrightarrow {}^1RH + {}^0RH$$

$$^3RH + {}^3RH \longrightarrow X + Y$$

where 0, 1 and 3 indicate ground, singlet and triplet states of the RH alkane molecule, respectively, and X and Y are alkyl radicals.

Electronically excited molecules and ions produced by ionizing radiation can lose their excess energy by light emission, by deactivation without decomposition (intersystem crossing) and by decomposition to ionic, radical and molecular fragments. Primary species may also transfer excitation energy or charge to surrounding molecules by various processes.

Hirayama et al. [56–58] have found the quantum yield of luminescence of alkane and monoalkene samples irradiated by photons with a wavelength of 147, 165 or 184.9 nm to be rather low ($\Phi \approx 10^{-5}$–10^{-2}). The luminescence yields and the shape of the spectrum are strongly influenced by the chemical structure [57]. The fluorescence quantum yields of irradiated aromatic compounds are much larger [59–61].

Early research revealed that radiation energy transferred to a given molecule does not necessarily cause the decomposition of this particular molecule, but there is a possibility, especially in condensed phases, that the energy or charge will be transferred within the system, from molecule to molecule. Charge transfer is generally assumed [62–64] to occur from molecules with higher ionization potential to ones with lower ionization potential. Little is known as yet about excitation transfer, but it can be stated that very complicated correlations do exist, and the influence of state of aggregation, composition, molecular structure and excitation levels should be considered.

One of the central problems of radiation chemistry is the elucidation of the decomposition mechanisms, and a good deal of relevant information has accumulated during the past few years.

A fraction of ions suffer bond rupture prior to neutralization. It has been observed that significant ionic decomposition can occur generally in low-pressure gases (where collisional deactivation is not important) [65] and specifically in the liquid-phase radiolysis of branched chain alkanes [66, 67] with low bond dissociation energies. An inspection of the mass spectrometric fragmentation patterns of various branched chain alkanes shows that the relative intensity of the parent ion is very low (or practically zero, e.g. for neopentane), as compared with the high relative intensities of some fragment ions [68].

The relative rates of competitive unimolecular decompositions are determined by differences in the energies and entropies of activation. This latter is related to the rearrangement of the energized ion needed for the decomposition to take place. For example, it has been shown that, with increasing internal energy of the ion, the role of so-called multicentre decompositions (low or negative entropy of activation) leading to a molecular ion and a neutral molecule decreases and, at the same time, that of the so-called direct decompositions (high entropy of activation) resulting in a free radical and a radical-ion increases. Among other things this fact also indicates that in radiolytic processes where the internal energy of primary ionic species may extend over a very wide range, several competing ionic decomposition reactions should be taken into consideration [69]. Although the importance of collisional deactivation is considerable in the gas phase at the pressures usually applied, as well as in the liquid phase, there are some indications of the occurrence of the consecutive ionic decomposition series well-known in mass spectrometry [69, 70], such as:

$$C_3H_8^+ \xrightarrow{-H} C_3H_7^+ \xrightarrow{-H_2} C_3H_5^+$$

$$C_2H_6^+ \xrightarrow{-H_2} C_2H_4^+ \xrightarrow{-H_2} C_2H_2^+$$

The thermodynamically most favourable reaction for some molecule ions is skeletal isomerization [70–72] (e.g.: n-$C_4H_{10}^+ \longrightarrow$ iso-$C_4H_{10}^+$).

Several qualitative and quantitative hypotheses have been developed for estimating the rate constants of various decomposition reactions of positive ions in the ground or vibrationally excited states, based mainly on developing mass-spectrometric methods [65, 73–76]. These allow one to interpret and, in many cases, even to calculate the mass spectrometric fragmentation pattern of relatively simple compounds. Of the various theories the so-called 'quasi-equilibrium theory' developed by Rosenstock, Wallenstein, Wahrhafting and Eyring has been most extensively applied

[65, 74]. In this theory the ion is treated as an isolated system in a state of internal equilibrium. It is assumed that the dissociation is so slow that the energy in the reactant ion has time to randomize over all internal degrees of freedom of the ion; the rate theory of Eyring applies to such an isolated molecular ion.

For the radiation chemist interested in speculations about the detailed mechanism of radiolysis, the recent work of Lossing and Maccoll [76A] presents a number of data on the ionization potentials of alkyl radicals and the heats of formation and hydride ion affinities of radical ions. The spectra of some C_6-C_{10} alkyl radical ions were published by Mehnert et al. [76B].

Little is known of the electronic structure of neutral, electronically excited molecules and the potential energy surfaces connected with their fragmentation processes. Therefore few correlations between decomposition reactions and the excited states involved in them have been derived. It is possible, however, to interpret fairly well the decomposition reactions of methane by using very approximate quantum mechanical calculations based on the structure of electronically excited states of low energy [77–79].

In connection with the decomposition pattern of electronically excited larger alkane molecules, the authors often refer to the work of Raymonda and Simpson, who measured and interpreted the absorption spectra of a large number of alkane molecules in the vacuum-UV region [80]. Partridge developed the theory further and used it to interpret the radiation chemical results obtained with alkane polymers [81]. According to this theory the excitations of the electrons in the C—C and C—H bonds occur separately ('exciton': see ref. [81] for details). If a certain C—H bond is excited, the excitation will be transferred predominantly between this bond and the other C—H bond attached to the same carbon atom, and only occasionally will it jump to any other C—H bond. The excitations in the C—C bonds, on the other hand, can be passed rapidly from one C—C bond to another along the carbon chain. The localization of C—H bond excitation generally leads to scission of one or both members of each C—H bond pair, ejecting H or H_2, while the C—C bond excitations lead mainly to C—C scission. As to the interpretation of absorption spectral features of alkanes found by Raymonda and Simpson at low energies in normal and cyclic alkanes, the electrons in the C—C bonds are excited and the C—H electron excitation onset lies at higher energies. At low energy we would thus expect considerable C—C bond scission which would decrease in importance as the probability of C—H scission increased with increasing photon energy. The experimental results, however, showed just the opposite trend [82–85]. It would appear that other less popular theories that describe the excitation and the excited states in terms of delocalized orbitals are in better agreement with these experimental findings [78, 84, 85].

In the decomposition processes of excited ions and electronically excited molecules involving multicentre transition complexes, stable molecular products can be formed. Mass spectrometric experiments on the decomposition of ions as well as vacuum-UV photolytic measurements on neutral species, indicate that the importance of decomposition processes involving rearrangements decreases with increasing internal energy (and consequently, with decreasing lifetime) of the primary species and, in turn, the probability of direct rupture leading to radical formation increases [65, 77, 79].

In alkanes (and also mostly in alkenes) one of the most important decomposition reactions of excited molecules produced in radiolysis is $C-H$ bond dissociation leading to hydrogen gas formation. If both H atoms of a H_2 molecule come from a single hydrocarbon molecule this is called 'unimolecular hydrogen formation', whereas in 'atomic hydrogen formation' the two atoms of a H_2 molecule originate from two distinct molecules.

1.2.2. Secondary reactions of free radicals

Radicals produced by decomposition of ions and/or excited molecules possess considerable excess energy at the moment of their formation, i.e. they are 'hot'. Consequently, mainly in low pressure gases with small collision probability, they may undergo further decomposition. On the other hand, this is less probable in liquids where the frequency of collisions is about 10^{13} s^{-1} and the thermalization of hot radicals is very rapid. The frequency of collisions (Z) in liquids can be calculated as follows:

$$Z = 0.892 \cdot 3\pi\sigma\eta/\mu$$

where σ is the collision diameter of the radicals, η is the viscosity of the medium and μ the reduced mass of a radical–solvent molecule pair [86, 87].

One of the most important reactions of radicals is hydrogen atom abstraction leading to radical exchange:

$$X + HR \xrightarrow{k_a} XH + R$$

where X may be either a hydrogen atom or hydrocarbon radical and R a hydrocarbon radical. Papers dealing with the kinetics of formation of hydrogen gas distinguish between 'hot' and 'thermal' hydrogen atoms, the distinction being more or less arbitrary.

It is very probable that 'hot' hydrogen atoms abstract another hydrogen atom as soon as they collide with one of the surrounding molecules. The yield of hydrogen gas in these cases is not greatly influenced by radical scavengers present at low concentrations ($<$ 100 mM), whereas 'thermal' hydrogen atoms are very sensitive to radical scavengers even at very low concentrations ($<$ 1 mM) thus permitting estimation of their yield.

It has to be noted, however, that our total knowledge of the participation of hot hydrogen atoms in radiation chemical processes is based on indirect evidence as they react too rapidly to be studied directly.

Hydrogen gas that is not affected by the presence of radical scavengers in low concentration is called molecular hydrogen. The molecular H_2 results from hot hydrogen atom reactions and unimolecular hydrogen formation.

It has sometimes been suggested in the interpretation of radiation chemical yields that the relative values of rate constants of hydrogen abstraction, combination and disproportionation reactions for radicals produced in radiolysis are approximately the same as those determined by other (e.g. thermal or photochemical) techniques [87].

Several papers have suggested that empirical correlations for hydrogen abstraction reactions of radicals determined by thermal or photochemical initiation [88–90] are also valid for radiation chemical reactions. Thus, the energy of activation increases with increasing carbon number of the attacking radical and decreases with the increasing number of carbon atoms in the molecule attacked. The presence of a tertiary C—H bond in the latter also decreases the activation energy (Table 1.3). Baldwin and Walker [88] pro-

Table 1.3

Activation energies of hydrogen atom abstraction
Data taken from ref. [88]

RH H + RH	E_a kJ mol^{-1}
CH_4	49.8
C_2H_6	40.6
C_3H_8	34.8
iso-C_4H_{10}	29.3
CH_3 + RH	
CH_4	53.6
C_2H_6	43.5
C_3H_8	37.7^a
iso-C_4H_{10}	31.4
neo-C_5H_{12}	50.2
CH_3CH_2 + RH	
CH_4	74.1
C_2H_6	54.4
C_3H_8	42.7^a
iso-C_4H_{10}	36.8^a

^{a} Refers to the weakest C—H bond

posed the following expression for the estimation of the energy of activation of hydrogen abstraction reactions (E_a in kJ mol^{-1}) in gas-phase hydrocarbon systems if the reaction enthalpy (ΔH), is known:

$$E_a = 51 + 0.55\,\Delta H + (\Delta H)^2/(167 + 2.2\,|\Delta H|)$$

On the basis of results obtained with thermal reactions it is widely accepted that hydrogen atoms, methyl radicals and a small fraction of ethyl radicals produced in liquid phase radiolysis of alkanes are stabilized in hydrogen abstraction reactions whereas combination and disproportionation reactions are more important for higher radicals. This suggestion is supported by the fact that whereas methane and ethane yields are, as a rule, strongly influenced by the dose rate (due to the competition between abstraction and combination processes), the yields of other hydrocarbon fragments are practically invariant as a function of the dose rate [91].

Two extensively investigated processes of radical disappearance (chain termination) are disproportionation and combination:

$$2R-CH_2-CH_2 \xrightarrow{k_d} R-CH_2-CH_3 + R-CH=CH_2$$
$$\xrightarrow{k_c} R-CH_2-CH_2-CH_2-CH_2-R$$

Both of these reactions are highly exothermic ($\Delta H \approx 300$ kJ mol^{-1}) and generally their rates are very high. For example, the following rates have been found for combination of radicals in the gas phase (142°C): methyl radicals, $10^{10.5}$, ethyl radicals, $10^{9.6}$, isopropyl radicals, $10^{8.6}$ and tert-butyl radicals, $10^{5.6}$ M^{-1}s^{-1} (189°C), respectively [92]. The 'collision frequency' of methyl radicals, is 10^{11} M^{-1}s^{-1}, thus, every fourth to sixth collision leads to reaction. The rate of termination in liquids is close to the rate determined by diffusion, although some significant deviations have also been reported.

The ratio of disproportionation and combination rates, i.e. the value of k_d/k_c (Table 1.4) is characteristic of the structure of the radicals participating. They are independent of the method of radical initiation and they are

Table 1.4

Ratios of disproportionation and combination reactions in the gas and liquid phase at about 25°C
Data taken from ref. [92][a]

Radical	k_d/k_c values	
	in gas phase	in liquid phase
CH_3CH_2	0.13	0.12–0.26
$CH_3CH_2CH_2$	0.15	0.13–0.15
$C_2H_5CH_2CH_2$	0.14	0.13; 0.14
$(CH_3)_2CHCH_2$	0.076	
$(CH_3)_2CH$	0.66	1.2
$CH_3CH_2CHCH_3$	0.63; 0.77	1.0; 1.1
$(CH_3)_3C$	2.3; 3.1; 2.7	4.5
cyclo-C_5H_9	~1.0	~1.0[b]
cyclo-C_6H_{11}	~0.5	1.1

[a] Determined by a wide variety of methods, see ref. [92] for details
[b] Ref. [99]

less dependent on the experimental conditions, on the phase (although there are indications that k_d/k_c values for radical reactions are higher by about 30–40% in the liquid phase than in the gas phase [92]), and on the pressure and temperature (according to recent data [92–94], the ratio k_d/k_c decreases with increasing temperature for both liquid and gas phases, a difference of 100°C bringing about a two- or three-fold change).

The mechanism of combination and disproportionation has been discussed extensively in the literature and the debate is still unfinished. Two basically different hypotheses have been put forward: one of them assumes a common transition state for both reactions, whereas the other suggests different acitvated complexes for disproportionation and for combination [92, 93]. The temperature dependence of k_d/k_c found for a number of radicals during the recent years seems to support the latter suggestion, although the differences in the energies of activation calculated on this basis are too low ($E_{a(c)}-E_{a(d)} = 1$–4 kJ mol^{-1}) to be regarded as unambiguous evidence for different transition states.

The k_d/k_c ratio is strongly dependent on the chemical structure [92, 95–98]. Its value for auto- (both radicals are of the same type) disproportionation and combination of primary radicals (terminal, chain-end) is 0.1–0.2, for the same reaction of secondary radicals it is 0.5–1, and for tertiary ones it is about 4. This shift may be attributed partly to the increasing number of hydrogen atoms in the β-position, i.e. in positions where attack during a disproportionation reaction is possible, thus increasing the probability of such an attack [100]. At the same time, the steric hindrance of combination, acting in the same direction, also increases the k_d/k_c-value.

If the partner radicals are of different types (cross disproportionation and combination), the ratio k_d/k_c lies between the auto-ratios of the two partners, being, to a first approximation, the geometric mean.

Two aspects of the importance of k_d/k_c values in radiation chemistry should be mentioned. First, it is relatively easy to generate radicals by using high-energy radiation; thus, in addition to other methods, such as photochemistry, pyrolysis etc., radiation chemistry can make a contribution to the collecting of data on elementary radical combination and disproportionation reactions. Second, data obtained using other methods can assist in the evaluation of radiation chemical measurements.

1.2.3. Secondary reactions of ions

Ion-molecule reactions represent a very important reaction possibility for ions (mainly of fragments). Such processes have been studied for more than 20 years. The interpretation of radiation chemical observations can often be based on the results obtained by the newer mass-spectrometric techniques, such as high-pressure mass spectrometry, ion-cyclotron resonance spectroscopy, and chemical ionization mass spectrometry (see Section 1.5.1 for details). Using these methods, it was possible to determine kinetic parameters for several ion-molecule reactions. Thus, calculations of critical energy values for such processes can be based on adiabatic ionization and appearance potentials obtained by mass-spectrometry.

Ion-molecule reactions characteristic of individual hydrocarbons will be discussed in detail in the later sections; here we show only a few important reaction types [101–104]:

H^--ion transfer:

$$C_3H_7^+ + C_6H_{14} \longrightarrow C_3H_8 + C_6H_{13}^+ \qquad (1.1)$$

H_2^--ion transfer:

$$\text{cyclo-}C_3D_6^+ + \text{cyclo-}C_6H_{12} \rightarrow CD_2HCD_2CD_2H + \text{cyclo-}C_6H_{10}^+ \qquad (1.2)$$

H_2-transfer:

$$\text{cyclo-}C_6H_{12}^+ + \text{cyclo-}C_3D_6 \rightarrow \text{cyclo-}C_6H_{10}^+ + CD_2HCD_2CD_2H \qquad (1.3)$$

H-transfer:

$$C_3H_8^+ + C_2H_4 \longrightarrow C_3H_7^+ + C_2H_5$$

The yield of ion-molecule reactions generally increases with increasing exothermicity of the process, but the rates are also strongly influenced by steric factors [105].

A special type of ion-molecule reactions is the so-called ion-molecule condensation. It can be assumed, by analogy with mass spectrometric studies that such processes are important not only in alkene [106] but also in alkane radiolysis [107]. For example:

$$R-CH_2-\overset{+}{C}H-CH_2 + R-CH_2-CH=CH_2 \rightarrow$$

$$\rightarrow R-CH_2-\overset{+}{C}H-CH_2-CH_2-CH-CH_2-R$$

and

$$C_2H_6^+ + C_2H_6 \longrightarrow C_4H_{12}^+$$

1.3. EFFECT OF EXPERIMENTAL CONDITIONS ON THE YIELDS OF END-PRODUCTS

1.3.1. Dose effects

The yields of some of the products from irradiated saturated hydrocarbons remain unchanged, whereas those of other ones increase or decrease with increasing dose (Fig. 1.3). The most common explanation is to attribute this phenomenon to the secondary reactions of molecular products (first of all, unsaturated hydrocarbons) accumulating during irradiation.

Holroyd, investigating the radiolysis of neopentane [91], found that increasing the dose from $1.0 \cdot 10^{19}$ to $43.8 \cdot 10^{19}$ eV g^{-1} decreased the G-value of hydrogen from 2.2 to 1.33 and that of isobutene from 2.34 to 1.37. He attributed this to the partial hydrogenation of isobutene:

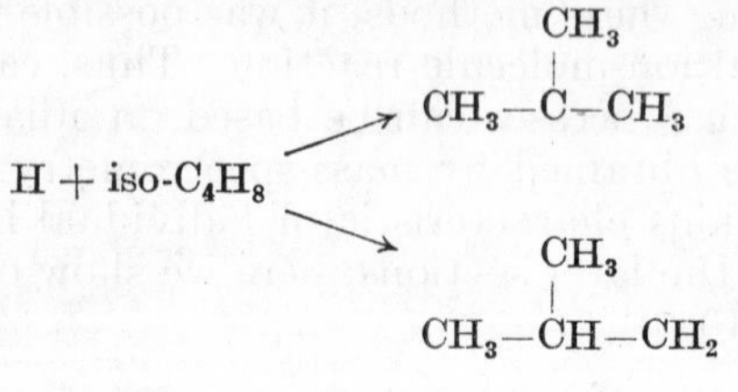

$$
H + \text{iso-}C_4H_8
\begin{cases}
\nearrow \quad \underset{\underset{\displaystyle CH_3-C-CH_3}{\mid}}{CH_3} \\
\\
\searrow \quad \underset{\underset{\displaystyle CH_3-CH-CH_2}{\mid}}{CH_3}
\end{cases}
$$

The yields of combination products of iso- and tert-butyl radicals (e.g. 2,2,4-trimethylpentane) increased with larger doses.

$$CH_3-\overset{\displaystyle CH_3}{\underset{\displaystyle CH_3}{\overset{|}{\underset{|}{C}}}}-CH_3 + CH_2-\overset{\displaystyle CH_3}{\overset{|}{CH}}-CH_3 \longrightarrow CH_3-\overset{\displaystyle CH_3}{\underset{\displaystyle CH_3}{\overset{|}{\underset{|}{C}}}}-CH_2-\overset{\displaystyle CH_3}{\overset{|}{CH}}-CH_3$$

Figure 1.4, depicts the yields of some radiolysis products of 2,2-dimethyl-butane as a function of the dose [109]. G-values of hydrogen and unsaturated products decrease rapidly with increasing dose, a slower decrease is observed for saturated hydrocarbon products. The different character of dose dependence of these yields indicates that the addition of hydrogen atoms to the π-bonds of product alkene takes place with a considerable rate.

Experiments indicate no significant dose dependence in hydrogen yields from irradiated alkenes and aromatics [110–112] in contrast to the case of alkanes. At the same time, the G-values of certain hydrocarbon products change markedly with the dose. A unified treatment of dose effects seems impossible in the radiolysis of unsaturated components. Several reports have indicated the molecular character of the formation of hydrogen and lower molecular weight products because thermal hydrogen atoms and alkyl radicals undergo addition reactions to the π-bonds. Some products are also formed from abstraction reactions with the participation of 'hot' hydrogen atoms and radicals.

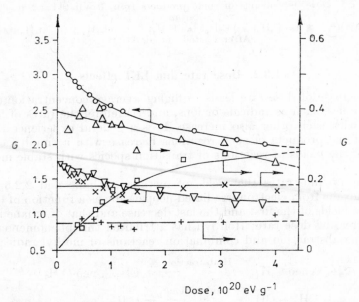

Fig. 1.3. Dose dependence of some products from irradiated cyclohexane. ○ Cyclohexene; ▽ bicyclohexane; △ hex-1-ene; × methylcyclopentane; + n-hexane; □ cyclohexylcyclohexene. After Ho and Freeman [108]

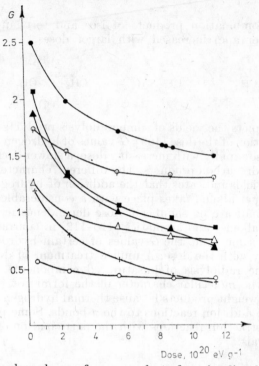

Fig. 1.4. Dose dependence of some products from irradiated 2,2-dimethyl-
butane.
● H_2; ◑ CH_4; ■ iso-C_4H_8; △ C_2H_6; ▲ C_2H_4; ○ iso-C_4H_{10}; + $(CH_3)_3CCH=CH_2$
After Castello et al. [109]

1.3.2. Dose rate and LET effects

The increase of dose rate leads to higher average concentration of transition species, such as radicals or ions, and therefore, the rates of their reactions with each other also increase. It is a general experience that the
yields of the products of such reactions increase with a simultaneous decrease of the yields of reactions of transition species with stable molecules
[113].

Figure 1.5 illustrates the yields of methane, ethane and 2,2,5,5-tetra-
methylhexane from electron irradiated neopentane as a function of the dose
rate. The yields of the first and the last decrease and that of ethane increase
with increasing dose rate; Holroyd has attributed this phenomenon to the
competing abstraction and combination reactions of methyl radicals [91]:

$$CH_3 + neo\text{-}C_5H_{10} \xrightarrow{\text{H-abstraction}} CH_4 + neo\text{-}C_5H_{11}$$

$$CH_3 + CH_3 \xrightarrow{\text{combination}} C_2H_6$$

$$neo\text{-}C_5H_{11} + neo\text{-}C_5H_{11} \xrightarrow{\text{combination}} 2,2,5,5\text{-tetramethylhexane}$$

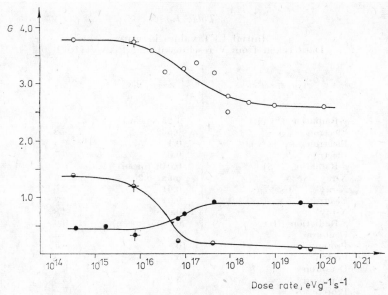

Fig. 1.5. Dose rate dependence of some products from electron irradiated neopentane.
○ CH_4; ● C_2H_6; ◐ 2,2,5,5-tetramethylhexane. After Holroyd [91]

The dependence of the experimental G-values on the dose rate and concentration of scavengers has been described reasonably well by inhomogeneous kinetics. Of these studies, those of Ganguly and Magee [114], Kuppermann and Belford [115–117], Freeman [36, 118] and Burns et al. [119, 120] may be mentioned. Typical LET values (see Section 1.2.1) for various radiations are summarized in Table 1.5. Corpuscular radiation with a high LET value transfers its energy along short paths, and the activated species are produced in a volume with cylindrical symmetry along the path. Energy transfer from electron and electromagnetic radiation with low LET values occurs along a longer pathway, and the activated species are produced in small packages (so-called 'spurs', 'blobs'). The exact definition of these concepts has been disputed very extensively in the literature; detailed discussion would far exceed the scope of the present book [36, 55].

With radiation of high LET values, the spurs overlap, forming continuous tracks; within these the reactive intermediates (excited and/or ionized species, radicals) can be found in very high concentrations, and hence, reactions between these species cannot be ignored (the expression 'spur' will be applied here for conditions where transition species are sufficiently close to influence their reactions [36]). At low LET values, the active species interact predominantly with the molecules of the medium. The higher the density of the medium, the more pronounced the LET effect, i.e. LET effects are observed first of all in high-pressure gases and liquids. Spur and LET effects are discussed in a number of reviews and data collections (e.g. [55, 119]); the inhomogeneous kinetic models already mentioned [114–120] can also be applied here.

3

Table 1.5

Initial LET values in water
Data taken from Vereschinskii and Pikaev [101]

Type of radiation	Energy, MeV	LET, keV μm^{-1}
γ-Radiation (^{60}Co)	1.25 (mean)	0.2
Electron	0.5–2	0.2
Electron	0.1	0.4
Electron	0.05	0.7
β-Radiation (^{35}S)	0.046 (mean)	0.7
X-ray	max. 0.25	1.0
X-ray	0.01	2.0
Electron	0.01	2.3
X-ray	0.008	2.8
β-Radiation (^{3}H)	0.0055 (mean)	3.6
Deuteron	20	4.5
Proton	10	4.7
Proton	5	8.2
Deuteron	8	10
Deuteron	5.2	13
Proton	2	17
Helion	38	22
Proton	1	27.7
Helion	12	50
Proton	0.3	54
α-Particles (^{210}Po)	5.3	88
Products of ^{6}Li(n, α)^{3}H reaction	2.05(α); 2.73 (^{3}H)	~100
α-Particles	3.4	120
Products of ^{10}B(n, α)^{7}Li reaction	1.50(α); 0.85 (^{7}Li)	~170

1.3.3. Temperature and phase effects

The yields of products from γ-irradiated liquid alkanes increase with increasing temperature: e.g. Dewhurst found about a $+3\%$ increment per 10°C for the G-value of H_2 from n-hexane, in the temperature range between -78 and $+30$°C. This corresponds to an energy of activation of about 12.5 kJ mol^{-1} which is characteristic of diffusion controlled processes [121].

Koob and Kevan have observed the overall decomposition yields of propane radiolysis by applying a $6 \cdot 10^{19}$ eV g^{-1} dose at -130, -78 and $+35$°C to be 6.4, 9.3 and 10.4 molecules per 100 eV, respectively [122]. They have shown that the variation of temperature affects the rate of free radical reactions but has practically no influence on ionic and molecular decomposition processes having a temperature-independent overall yield of $G = 1.6$.

It is generally observed that higher temperatures are favourable for reactions with higher energies of activation. Thus, according to Koob and Kevan, at lower temperatures addition of hydrogen atoms to the π-bond of the propene product ($E_a = 9.2$ kJ mol^{-1}) and the combination of the propyl radicals thereby produced predominate [122]. The abstraction reaction leads to a higher overall $G(-C_3H_8)$ values and consequently, to a higher overall product yield, since reactions (1.4) and (1.5) compete with each other:

$$H + C_3H_8 \longrightarrow H_2 + C_3H_7 \tag{1.4}$$

$$H + C_3H_6 \longrightarrow C_3H_7 \tag{1.5}$$

$$C_3H_7 + C_3H_7 \longrightarrow C_6H_{14}$$

The experiments of Koch et al. with n-pentane [123] indicate that the G-value of alkene products increases rapidly between -120 and $0°C$, presumable owing to the relative acceleration of hydrogen abstraction at the cost of addition. Investigating the approximately complete product spectrum, they have measured the $G(C-C$ bond rupture) values at -116, $+20$ and $+95°C$ to be 1.0, 1.5 and 1.7, respectively.

Gäumann et al. have studied the temperature dependence of yields of products formed from irradiated C_5-C_{16} n-alkanes in the temperature range between -50 and $+100°C$ [124]. Different apparent activation energy values were obtained (as determined from the ln G vs. T^{-1} plots) for different kinds of products. Thus, for H_2 formation $E_a = 1.5$ kJ mol^{-1}, for the formation of C_1-C_{n-1} n-alkane products from C_nH_{2n+2} alkanes $E_a = 2.1$, for C_2-C_{n-1} alkenes $E_a = 1.6$ and for C_nH_{2n} alkenes $E_a \approx 0$ kJ mol^{-1}, independently of the value of n. The G-values of $C_{n+1}-C_{2n-1}$ products decrease with increasing temperature ($E_a = -6.3$ kJ mol^{-1}) and for $C_{2n}H_{4n+2}$ dimers $E_a = = 2.8$. Apparent energies of activation do not give much information about the character of the processes and on the transition species involved, nevertheless they are useful in estimating G-values for various irradiation temperatures.

Földiák et al. [125] have found that the overall rates of gas-phase decomposition of hexadecane and liquid phase decomposition of liquid paraffin are practically the same up to $250°C$, but a sharp increase can be observed above this temperature: a chain reaction starts with an apparent energy of activation of about 60 kJ mol^{-1}. Thermal decomposition sets in at about $310-320°C$ and above this threshold value the rate of radiation chemical initiation and the importance of radiation chemical processes in determining product formation become negligible.

On comparing product yields observed in vapour-phase and liquid-phase radiolysis, it is possible to draw conclusions concerning the mechanisms of basic radiation chemical processes.

Overall G-values in vapour-phase and liquid-phase radiolysis are: for propane, 12.1 and 10.4 [122]; for neopentane, 6.6 and 5.9 [126, 127]; for n-hexane, 8.8 and 6.5 [128, 129], respectively. The difference has been interpreted by Koob and Kevan [122] in terms of the average excitation energy content of activated molecules being smaller in the liquid phase, owing to collisional deactivation.

Judging by the papers dealing with solid phase radiolysis of hydrocarbons, the processes in this case can be regarded as similar to those observed in gas or liquid phases [11, 130–133]. Some products, however, are formed in lower yields than in fluid phases. Freeman [36] attributed this fact to the smaller free volume between molecules in the solid phase, so that reactions requiring a positive activation volume (such as diffusion, fragmentation, some isomerization processes) slow down. This can explain the presence of electrically conductive ion pairs and free radicals in the sample even after irradiation.

1.4. A SHORT SUMMARY OF EXPERIMENTAL METHODS

Several chemical and physical methods have been developed during the past 20 years for the study of the reaction mechanisms of radiation chemical processes, as well as for determination of the yields of short-lived transition species such as radicals and ions.

1.4.1. Chemical methods

The application of additives, the so-called scavengers (S, Table 1.6), is a widespread method for investigating the mechanisms of product formation. The yields of some products decrease, those of others increase or remain unchanged in the presence of such substances, and new components appear

Table 1.6

Physical properties of some scavengers in the gas phase.
Data taken from refs [19–21, 194–196]

Scavengers	Photoionization potential, eV	Electron affinity, eV	Proton affinity, eV
Electron scavengers			
CH_3Cl	11.28		
CH_3Br	10.50		
C_2H_5Br	10.24		
N_2O	12.82	−0.15	
SF_6		1.29	
CCl_4	11.47		
C_6H_5Cl	9.07		
C_6H_5Br	8.98		
Positive ion scavengers			
cyclo-C_3H_6	10.06		
C_6H_6	9.25	−1.59...0	
NH_3	10.25		9.20
C_2H_5OH	10.60		8.60
Radical scavengers			
O_2	12.20	0.44	
NO	9.25	0.91	
p-$C_6H_4O_2$	9.67	1.37	
I_2	9.40	1.80	
SO_2	12.34	1.1	
H_2S	10.46		
C_2H_4	10.52		

among the products, as a result of the scavenging processes. On the basis of such experiments, and considering the chemical properties of the scavenger itself, the yields of some products can be estimated, the approximate ratio of parallel reaction possibilities can be evaluated and some reaction steps can be excluded from the mechanism.

A general problem with scavenger experiments is the practical impossibility of finding specific additives that react with only one type of intermediates among the several types present during radiolysis (excited molecules, positive and maybe negative ions, electrons and free radicals). For example, substances generally used as radical scavengers can also react with electrons, as both radical and electron scavenging is related to the electron affinity of the substance in question. In order to avoid undesirable secondary reactions as far as possible, it is advisable to use scavenger concentrations only just great enough to bring about the necessary effect. This is, however, sometimes difficult: for example, an iodine concentration as low as 10^{-4} M is sufficient to capture all alkyl radicals produced that escape the spur during alkane radiolysis [134, 135]. However, in order to avoid the consumption of iodine, a very low dose has to be applied and this requirement puts the analyst in a difficult position. The G-value of alkyl iodides obtained in low yields can be measured by applying radioactive I_2.

Most scavenger studies reported in the literature have been performed with alkanes, particularly with cyclohexane. The mechanism of scavenging is even more complex in systems containing alkenes and aromatics: this may be the reason why so few studies of this kind have been published [136].

Investigation of charged species

Several substances capture electrons produced by radiation and thus inhibit charge neutralization processes. As the lifetime of positive ions becomes thus longer, the probability of ion-molecule reactions increases. Electron scavengers used generally are: nitrous oxide, sulphur hexafluoride, carbon tetrachloride, methyl and ethyl halides, chlorobenzene, benzyl chloride and bromobenzene. Anions formed from the additive molecules can be stabilized via electron detachment or dissociation. The decomposition of activated molecules formed in charge neutralization between hydrocarbon molecule ions and scavenger anions will alter the product spectrum compared with that obtained from pure hydrocarbons. Charge scavenging can be followed either by the decrease of yields of products originating from the charge neutralization reactions or by the G-values of products of charge scavenging.

Several formulae have been developed to describe concentration dependence [137, 138]; the most frequently applied of these is that derived by Hummel [139], and developed further empirically by Warman et al. [140, 141]:

$$G(\mathrm{P})_\mathrm{S} = \left(G_{\mathrm{fi}} + G_{\mathrm{gi}} \frac{\sqrt{\alpha[\mathrm{S}]}}{1 + \sqrt{\alpha[\mathrm{S}]}} \right) \varepsilon \qquad (1.6)$$

where ε expresses the efficiency of production of P during or after the charge scavenging: G_{fi} is the yield of the so-called 'free' ions or electrons, G_{gi} is that of the so-called 'geminate' ions.

The evaluation of ε can be illustrated by reference to the scavenging of positive cyclohexane ions by added cyclopropane:

$$\text{cyclo-C}_6\text{H}_{12}^+ + \text{cyclo-C}_3\text{H}_6 \longrightarrow \text{cyclo-C}_6\text{H}_{10}^+ + \text{C}_3\text{H}_8$$

Propane is produced with 63% efficiency, therefore $\varepsilon = 0.63$ for propane formation [142]. For many scavengers, ε is equal to or close to unity.

This expression has so far been found to be valid in the scavenger concentration range between 10^{-5}–$5 \cdot 10^{-1}$ M. With low concentrations of scavenger the denominator of the expression is very close to unity, therefore:

$$G(\text{P})_\text{S} = (G_{fi} + G_{gi}\sqrt{\alpha[\text{S}]})\varepsilon$$

The intercept with the ordinate axis gives the value of G_{fi} when $\varepsilon = 1$. Values determined in this way are generally rather close to those determined by conductivity measurements. Values of α and G_{gi} can be determined by fitting equation (1.6) to the experimental points. The sum of G_{fi} and G_{gi} is, in most cases, equal to the total ion yield, $G \approx 4.4$, calculated on the basis of the W value in the gas phase [143].

Table 1.7

α-Values measured in cyclohexane systems for some electron and positive ion scavengers

Electron scavengers	α, M^{-1}	References
CH_3Cl	5.0–5.4	141, 144, 185
CH_3Br	16	141, 144, 185
CH_3I	22	185
$\text{C}_2\text{H}_5\text{Br}$	7.8, 10,16	141, 144, 185
N_2O	8, 10, 16	144, 159, 185
SF_6	16—18	144, 185
CCl_4	12	185
$(\text{C}_6\text{H}_5)_2$	15	185
$\text{cyclo-C}_4\text{F}_8$	14	185
$\text{cyclo-C}_6\text{F}_{12}$	21, 30	170, 185
CO_2	8, 13	144, 158
$\text{C}_6\text{H}_5\text{Cl}$	9	162
$\text{C}_6\text{H}_5\text{CH}_2\text{Cl}$	8	162
$\text{C}_6\text{H}_5\text{Br}$	10	143, 165
o-Carborane	33	186
1-Ethylcarborane	11	186
Positive ion scavengers		
$\text{cyclo-C}_3\text{H}_6$	0.40	142
C_6H_6	1.2	142
ND_3	0.85	178

The value of α is equal to or greater than 10 M^{-1} for good electron scavengers, but it hardly reaches $\alpha = 1 \text{ M}^{-1}$ for even the best positive ion scavengers (Table 1.7). By comparing α-values for different additives it is possible to compare the charge-scavenging abilities of individual scavengers. By expressing yields as a function of $(\alpha_2/\alpha_1) [S]$ (where α_1 is the α value for a scavenger taken as standard, α_2 is the same for any other scavenger), the results obtained with the same hydrocarbon and various scavengers can be plotted as a single curve [140, 144].

The variation of yields of individual radiolysis products of hydrocarbons in the presence of scavengers may give information on the different pathways of product formation. The formation of the three main products of cyclohexane radiolysis (hydrogen, cyclohexene and bicyclohexane) has been studied in great detail. The yield of hydrogen generally decreases very considerably in the presence of scavengers (with the exception of NH_3 and C_2H_5OH), whereas no unambiguous correlation could be found between the yields of bicyclohexane and cyclohexene and the concentration of charge scavengers. Some scavengers (such as N_2O [145] and cyclo-C_6F_{12} [146]), while they prevent the production of cyclohexene and bicyclohexane by inhibiting the formation of their common precursor, namely cyclohexyl radicals, however, they produce cyclohexyl radicals in the scavenging process itself (see later, the scavenging reactions of N_2O and cyclo-C_6F_{12}). In these cases the yield of bicyclohexane generally increases and that of cyclohexene increases or decreases somewhat with increasing scavenger concentration. This can be attributed to the fact that cyclohexene can also be produced in unimolecular reactions: the scavenger prevents unimolecular cyclohexene formation following electron–ion neutralization and swaps its precursor for cyclohexyl radicals.

Particular attention has been paid to the use of nitrous oxide as an electron scavenger. Studies with this compound exemplify the contradictions and unsolved problems the solution of which could revitalize research in the field of the radiolysis of liquid hydrocarbons. The addition of N_2O decreases the yield of hydrogen from cyclohexane considerably and increases somewhat the yield of cyclohexene and bicyclohexane [145]. Decomposition of N_2O gives N_2, and in stoichiometric ratio, H_2O plus cyclo-$C_6H_{11}OH$, the amount of water being about 15 times greater than that of cyclohexanol [147–150]. The ratio of N_2 formation to the decrease of hydrogen yield from cyclohexane is about 1.6, and is independent of concentration in the range 10^{-3}–10^{-1} M, in spite of the widely accepted assumption that every neutralization process without scavenger leads to the formation of H_2. Therefore, it is more or less generally accepted that, apart from reactions (1.7)–(1.10):

$$e^- + N_2O \longrightarrow N_2 + O^- \tag{1.7}$$

$$e^- + N_2O \longrightarrow N_2O^- \tag{1.8}$$

$$N_2O^- + \text{cyclo-}C_6H_{12}^+ \longrightarrow N_2 + OH + \text{cyclo-}C_6H_{11} \tag{1.9a}$$

$$\longrightarrow N_2 + H_2O + \text{cyclo-}C_6H_{10} \tag{1.9b}$$

$$N_2O^- + \text{cyclo-}C_6H_{12} \longrightarrow N_2 + OH^- + \text{cyclo-}C_6H_{11} \tag{1.10}$$

Other nitrogen producing reactions may also occur [33, 34, 51, 147–156] such as:

$$O^- + N_2O \longrightarrow N_2O_2^-$$

$$N_2O_2^- + \text{cyclo-}C_6H_{12}^+ \longrightarrow N_2 + \text{products}$$

$$\text{cyclo-}C_6H_{12}^* + N_2O \longrightarrow N_2 + \text{products}$$

where $\text{cyclo-}C_6H_{12}^*$ denotes an electronically excited cyclohexane molecule. The intermediates OH, OH^-, O^- and $\text{cyclo-}C_6H_{11}$ can participate in many further reactions leading ultimately to the formation of water, cyclohexanol, cyclohexene and bicyclohexane. For instance:

$$OH + \text{cyclo-}C_6H_{12} \longrightarrow H_2O + \text{cyclo-}C_6H_{11}$$

$$OH^- + \text{cyclo-}C_6H_{12}^+ \longrightarrow H_2O + \text{cyclo-}C_6H_{11}$$

$$O^- + \text{cyclo-}C_6H_{12} \longrightarrow OH^- + \text{cyclo-}C_6H_{11}$$

$$OH + \text{cyclo-}C_6H_{11} \longrightarrow H_2O + \text{cyclo-}C_6H_{10}$$

$$\searrow \text{cyclo-}C_6H_{11}OH$$

$$2\,\text{cyclo-}C_6H_{11} \quad \nearrow \text{cyclo-}C_6H_{10} + \text{cyclo-}C_6H_{12}$$

$$\searrow (\text{cyclo-}C_6H_{11})_2$$

The upper limit of $G(N_2)$ from N_2O dissolved in cyclohexane and some other alkanes has been found by Freeman and Sambrook to be 5.2 ± 0.3 [33], using a correction for the yield of N_2 from the direct radiolysis of N_2O. They assumed that this yield originates almost entirely from reactions (1.7)–(1.10), i.e. the capture of one electron produces one molecule of nitrogen. Consequently, the ion yield in the liquid phase must be higher than the value of $G(\text{ion})$ in the gas phase, and it even exceeds the yield calculated from equation (1.6). This difference between the yields in the two phases has been verified by Robinson and Freeman [156] using direct conductivity measurements in the radiolysis of Ar, Kr and Xe. If the conclusions of Freeman et al. [33, 152, 156] are correct, then several experimental facts have to be reconsidered concerning the role of ions.

A considerable number of studies have been carried out using alkyl halides, such as CH_3Br, C_2H_5Br, CH_3Cl, as additives in the presence of 10^{-4}– 10^{-3} M of I_2. This method proved to be very useful in the determination of yields of electrons. The added iodine serves for the measurement of alkyl radicals produced in the dissociative electron capture process [reaction (1.11)]; the scavenging of these radicals is often realized by using ^{131}I, when the yields can be measured by radiochemical methods:

$$e^- + RX \longrightarrow R + X^- \tag{1.11}$$

$$R + I_2 \longrightarrow RI + I$$

A small fraction (5–10%) of R radicals does not give alkyl halides but produces alkanes by hydrogen abstraction [135, 141, 144, 157–159].

CCl_4 is also used frequently as an electron scavenger: it captures an electron to give Cl^- and CCl_3. The stabilization of these two species produces trichloromethylated and chlorinated hydrocarbons as well as hydrochloric acid [99, 160, 161]. Benzyl chloride and chlorobenzene also gives hydrochloric acid following electron capture, the determination of which can be used to follow the scavenging reaction [162].

In solutions containing bromobenzene the capture of electrons leads to the production of phenyl radicals, which then abstract hydrogen to form benzene [163–165]:

$$e^- + C_6H_5Br \longrightarrow C_6H_5 + Br^-$$

$$C_6H_5 + RH \longrightarrow C_6H_6 + R$$

Several papers indicate that the halogene compounds mentioned above also participate in excitation transfer processes [52, 166, 167].

Some recent publications have reported the advantageous electron-scavenging properties of perfluorocyclohexane [146, 168–172]. The analysis of its decomposition product (cyclo-$C_6F_{11}H$) is a convenient way of following the scavenging processes.

Three suggestions have been put forward to explain the formation of cyclo-$C_6F_{11}H$. Two of these follow Sagert's suggestion and assume the formation of cyclo-C_6F_{11} radicals [146, 170]:

$$\text{cyclo-}C_6F_{12}^- + RH^+ \longrightarrow \text{cyclo-}C_6F_{11} + R + HF$$

$$\text{cyclo-}C_6F_{12}^- \longrightarrow \text{cyclo-}C_6F_{11} + F^-$$

$$\text{cyclo-}C_6F_{11} + RH \longrightarrow \text{cyclo-}C_6F_{11}H + R$$

Rajbenbach's mechanism proposes a cyclo-$C_6F_{11}^-$ anion intermediate [168, 169, 172]:

$$\text{cyclo-}C_6F_{12}^- + RH \longrightarrow \text{cyclo-}C_6F_{11}^- + HF + R$$

$$\text{cyclo-}C_6F_{11}^- + RH^+ \longrightarrow \text{cyclo-}C_6F_{11}H + R$$

The electron-scavenging abilities of additives are highly solvent-dependent, and are also dependent on the temperature [43, 173–176]. This phenomenon can be traced back mainly to changes in the mobility of electrons and thus to changes in the rate constant of the scavenging reaction [173, 174]. Allen et al. [174] have shown that the rate constants of the reactions of the electron with various scavengers change with the conduction band energy (V_0) in the solvent according to a maximum function. This phenomenon is not yet understood.

One of the best positive ion scavengers is ammonia, often used in its deuterated form. Its addition leaves the total hydrogen yield virtually unchanged [177, 178], e.g. for cyclohexane:

$$\text{cyclo-}C_6H_{12}^+ + ND_3 \longrightarrow \text{cyclo-}C_6H_{11} + ND_3H^+$$

$$ND_3H^+ + e^- \longrightarrow ND_3 + H$$
$$\longrightarrow ND_2H + D$$

$$H + \text{cyclo-}C_6H_{12} \longrightarrow H_2 + \text{cyclo-}C_6H_{11}$$

$$D + \text{cyclo-}C_6H_{12} \longrightarrow HD + \text{cyclo-}C_6H_{11}$$

According to the measurements of Asmus [178], the elimination of H-and D-atoms during neutralization occurs in a random way, in the ratio of 1 : 3; thus no isotope effect is apparent.

Bansal and Rzad have investigated the radiolysis of 2,2,4-trimethylpentane and have concluded that ND_3 does not react with the free ions. The reason for this phenomenon is as yet unknown [176].

Positive-ion scavenging by cyclopropane produces propane [179–181]. As the radiolysis of pure cyclohexane or pure cyclopropane yields very low amounts of propane ($G = 0.011$ [182], and $G = 0.27$ [183], respectively), the determination of propane is a convenient way of following the reaction. The use of cyclopropane for the study of positive ions is now a standard technique. For example, an unambiguous picture of positive-ion capture by benzene dissolved in cyclohexane can be obtained by comparing the propane yields from the cyclohexane–cyclopropane and the cyclohexane–cyclopropane–benzene systems [142].

Benzene, ethylene and other alkenes also capture hydrogen atoms, in addition to positive ions [142, 144, 184].

Investigation of radical processes

Radical scavengers have been used for the study of hydrocarbon radiolysis for some time, but quantitative evaluation has only recently been achieved. This is due to the fact that a scavenger is generally both an electron scavenger, and a radical scavenger. Therefore the correlation between the depression of yields following the addition of a scavenger, the yield of the products of scavenging and the radical processes is very complicated. Another problem arises from the ability of the primary products of scavenging to participate in secondary reactions. Therefore, reliable studies can be performed with sufficiently low doses only.

The detection of free radicals and the determination of their yields can be achieved by using as scavengers unsaturated hydrocarbons (e.g. ethylene, pent-1-ene, hex-3-ene) or substances with unsaturated electron orbitals or with unpaired electrons (iodine, oxygen, nitrogen monoxide, sulphur dioxide, hydrogen sulphide, benzoquinone, propanethiol, DPPH etc.) [187, 188].

The most important expression for describing the yield of products of hydrogen-atom capture (SH) is equation (1.12):

$$G(\text{SH}) = \frac{G_{\text{H, therm}}}{1 + \dfrac{k_a[\text{RH}]}{k_b[\text{S}]}} \qquad (1.12)$$

where k_a and k_b are, respectively, the rate constants of the hydrogen atom abstraction process from RH hydrocarbon and of the reaction occurring between the hydrogen atom and the scavenger. $G_{H,therm}$ means the yield of 'thermal' hydrogen atoms, i.e. the theoretical maximum of scavengeable species. The ratio k_a/k_b for some additives has been summarized in Table 1.8.

Table 1.8

k_a/k_b values for some hydrogen atom scavengers

Scavengers	k_a/k_b	References
C_2H_4	0.00475	144
I_2	0.00022	135
C_3H_7-SD	0.008	191
C_6H_6	0.028	142

One of the most efficient scavengers of hydrogen atoms is ethylene: ethyl radicals are produced and can be determined, for example, by iodine. But ethylene reacts also with cyclo-$C_6H_{12}^+$ ions, and ethyl radicals are also produced in this process. Schuler et al. [144] have developed a mathematical treatment that describes scavenging phenomena exactly.

Iodine is 22 times more efficient than ethylene for the scavenging of hydrogen atoms, but it also reacts more readily with positive ions (Table 1.8) [135]. Iodine is one of the best scavengers for alkyl radicals. Cyclohexyl radicals that escape the spur produced from irradiated cyclohexane are captured quantitatively by iodine applied in concentrations as low as 10^{-4} M ($G = 5.35$); at the same time, iodine in such low concentrations participates only to a limited extent in ionic reactions ($G \approx 0.16$) and hydrogen atom capture ($G \approx 0.07$), and correction factors can be applied for these effects. At higher concentrations, the yields of the last two reactions increase and so does the yield of iodine-containing substances of high boiling point [135].

The reactions between cycloalkyl radicals and iodine are diffusion-controlled; the rate constant for the process cyclohexyl $+ I_2$ is $1.2 \cdot 10^{10}$ M^{-1}s^{-1} at room temperature [188A].

Iodine is less advantageous for studies with branched-chain alkanes. Because the stability of tertiary alkyl iodides is rather low, the scavenging of tertiary radicals by iodine gives unreliable results [187].

Oxygen is also frequently used as a free radical scavenger. Its presence in irradiated cyclohexane leads to the formation of cyclohexanone and cyclohexanol ($G = 2.63$ and 3.17, respectively [108]). The slight electron capturing and triplet quenching properties of oxygen also influence the yields obtained in its presence.

The disadvantage of NO (although it is a good radical scavenger and reasonably soluble in hydrocarbons) is that its ionization potential is low (9.20–9.25 eV [19]). Positive charge transfer may thus also take place from hydrocarbons to NO [189].

Stone et al. have reported the favourable scavenger properties of propanethiol in cyclohexane [190, 191], whereas the properties of ethanethiol, propane disulphide, hydrogen sulphide [191, 192] and thiophenol [193] seem to be less attractive.

Reaction of deuterated propanethiol with hydrogen atoms leads to the formation of HD:

$$C_3H_7-SD + H \longrightarrow C_3H_6-\dot{S} + HD$$

There is a possibility that propanethiol may also take part in ionic reactions, but such processes can be neglected at low concentrations [190, 191].

1.4.2. Physical methods

Physical methods can yield somewhat more direct information on primary processes than chemical ones.

Pulse radiolysis

The main field of application of pulse radiolysis is the study of intermediates, although it can be used to advantage even in cases where the analysis of end-products is the only method of evaluation, when e.g. the very high dose rate of pulse devices is used to produce very high concentrations of radicals.

Pulse radiolysis offers a convenient method for the study of radical–radical reactions.

Pulse radiolysis has been used from the late 1950s for the study of short-lived transition species. The method involves the passage of an electron beam of very high intensity (produced by devices such as the microwave linear accelerator or the Van de Graaff accelerator) through a sample. The decay kinetics of transition species are then measured. Early instruments produced pulses of a few microseconds; later the pulse duration was decreased further and nowadays there are devices with pulses as short as ~10 ps. This development permitted to study the kinetics of very rapid processes. However, an 'infinite' decrease of the pulse duration is limited not only by electronics, but also by the analytical equipment. The lower detection limit of a product is determined by the absolute number of species formed. This, in turn, depends on the absorbed dose, being the product of the pulse duration and the dose rate. Thus an 'infinite' decrease of the pulse duration would be possible only with an 'infinite' increase of the dose rate.

An early method of detection used spectrophotographic recording (borrowed from flash photolysis). More recently, spectrophotoelectric recording has been the method of choice [197, 198]. The irradiation cell (or two of its walls) is made of high-purity silica glass for both methods, and the cell is illuminated by analysing light with adjustable wavelength. The light absorption at a given moment is recorded photographically in the spectrophotographic method, whereas the spectrophotoelectric method uses a photomultiplier completed with an amplifier and an oscilloscope enabling thus to follow up time dependence in one single shot. The transient signal can be stored in digital form in a computer memory and the results from a large number of shots can be averaged. This method provides especially high sensitivity of detection and easy handling of the data [198A]. Light absorption can be used to study some free radical species (e.g. the cyclo-hexadienyl radical [199]), some ions (e.g. ions of amines) and solvated electrons [200], as well as some triplet states (e.g. in aromatics). Other methods utilize the recording of radiation-induced luminescence [201, 202].

Another detection method is the coupling of the pulse radiolysis system to an electron spin resonance instrument that permits the identification and quantitative determination of radicals [203, 204]. The advantage of this method is that radicals can be studied that absorb light very weakly or not at all in the practical wavelength range (e.g. alkyl radicals).

Special detection methods include the measurement of the electric conductivity of the irradiated system and the collection of the ions produced. Such methods render possible the determination of the yield of free ions (e.g. clearing field method [41–43, 205]) or the study ionic reactions such as electron capture [206].

Electron spin resonance spectroscopy

Electron spin resonance (ESR) spectroscopy is widely used in the determination of the structure and the concentration of free radicals. The method is based on the paramagnetic properties of free radicals: in an applied magnetic field, their vector of magnetic moments will be oriented parallel or antiparallel to the direction of the field. Energy impact via electromagnetic radiation may lead to the mutual transformation of these two states having somewhat different energies. The energy necessary for this (or the magnetic field when the electromagnetic radiation energy is constant) is influenced by the environment of the radical site i.e. its chemical structure. This results in the fine structure of the ESR spectrum, which is characteristic of the chemical structure of the radical.

The ESR technique is generally used in the solid state because the lifetimes of radicals are then relatively long, and so the sensitivity and ease of measurement are enhanced.

Liquid-phase ESR spectra can be obtained during irradiation or combined with pulse radiolysis, after short irradiation pulses. The lower detection limit of modern instruments is about 10^{12} radicals per gram, provided that the lifetime of the radical is longer than 10^{-10}–10^{-9} s. These severe preconditions explain why most such reports (e.g. [207, 208]) to date have dealt with relatively unreactive high molecular weight tertiary or secondary radicals (e.g. cyclohexyl, tert-pentyl). Other studies, such as the detection of hydrogen atoms with a mean lifetime of 10^{-7} s in irradiated liquid methane and ethane [54], can be regarded as curiosities.

The irradiation of cyclohexane in a gas flow followed by its very rapid cooling has been reported: the frozen radicals were studied by ESR [209].

The investigation of spin–spin couplings has led to the observation of excited naphthalene molecules in the triplet state with a relatively long lifetime (about 10^{-3} s) in irradiated liquid naphthalene [210].

Conductivity measurements

Conductivity measurements have been applied over the past 70 years for determining the yields of charged species produced during irradiation, and for studying their mobility and their reactions with ion scavengers. Such measurements include the generation of charge carriers by external irradiation (by stationary or pulse-type radiation sources) or by using internal radiation sources dispersed in the hydrocarbons to be studied or applied as a layer on the electrodes [43]. A special (although not 'pure' radiation chemical) method of generation of charge carriers is to induce excess electrons by UV light from a metal layer with low work function placed in the hydrocarbons [211, 212].

The potential applied to the electrodes of the irradiation cell causes the charge carriers (or some of them) to move towards the electrodes.

The detailed description of the various devices is outside the scope of this book; the subject has recently been treated extensively by Hummel and Schmidt [43].

Conductivity is the sum of electron and positive ion conduction. The mobility of the former species is some 2–3 orders of magnitude higher than that of the latter. If an electron scavenger is present in the system, it decreases the mobility of the electrons to about that of positive ions (by transforming them into negative ions). Thus the conductivity becomes lower and electron-capture reactions can be followed. Impurities in the sample (e.g. alcohols) also induce electron capture, even in very low concentrations.

An interesting conductivity method is the coupling of pulse-induced conductivity with a special pulse of light that follows the high-energy radiation pulse after an interval. If there is an electron scavenger present in the hydrocarbon sample, the light pulse may cause photodetachment of the electron from the P_e^- ion produced during the electron scavenging. Lukin and Yakovlev applied this method in a study of the formation and decay of the N_2O^- ion in a solution of 2,2,4-trimethylpentane [213].

1.5. AUXILIARY TECHNIQUES

The techniques described in this section do not belong to radiation chemistry in the narrow sense. Their inclusion is justified by the fact that the kinetic parameters of reactions of intermediates, their structures and reactivity can be studied more precisely and more conveniently by these techniques.

Some of the techniques discussed in this section are similar to radiation chemistry as far as the physical essence of the basic phenomenon (e.g. the method of energy impact) is concerned: mass spectrometry or some branches of photochemistry belong to this group. In other cases, such as pyrolysis, photochemistry, plasmochemistry, the analogy refers rather to the final results; such techniques are useful in the interpretation of particular results (e.g. the yields of radical processes).

Naturally, experimental details of individual techniques cannot be included in this book: we shall keep to the essence of each method, emphasizing the fields of application in radiation chemistry. More attention will be paid to some methods that can be regarded — at present or presumably in the near future — as especially important for a better interpretation of radiation chemical results. These are, among others, vacuum-UV photolysis and ion-cyclotron resonance spectroscopy. The main purpose of this short summary is to give the reader some idea of the very wide variety of modern techniques, and to include in the list of references works or monographs that can give more detailed information on the subject.

1.5.1. Mass spectrometry

Much additional information on important processes in radiation chemistry (mechanism of ion reactions) can be obtained from mass spectrometry. The gas (vapour) phase sample is introduced through an inlet system into the conventional analytical mass spectrometer [214–216]. It is ionized in the ionization chamber by electrons emitted by a hot wire and accelerated by an electric field. The energy of the electrons varies between 5 and 100 eV in most modern instruments. As the electron energy generally greatly exceeds the first ionization energy of most substances (7–13 eV), primary ions with a lot of excess energy undergo various consecutive decomposition and rearrangement processes. Owing to the very low pressure in the system $(10^{-10}-10^{-8}$ bar) the mean free path is large and no collisions or other bimolecular reactions between parent and daughter ions need to be considered. A low repeller potential promotes the transfer of ions from the ionization chamber into the analyser unit where, on passing between accelerating electrodes, every ion is given the same kinetic energy. Single-focusing devices separate these ions in a homogeneous magnetic field, according to their mass/charge (m/e) ratio. Double-focusing systems offer a much higher resolution, the ion beam, which initially has a slightly variable kinetic energy (owing to the Boltzmann distribution and inhomogeneity in the ion source field) passes through an electrostatic field and a slit prior to reaching the magnetic analyser. Thus, the kinetic energy of the ion beam is homogenized before separation into components of different m/e ratios. The relative amounts of individual components are determined by using photosensitive plates, photomultipliers or collectors.

Time-of-flight mass spectrometers use the principle that the velocities of components with different masses in an ion beam with the same kinetic energy are different, so that they travel the same distance in different times. Thus, the components arrive at the detector in the order of increasing mass. Such instruments operate in the pulse mode.

A high mass resolution can be achieved by using a quadrupole mass filter [216]. The quadrupole force field generated by four electrodes in parallel arrangement allows only ions with strictly defined mass to enter simultaneously into the detecting device.

These are the basic types of the electron-beam apparatus that are most widely used in analysis and research. In addition, instruments with other ionization sources are also known. For special purposes, radioactive radiation or light from discharge tubes are also used for ionization. Field ionization (FI) mass spectrometers represent a basically different type that uses a very high electric field ($\sim 10^{10}$ V m^{-1}) for ionization of the gas sample [216]. Owing to this high electric field, the residence time of ions in the ion chamber is very short ($\sim 10^{-12}$ s). This fact, together with the relatively low excess energy of parent ions, prevents extensive fragmentation, so that FI spectroscopy can be used for the investigation of thermally unstable materials.

Secondary decomposition processes of fragment ions can be studied especially easily by generating collision-induced metastable states [216]. By allowing ions with high kinetic energy (a few keV) to collide with molecules of an inert gas, it is possible to transform considerable fraction of the translational energy into internal energy of the ion. Under such conditions the rate of decomposition (mainly direct bond rupture) processes increases.

A considerable part of our knowledge of decomposition processes of hydrocarbon ions has originated from modern mass spectrometry. Experimental techniques described so far are useful for determining not only the unimolecular fragmentation pattern of ions but also some important thermochemical parameters such as heats of formation of radicals and ions, bond energies etc. (For recent results see refs 217 and 218.) Such data can be calculated from appearance and ionization potentials which can be measured with high sensitivity using modern instruments. (The appearance potential of an ion with respect to a given decomposition is the lowest electron beam energy required to produce the ion in question.) For example, the reaction enthalpy of the following processes

$$R_1R_2 + e^- \longrightarrow R_1^+ + R_2 + 2e^-$$

is equal to the energy corresponding to the appearance potential of R_1^+ ion, and can be calculated from the following equation:

$$A(R_1^+) = \Delta H_f(R_1^+) + \Delta H_f(R_2) - \Delta H_f(R_1R_2)$$

where A stands for the energy equivalent of the appearance potential and ΔH_f represents the enthalpy of formation of the species shown in the brackets. Enthalpies of formation of radical ions are generally calculated this way, from appearance potential data and heats of formation taken from appropriate tables. The following equation, based on Hess's law, is frequently used:

$$A(R_1^+) = I(R_1) + D(R_1 - R_2)$$

This expression gives a correlation between the ionization potential I of the radical R_1, the appearance potential of the corresponding radical ion and the dissociation energy D of the given bond in the original molecule. Knowledge of appearance potentials is very important for the evaluation of competing ionic decomposition processes because the difference between the adiabatic appearance potential and the adiabatic ionization potential of the molecule gives the approximate value of the energy of dissociation.

Modern mass spectrometry makes possible the study of second-order kinetic processes of ions, the so-called ion-molecule reactions. There are two possible modifications of the analytical mass spectrometer that permit the study of reactions of the following type [219, 220]:

$$A^+ + B \longrightarrow C^+ + D \qquad (1.13)$$

The reaction can be observed if the ion undergoes several collisions within the source. One possibility is to increase the pressure, the other is to prolong the residence time of ions in the ionization chamber. Techniques for the study of ion-molecule reactions can thus be divided into two groups: one is known as high-pressure mass spectrometry, the other is ion-cyclotron resonance spectroscopy (ICR).

Kinetic parameters and the mechanism of ion-molecule reactions can be evaluated on the basis of the pressure dependence of the intensity of the ion current (between 10^{-6}–10^{-3} bar) in high-pressure mass spectrometry of pure substances. The current intensity of primary ions is proportional to the first power of pressure, that of secondary ones to its second power, and so on. The slopes of the pressure dependences can be used to estimate the effective cross-sections of the reactions, together with their rate constants. Several studies have been carried out with variation of the accelerating potential in order to get more information on the factors controlling the kinetic parameters of the reactions. Various experimental methods are available for correlating the rates of certain ion-molecule reactions (among others, the so-called stripping reactions) and the translational energy of the ions. The mechanism of ion-molecule reactions is discussed in several papers based on experiments where the direction of movement of primary and product ions has also been determined in order to elucidate the structure of the transition complex (the so-called 'beam' procedures).

A large number of studies have been carried out in this field with so-called tandem devices, which are two mass spectrometers coupled together [221]. One of them produces ions by electron impact. These are shot into the second spectrometer, which contains a flow of the target gas: the ions produced are determined in the analyser unit of the second instrument. This technique is excellent for the study of one-stage or two-stage processes; for more complex sequences the ICR technique is more useful.

High-pressure mass spectrometry is a variation of chemical ionization mass spectrometry [222, 223]. This latter technique consists of injecting a low-pressure sample (10^{-7} bar) together with a larger volume of reactive gas with a higher ionization potential (methane, isobutane, noble gases, etc.) into the ionization chamber, which is kept at a constant pressure. Ionization of the sample occurs under such conditions exclusively by means of primary ions produced from the reactive gas under electron impact (such as CH_5^+ from methane, tert-$C_4H_9^+$ from isobutane). The conventional low pressure mass spectrum is very different from that obtained by chemical ionization, as practically no collisions occur in the conventional mass spectrometer. At higher pressures, frequent collisions may bring the system into thermodynamic equilibrium, so that even the Arrhenius equation can be applied in kinetic studies.

One of the newest and most promising methods for studies of ion-molecule reactions and ion structure is ion-cyclotron resonance spectroscopy (ICR) [105, 216, 224]. This technique does not require higher pressures than conventional mass spectrometry, as the higher collision number required for second-order reactions is achieved by prolonging the residence time of ions in the ion source (up to 10^{-4}–10^{-3} s). A strong homogeneous magnetic field is applied to the ion source (or reaction chamber), so that the ions are forced into a circular orbit perpendicular to the magnetic field. The angular frequency, w_c, of the circular movement and the radius of the orbit, r, are given by the following equations:

$$w_c = \frac{eH}{m} \qquad r = \frac{v_p}{w_c}$$

where e is the charge of the ion, m is its mass, H is the intensity of the magnetic field and v_p is the velocity component perpendicular to the magnetic field. For a given value of H, the angular frequency is a function of e/m. In an electric field perpendicular to the magnetic field oscillating with frequency w_1, the ions will absorb energy if $w_1 = w_c$; this can be detected by oscillator detectors similar to those used in electron spin and nuclear magnetic resonance (NMR) techniques. In practice, w_1 is fixed and the spectrum is obtained by varying H. This so-called ion-cyclotron single resonance technique is suitable for the detection of the A^+ and C^+ components in reaction (1.13). The double-resonance technique involves another electric field with a frequency w_2. If w_2 is equal to the angular frequency of A^+, this leads to an increase in the translational energy of A^+; consequently the rate of reaction (1.13) also increases and this can be detected by a dramatic change in the energy absorption of C^+ product ions.

Single and multiple resonance methods are useful not only for determining cross sections of individual ion-molecule reactions but also for investigating the details of complex competitive–consecutive reaction mechanisms. The more general use of this technique should contribute significantly to the elucidation of problems of ionic reaction mechanisms and ion structures.

1.5.2. Photolysis and electron impact

The frequent use of photochemical methods as auxiliary techniques in radiation chemical experiments is justified by the fact that the electrons of the molecules in the system react with quanta of well-defined energy. Careful choice of the light frequency may thus ensure the selective formation of activated (electronically excited and/or ionized) species with given energy.

The absorption of monochromatic light by a homogeneous system is described by Beer's law:

$$\log (I_0/I) = \varepsilon[c]\,l$$

where I_0 is the intensity of the incident light, I is the intensity of the light after it has passed through a layer of thickness l, ε is the molar absorption coefficient, and $[c]$ is the molar concentration. The main factor governing the extent of light absorption is the absorbance, which depends on the wavelength and so should be taken into account when a light source is being selected. Alkenes and aromatic compounds have absorption maxima in the wavelength range above 180 nm, and the spectrum of mercury vapour lamps is suitable for exciting the $\pi \rightarrow \pi^*$ transition. Saturated hydrocarbons are practically transparent in this spectral range and hence light sources emitting in the vacuum-UV range (< 180 nm) have to be used [80, 225, 226]. Such light sources (Table 1.9) are either discharge tubes filled with hydrogen

Table 1.9

Wavelengths and energies of most frequently used vacuum-UV lamps

Gas	Wavelength, nm	Energy, eV
Helium	58.4	21.22
Neon	73.6	16.85
	74.4	16.67
Argon	104.8	11.83
	106.7	11.62
Hydrogen	121.6	10.2
Krypton	123.6	10.03
Xenon	147.0	8.44
Bromine	163.4	7.6
Nitrogen	174.3	7.11
	174.5	7.10

or noble gases, or resonance lamps filled with noble gases, nitrogen or bromine vapour [167, 226–229]. The field of application of mercury lamps in alkane photolysis has been limited to so-called mercury-sensitized photolysis: low-pressure mercury lamps emitting light at 184.9 and 253.7 nm are the best light sources for this purpose [226]. This technique is very useful in studying kinetic parameters of radical reactions as indicated by the large number of papers in this field.

Recently several papers have appeared on the use of cadmium resonance lamps in connection with cadmium photosensitized ($Cd(^3P_1)$) reactions [230–235]. This lamp gives a relatively long wavelength radiation (326.1 nm) and is applied mainly to the photolysis of such compounds as alcohols or amines, although a few papers dealing with cadmium sensitized reactions of hydrocarbons have also been published [231, 232].

Our understanding of processes in hydrocarbon radiation chemistry has been significantly increased by studies with resonance lamps [236]. These consist of various gases in glass tubes (Table 1.9), excited usually by a 2450 MHz microwave generator. The great advantage of resonance lamps is that they produce well-defined, relatively high photon energies that may exceed the ionization potential of hydrocarbons (9–11 eV), and they can emit high light intensities (up to 10^{16} s^{-1}). The energy of the Xe resonance lamp is somewhat lower, and that of the Ar lamp somewhat higher than the ionization potential of most hydrocarbons (Table 1.1), whereas that of the Kr lamp is in the same energy range as the ionization potential. Thus, such lamps can be used in combination to determine which products are formed in ionic reactions or subsequent to electron–ion recombination, and which have electronically excited precursors. Apart from the relatively easy experimental technique, this is the reason why reports featuring these lamps appear so frequently in the literature [16, 51–53, 82, 236–240].

Resonance lamps are usually applied in gas-phase studies. Experiments in the liquid phase are difficult to perform as the radiation is absorbed within a few micrometres, causing severe polymerization in this layer. The polymer covers the window and decreases its transparency, thus halting the reaction. This can be eliminated by thorough mixing of the sample [167, 228].

Resonance lamps have also been applied in some cases to solid-state photolysis [238, 239].

A special field of photochemistry is the photolysis of azo and diazo compounds, ketones and aldehydes that readily produce radicals [241]. Such studies may be useful for the investigation of combination, disproportionation and abstraction reactions of several hydrocarbon radicals. This method is advantageous as it does not involve high temperatures and thus the chain lengths involved are short. Such processes are generally easier to interpret than thermally initiated ones.

Recently great interest has been attached to the application of lasers for studying chemical reactions [242, 243]. The multiphoton excitation brought about by lasers provides the chemists with a means for elucidating the decomposition mechanism of excited molecules in their ground electronic state and at selected vibrational levels.

Studies of the chemical effects induced by the impact of low energy electrons might be invaluable in discovering the role of particular intermediates in radiolysis. Work of this kind has recently been performed on solid and gaseous hydrocarbons.

Hamill et al. [244, 245] studied the product distribution and fluorescence phenomena brought about by 1.5–15 eV electrons in thin layers of solid hydrocarbons. Derai et al. [246–249] investigated the dependence of product yields on the electron beam energy in the range 3.5–15 eV, in flowing hydrocarbon gases, at low pressure (10^{-5} bar). The product distribution was determined in both methods by gas chromatography.

On the basis of the appearance potentials and the energy dependence of the product yields, the precursors of certain products have been identified and conclusions drawn on the mechanism of the radiolysis.

1.5.3. Other methods

The vast majority of our knowledge of radical reactions has originated from sources of thermal kinetics [250, 251]. Kinetic parameters of radical reactions of hydrocarbons have been measured, as a rule, by investigating thermal cracking, pyrolysis, and decomposition of radical-producing substances of low thermal stability. The detailed description of the methods used in this field would fill several volumes, so we shall confine ourselves to the mentioning of a few experimental methods and the listing of some references.

The shock tube is used to study a shock wave, moving forward with high velocity (see for instance refs [252–255]). The method can be used to study radical-type chain reactions, as can the method of adiabatic compression which was developed for the study of hydrocarbon combustion.

The mechanisms of reactions and the character of transient species in irradiated systems are very similar to those in plasmochemistry. In the plasmochemical reactors, e.g. a high-voltage silent electric discharge produces ion-electron pairs and excited molecules in a similar manner to irradiation [256]. The high conversion usually achieved in such systems facilitates end-product analysis; however, this method generally offers no advantages over radiation chemistry as far as intermediates are concerned. High conversions and the fact that relatively large quantities can be transformed rather easily have facilitated the introduction of plasmochemical reactors into industry.

REFERENCES

1. HAISSINSKY, M., *Nuclear Chemistry and its Application*, Addison-Wesley Publishing Co. Reading, Massachusetts, 1964
2. BURTON, M., FUNABASHI, K., HENTZ, R. R., LUDWIG, P. K., MAGEE, J. L. and MOZUMDER, A., *Transfer and Storage of Energy by Molecules* (Eds BURNETT, G. M. and NORTH, A. M.), Wiley-Interscience, London, 1969, Vol. 1, p. 161
3. TOPCHIEV, A. V., *Radiolysis of Hydrocarbons*, Elsevier, Amsterdam, 1964
4. SWALLOW, A. J., *Radiation Chemistry*, Longman, London, 1973
5. KLOTS, C. E., Energy Deposition Processes, in *Fundamental Processes in Radiation Chemistry* (Ed. AUSLOOS, P.), Interscience, New York, 1968
6. PLATZMAN, R. L., *The Vortex*, **23**, 372 (1962)
7. PLATZMAN, R. L., in *Radiation Research* (Ed. SILINI, G.), North-Holland, Amsterdam, 1967, p. 20
8. HATANO, Y., *Bull. Chem. Soc. Jpn.*, **41**, 1126 (1968)
9. SHIDA, S. and HATANO, Y., *Int. J. Radiat. Phys. Chem.*, **8**, 171 (1976)
10. GRACHOVA, T. A., MAKAROV, V. I., POLAK, L. S. and AVDONINA, E. N., *Radiat. Eff.*, **10**, 157 (1971)
11. MAKAROV, V. I. and POLAK, L. S., *Int. J. Radiat. Phys. Chem.*, **8**, 187 (1976)
12. DE HEER, F. J., *Int. J. Radiat. Phys. Chem.*, **7**, 137 (1975)
13. KLEIN, G. and VOLTZ, R., *Int. J. Radiat. Phys. Chem.*, **7**, 155 (1975)

14. KOCH, E. E. and OTTO, A., *Int. J. Radiat. Phys. Chem.*, **8**, 113 (1976)
15. MAKAROV, V. I. and POLAK, L. S., *Khim. Vys. Energ.*, **4**, 3 (1970)
16. AUSLOOS, P. and LIAS, S. G., *Chemical Spectroscopy and Photochemistry in the Vacuum Ultraviolet* (Eds SANDORFY, C., AUSLOOS, P. and ROBIN, M. B.), Reidel, Dordrecht, Netherlands, 1974, p. 465
17. COOPER, R. and MOORING, R. M., *Australian J. Chem.*, **21**, 2417 (1968)
18. DEMEO, D. A. and EL-SAYED, M. A., *J. Chem. Phys.*, **52**, 2622 (1970)
19. VEDENEYEV, Y. I., GURVICH, L. V., KONDRAT'YEV, V. N., MEDVEDEV, A. A. and FRANKEVICH, YE. L., *Bond Energies, Ionization Potentials and Electron Affinities*, Edward Arnold Publishers Ltd., 1966
20. FRANKLIN, J. L., DILLARD, J. G., ROSENSTOCK, H. M., HERRON, J. T., DRAXL, K. and FIELD, F. H., *Nat. Stand. Ref. Data Ser.*, Nat. Bur. Stand., NSRDS-NBS 26, 1969
21. BLAUNSTEIN, R. P. and CHRISTOPHOROU, L. G., *Radiat. Res. Rev.*, **3**, 69 (1971)
22. STONEHAM, T. A., ETHRIDGE, D. R. and MEISELS, G. G., *J. Chem. Phys.*, **54**, 4054 (1971)
23. LAUB, R. J. and PECSOK, R. L., *Anal. Chem.*, **46**, 1214 (1974)
24. MAIER, J. P. and TURNER, D. W., *Disc. Faraday Soc.*, **54**, 149 (1972)
25. HERNDON, W. C., *J. Am. Chem. Soc.*, **98**, 887 (1976)
26. ROBIN, M. B., *Higher Excited States of Polyatomic Molecules*, Academic Press, New York, 1974, Vol. 1
27. KLEIN, G., VOLTZ, R. and SCHOTT, M., *Chem. Phys. Lett.*, **19**, 391 (1973)
28. MOLLER, W. M. and POPE, M., *J. Chem. Phys.*, **59**, 2760 (1973)
29. SWENBERG, C. E., RATNER, M. A. and GEACINTOV, N. E., *J. Chem. Phys.*, **60**, 2152 (1974)
30. RZAD, S. J. and BANSAL, K. M., *J. Phys. Chem.*, **76**, 2374 (1972)
31. RZAD, S. J. and BAKALE, G., *J. Chem. Phys.*, **59**, 2768 (1973)
32. RZAD, S. J., KLEIN, G. W. and INFELTA, P. P., *Chem. Phys. Lett.*, **24**, 33 (1974)
33. FREEMAN, G. R. and SAMBROOK, T. E. M., *J. Phys. Chem.*, **78**, 102 (1974)
34. BUSI, F., FLAMIGNI, L. and RODA, A., *Int. J. Radiat. Phys. Chem.*, **7**, 589 (1975)
35. ONSAGER, L., *Phys. Rev.*, **54**, 554 (1938)
36. FREEMAN, G. R., *Radiat. Res. Rev.*, **1**, 1 (1968)
37. DODELET, J.-P., FUOCHI, P. G. and FREEMAN, G. R., *Can. J. Chem.*, **50**, 1617 (1972)
38. DODELET, J.-P., SHINSAKA, K., KORTSCH, U. and FREEMAN, G. R., *J. Chem. Phys.*, **59**, 2376 (1973)
39. SHINSAKA, K. and FREEMAN, G. R., *Can. J. Chem.*, **52**, 3495 (1974)
40. ABELL, G. C. and FUNABASHI, K., *J. Chem. Phys.*, **58**, 1079 (1973)
41. SCHMIDT, W. F. and ALLEN, A. O., *J. Chem. Phys.*, **52**, 2345 (1970)
42. SCHMIDT, W. F. and ALLEN, A. O., *J. Phys. Chem.*, **72**, 3720 (1968)
43. HUMMEL, A. and SCHMIDT, W. F., *Radiat. Res. Rev.*, **5**, 199 (1974)
44. YAMAGUCHI, Y. and NISHIKAWA, M., *J. Chem. Phys.*, **59**, 1298 (1973)
45. NISHIKAWA, M., YAMAGUCHI, Y. and FUJITA, K., *J. Chem. Phys.*, **61**, 2356 (1974)
46. NISHIKAWA, M., *Can. J. Chem.*, **53**, 3075 (1975)
47. SHINSAKA, K. and FREEMAN, G. R., *Can. J. Chem.*, **52**, 3556 (1974)
48. SHINSAKA, K., DODELET, J.-P. and FREEMAN, G. R., *Can. J. Chem.*, **53**, 2714 (1975)
49. DODELET, J.-P., SHINSAKA, K. and FREEMAN, G. R., *Can. J. Chem.*, **54**, 744 (1976)
50. BABA, M. and FUEKI, K., *Bull. Chem. Soc. Jpn.*, **48**, 2240 (1975)
51. WADA, T. and HATANO, Y., *J. Phys. Chem.*, **79**, 2210 (1975)
52. WADA, T. and HATANO, Y., *J. Phys. Chem.*, **81**, 1057 (1977)
53. WOJNÁROVITS, L., WADA, T. and HATANO, Y., unpublished results
54. HOLROYD, R. A., in *Fundamental Processes in Radiation Chemistry* (Ed. AUSLOOS, P.), Wiley, New York, 1968
55. *Charged Particle Tracks in Solids and Liquids*, Proc. 2nd L. A. Gray Conf., *Cambridge, April 1969* (Eds ADAMS, G. E., BEWLEY, D. K. and BOAG, J. W.), Institute of Physics and Physical Society, Conference Series No. 8. Adlard and Son, Bartholomew, Dorking, 1970
56. HIRAYAMA, F., ROTHMAN, W. and LIPSKY, S. ,*Chem. Phys. Lett.*, **5**, 296 (1970)

57. ROTHMAN, W., HIRAYAMA, F. and LIPSKY, S., *J. Chem. Phys.*, **58**, 1300 (1973)
58. HIRAYAMA, F. and LIPSKY, S., *J. Chem. Phys.*, **62**, 576 (1975)
59. BIRKS, J. B., *Photophysical Processes of Aromatic Compounds*, Wiley, New York, 1970
60. HIRAYAMA, F. and LIPSKY, S., in *Organic Scintillators* (Ed. HORROCKS, D. L.), Academic Press, New York, 1970
61. NOYES, W. A. Jr. and AL-ANI, K. E., *Chem. Rev.*, **74**, 29 (1974)
62. MANION, J. P. and BURTON, M., *J. Phys. Chem.*, **56**, 560 (1952)
63. HARDWICK, T. J., *J. Phys. Chem.*, **66**, 2132 (1962)
64. KROH, J. and KAROLCZAK, S., *Radiat. Res. Rev.*, **1**, 411 (1969)
65. VESTAL, M. L., in *Fundamental Processes in Radiation Chemistry* (Ed. AUSLOOS, P.) Wiley, New York, 1968
66. TANNO, K., SHIDA, S. and MIYAZAKI, T., *J. Phys. Chem.*, **72**, 3496 (1968)
67. TANNO, K. and SHIDA, S., *Bull. Chem. Soc. Jpn.*, **42**, 2128 (1969)
68. FIELD, E. H. and LAMPE, F. W., *J. Am. Chem. Soc.*, **80**, 5587 (1958)
69. BONE, L. I., SIECK, L. W. and FUTRELL, J. H., *The Chemistry of Ionization and Excitation* (Eds JOHNSON, G. R. A. and SCHOLES, G.), Taylor and Francis London, 1967, p. 223
70. MCLAFFERTY, F. W., *Interpretation of Mass Spectra*, Benjamin, New York, 1967
71. BROWN, P. and DJERASSI, C., *Angew. Chem. Internat. Edit.*, **6**, 477 (1967)
72. COOKS, R. G., *Org. Mass. Spectrom.*, **2**, 481 (1969)
73. LEVINE, R. D., *J. Chem. Phys.*, **44**, 2046 (1966)
74. ROSENSTOCK, H. and KRAUSS, M., in *Mass Spectrometry of Organic Ions* (Ed. MCLAFFERTY, F. W.), Academic Press, New York, 1963
75. LORQUET, J. C., *Mol. Phys.*, **9**, 101 (1965)
76. LORQUET, J. C. and HALL, G. G., *Mol. Phys.*, **9**, 29 (1965)
76A. LOSSING, F. P. and MACCOLL, A., *Can. J. Chem.*, **54**, 990 (1976)
76B. MEHNERT, R., BREDE, O. and BOES, J., *Z. Chem.*, **17**, 268 (1977)
77. SIECK, L. W., in *Fundamental Processes in Radiation Chemistry* (Ed. AUSLOOS, P.), Wiley, New York, 1968
78. GORDON, M. S., *Chem. Phys. Lett.*, **52**, 161 (1977)
79. RICE, S. A., in *Excited States*, Vol. **2**. (Ed. LIM, E. C.), Academic Press, New York, 1975
80. RAYMONDA, J. W. and SIMPSON, W. T., *J. Chem. Phys.*, **47**, 430 (1967)
81. PARTRIDGE, R. H., *J. Chem. Phys.*, **52**, 2485 (1970)
82. AUSLOOS, P. J. and LIAS, S. G., Photochemistry in the Far Ultraviolet, *Ann. Rev. Phys. Chem.*, **22**, 85 (1971)
83. OBI, K., AKIMOTO, H., OGATA, Y. and TANAKA, I., *J. Chem. Phys.*, **55**, 3822 (1971)
84. REBBERT, R. E., LIAS, S. G. and AUSLOOS, P., *J. Photochem.*, **4**, 121 (1975)
85. SAATZER, P. M., KOOB, D. R. and GORDON, M. S., *J. Am. Chem. Soc.*, **97**, 5054 (1975)
86. HIROKAMI, S. and CVETANOVIC, R. J., *J. Phys. Chem.*, **78**, 1254 (1974)
87. MOELWYN-HUGHES, E. A., *Chemical Statics and Kinetics of Solutions*, Academic Press, London, 1971, p. 118
88. BALDWIN, R. R. and WALKER, R. W., *J. Chem. Soc., Perkin Trans.* **2**. 361 (1973),
89. KAGIYA, T., SUMIDA, Y., INOUE, T. and DYACHKOVSKII, F. S., *Bull. Chem. Soc. Jpn.*, **42**, 1812 (1969)
90. ZAVITSAS, A. A., *J. Am. Chem. Soc.*, **94**, 2779 (1972)
91. HOLROYD, R. A., *J. Phys Chem.*, **65**, 1352 (1961)
92. GIBIAN, M. J. and CORLEY, R. C., *Chem. Rev.*, **73**, 441 (1973)
93. GILLIS, H. A., *Can. J. Chem.*, **49**, 2861 (1971)
94. TILQUIN, B., ALLAERT, J. and CLAES, P., *J. Phys. Chem.*, **82**, 277 (1978)
95. HOLROYD, R. A. and KLEIN, G. A., *J. Phys. Chem.*, **67**, 2273 (1963)
96. LARSON, C. W., RABINOVITCH, B. S. and TARDY, D. C., *J. Chem. Phys.*, **47**, 4570 (1967)
97. GEORGAKAKOS, H. H., RABINOVITCH, B. S. and LARSON, C. W., *Int. J. Chem. Kinet.*, **3**, 535 (1971)
98. KLEIN, R. and KELLEY, R., *J. Phys. Chem.*, **79**, 1780 (1975)
99. KESZEI, Cs., WOJNÁROVITS, L. and FÖLDIÁK, G., *Acta Chim. Acad. Sci. Hung.*, **92**, 329 (1977)

100. THOMMARSON, R. L., *J. Phys. Chem.*, **74**, 938 (1970)
101. VERESCHINSKII, I. V. and PIKAEV, A. K., *Introduction to Radiation Chemistry* (in Russian), Akad. Izd., Moscow 1963
102. FUTRELL, J. H., *J. Am. Chem. Soc.*, **81**, 5921 (1959)
103. AUSLOOS, P., LIAS, S. G. and SCALA, A. A., *Advan. Chem. Ser.*, **58**, 264 (1966)
104. COLLIN, G. J. and AUSLOOS, P., *J. Am. Chem. Soc.*, **93**, 1336 (1971)
105. AUSLOOS, P. and LIAS, S. G., *Ion-Molecule Reactions and their Role in Radiation Chemistry*, American Chemical Society, Washington D. C., 1975
106. WAGNER, C. D., *J. Phys. Chem.*, **71**, 3445 (1967)
107. AUSLOOS, P., REBBERT, R. E. and SIECK, L. W., *J. Chem, Phys.*, **54**, 2612 (1971)
108. HO, S. K. and FREEMAN, G. R., *J. Phys. Chem.*, **68**, 2189 (1964)
109. CASTELLO, G., GRANDI, F. and MUNARI, S., *Radiat. Res.*, **45**, 399 (1971)
110. HATANO, Y. and SHIDA, S., *J. Chem. Phys.*, **46**, 4784 (1967)
111. KOVÁCS, A., CSERÉP, GY. and FÖLDIÁK, G., *Radiochem. Radioanal. Lett.*, **22**, 221 (1975)
112. HOIGNÉ, J., in *Aspects of Hydrocarbon Radiolysis* (Eds GÄUMANN, T. and HOIGNÉ, J.), Academic Press, London, 1968, p. 61
113. BURNS, W. G. and BARKER, R., in *Aspects of Hydrocarbon Radiolysis* (Eds GÄUMANN, T. and HOIGNÉ, J.), Academic Press, London, 1968, p. 33
114. GANGULY, A. K. and MAGEE, J. L., *J. Chem. Phys.*, **25**, 129 (1956)
115. KUPPERMANN, A. and BELFORD, G. G., *J. Chem. Phys.*, **36**, 1427 (1962)
116. KUPPERMANN, A., in *Radiation Research* (Ed. SILINI, G.), North-Holland, Amsterdam, 1967, p. 212
117. KUPPERMANN, A., in *The Chemical and Biological Action of Radiations* (Ed. HAISSINSKY, M.), Academic Press, London, 1961, Vol. 5, p. 85
118. FREEMAN, G. R., *J. Chem. Phys.*, **46**, 2822 (1967)
119. BURNS, W. G. and BARKER, R., *Progr. Reaction Kinetics*, **3**, 303 (1965)
120. BURNS, W. G. and REED, C. R. V., *J. Chem. Soc., Faraday Trans. I*, **68**, 67 (1972)
121. DEWHURST, H. A., *J. Phys. Chem.*, **62**, 15 (1958)
122. KOOB, R. D. and KEVAN, L., *Trans. Faraday Soc.*, **64**, 422 (1968)
123. KOCH, R. O., HOUTMAN, J. P. W. and CRAMER, W. A., *J. Am. Chem. Soc.*, **90**, 3326 (1968)
124. GÄUMANN, T., RAPPOPORT, S. and RUF, A., *J. Phys. Chem.*, **76**, 3851 (1972)
125. FÖLDIÁK, G., MEDER, W., and SCHENCK, G. O. *Radiochim. Acta*, **2**, 105 (1963)
126. LAMPE, F. W., *J. Phys. Chem.*, **61**, 1015 (1957)
127. TAYLOR, W. H., MORI, S. and BURTON, M., *J. Am. Chem. Soc.*, **82**, 5817 (1960)
128. DEWHURST, H. A., *J. Am. Chem. Soc.*, **83**, 1050 (1961)
129. WIDMER, H. and GÄUMANN, T., *Helv. Chim. Acta*, **46**, 2766 (1963)
130. STONE, J. A., *Can. J. Chem.*, **46**, 1267 (1968)
131. SAGERT, N. H., *Can. J. Chem.*, **46**, 89 (1968)
132. BOUILLOT, M. S., *Int. J. Radiat. Phys. Chem.*, **2**, 117 (1970)
133. MIYAZAKI, T., *Int. J. Radiat. Phys. Chem.*, **8**, 57 (1976)
134. DAUPHIN, J., *J. Chim. Phys.*, **59**, 1207, 1223 (1962)
135. BANSAL, K. M. and SCHULER, R. H., *J. Phys. Chem.*, **74**, 3924 (1970)
136. KOVÁCS, A., CSERÉP, GY. and FÖLDIÁK, G., *Acta Chim. Acad. Sci. Hung.*, **92**, 223 (1977)
137. RZAD, S. J. and SCHULER, R. H., *J. Phys. Chem.* **72**, 228 (1968)
138. SATO, S., TERAO, T., KONO, M. and SHIDA, S., *Bull. Chem. Soc. Jpn.*, **40**, 1818 (1967)
139. HUMMEL, A., *J. Chem. Phys.*, **48**, 3268 (1968)
140. WARMAN, J. M., ASMUS, K.-D. and SCHULER, R. H., *Advan. Chem. Ser.*, **82**, 25 (1968)
141. WARMAN, J. M., ASMUS, K.-D. and SCHULER, R. H., *J. Phys. Chem.*, **73**, 931 (1969)
142. KENNEDY, M. G. and STONE, J. A., *Can. J. Chem.*, **51**, 149 (1973)
143. TANAKA, M. and FUEKI, K., *J. Phys. Chem.*, **77**, 2524 (1973)
144. ASMUS, K.-D., WARMAN, J. M. and SCHULER, R. H., *J. Phys. Chem.*, **74**, 246 (1970)
145. SAGERT, N. H. and BLAIR, A. S., *Can. J. Chem.*, **45**, 1351 (1967)
146. SAGERT, N. H., *Can. J. Chem.*, **46**, 95 (1968)

147. TAKAO, S., HATANO, Y. and SHIDA, S., *Bull. Chem. Soc. Jpn.*, **44**, 873 (1971)
148. TAKEUCHI, K., SHINSAKA, K., TAKAO, S., HATANO, Y. and SHIDA, S., *Bull. Chem. Soc. Jpn.*, **44**, 2004 (1971)
149. HATANO, Y., TAKEUCHI, K. and TAKAO, S., *J. Phys. Chem.*, **77**, 586 (1973)
150. HATANO, Y., ITO, K. and TAKAO, S., *Int. J. Radiat. Phys. Chem.*, **7**, 39 (1975)
151. SATO, S., YUGETA, R., SHINSAKA, K. and TERAO, T., *Bull. Chem. Soc. Jpn.*, **39**, 156 (1966)
152. SAMBROOK, T. E. M. and FREEMAN, G. R., *J. Phys. Chem.*, **78**, 32 (1974)
153. HIROKAMI, S., WOJNÁROVITS, L. and SATO, S., *Bull. Chem. Soc. Jpn.*, **52**, 299 (1979)
154. HOLROYD, R. A., *Advan. Chem. Ser.*, **82**, 488 (1968)
155. MENGER, A. and GÄUMANN, T., *Helv. Chim. Acta*, **52**, 2477 (1969)
156. ROBINSON, M. G. and FREEMAN, G. R., *Can. J. Chem.*, **51**, 641 (1973)
157. RZAD, S. J., SCHULER, R. H. and HUMMEL, A., *J. Chem. Phys.*, **51**, 1369 (1969)
158. INFELTA, P. P. and SCHULER, R. H., *J. Phys. Chem.*, **76**, 987 (1972)
159. INFELTA, P. P. and SCHULER, R. H., *Int. J. Radiat. Phys. Chem.*, **5**, 41 (1973)
160. HARDWICK, T. J., *J. Phys. Chem.*, **66**, 2246 (1962)
161. HENGLEIN, A., HECKEL, E., OJIMA, Y. and MEISSNER, G., *Ber. Bunsenges. Phys. Chem.*, **67**, 988 (1964)
162. HOROWITZ, A. and RAJBENBACH, L. A., *Int. J. Radiat. Phys. Chem.*, **5**, 163 (1973)
163. KIMURA, T., FUEKI, K. and KURI, Z., *Bull. Chem. Soc. Jpn.*, **43**, 3090 (1970)
164. BABA, M. and FUEKI, K., *Bull. Chem. Soc. Jpn.*, **48**, 3039 (1975)
165. KARASAWA, H., KIM, E. R. and SATO, S., *Bull. Chem. Soc. Jpn.*, **50**, 1670 (1977)
166. BECK, G. and THOMAS, J. K., *J. Phys. Chem.*, **76**, 3856 (1972)
167. NAFISI-MOVAGHAR, J. and HATANO, Y., *J. Phys. Chem.*, **78**, 1899 (1974)
168. RAJBENBACH, L. A., *J. Am. Chem. Soc.*, **88**, 4275 (1966)
169. RAJBENBACH, L. A., *J. Phys. Chem.*, **73**, 356 (1969)
170. SAGERT, N. H., REID, J. A. and ROBINSON, R. W., *Can. J. Chem.*, **47**, 2655 (1969)
171. KENNEDY, G. A. and HANRAHAN, R. J., *J. Phys. Chem.*, **78**, 360 (1974)
172. KENNEDY, G. A. and HANRAHAN, R. J., *J. Phys. Chem.*, **78**, 366 (1974)
173. ALLEN, A. O. and HOLROYD, R. A., *J. Phys. Chem.*, **78**, 796 (1974)
174. ALLEN, A. O., GANGWER, T. E. and HOLROYD, R. A., *J. Phys. Chem.*, **79**, 25 (1975)
175. ITO, K. and HATANO, Y., *J. Phys. Chem.*, **78**, 853 (1974)
176. BANSAL, K. M. and RZAD, S. J., *J. Phys. Chem.*, **76**, 2381 (1972)
177. WILLIAMS, F., *J. Am. Chem. Soc.*, **86**, 3954 (1964)
178. ASMUS, K.-D., *Int. J. Radiat. Phys. Chem.*, **3**, 419 (1971)
179. AUSLOOS, P. and LIAS, S. G., *J. Chem. Phys.*, **43**, 127 (1965)
180. DOEPKER, R. D. and AUSLOOS, P., *J. Chem. Phys.*, **42**, 3746 (1965)
181. AUSLOOS, P., SCALA, A. A. and LIAS, S. G., *J. Am. Chem. Soc.*, **89**, 3677 (1967)
182. WOJNÁROVITS, L. and FÖLDIÁK, G., *Acta Chim. Acad. Sci. Hung.*, **82**, 285 (1974)
183. HORVÁTH, Zs. and FÖLDIÁK, G., *Acta Chim. Acad. Sci. Hung.*, **85**, 417 (1975)
184. VAN INGEM, J. F. W. and CRAMER, W. A., *Trans. Faraday Soc.*, **66**, 857 (1970)
185. KLEIN, G. W. and SCHULER, R. H., *J. Phys. Chem.*, **77**, 978 (1973)
186. THIBAULT, R. M., HEPBURN, D. R. Jr. and KLINGEN, T. J., *J. Phys. Chem.*, **78**, 788 (1974)
187. HOLROYD, R. A., in *Aspects of Hydrocarbon Radiolysis* (Eds GÄUMANN, T. and HOIGNÉ, J.), Academic Press, London, 1968, p. 1
188. SHINSAKA, S. and SHIDA, S., *Bull. Chem. Soc. Jpn.*, **43**, 3728 (1970)
188A. FÖLDIÁK, G. and SCHULER, R. H., *J. Phys. Chem.*, **82**, 2756 (1978)
189. HORACEK, K. and GETOFF, N., *Int. J. Appl. Radiat. Isotop.*, **20**, 43 (1969)
190. HARRIS, M., ESSER, J. and STONE, J. A., in *Proc. 3rd Tihany Symp. Rad. Chem.* (Eds DOBÓ, J. and HEDVIG, P.), Akadémiai Kiadó, Budapest, 1972, p. 347
191. ESSER, J. and STONE, J. A., *Can. J. Chem.*, **51**, 192 (1973)
192. STONE, J. A. and ESSER, J. *Can. J. Chem.*, **52**, 1253 (1974)
193. BERDOLT, A. and SCHULTE-FROHLINDE, D., *Z. Naturforsch.*, **22b**, 270 (1967)
194. CHRISTOPHOROU, L. G., *Chem. Rev.*, **76**, 409 (1976)
195. NENNER, I. and SCHULZ, G. J., *J. Chem. Phys.*, **62**, 1747 (1975)
196. COMPTON, R. N. and COOPER, C. D., *J. Chem. Phys.*, **59**, 4140 (1973)
197. *Pulse Radiolysis, Proc. of the Int. Symp. held at Manchester, April 1965* (Eds EBERT, M., SWALLOW, A. J., KEENE, J. P. and BAXENDALE, J. H.), Academic Press, London, 1965

198. MATHESON, M. S. and DORFMAN, L. M., *Pulse Radiolysis*, The M. I. T. Press, Cambridge, Massachusetts, 1969
198A. PATTERSON, L. K. and LILIE, J., *Int. J. Radiat. Phys. Chem.*, **6**, 129 (1974)
199. SAUER, M. C. Jr. and MANI, I., *J. Phys. Chem.*, **72**, 3856 (1968)
200. GILLIS, H. A., KLASSEN, N. V., TEATHER, G. G. and LOKAN, K. H., *Chem. Phys. Lett.*, **10**, 481 (1971)
201. BAXENDALE, J. H. and WARDMAN, P., *Trans. Faraday Soc.*, **67**, 2997 (1971)
202. HENRY, M. S. and HELMAN, W. P., *J. Chem. Phys.*, **56**, 5734 (1972)
203. FESSENDEN, R. W., *J. Phys. Chem.*, **68**, 1508 (1964)
204. FESSENDEN, R. W. and SCHULER, R. H., in *Advances in Radiation Chemistry* (Eds BURTON, M. and MAGEE, J. L.), Wiley, New York, 1970, Vol. 2
205. SCHMIDT, W. F. and ALLEN, A. O., *Science*, **160**, 301 (1968)
206. HUMMEL, A. and LUTHJENS, L. H., *J. Chem. Phys.*, **59**, 654 (1973)
207. FESSENDEN, R. W. and SCHULER, R. H., *J. Chem. Phys.*, **39**, 2147 (1963)
208. KANICK, S. W., LINDER, R. E. and LING, A. C., *J. Chem. Soc. A.*, 2971 (1971)
209. SMITH, D.R. and TOLE, J. C., *Can. J. Chem.*, **45**, 779 (1967)
210. DORFMAN, L. M. and MATHESON, M. S., *Progress in Reaction Kinetics* (Ed. PORTER, G.), Pergamon Press, Oxford, 1965, Vol. 3
211. HOLROYD, R. A., DIETRICH, B. K. and SCHWARZ, H. A., *J. Phys. Chem.*, **76**, 3794 (1972)
212. ALLEN, A. O. and HOLROYD, R. A., *J. Phys. Chem.*, **78**, 796 (1974)
213. LUKIN, L. V. and YAKOVLEV, B. S., *Int. J. Radiat. Phys. Chem.*, **7**, 667 (1975)
214. BENZ, W. and HENNEBERG, D., *Massenspektrometrie Organischer Verbindungen*, Akademische Verlagegesellschaft, Frankfurt am Main, 1969
215. BEYNON, J. H., SAUNDERS, R. A. and WILLIAMS, A. E., *The Mass Spectra of Organic Molecules*, Elsevier, Amsterdam, 1968
216. WILLIAMS, D. H. and HOWE, I., *Principles of Organic Mass Spectrometry*, McGraw-Hill, London, 1972
217. LOSSING, F. P. and TRAEGER, J. C., *Int. J. Mass. Spectrom. Ion Phys.*, **19**, 9 (1976)
218. SOLOMON, J. J. and FIELD, F. H., *J. Am. Chem. Soc.*, **97**, 2625 (1975)
219. FUTRELL, J. H. and TIERNAN, T. O., in *Fundamental Processes in Radiation Chemistry* (Ed. AUSLOOS, P.), Wiley-Interscience, New York, 1968
220. McDANIEL, E. W., CERNAK,V., DALGARNO, A., FERGUSON, E. E. and FRIEDMAN, L., *Ion-Molecule Reactions*, Wiley-Interscience, New York, 1970
221. LINDHOLM, E., *Adv. Chem. Ser.*, **58**, 1 (1966)
222. FIELD, F. H., MUNSON, M. S. B. and BECKER, D. A., *Adv. Chem. Ser.*, **58**, 167 (1966)
223. HYATT, D. J., DODMAN, A. E. and HENCHMAN, M. J., *Adv. Chem. Ser.*, **58**, 131 (1966)
224. LIAS, S. G., EYLER, J. R. and AUSLOOS, P., *Int. J. Mass. Spectrom. Ion. Phys.*, **19**, 219 (1976)
225. BAMFORD, C. H. and WAYNE, R. P., in *Photochemistry and Reaction Kinetics* (Eds ASHMORE, P. G., DAINTON, F. S. and SUGDEN, T. M.), Cambridge, University Press, 1967
226. CALVERT, J. G. and PITTS, J. N. Jr., *Photochemistry*, Wiley, New York, 1966
227. McNESBY, J. R. and OKABE, H., *Adv. Photochem.*, **3**, 157 (1964)
228. RADNOTI, D., EISEL, E. and YANG, J. Y., *Rev. Sci. Instr.*, **37**, 970 (1966)
229. GORDEN, R. Jr., REBBERT, R. E. and AUSLOOS, P., NBS Tech. Note 496, *Natl. Bur. Stand.* (U. S.), 1969
230. KALRA, B. L. and KNIGHT, A. R., *Can. J. Chem.*, **48**, 1333 (1970)
231. KALRA, B. L. and KNIGHT, A. R., *Can. J. Chem.*, **50**, 2010 (1972)
232. KALRA, B. L. and KNIGHT, A. R., *Can. J. Chem.*, **54**, 77 (1976)
233. YAMAMOTO, S., TAKAOKA, M., TSUNASHIMA, S. and SATO, S., *Bull. Chem. Soc. Jpn.*, **48**, 130 (1975)
234. YAMAMOTO, S. and SATO, S., *Bull. Chem. Soc. Jpn.*, **48**, 1382 (1975)
235. YAMAMOTO, S., TANAKA, K. and SATO, S., *Bull. Chem. Soc. Jpn.*, **48**, 2172 (1975)
236. AUSLOOS, P. and LIAS, S. G., *Radiat. Res. Rev.*, **1**, 75 (1968)
237. AUSLOOS, P., *Mol. Photochem.*, **4**, 39 (1972)
238. SCALA, A. A. and AUSLOOS, P., *J. Chem. Phys.*, **49**, 2282 (1968)

239. GORODETSKII, I. G., SKURAT, V. E. and TAL'ROSE, V. L., *Khim. Vys. Energ.*, **10**, 132 (1976)
240. COLLIN, G. J., *J. Chim. Phys.*, **74**, 302 (1977)
241. STRAUSZ, O. P., LOWN, J. W. and GUNNING, H. E., in *Comprehensive Chemical Kinetics* (Eds BAMFORD, C. A. and TIPPER, C. F. H.), Decomposition and Iso-merization of Organic Compounds. Elsevier, Amsterdam, 1972, Vol. 5
242. MUKAMEL, S. and JORTNER, J., *Chem. Phys. Lett.*, **40**, 150 (1976)
243. LESIECKI, M. L. and GUILLORY, W. A., *J. Chem. Phys.*, **66**, 4317 (1977)
244. MATSUSHIGE, T. and HAMILL, W. H., *J. Phys. Chem.*, **76**, 1255 (1972)
245. HUANG, T. and HAMILL, W. H., *J. Phys. Chem.*, **78**, 2081 (1974)
246. DERAI, R., NECTOUX, P. and DANON, J., *J. Phys. Chem.*, **80**, 1664 (1976)
247. DERAI, R. and DANON, J., *J. Chem. Phys.*, **15**, 331 (1976)
248. DERAI, R. and DANON, J., *J. Phys. Chem.*, **81**, 199 (1977)
249. DERAI, R. and DANON, J., *Chem. Phys. Lett.*, **45**, 134 (1977)
250. STEACIE, E. W. R., *Atomic and Free Radical Reactions*, Reinhold, New York, 1954
251. BENSON, S. W., *Thermochemical Kinetics: Methods for the Estimation of Thermo-chemical Data and Rate Parameters*, Wiley, New York, 1968
252. GLICK, H. S., SQUIRE, W. and HERTZBERG, A., *Symp. Int. Combust. Proc.*, **5**, 393 (1955)
253. BRADLEY, J. N. and FREND, M. A., *Trans. Faraday Soc.*, **67**, 1 (1971)
254. BRADLEY, J. N. and WEST, K. O., *J. Chem. Soc., Faraday Trans. I.*, **72**, 8 (1976)
255. BRADLEY, J. N. and WEST, K. O., *J. Chem. Soc., Faraday Trans. I.*, **72**, 558 (1976)
256. VENUGOPALAN, M. (Ed.), *Reactions under Plasma Conditions*, Wiley-Interscience, New York, 1971

2. ALIPHATIC ALKANES

Recent results in this field have been discussed in several monographs and reviews. Of these, the works of Topchiev [1], Ausloos [2], Allen [3], Hardwick [4] and Gäumann [5] deserve particular mention. This chapter concentrates on the results published since 1967–68, although, to make the picture complete, data from earlier papers are included to support our conclusions.

Systematic survey of the literature revealed the qualitative and quantitative heterogeneity of the material available. On the one hand, for example, radiolysis of n-hexane, neopentane and methane is discussed in a very large number of papers, but few papers deal with ethane, branched-chain hexanes and heptanes, for no apparent reason. On the other hand, and in connection with the above remark, the interpretation of the radiation chemical behaviour of individual hydrocarbons reflects the (sometimes somewhat biased) views and attitude of the most active research team. Therefore, the reaction mechanisms proposed cannot necessarily be regarded as fully elucidated nor as entirely verified.

2.1. INTRODUCTORY REMARKS

Under ordinary conditions (room temperature, 1 bar pressure) C_1-C_4 hydrocarbons are gaseous, C_5-C_{16} are liquid, and those with more carbon atoms solid. The values of melting and boiling points as well as density and refractive index increase with increasing carbon number [6, 7]. Of isomers with the same carbon number, branched chain species have generally lower melting and boiling points than corresponding n-alkanes (see e.g. Table 2.1).

Carbon atoms in the saturated aliphatic chain are sp^3 hybridized. As shown by X-ray diffraction studies, the tetrahedral bond angle (109.5°) is

Table 2.1
Some physicochemical data of gaseous hexane isomers [13]

Hydrocarbon	Melting point, K	Boiling point, K	$\Delta H^0_{f,\,298\,K}$, kJ mol^{-1}	$S^0_{298\,K}$, J mol^{-1}	$\Delta G^0_{f,\,298\,K}$, kJ mol^{-1}
n-Hexane	177.8	341.89	−167.2	388.4	0.25
2-Methylpentane	119.49	333.42	−174.3	380.5	−5.02
3-Methylpentane	—	336.43	−171.6	379.8	−2.13
2,3-Dimethylbutane	144.61	331.14	−177.8	365.8	−4.10
2,2-Dimethylbutane	173.28	322.89	−185.6	358.2	−9.62

approximately valid for most straight and branched-chain alkanes; the deviations observed are due to repulsion between adjacent groups and are rather small (e.g. for n-butane, the C—C—C angle is 112.4°) [8–13].

A comparison of thermodynamical data for hexane isomers [13, 14] indicates that branched isomers are more stable at room temperature, the stability increasing with more methyl groups in the molecule (Table 2.1). As, however, the entropies of formation change in the opposite direction, the order of stability will be reversed at higher temperatures. Of course, the mixture cannot be equilibrated thermodynamically, owing to the very high energies of activation for hexane isomerization. Thermal decomposition of aliphatic alkanes commences at about 300—350°C, and involves a radical mechanism (for details, see reviews in refs [15–18]).

As discussed in Chapter 1, radiolysis generates ion-electron pairs in the medium, in addition to excited particles. In the gas phase, the total yield of ionization of alkanes is $G \approx 4.3$ [19–28]. Data on ion yields in the liquid phase are rather contradictory: they may be equal to [29], lower than [30] or higher than [31] those observed in the gas phase. Recent results on liquid xenon have shown that the average energy expended on the formation of an ion-pair (W) is lower in the liquid phase than in the gaseous phase [198]. This would imply that the total yield of ionization is greater in the liquid than in the gaseous phase. Some calculations indicate a difference between the total yield of ionization and that of scavengable electrons: $G = 4.65$ and 4.10, respectively [32]. N_2O, the electron scavenger used most frequently for the determination of ion yield, also reacts with excited molecules [33–36]. Thus, $G(N_2) = 5.2 \pm 0.2$ was measured for the radiolysis of ethane, propane and neopentane [35–37]; this high yield was attributed, on the basis of photolysis studies (147 nm), to the reaction

$$RH^* + N_2O \longrightarrow N_2 + \text{products}$$

taking place in addition to electron capture.

Several papers have dealt with the correlation between the molecular structure of alkanes and their free ion yields, as well as the mobility of electrons in them. Steady-state and pulse radiolysis, charge scavenging [38–45] and physical (e.g. conductivity) studies [46, 47] are used. The good agreement between data obtained by various methods applied indicates that the free ion yield as well as the electron mobility increases with increasing spherical symmetry of the alkane molecule. Thus, alkanes containing tert-butyl groups give more free ions than the corresponding n-alkane (Tables 2.2 and 2.3). The free ion yield and the electron mobility do not depend on the dielectric constant (ε) or dipole moment of the medium (which are nearly identical for all hydrocarbons), or on the viscosity, but they increase sharply with increasing temperature. Free ion yield measured in the gas phase changes with changing pressure: with increasing pressure it drops rapidly and, in the limit, approximates the free ion yield measured in the liquid phase (Fig. 2.1). The free ion yield and the mean penetration range of secondary electrons (b) calculated from it decrease with increasing pressure in propane and propene radiolysis. The value of the latter multiplied by the

Table 2.2

Free ion yield (G_{fi}^0), most probable electron range (b_{gp}) and this range normalized for the density $(b_{gp} \cdot \varrho)$, for some alkanes

Hydrocarbon	G_{fi}^0	b_{gp}, [55] nm	$b_{gp} \cdot \varrho$, [55] 10^{-5} g cm^{-2}
n-Butane	0.22 [52]; 0.225 [41]		
2-Methylpropane	0.37 [52]; 0.31 [50]		
n-Pentane	0.12 [49]; 0.145 [55]	7.15	44.6
2-Methylbutane	0.170 [55]	7.6	46.9
2,2-Dimethylpropane	0.857 [55]; 0.81 [49]	17.84	104.9
n-Hexane	0.131 [55]; 0.11 [49]	6.74	44.2
3-Methylpentane	0.146 [55]	6.96	46.0
2,3-Dimethylbutane	0.192 [55]	7.49	49.4
2,2-Dimethylbutane	0.304 [55]; 0.40 [49]	9.2	59.5
n-Octane	0.124 [55]	6.42	44.9
Iso-octane isomers	0.332 [55]; 0.22 [53]		
	0.36 [34]; 0.34 [39]		
	0.35 [48]	9.5	65.4
n-Hexadecane	0.11 [48]	—	—

Table 2.3

Dependence of G_{fi}^0 and most probable electron range (b_{gp}) on branching. After Dodelet and Freeman [51]

Hydrocarbons	Temperature, K	G_{fi}^0	ε	b_{gp}, nm	$b_{gp} \cdot \varrho$, 10^{-5} g cm^{-2}
Neopentane	294	1.09	1.80	21.5	128
2,2,4,4-Tetramethylpentane	295	0.83	1.98	15.8	117
2,2,5,5-Tetramethylhexane	293	0.67	1.97	13.8	102
2,2,6,6-Tetramethylheptane	293	0.47	1.97	11.3	81
2,2,7,7-Tetramethyloctane	316	0.34	2.13	8.3	70
2,2,3,3-Tetramethylpentane	295	0.42	2.05	10.2	74

density $(b\varrho$, g cm$^{-2})$ is constant for a wide range of density and temperature [22, 23], in spite of the strong dependence of energy loss processes of electrons on the state of the medium [52]. The free ion yield and the electron mobility are correlated closely with the b value. The mean penetration range of electrons can be measured experimentally either by irradiation with X-rays [47, 55, 56, 263, 281, 288, 289] or by applying the photoelectric effect [54]. The photoelectric effect is used also for the determination of the energy of the conduction band (V_0) in condensed hydrocarbons. (V_0 is the energy of an excess electron in the lowest quasifree level in a medium, in relation to its energy in vacuum.)

So-called solvated electrons can be observed relatively easily in polar substances (e.g. by conductivity or optical absorbance measurements) as their lifetime is sufficiently long. Electrons generated in non-polar hydro-

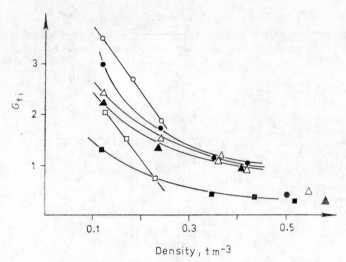

Fig. 2.1. Effect of density on free ion yields (G_{fi}) in various gases.
○ C_2H_6; □ C_2H_4; ● C_3H_8; ■ C_3H_6; ▲ n-C_4H_{10}; △ iso-C_4H_{10} (radiation: ^{60}Co-γ)
After Yamaguchi and Nishikawa [52, 200]

carbons were first observed some time ago in the glassy state at 77 K, by
means of optical or ESR techniques (see e.g. [62]), but only the latest technical developments have made it possible to trace electrons and their reactions directly in liquids of low viscosity at room temperature [63, 64].
Initially the kinetics of recombination in the spur having a higher yield
could be followed; later on, with the appearance of still more sensitive techniques, the recombination of 'free' electrons with low yields became measurable [63]. The absorption spectrum of the solvated electron has been determined in glassy 3-methylhexane at 77 K, and the time dependence of the
spectrum has been measured [64]. The presence of solvated or 'trapped'
electrons has been shown spectroscopically ($\lambda_{max} = 1500$ nm) at 193 K;
at the same time, the rate constant of a very important reaction between
electron and electron scavenger has also been measured: the rate of electron
capture by oxygen or carbon tetrachloride has been found to be about 10
times greater than the average rate of diffusion-controlled ionic reactions
[63, 64]. The rate constants of charge neutralization in various alkanes at
20°C have been determined by pulse radiolysis coupled with kinetic spectrophotometry [64]: the values are $1.1 \cdot 10^{14}$ and $5.0 \cdot 10^{15}$ $M^{-1}s^{-1}$ for n-
hexane and isopentane, respectively. The rate constant of neutralization in
3-methylhexane increased from $9.3 \cdot 10^{10}$ measured at 143 K to $4.4 \cdot 10^{12}$
$M^{-1}s^{-1}$ at 283.5 K [66].

There is a relatively close correlation between the shape of molecules and
the mobility of thermal electrons, their mean penetration range and the
yield of free ions [49–51, 331]. The rate of energy loss and the penetration
range of epithermal and thermal electrons is determined by the non-elastic
scattering interactions. The cross-section of the non-elastic scattering

in alkanes is influenced by the anisotropy of molecular polarization, which is inversely proportional to the sphericity of the molecules [51]. It is supposed that the electron mobility in the medium is determined by the various electron-accepting abilities (in fact, the relative inductive effect) of functional groups with different geometries in the molecule [51]. Thus, a much higher electron mobility was observed in methane, which has full spherical symmetry, than in ethane or propane; free ion yields for these three compounds are 0.8, 0.13 and 0.076, respectively [65].

It has been shown several times by pulse radiolysis experiments [49, 65] that high free ion yields are concomitant with the formation of an electrically conductive transition species, identified tentatively with quasi-free electrons. With low dose rates, the decay of the electron current in irradiated alkanes follows an exponential function (half-life: 5–400 ns), whereas with high dose rates a non-exponential second-order decay was found. The phenomenon was interpreted in terms of the predominance of electron-attachment process in the former instance and ion-pair recombination in the latter. This observation has been supported by steady-state irradiation studies; it was found that about 95% of the geminate n-alkane ion pairs recombine within 1 ns. Neutralization in n-alkanes is accompanied mainly by hydrogen formation: this reaction becomes less important with increasing branching of the carbon chain [25].

The correlation of geminate recombination, free ion yield and electron mobility with molecular structure sparked off several disputes in the radiation chemical literature. In spite of the growing number of experimental results, and hypotheses derived from them, the problem is still far from being unambiguously solved. The essence of the question lies in the fact that the mobility of negative charge carriers generated by radiation is higher by about two orders of magnitude than that of the ordinary large negative ions, but lower than that of quasi-free electrons ought to be (Table 2.4). It is convenient to assume partially localized electrons, whose motion is slightly hindered by scattering or trapping [57, 61, 67]. It is surprising that there are mobility differences of 2–3 orders of magnitude between alkanes with very similar physical properties, such as isomers. The

Table 2.4

Mobilities of electrons (μ) and V_0 values
of various alkanes

Hydrocarbon	μ, cm^2 V^{-1}s^{-1}	V_0, eV [61]
Tetramethylsilane	90 $\pm$ 5 [57]	
Neopentane	55 $\pm$ 5 [57]; 54 [58]; 70 [59]; 50 [51]	-0.43
2,2-Dimethylbutane	10 $\pm$ 1 [57]	
2,2,4-Trimethyl-		
pentane	7 $\pm$ 2 [57]	-0.18
n-Butane	0.4 $\pm$ 0.1 [57]	
n-Pentane	0.16 $\pm$ 0.01 [57]; 0.14 [58]; 0.075 [60]	-0.01
n-Hexane	0.09 $\pm$ 0.01 [57]; 0.082 [58]; 0.08 [59]	$+0.04$

5

temperature coefficient of mobility is positive, with the highest values for the n-alkanes which exhibit lowest absolute values of mobility [57].

Table 2.4 also contains V_0 values determined by photoelectron injection. The V_0 value expresses the energy level of excess electrons introduced into the system, or that of the 'conduction band'. The lower the value of V_0 the higher is the 'electron affinity' (and the electron mobility) in the medium [61]. Localization of the electron by the medium will occur if the average energy (E_t) of the electron to be localized is lower than the average energy of quasi-free electrons (V_0), i.e.: $E_t < V_0$ [67]. It has been verified on the basis of the statistical mechanics of liquids that, considering the very significant polarizing effect of the electrons, the localizing of electrons in so-called 'bubbles' is energetically favourable for most alkanes in spite of the value of V_0 being equal to or lower than zero [67]. On this basis it was assumed [67] that quasi-free and localized electrons are in equilibrium in hydrocarbons. This equilibrium is governed by local fluctuations around the average energy level of the medium. Apart from transient energy fluctuations, the dipole character of C—H bonds may also have a role in directing localization or delocalization, inasmuch as molecular movements may produce short-lived potential 'valleys' [57]. Recent works on reactions of electron scavengers in liquid alkanes revealed that there is a correlation between the electron scavenging rate constant (k_s) and the V_0 value of the medium. Allen et al. [329] have shown that k_s has a maximum at a certain V_0. This phenomenon is not yet understood.

The localized electron model can be developed further. The G_{fi} vs. V_0 and G_{fi} vs. μ functions calculated on the basis of Onsager's recombination law are in good agreement with experimental data [67]. The discussion of approximative theories on recombination would, however, be beyond the scope of the present work. Similarly, the theoretical and experimental papers dealing with the fate of the ion-electron pairs from the point of view of scavenger studies cannot be reviewed here either.

Much less information is available on the nature of electronically excited molecules produced in alkane radiolysis. The decomposition pathways of electronically excited states can be followed up by vacuum-UV photolysis: a few reviews on this subject should be mentioned here [69–72]. Experimental techniques developed during the last few years, such as radiation-induced fluorescence, and low-energy electron impact methods, have not yet afforded sufficient data to clarify the picture about the reactivity of excited states.

As stated in Chapter 1, electron-ion recombination and inelastic collisions of secondary electrons lead to the formation of electronically excited molecules. Scavenger studies (discussed below) as well as experiments with lasers as radiation sources [73] indicate that the neutralization reaction produces the vast majority of the excited species. Both singlet (S) and triplet (T) states can be formed, so the optical approximation cannot be accepted for excitation by inelastic collisions with relatively low energy secondary electrons (as triplet formation is optically forbidden). Triplets may be formed under the effect of electron impact, and during electron-ion recombination, as well as in intersystem crossing processes [74]. For example, in a system with total spin relaxation, the triplet/singlet ratio corresponds to the random distribution and is equal to 3/1 [74]. With liquid hydrocarbons at room temperature it is supposed that neutralization is much more rapid than spin relaxation, therefore the triplet/singlet ratio is expected to differ somewhat from this value. The T/S ratio depends on the LET value and on the size of the spur, too. It has been shown by statistical calculations that, in n-hexane, ion-electron recombination produces $T/S \approx 0.5$ [74]. This is due to the much slower spin relaxation ($\tau \approx 10^{-6}$ s) which cannot compete with the much more rapid ($\tau \approx 10^{-11}$ s) recombination. Neutralization in some hydrocarbons led to $T/S \approx 1$ values. Direct [75–77] and cross-over [78] formation of electronically excited molecules has been discussed in a few theoretical papers.

The main sources of information on the electronically excited states of alkanes are spectroscopy and photochemistry. Saturated hydrocarbons show characteristic absorption spectra in the infrared region that can be used for analysis and chemical structure determination. They are transparent in the visible and near ultraviolet region and show a noticeable absorption again only in the vacuum-UV region ($\lambda < 150$ nm), which has only recently become accessible. The sophisticated experimental technique is part of the reason why vacuum-UV spectral investigation is not very widespread. On the basis of molecular orbital theory, it is assumed that two types of transition may take place. One is the so-called $N \rightarrow V$ (or $\sigma \rightarrow \sigma^*$) transition involving the excitation of a bonding electron into an antibonding orbital: the other is called the $N \rightarrow R$ transition and leads to the formation of the so-called Rydberg states which are important in ionization. Methane and ethane show $\sigma \rightarrow \sigma^*$ transitions at wavelengths of 125 and 135 nm respectively, in the near vicinity of the sharp Rydberg bands pointing to ionization [79] (see also Section 2.2).

Although more and more UV spectra of hydrocarbons are being published, very little is known about the energetics and structure of individual excited states. A hypothesis based on quantum mechanical considerations is widely accepted: this treats the absorption spectra of alkanes as those of systems containing separate C—C and C—H bonds. The experimental justification of this is the fact that the $\sigma \rightarrow \sigma^*$ excitation of C—C bonds takes place in the region above 140 nm, whereas that of C—H bonds below 120 nm [79–81]. It is to be hoped that with the spreading of tunable lasers and photoelectron spectroscopy, more and more reliable quantitative information will be collected on the electronically excited states and reactivity of alkanes.

Irradiation of alkanes with vacuum UV photons or X-rays brings about their fluorescence [82]. The fluorescence lifetime grows from a few tenth of a nanosecond to several nanoseconds as the carbon number increases from C_6 to C_{17} (n-alkanes) together with a similar increase in the quantum yield ($2.10^{-4} \rightarrow 7.6 \cdot 10^{-3}$) [82]. This yield decreases with increasing methyl substitution, and is lowest with gem-dimethylalkanes. The differences in quantum yields of various alkanes are due to the varying rates of other processes (such as decomposition) competing with fluorescence [82]: the quantum yield of less stable, more easily decomposing 2-methylalkanes is much less than that of 3- and 4-methylalkanes. The radiating state of branched chain alkanes is characterized by the strong localization of electron excitation energy in the vicinity of the branching, resulting in the formation of a 'chromophorous group' in the molecule. Energy localization distorts the molecule geometrically in the vicinity of the branching, thus increasing the probability of processes of non-radiative character (e.g. decomposition) [82].

Experiments with low-energy electron impact indicate that alkanes possess a considerable oscillator strength in the energy ranges 3–5 and 7–14 eV [83–86]. Low-lying triplet states can be found in the former range, e.g. irradiation of n-hexane produces an unstable triplet molecule with an energy of about 3.6 eV [84]. The threshold energy for ionization is about 10 eV in alkanes. With energies as high as 16–18 eV, the efficiency of ionization reaches unity [75]. The considerable energy absorption observed between 22 and 24 eV can be attributed to the collective excitation of the molecules of the medium [85].

As stated above, the lower energy limit of ionization of alkanes is near to 10 eV; this value is not particularly structure-dependent. The cross-section of total energy absorption above the ionization potential is higher than that of ionization, except near 16 eV; the difference must be due to the formation of superexcited states having excitation energies exceeding the ionization potential. The decay of such states involves both physical (internal conversion, intersystem crossing, photon emission, autoionization) and chemical (predissociation, dissociation) processes. Very little is known about the relative rates of energy dissipation through these different channels in irradiated systems.

Some data indicate that the formation of 'hot' hydrogen atoms with high kinetic energy must be due to the decomposition of superexcited molecules. Polyakova et al. [380, 381] determined the velocity distribution of hydrogen atoms formed in methane and ethane upon collision with 300 eV electrons. They found that the distribution was close to Maxwellian with a maximum energy around 15–30 eV. This suggests that a large fraction of the decomposing molecules has very high internal energy. Even less is known about the distribution of energy and the mechanism of energy loss in polyatomic fragments formed upon radiolysis. Some data suggest that the distribution of energy on translational, vibrational and rotational degrees of freedom of the fragments is nearly equilibrium [382, 383, 386, 388]. This subject, however, needs extensive further study.

Both inter- and intra-molecular excitation energy transfer, as well as the decomposition of excited states, were interpreted by means of the exciton model [87].

2.2. INDIVIDUAL COMPOUNDS

The available data will be reviewed in order of increasing molecular mass; straight and branched chain hydrocarbons will be treated separately.

2.2.1. Methane

The large number of papers dealing with the radiation chemistry of methane is explained by the simple structure of the molecule, which decomposes into only a few radiolysis products easily identifiable by gas chromatography.

The radiation chemical yields summarized in Table 2.5 have been deter-

Table 2.5

Yields from gaseous methane

Radiation	^{60}Co-γ	Reactor	
Reference	88	129	129*
Product	G, molecule/100 eV		
Hydrogen	5.73		
Ethane	2.20		1.33
Ethylene	0.004		
Acetylene			0.30
Propane	0.36	0.36	2.02
Propene			0.37
n-Butane	0.114		2.03
Isobutane	0.040	0.24	
Butenes			0.35
Pentanes		0.07	0.90
Pentenes	0.03		0.31
Hexanes		0.03	0.34
'Polymers'	2.1		

* In the presence of 1–2% C_2H_4

mined with a total conversion of about 0.20–0.25%. The product G-values are independent of pressure in the range between 0.12 and 1.2 bar.

Owing to its relatively high ionization potential and stable C—H bonds, methane is more resistant to radiation than several products of its irradiation: therefore the extent of secondary reactions may be considerable even with very low doses [89]. These processes contribute to the strong dose-dependence of the product yields.

Contradictory data can be found in the literature concerning the yield of ethylene during methane radiolysis. Hummel's value of $G(C_2H_4) = 1.4$ [98], measured in the presence of propene added in order to eliminate the ethylene-consuming H-atom addition reaction (2.1), was later criticized [99]. Gorden and Ausloos regarded this value as too high

$$H + C_2H_4 \longrightarrow C_2H_5 \qquad (2.1)$$

$$C_2H_5^+ + C_3H_6 \longrightarrow C_2H_4 + C_3H_7^+ \qquad (2.2)$$

owing to the proton transfer process (2.2) [99]. Repeated experiments using extremely pure methane have shown, however, that the ethylene yield at

1.2 bar pressure and room temperature with a methane conversion as low as $\sim 2 \cdot 10^{-5}\%$, is indeed $G(C_2H_4) = 1.4 \pm 0.3$, extrapolated to zero irradiation dose [100]. Thus reaction (2.2) is not fast enough to influence the ethylene yield noticeably under the experimental conditions applied by Hummel. Although these latter studies gave a reliable result with respect to the initial ethylene yield, the complete neglect of the proton transfer processes in this and similar systems would be just as doubtful as the overestimation of their role. They were actually shown in the radiolysis of propane–propene [107] and methane–isobutene [106] mixtures, for example.

Klassen [117] studied the argon-sensitized radiolysis of liquid methane. He showed that a large part of the radiation energy absorbed by argon is transferred to the hydrocarbon and brings about its decomposition. Ethylene formation was observed only in the presence of NO as a radical scavenger: the yields of ethane and propane decreased in the presence of NO. These studies indicate that ethane and propane are formed mainly in radical reactions, the radical intermediates being consumed by ethylene in the absence of an effective radical scavenger such as NO.

No acetylene formation was observed following ^{60}Co γ-radiation with a dose rate of $2.8 \cdot 10^{15}$ eV g^{-1} s^{-1} and a dose of $46 \cdot 10^{19}$ eV g^{-1} [113], but acetylene does form when the irradiation is carried out with 1% propene additive, as propene 'protects' acetylene against attack by free radicals and H-atoms. Pulse radiolysis experiments have shown that small doses and high dose rates are favourable for acetylene formation [113].

The yield of polymer formation from methane in the gas phase is about $G(-CH_4) = 2.1$ [88], and in the solid phase $G(-CH_4) = 0.3$ [112, 127]. The composition of the polymer corresponds to the formula $C_{20}H_{40}$ and is independent of the absorbed dose up to about $936 \cdot 10^{19}$ eV g^{-1}. NMR spectroscopy has been used to show that, apart from methylene and methylidene groups, the polymer contains eight methyl groups [112, 127].

Earlier papers on methane radiolysis generally assumed a radical mechanism of decomposition. Sieck and Johnsen [88] attributed the decrease of $G(H_2)$ and the increase of $G(C_2H_6)$ with the absorbed dose to the competition of reaction (2.1) with the following:

$$H + H + CH_4 \longrightarrow H_2 + CH_4$$

$$H + C_2H_5 \longrightarrow C_2H_6$$

where ethylene produced during radiolysis acts as an 'internal scavenger' and inhibits hydrogen formation. The importance of radical processes is underlined by the increased yields of unsaturated products (ethylene and propene) in the presence of NO as a radical scavenger, with a concomitant decrease in the yields of hydrogen and saturated hydrocarbons — presumably owing to the capture of the H-atom by the additive in competition with reaction (2.1) [88]. The addition of about 1–2% of ethylene as radical scavenger to methane increases the yield of saturated hydrocarbon products by several orders of magnitude [129], indicating the acceleration of reaction (2.1) and the enhanced steady-state concentration of ethyl radicals.

The recent widespread application of positive ion and electron scavengers as well as several new instrumental techniques has directed attention to the study of ionic decomposition and ion-molecule reactions. Freeman was the first to propose a reaction mechanism for the radiolysis of methane involving ionic processes as well as radical reactions [89]:

$$CH_4 \xrightarrow{\;\;\sim\!\!\!\sim\;\;} CH_4^+ + e^-$$

$$\xrightarrow{\;\;\sim\!\!\!\sim\;\;} CH_4^{+*} + e^-$$

$$\xrightarrow{\;\;\sim\!\!\!\sim\;\;} CH_4^*$$

$$CH_4^{+*} \longrightarrow CH_3^+ + H$$

$$\longrightarrow CH_2^+ + H_2$$

$$\longrightarrow CH^+, C^+, H_2, H$$

$$CH_4^* \longrightarrow CH_3 + H \qquad (2.3a)$$

$$\longrightarrow CH_2 + H_2 \qquad (2.3b)$$

$$CH_4^+ + CH_4 \longrightarrow CH_5^+ + CH_3 \qquad (2.4)$$

$$CH_3^+ + CH_4 \longrightarrow C_2H_7^{+*} \qquad (2.5)$$

$$C_2H_7^{+*} \longrightarrow C_2H_5^+ + H_2 \qquad (2.6)$$

$$CH_2^+ + CH_4 \longrightarrow C_2H_6^{+*}$$

$$C_2H_6^{+*} \longrightarrow C_2H_4^+ + H_2$$

$$CH_2 + CH_4 \longrightarrow C_2H_6^* \qquad (2.7)$$

$$C_2H_6^* \longrightarrow 2 CH_3 \qquad (2.8)$$

$$C_2H_6^* + M \longrightarrow C_2H_6 + M \qquad (2.9)$$

With increasing dose the following secondary reactions contribute significantly to the formation of the main products:

$$CH_5^+ + C_2H_6 \longrightarrow CH_4 + C_2H_7^{+*}$$

$$C_2H_7^{+*} \longrightarrow C_2H_5^+ + H_2$$

$$C_2H_5^+ + C_3H_8 \longrightarrow C_2H_6 + C_3H_7^+$$

The methylene biradical has been observed during irradiation of $CH_4 - CD_4$ and $CH_4 - C_6H_6$ systems, with yields of $G(CH_2) = 1.3$ and 1.0, respectively [95]. For its direct detection, Hummel [96] proposed its addition reaction to the π-bond of tert-but-2-ene, leading to dimethylcyclopropane. Scavenger studies have shown that about half of the biradicals in the singlet and triplet states are formed by decomposition of ionic species, the other half originates from excited intermediates, e.g. via reaction (2.3). As indicated by results from radiolytic and photolytic studies of CH_4–CD_4 gas mixtures, reactions (2.7) and (2.9) are important processes of ethane formation [135, 136]. The reactivity of CH_2 biradicals has extensively been studied [400, 402, 403], mostly by photochemical methods.

Ethane is also formed by the combination of methyl radicals as well as from reactions (2.7) and (2.9). By investigating the pulse radiolysis of aqueous methane solutions, the rate constant of this combination reaction has been found to be $k = 1.24 \pm 0.2 \cdot 10^9$ M^{-1} s^{-1} [134]. The H-atom abstraction reaction contributes little to the formation of methyl radicals as the C—H bond is very strong (427 kJ mol^{-1}). Data on gas-phase radiolysis indicate that $G(H_2)$ is constant between 32 and 100°C; above this temperature it increases and, above about 150°C, the reaction

$$H + CH_4 \longrightarrow H_2 + CH_3$$

also contributes to hydrogen formation [104].

Vacuum-UV photolysis of methane carried out with He (21.2 eV) and Ne (16.7–16.8 eV) resonance lamps involves several reactions of great importance in radiolysis. Such experiments have confirmed that proton transfer from the $C_2H_5^+$ ion produced in reaction (2.6) is important in ethylene formation [122]. Rebbert and Ausloos did not observe the formation of C, CH and CH_2 species during the irradiation of methane with 16.7–16.8 eV photons, these species being the precursors of ethylene. The quantum yield of ionization in this energy range is $\Phi = 1$. These experiments provide indirect proof of the formation of most of the above-mentioned fragments via excited or superexcited methane, involving perhaps consecutive decomposition processes [122].

Some of the ionic reactions involved in methane radiolysis can be observed directly, and their kinetic parameters have been determined by mass spectrometry. In a high-pressure mass spectrometer the formation of C_1–C_6 alkane ions from methane has been observed [90]. What is more, the formation of polymers with significant $G(-CH_4)$ values was attributed to consecutive ion-molecule reactions [88, 91]. The rate constant of reaction (2.4) was found by ion-cyclotron resonance spectroscopy to be $k = 6.6 \cdot 10^{11}$ M^{-1} s^{-1} [92]. It has been shown by experiments using crossed molecular beams [93] that reactions (2.5) and (2.6) give hydrogen directly without the formation of $C_2H_7^+$ transition species. Field and Beggs, as well as Hiraoka and Kebarle have investigated methane decomposition in a high-pressure 0.1—1 mbar) mass spectrometer [94, 237]; they found that the following reactions lead to equilibrium:

$$CH_5^+ + CH_4 \rightleftharpoons C_2H_9^+$$

$$C_2H_5^+ + CH_4 \rightleftharpoons C_3H_9^+$$

$$C_2H_9^+ + CH_4 \rightleftharpoons C_3H_{13}^+$$

The latter authors observed the formation of tert-butyl ions in the ion chamber of the mass-spectrometer [256]. The relative intensities of ion-molecule condensation products decrease with increasing temperature, whereas those of CH_5^+, $C_2H_4^+$, $C_2H_5^+$ and $C_2H_9^+$ ions increase significantly [94]. The ions observed with highest intensity during methane irradiation are CH_5^+ and $C_2H_5^+$: the rate constants of all their second-order reactions so far observed are close to $6 \cdot 10^{11}$ M^{-1} s^{-1}, i.e. practically every collision leads to reaction. These two ions react with saturated and unsaturated products, present in higher concentrations with increasing dose, by processes such as proton transfer and condensation [101].

The energy dependence of certain ion-molecule reactions in methane are discussed in a number of works [242, 245, 255, 257, 441, 442, 445, 449].

Pulse radiolysis and photolysis have been used to show that the C, CH and CH_2 fragments participating in reactions (2.10) and (2.11) are formed in high yields and are important precursors of ethylene and acetylene [105, 118–120]:

$$CH(^2\pi) + CH_4 \longrightarrow C_2H_5^* \longrightarrow H + C_2H_4 \qquad (2.10)$$

$$C(^1D) + CH_4 \longrightarrow C_2H_4^* \longrightarrow C_2H_2 + H_2 \qquad (2.11)$$

$$(k_{2.10} = 1.5 \cdot 10^9 \text{--} 4.8 \cdot 10^{10} \text{ } M^{-1} \text{ } s^{-1}; \quad k_{2.11} = 1.9 \cdot 10^{10} \text{ } M^{-1} \text{ } s^{-1})$$

Combination reactions between relatively inert species may also contribute to ethylene and acetylene formation, especially at higher pressures and dose rates [121]:

$$C(^3P) + CH_3 \longrightarrow C_2H_2 + H$$

$$^1CH_2 + M \longrightarrow {}^3CH_2 + M$$

$$^3CH_2 + {}^3CH_2 \longrightarrow C_2H_2 + H_2$$

$$^3CH_2 + CH_3 \longrightarrow C_2H_4 + H$$

Several contradictory statements relating to the mechanism of acetylene formation have been published in the literature. By changing the energy of the irradiating electrons it has been shown [108–111] that the threshold energy of acetylene formation is close to the ionization potential (i.e. presumably ionic intermediates play a role in its formation), whereas ethane and ethylene can also be produced at much lower electron energies (about 7.5 eV). Comparison of data obtained using pulse or steady-state irradiations, as well as those obtained with stable isotopic tracers [98, 105, 110, 113, 114, 115, 126], reveals that the G-values given for acetylene show a rather large scatter, indicating a considerable dependence on dose and dose rate (Table 2.6). Rebbert and Ausloos have concluded, on the basis of scavenger and pressure-dependence studies, that with steady-state irradiation (e.g. γ-radiation) the reactions of species formed by direct excitation, such as $C(^1S)$ and $C(^1D)$, represent the main pathway of acetylene formation, but that with pulse radiolysis, ionic species (CH_3^+, $C_2H_5^+$) are more important [126].

Table 2.6

Yields of C_2-products from methane

Radiation	Electron (pulse: 20 ns)	Electron (pulse: 20 ns)	^{60}Co-γ	^{60}Co-γ
Dose rate, eV g^{-1} s^{-1}	$5.6 \cdot 10^{26}$	$1.6 \cdot 10^{27}$		
Dose, 10^{19} eV g^{-1}	1.12	3.2	100<	15.6
Reference	114	115	88	114
Product	*G*, molecule/100 eV			
Ethane	0.7	2.1	2.20	2
Ethylene	0.7	1.05	0.004	<0.008
Acetylene	0.5	0.85		<0.002

Several possible explanations of polymer formation can be found in the literature. Hamlet et al. [112] assumed that this polymer is produced in so-called 'blobs' with a diameter of about 10 nm under the effect of secondary electrons with energies between 100 and 500 eV, in regions where the local concentration of ionic intermediates CH_n^+ ($n = 0, 1, 2$ or 3) is very high. The ionic pathway of formation is also supported by experiments indicating high yields of polymer under the effect of light with a wavelength of 58.4 nm (21 eV), as well as by the yield of polymer $G(-CH_4)$ being 1.3 in the liquid phase in an argon matrix, where a highly effective charge transfer takes place from primary Ar^+ ions to hydrocarbon molecules [127]. Several authors have suggested the following reaction for the chain build-up [112, 127, 128]:

$$CH_4 + C_nH_{2n+1}^+ \longrightarrow H_2 + C_{n+1}H_{2n+3}^+$$

Rate constants of reactions between CH_5^+, $C_2H_5^+$, CH_3^+, CH_2^+ etc. and methane have been given in several reviews [102, 103]. Thus whereas at low doses CH_5^+ in pure methane suffers neutralization and gives H_2 and a radical fragment, it reacts with the propane product in an exothermal process at high doses [130–133]:

$$CH_5^+ + CH_3-CH_2-CH_3 \longrightarrow C_2H_5^+ + 2CH_4 \quad \Delta H = -54 \text{ kJ mol}^{-1}$$

Huntress has verified the proton transfer and hydrogen abstraction processes of CH_4^+ and CH_3^+ ions by studying deuteromethane in a high-pressure mass spectrometer [116]. He suggested that the direction and rate of the reactions, as well as the lifetime of the transition species ($0.1–3.6 \cdot 10^{-12}$ s) are significantly influenced by the kinetic energy of the ions.

It is clear from the above discussions that the radiolytic decomposition of methane is a very complex process that has not yet been fully elucidated.

2.2.2. Ethane

Holland and Stone [165] have shown that the main products of the radiolysis of liquid ethane (195 K) are hydrogen, methane, ethylene, propane and n-butane. Table 2.7 summarizes G-values for ethane radiolysis carried out in the liquid and gas phases.

Table 2.7

Yields from ethane (radiation: ^{60}Co-γ)

Dose rate, eV $g^{-1} s^{-1}$	$2.8 \cdot 10^{16}$	$1.8 \cdot 10^{16}$	$1.1 \cdot 10^{15}$
Dose, 10^{19} eV g^{-1}	1	1	15
Temperature, K	313	195	195
Pressure, bar	1.45		
Reference	168	165	157
Phase	Gaseous	Liquid	
Product	G, molecule/100 eV		
Hydrogen	8.7	3.68 ± 0.05	3.96
Methane	0.75	0.44 ± 0.05	0.39
Ethylene	1.2	0.67 ± 0.05	0.74
Propane	0.75	0.55 ± 0.05	0.52
n-Butane	2.5	2.6 ± 0.1	2.40
$G(H)^a$		2.3 ± 0.2	2.3 ± 0.2
$G(C_2H_5)^a$		5.81 ± 0.6	—
$G_0(H_2)^b$		5.97 ± 0.25	6.08
$G_0(C_2H_4)^c$		2.96 ± 0.25	—

^{a} Yield of the species given in parentheses
^{b} Yield extrapolated to zero dose, determined by NO radical acceptor
^{c} The yield was determined in the presence of 6 mole% NO

Whereas the yields of methane and propane are independent of the dose, the G-values of hydrogen and ethylene decrease, that of n-butane increases with increasing dose in the range 0.3–$6 \cdot 10^{19}$ eV g^{-1} [165]. With a very low dose $G(C_2H_4) = 2.96$, but with doses higher than $4 \cdot 10^{19}$ eV g^{-1} $G(C_2H_4) = 0.7$. According to the hypothesis of Holland and Stone, the difference is due to the addition of thermal hydrogen atoms to ethylene: it is surprisingly about equal to the yield of thermal hydrogen atoms, as determined by the use of radical scavengers: $G(H_{therm}) = 2.3$ [165].

It should be noted that Gawlowski et al. [142] obtained 1.07 for the G-value of n-C_4H_{10} in the gas phase, as opposed to $G = 2.5$ cited in Table 2.7. They carried out their experiments with extremely pure ethane and concluded that the much higher yields of butane reported in the literature were due to the presence of traces of water in the samples [142, 208].

Fessenden and Schuler [138] have observed ethyl, methyl and vinyl radicals in the liquid phase radiolysis of ethane, using ESR spectroscopy.

G-values of gas phase radiolysis of ethane show a slight dependence on pressure (and density). With increasing density, the yields approach the values measured in the liquid phase, e.g. with $\varrho = 0.529$ g cm^{-3} the observed G-values are equal to those determined in the liquid phase at 195 K [168]. Scavenger studies indicate that the mechanism of radiolysis changes with increasing density: primary decomposition of excited ethane and ion-molecule condensation reactions slow down whereas the probability of charge neutralization increases. Because the latter process, as shown by scavenger and tracer studies, leads mainly to the formation of ethyl radicals and ethylene, it is no wonder that the yields of butane and ethylene increase with increasing density while a simultaneous decrease of $G(H_2)$, $G(CH_4)$ and $G(C_3H_8)$ is also observed [168]. These conclusions are based on the experimental facts that the free ion yield [168] and the ionic fragment yield [146, 168, 170] both decrease with increasing gas density.

According to studies carried out in the temperature range between 16 and 34°C, under critical conditions ($T_{cr} = 32.2$°C, $\varrho_{cr} = 0.209$ g cm^{-3}), the overall yield of C_1-C_4 products of γ-radiolysis of ethane increases abruptly: moving away from the critical temperature, G-values decrease [140]. This phenomenon is as yet unexplained.

Holland and Stone [165] found that G-values for hydrogen and butane decrease, whereas that for ethylene increases in the presence of cyclopentene radical scavenger (max. 1.7 mole%). Systems containing electron scavengers (such as SF_6) exhibited an approximately twofold decrease of the yields of thermal hydrogen atoms and ethyl radicals, indicating that about half of both these species are formed in ionic processes [166]. Similar results have been obtained using ND_3 as a positive ion scavenger. Holland and Stone have also shown that some ethylene is formed in ionic reactions.

Bakale and Gillis [157] have determined the value of $G(H_{therm})$ by adding C_2D_4 radical scavenger to ethane: they found the same value as determined on the basis of dose dependence [$G(G) = 2.3$]. From dose dependence and scavenger studies, as well as from a consideration of the reaction mechanism (2.12)–(2.16) the ethylene yield extrapolated to zero dose has been estimated to be $G_0(C_2H_4) = 2.8 \pm 0.2$. The ratio of the rate constants of addition and abstraction reactions of H atoms has also been calculated: $k_{add}/k_{abs} = 5 \pm 3 \cdot 10^5$. Assuming the same A factors for these two reactions, the difference in activation energies has been estimated to be: $E_{abs} -$ $- E_{add} = 30 \pm 1$ kJ mol^{-1} (the value obtained in gas-phase thermal decomposition experiments was 18 kJ mol^{-1}). The ratios of rate constants of the disproportionation and combination reactions, k_d/k_c, for individual radical reactions have also been calculated on the basis of radiolysis yields in the presence of scavengers: $k_d/k_c = 0.13 \pm 0.03$ and 0.41 ± 0.06 for the reactions methyl–methyl and ethyl–ethyl, respectively [157]. These values are considerably higher than those obtained from gas phase pyrolysis experiments but are in agreement with data related to the liquid phase. The phase effect on k_d/k_c has been explained by several hypotheses, none of which gives an unambiguous interpretation of the 30–40% difference between k_d/k_c values measured in the gas and the liquid phase at room temperature [196].

Holland and Stone have put forward a radical mechanism explaining the experimental results [165], with the following main steps:

$$C_2H_6 \quad \longrightarrow\!\!\!\wedge\!\!\wedge\!\!\longrightarrow \quad H, CH_3, C_2H_5 \qquad\qquad (2.12)$$

$$H + C_2H_6 \quad \longrightarrow \quad H_2 + C_2H_5 \qquad\qquad (2.13)$$

$$H + C_2H_4 \quad \longrightarrow \quad C_2H_5 \qquad\qquad (2.14)$$

$$2\,C_2H_5 \quad \Bigg\langle \quad
\begin{array}{ll}
C_2H_4 + C_2H_6 & (2.15\mathrm{a}) \\[1em]
\text{n-}C_4H_{10} & (2.15\mathrm{b})
\end{array}$$

$$CH_3 + C_2H_5 \quad \Bigg\langle \quad
\begin{array}{ll}
CH_4 + C_2H_4 & (2.16\mathrm{a}) \\[1em]
C_3H_8 & (2.16\mathrm{b})
\end{array}$$

It can be assumed on the basis of this reaction sequence that the effect of radical scavengers (e.g. cyclopentene) is due to the 'protection' of ethylene from attack by radicals or H-atoms.

Several papers deal with the yields of various pathways of hydrogen formation (thermal, 'hot', molecular) in ethane radiolysis as well as with the G-values of methyl and ethyl radicals. It has been stated that about 60% of the hydrogen $[G(H_2) = 8.3\text{--}8.8]$ formed at room temperature and atmospheric pressure is produced in reactions of thermalized H-atoms [143–145]. The remainder $[G(H_2) \approx 3.3]$ can be attributed to abstraction reactions of 'hot' hydrogen atoms and molecular hydrogen elimination: these latter reactions are not affected by ethylene radical scavengers or by increasing temperature, as shown by Lias et al. [139]: they have studied the radiolysis of $C_2H_6\text{--}C_2D_6\text{--}C_2H_4$ mixtures at a pressure of 0.4 bar. Scavenger studies with NH_3 and SF_6 indicate that this fraction of the hydrogen is formed via decomposition of electronically excited ethane molecules. These reactions must be extremely rapid, as Pearson and Innes have shown by optical spectroscopy that the dissociative lifetime of the first electronically excited level of ethane is $\tau \approx 10^{-12}\text{--}10^{-11}$ s [148]. Possible reactions of 'hot' hydrogen atoms have been studied by tritium 'recoil' studies. Decomposition of ethane initiated by $^3\mathrm{He}(n, p)\mathrm{T}$ nuclear reaction leads to several abstraction and substitution reactions, with HT, CH_3T, C_2H_5T and C_2H_3T as the main products [159].

G-values for thermal, 'hot' and molecular hydrogen yields in the radiolysis of gaseous ethane are: $G(H_2) = 5.0\text{--}5.5$, 2.7 and 0.6, respectively [145]. In the liquid phase, $G_0(H_2) = 5.7\text{--}5.8$ (extrapolated to zero dose), whereas the yield of thermal H-atoms is $G(H_{\text{therm}}) = 2.3\pm0.2$ [157, 165].

Gawlowski et al. [142] have established that neutralization followed by molecular H_2 elimination is the main process of hydrogen formation in the gas-phase radiolysis of ethane.

The $^{14}C_2H_5$ radical sampling technique has been used to determine the yield of methyl radicals as $G(CH_3) = 0.6$ [176]. The ratio of the yields of methyl and ethyl radicals was reported to be $G(CH_3)/G(C_2H_5) \leq 0.10$: using ESR spectroscopy the relative abundance of methyl, ethyl and vinyl radicals was found to be $4 : 93 : 3$ [138]. Thus, the probability of C—C relative to C—H bond rupture is in the range 0.15–0.3 [138].

Ionic reactions can be studied using electron and positive ion scavengers and by high-pressure mass spectrometry and ion-cyclotron resonance spectroscopy. The dependence of relative intensities on the pressure permits the elucidation of complex reaction mechanisms consisting of consecutive ionic decomposition processes and ion–molecule reactions. Thus, it was found that $C_2H_6^+$, $C_2H_4^+$, $C_2H_3^+$, $C_2H_2^+$, and CH_3^+ are the products of primary decomposition reactions, whereas $C_2H_5^+$, $C_3H_3^+$, $C_3H_5^+$, $C_3H_7^+$ and $C_4H_7^+$ are the products of secondary ion-molecule processes [153, 158]. With increasing pressure the rates of condensation-type reactions (such as the combination of $C_2H_4^+$, $C_2H_5^+$ and $C_2H_6^+$ ions with ethane molecules leading to $C_3H_7^+$, $C_3H_9^+$, $C_4H_9^+$ ions) increase:

$$C_2H_6^+ + C_2H_6 \longrightarrow [(C_2H_6)^{+*}]_2 \underset{M}{\overset{}{\nearrow}} \begin{array}{l} C_3H_9^+ + CH_3 \quad (2.17a) \\ (C_2H_6)_2^+ \quad\quad (2.17b) \end{array}$$

Ausloos et al. considered the ethyl, ethylene, ethane and vinyl ions to be of greatest importance in ethane radiolysis [153]. $C_4H_9^+$ and $C_3H_7^+$ ions, formed in reactions (2.18a) and (2.18b)

$$C_2H_5^+ + C_2H_6 \overset{k}{\longrightarrow} [C_4H_{11}^{+*}] \begin{array}{l} \nearrow C_4H_9^+ + H_2 \quad (2.18a) \\ \searrow C_3H_7^+ + CH_4 \quad (2.18b) \end{array}$$

($k = 1.7 \cdot 10^{11}$ M^{-1} s^{-1} [156]) have a thermodynamically more stable secondary structure as shown by deuterium tracer studies [131, 153].

The yield of ethylene ions is much lower than that of ethyl ions [153]. The reaction of ethylene ions with ethane molecules is very slow ($k = 6 \cdot 10^9$–$1.8 \cdot 10^{10}$ M^{-1} s^{-1}) [161]. Vinyl ions are even less important: e.g. in the radiolysis of a $1 : 1$ mixture of C_2H_6 and C_2D_6 only about 10% of ethylene consists of C_2H_3D and C_2HD_3.

Harrison and co-workers have stated that the relative intensities of all ionic species approach a saturation value or decrease with increasing pressure in the range below 0.16 μbar. The only exception is $C_4H_9^+$, with a rapidly increasing relative intensity [131]. Although there are no direct evidence for the formation of ions with more carbon atoms, the detailed investigation of this phenomenon could provide indirect support for the ionic mechanism of polymerization.

As stated by Ausloos et al. [153], the rate of collisional stabilization of $C_4H_{11}^{+*}$ ions becomes higher with increasing pressure and, consequently, the yield of $C_4H_9^+$ decreases. Stabilized $C_4H_{11}^+$ is not a reactive ion; its most probable reaction is neutralization followed by decomposition to radical and molecular products.

The formation of a polymer with an average composition of $C_{23}H_{46}$ has been observed in the radiolysis of solid state ethane at 77 K. The yield was reported to be $G(-C_2H_6) = 0.2$ [127] whereas in the liquid phase in an argon matrix $G(-C_2H_6) = 2.1$. Libby et al. have proposed an ionic mechanism for polymerization [127, 128] that would involve reaction (2.19)

$$C_2H_5^+ + C_2H_6 \longrightarrow C_4H_9^+ + H_2 \qquad (2.19)$$

as its most important step.

Table 2.8 contains kinetic parameters of important ion-molecule reactions of ethane radiolysis, as well as collision rate constants (k_{calcd}) calculated on the basis of the ion-induced dipole model, considered in refs [131,

Table 2.8

Rate constants for some reactions of ions with
ethane molecules

Ion	k_{obs}, 10^{10} M^{-1} s^{-1}	k_{calc}, 10^{10} M^{-1} s^{-1}
$C_2H_2^+$	78 $\pm$12 [131]; 126 [161]	80.4 [131]
$C_2H_3^+$	22.2$\pm$ 1 [131] 29 $\pm$ 5 [161]	79.8 [131]
$C_2H_4^+$	0.3 [131]; 0.6 $-$ 1.8 [161]; 5.4 $-$12 [154]	79.2 [131]
$C_2H_5^+$	3 $-$15 [154]	
$C_2H_6^+$	1.1$-$1.92 [131]; 4.2$-$10.8 [161]; 3 $-$12 [154]	78.0 [131]

154, 158, 161 adn 162]. Considerable differences are observed between rate constants obtained from different sources (owing presumably, to the differences of experimental technique, temperature and pressure values), however, different reactivities of individual ions are obvious.

Klassen has determined the following values for the yields of main products in the liquid phase in an argon matrix: $G(CH_4) = 0.33$; $G(C_2H_4) = 0.89$; $G(C_3H_8) = 0.24$; $G(n\text{-}C_4H_{10}) = 1.81$. These values are independent of ethane concentration in the range 0.05–1% [117]. Therefore, it has been concluded that the rate determining steps of processes leading to these products involve intermediates with relatively long lifetimes. The neutralization of Ar$^+$ ion competes with the positive charge transfer to ethane. However, yields are independent of alkane concentration, so it can be assumed that excitation transfer predominates.

Decomposition of the electronically excited ethane molecule has been studied by Ausloos et al. by vacuum-UV photolysis [139, 150]. The following important primary processes have been observed during irradiation with Ar resonance lines (11.6–11.8 eV) [139]:

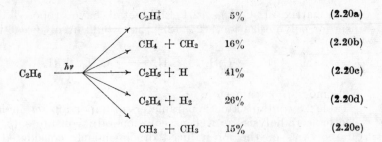

$$C_2H_6 \xrightarrow{h\nu}$$

$C_2H_6^+$	5%	(2.20a)
$CH_4 + CH_2$	16%	(2.20b)
$C_2H_5 + H$	41%	(2.20c)
$C_2H_4 + H_2$	26%	(2.20d)
$CH_3 + CH_3$	15%	(2.20e)

At pressures of less than 1 bar C_2H_5 and C_2H_4 suffer considerable secondary decomposition. If photolysis is carried out with Xe (8.4 eV), Kr (10.0 eV) and Ar (11.6–11.8 eV) lamps, the percentages of the bond-rupture processes (2.20c) and (2.20e) in the total yield were 13, 38 and 56%, respectively, whereas the importance of processes (2.20b) and (2.20d), involving partial rearrangement of the molecule, decreased with increasing photon energy. The phenomenon was interpreted in terms of a shorter lifetime of the excited species with increasing energy, thus decreasing the time available for re-arrangement [139].

The methylene biradical is also formed in good yield in the γ-radiolysis of ethane. 95% of the propane formed during radiolysis of a 1 : 1 mixture of C_2H_6–C_2D_6 consisted of C_3D_8, C_3H_8, $C_3H_6D_2$ and $C_3H_2D_6$ species [153], and the yield increased with increasing pressure. This fact can be explained by insertion of the methylene biradical into C–C and C–H bonds, followed by the collision stabilization of the excited propane molecule [153].

Several papers have been published on the decomposition kinetics of vibrationally excited ethane molecules. In some cases, the decomposition reaction rate constants calculated on the basis of transition-state theories and their quantum-statistical reformulation by Rice, Ramsperger, Kassel and Marcus (RRKM) have been correlated with values measured in photolytic, pyrolytic and chemically activated systems [125, 173, 175]. The detailed discussion of such results is beyond the scope of this book.

2.2.3. Propane

Radiation chemical yields measured in the gas or liquid phase radiolysis of propane are summarized in Table 2.9 [184, 187, 189]. All the G-values were obtained using relatively small doses of ^{60}Co γ-radiation (3.1–12.4 · 10^{19} eV g^{-1}).

The main radiolytic process is hydrogen formation. The G-value of C_1-C_2 fragments is relatively low as compared with the yields of hydrogen and propene. Zhitneva et al. [184] have attributed the formation of C_4-C_6 products to combination reactions of fragment and parent alkyl radicals.

Gawlowski et al. [142, 208] obtained much lower yields for hexane products than those given in Table 2.9, G(2,3-dimethylbutane) = 0.69, G(2-methylpentane) = 0.28 and G(n-hexane) = 0.04. They gave evidence

Table 2.9

Yields from propane (radiation: ^{60}Co-γ)

Temperature, °C	35	35	—78	—78
Reference	189	189	184	187
Phase	Gas (1 bar)	Liquid		
Product	G, molecule/100 eV			
Hydrogen			4.90	4.8
Methane	1.4	0.75	0.60	0.68
Ethylene	1.0	0.66	0.57	0.41
Ethane	2.2	0.50	0.56	0.37
Acetylene	0.31	0.09		
Propene	2.0	2.9	3.30	1.75
Isobutane	1.4	0.8		
n-Butane	0.33	0.3		
Isopentane	0.29	0.13		
n-Pentane	0.07	0.07	2.80	
2,3-Dimethylbutane	1.7	1.4		
2-Methylpentane	0.48	0.85		
n-Hexane	0.07	0.15		
$G(-C_3H_8)^a$	12.1	10.4		

^{a} These yields were estimated by taking into account the yields and carbon atom numbers of products

that the enhancement of the yields of these products is observed if traces of water are not properly removed from the propane gas sample.

Sieck et al. have observed a strong dose dependence in the gas-phase radiolysis of propane [185]. Wei and Bone [186] studied this phenomenon in C_3H_8–C_3D_8 mixtures, and showed that some radiolysis products, principally unsaturated hydrocarbons, act as internal scavengers. The effect is two-fold: they capture H-atoms, and inhibit product formation by scavenging positive ions.

The G-values of alkane and alkene products decrease with increasing dose in liquid-phase radiolysis, as shown by Koob and Kevan [189]; this phenomenon can be interpreted in the same way as the gas-phase process [143, 189, 197]. Bone et al. [199] underlined the role of proton or positive charge transfer reactions from $C_3H_7^+$ and $C_3H_8^+$ intermediates to alkene molecules (in addition to the radical scavenging effect of alkene products) among the processes leading to yield decrease.

The comparison of yields measured in the gas and liquid phases leads to conclusions concerning the mechanisms of radiation chemical processes. The phase effect in propane radiolysis has been studied at 35°C [188, 189]. The overall G-values of decomposition in the vapour and liquid phase were 12.1 and 10.4, respectively. Koob and Kevan explained this difference of about 14% by the lower average excitation energy content of activated molecules in the liquid phase, caused by the higher rate of collisional de-activation [189].

The major contribution to the decrement of the overall yields when going from the gaseous to the liquid phase is due to the non-radical processes: the decrease of the yield of radical processes is as low as 4% within the overall decrease of 14% [189]. This observation is in general agreement with other results [191–194], although the comparability of experimental data is doubtful in many cases because G-values at identical temperatures in the gas and liquid phases have been reported in very few papers.

Data on the low-pressure (below 1 bar) radiolysis of propane are rare in the literature, but the study of the medium pressure range (1–8 bar) shows that yields are not sensitive to pressure variation between 1 and 6 bar [178, 184, 189]. Although G-values are not generally dependent on pressure, there are some indications that the mechanism may depend on the pressure and density. Yamaguchi and Nishikawa irradiated propane gas in a supercritical state with γ-rays, and found that variation of the density between 0.12 and 0.42 g cm^{-3} was accompanied by a decrease of the free ion yield from $G = 3.0$ to $G = 1.0$ [200]. Freeman et al. have reported that the free ion yields are $G_{\mathrm{fi}} = 0.36$ and 0.11, respectively, in the liquid state at $+23$ and $-90°$C when the corresponding densities are 0.50 and 0.645 g cm^{-3}, respectively [201, 202]. The mean penetration range of secondary electrons has been calculated by using semiempirical formulae: in the density range between 0.12 and 0.645 g cm^{-3} this range varied between 106 and 9.3 nm [200, 202]. These data indicate that whereas the neutralization of ions produced in low-pressure gas has a low probability, neutralization is a favourable process at high pressures or in the liquid state.

The following relative probabilities of individual decomposition reactions have been calculated by Bone, Sieck and Futrell, who applied deuterium tracers and radical scavengers in the gas phase γ-irradiation of propane [205]:

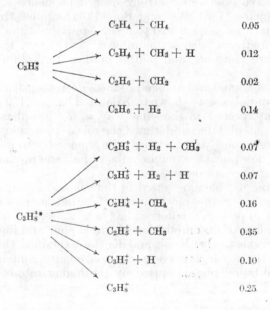

$$C_3H_8^* \begin{cases} C_2H_4 + CH_4 & 0.05 \\ C_2H_4 + CH_3 + H & 0.12 \\ C_2H_6 + CH_2 & 0.02 \\ C_3H_6 + H_2 & 0.14 \end{cases}$$

$$C_3H_8^{+*} \begin{cases} C_2H_3^+ + H_2 + CH_3 & 0.07 \\ C_3H_5^+ + H_2 + H & 0.07 \\ C_2H_4^+ + CH_4 & 0.16 \\ C_2H_5^+ + CH_3 & 0.35 \\ C_3H_7^+ + H & 0.10 \\ C_3H_8^+ & 0.25 \end{cases}$$

Ausloos and Lias [182] have concluded, on the basis of photolytic (10.0 eV) studies of the C_3D_8–H_2S system that the following reactions have also high probabilities:

$$C_3H_8^* \quad \longrightarrow \quad \begin{array}{l} CH_3 \dotplus C_2H_5 \\ \\ C_3H_7 + H \end{array}$$

C_2H_4, C_2H_5, C_3H_6 and C_3H_7 intermediates containing excess internal energy give H, CH_3, C_2H_2, C_2H_3 and C_2H_4 in a secondary decomposition process, or, alternatively, can be stabilized in collision.

Wada et al [258] studied the effect of phase on the decomposition and stabilization of the $C_3H_8^+$ excited cation. Their yields of $C_2H_5^+$, $C_3H_7^+$ and $C_3H_8^+$ were, in the liquid phase, 0.3, 0.7 and 2.5, respectively. These values should be compared with the yields of the same species in the gas phase [205], 1.30, 0.37 and 0.94, respectively. The differences between the yields in the two phases illustrate the great importance of collisional deactivation in the liquid phase.

Several papers have discussed the mechanism of propane radiolysis on the basis of scavenger and tracer studies. Fujisaki et al. have shown [180] using C_2H_4, N_2O and SF_6, that the hydrogen yield (the value of which was $G(H_2) = 5.6 \pm 0.5$ when the experiments were carried out at low doses corresponding to a conversion of less than 0.01%) results from the reactions of different precursors as shown in Table 2.10. The hydrogen yield due to

Table 2.10

Sources of the hydrogen formation from propane
(radiation: ^{60}Co-γ). After Fujisaki et al. [180]

Source of hydrogen	Phase	Direct excitation and ionization	Neutralization
		G, molecule/100 eV	
Thermal H		2.5	} 1.9
Hot H	Gaseous	0.9	
Molecular H_2		1.4	0.7
Thermal H		0.9	0.6
Hot H	Liquid	1.2	0.7
Molecular H_2		0.4	1.0

neutralization is higher in the liquid than in the gas phase: in turn, the yield of hydrogen originating from the decomposition of directly excited molecules and ions is higher in the gas phase [180]. These results do not contradict the conclusions drawn with respect to the free ion yield, namely that the yield of charge neutralization increases at higher pressures (or densities) but the yield of ionic decomposition and simultaneously with the overall yield of free ions both decrease.

6*

Tanno and Shida have proposed the following important reactions for ethane and ethylene formation in liquid systems containing SF_6 and NH_3 additives [187]:

$$C_3H_8 \longrightarrow C_2H_5^+ + CH_3 + e^-$$

$$C_2H_5^+ + e^- \longrightarrow C_2H_4 + H \tag{2.21}$$

$$C_3H_8^+ + e^- \longrightarrow C_2H_4 + CH_4$$

$$C_2H_5^+ + C_3H_8 \longrightarrow C_2H_6 + C_3H_7^+$$

The yield of ethane is enhanced in the presence of SF_6 as electron scavenger owing to competition between processes (2.21) and (2.22):

$$e^- + SF_6 \longrightarrow SF_6^- \tag{2.22}$$

In a system containing both SF_6 electron and NH_3 positive ion scavengers, the G-value of ethylene and propene increases two- to three-fold, whereas the ethane yield decreases by about 60%, owing to the competition between reactions (2.22), (2.23) and (2.24) [187]:

$$C_2H_5^+ + NH_3 \longrightarrow C_2H_4 + NH_4^+ \tag{2.23}$$

$$C_3H_7^+ + NH_3 \longrightarrow C_3H_6 + NH_4^+ \tag{2.24}$$

On the basis of scavenging plots, $G(C_2H_5^+) = 0.28$ and $G(C_3H_7^+) \approx 1.1$ [187]. Koob and Kevan used oxygen as a radical scavenger and showed that of the principal products, methane, ethane, ethylene and propene are formed mainly via non-radical processes, either unimolecular decomposition or ion-molecule reactions [188].

Deuterium tracer studies revealed that in addition to (2.21), reactions (2.25) and (2.26) have considerable yields [188]:

$$C_2H_4^+ + C_3H_8 \longrightarrow C_2H_6 + C_3H_6^+ \tag{2.25}$$

$$C_2H_3^+ + C_3H_8 \longrightarrow C_2H_4 + C_3H_7^+ \tag{2.26}$$

Koob and Kevan attributed about 25% of the total ethylene yield to reaction (2.26), the remaining 75% being formed by decomposition of excited propane molecules. The proposal of reaction (2.25) provoked several disputes: on the basis of theoretical considerations it has been concluded that the H_2^- transfer reaction should proceed very slowly and in very low yield [187].

Unimolecular and bimolecular reactions are responsible for 80 and 20% of methane formation, respectively, both in the gas and the liquid phase [188]. In the gas phase 64% of the methane is due to ionic and 36% to electronically excited intermediates, but in the liquid phase the unimolecular fraction of methane is produced entirely by ionic processes [188].

Koob and Kevan [189] have also studied the reactions of alkyl radicals produced during γ-radiolysis of propane with oxygen as radical scavenger.

The decrease in yield observed in the presence of oxygen has been interpreted in terms of the interception of reactions of thermal alkyl radicals. Thus, the yields of C_4-C_6 products become zero verifying the hypothesis that their formation involves exclusively radical combination processes. The values of k_d/k_c for 1-propyl and 2-propyl radicals calculated on the basis of scavengeable C_4–C_6 yields in the gas phase are in good agreement with literature data determined by other experimental methods, such as photolysis and thermal decomposition in the gas phase [189]. The k_d/k_c values in the liquid phase are higher by about 30% than those measured in the gas phase, in agreement with other sources [188]. It has been reported that the overall yields of the decomposition of propane molecules irradiated by $6.2 \cdot 10^{19}$ eV g^{-1} are $G(-C_3H_8) = 6.4$, 9.3 and 10.4 at -130, -78 and $+35°C$, respectively [189]. The temperature variation affected the rates of free radical processes, but had no influence on ionic and molecular decomposition reactions, which had a temperature-invariant yield of $G = 1.6$.

It is general experience that higher temperatures promote reactions with higher energies of activation. Thus, in the liquid phase radiolysis of propane, addition of hydrogen atoms onto the π-bond of the product propene ($E_a = 8.5$ kJ mol^{-1}) and the subsequent combination of propyl radicals ($E_a \approx 0$ kJ mol^{-1}) are predominant at lower temperatures, whereas at higher temperatures hydrogen atom abstraction ($E_a = 30$–40 kJ mol^{-1}) is favoured. The latter reaction pathway leads to higher overall $G(-C_3H_8)$ and product yield values owing to competition between reactions (2.27) and (2.28):

$$H + C_3H_8 \longrightarrow H_2 + C_3H_7 \qquad (2.27)$$

$$H + C_3H_6 \longrightarrow C_3H_7 \qquad (2.28)$$

$$C_3H_7 + C_3H_7 \longrightarrow C_6H_{14} \qquad (2.29)$$

Of the dimer products, the yields of 2-methylpentane and n-hexane, as a function of temperature, reach a maximum, whereas that of 2 3-dimethylbutane increases monotonically (Table 2.11). The former phenom-

Table 2.11

Yields of dimer products from liquid propane
(radiation: ^{60}Co-γ). After Koob and Kevan [189]

Temperature, °C	−130	−78	35
Product	G, molecule/100 eV		
2-Methylpentane	0.58	1.14	0.85
2,3-Dimethylbutane	0.34	0.61	1.4
n-Hexane	0.29	0.53	0.15

enon has been interpreted by Koob and Kevan [189] in terms of an increased relative rate of hydrogen atom abstraction by 1-propyl radicals at higher temperatures, as compared with that of the combination reaction, which has approximately zero energy of activation. Consequently, the rates of the reactions

$$1\text{-}C_3H_7 + 1\text{-}C_3H_7 \longrightarrow n\text{-}C_6H_{14}$$

$$1\text{-}C_3H_7 + 2\text{-}C_3H_7 \longrightarrow (CH_3)_2CHCH_2CH_2CH_3$$

decrease. The increase in the G-value of 2,3-dimethylbutane at higher temperatures has been attributed to the acceleration of the reactions

$$R + C_3H_8 \longrightarrow RH + 2\text{-}C_3H_7$$

$$2\text{-}C_3H_7 + 2\text{-}C_3H_7 \longrightarrow (CH_3)_2CHCH(CH_3)_2$$

where $R = CH_3$, $1\text{-}C_3H_7$, etc. [189]. On the basis of total product analysis of propane radiolysis the ratio of isopropyl to n-propyl radicals has been estimated to be $G(2\text{-}C_3H_7)/G(1\text{-}C_3H_7) = 1.4$, 1.4 and 4.0 at -130, -78 and $+35°C$, respectively, owing to the different temperature dependences of the rates of the abstraction and combination reactions [189]. A similar temperature dependence was found by Widmer and Gäumann for the ratio of 2-hexyl and 1-hexyl radicals [194].

The k_d/k_c ratios calculated for liquid phase radiolysis were 1.3, 1.5 and 1.7 times higher at $+35$, -78 and $-130°C$, respectively, than that observed at $+35°C$ in the gas phase [189].

The ratio of the rate constants of abstraction (2.27) and addition (2.28) reactions, k_{abs}/k_{add}, has been calculated on the basis of the temperature dependence of propene and dimer yields with and without radical scavengers. The value obtained is $k_{abs}/k_{add} = 3 \cdot 10^{-6}$, which is in agreement with other data obtained in the gas phase with different experimental methods [195].

Very few papers have been published on the high-temperature, so-called radiation-thermal decomposition of propane [203, 204]. Results show that both product distribution and overall energy of activation depend on the temperature. Polak et al. have suggested a radical mechanism for high-temperature radiolysis [204], which involves methyl radicals abstracting hydrogen and hydrogen atoms themselves acting as the chain carriers. 1-Propyl and 2-propyl radicals, produced by abstraction, decompose to give methyl radicals and hydrogen atoms:

$$1\text{-}C_3H_7 \longrightarrow CH_3 + C_2H_4 \qquad\qquad (2.30)$$

$$2\text{-}C_3H_7 \longrightarrow C_3H_6 + H \qquad\qquad (2.31)$$

The increasing rate of methane formation with increasing temperature indicates that the yield of 1-propyl radicals increases, whereas that of 2-propyl radicals decreases. Experimental values of energies of activation support this hypothesis [204]: the activation energy measured in radiation-thermal experiments between 350 and 480°C is 100 ± 4 kJ mol^{-1}, which is exactly the same as that reported for reaction (2.30) in purely thermal systems. The activation energy given by Polak et al. for the temperature range 523–576°C is $E_a = 267\pm4$ kJ mol^{-1}, that is higher than that of reaction (2.31) measured in ordinary thermally induced system s, $E_a = 170$ kJ mol^{-1}.

The primary processes of propane radiolysis, their yields as determined by extensive scavenger and tracer studies, and their dependence on temperature and state [188, 189] are summarized in Table 2.12.

The photolysis [160, 169, 174, 177, 181, 183, 218], mass spectrometric reactions [205] and noble-gas-sensitized radiolysis of propane [205, 206] have been studied by several research groups but detailed discussion of these subjects cannot be included in this volume.

Table 2.12

Yields of primary decomposition from propane
(radiation: ^{60}Co-γ). After Koob and Kevan [188–189]

Temperature, °C	35	35	—78	—130
Phase	Gas (1 bar)	Liquid		
Reaction	G, molecule, radical or ion/100 eV			
$C_3H_8^{+*} \rightarrow C_2H_4^+ + CH_4$	0.34	0.15	0.19	0.16
$C_3H_8^{+*} \rightarrow C_2H_5^+ + CH_3$	1.3	0.27	0.14	0.09
$C_3H_8^{+*} \rightarrow C_3H_5^+ + H_2 + H$	0.28	0.13		
$C_3H_8^{+*} \rightarrow C_2H_3^+ + CH_3 + H_2$	0.27	0.13	0.12	0.08
$C_3H_8^* \rightarrow C_3H_6 + H_2$	0.60	0.75		
$C_3H_8^* \rightarrow C_2H_4 + CH_4$	0.16	0.10	0.10	0.07
$C_3H_8 \rightarrow C_2H_4 + CH_3 + H$	0.45	0.28	0.26	0.18

2.2.4. n-Butane

The radiation chemical yields of n-butane determined under various experimental conditions are collected in Table 2.13.

The hydrogen yield in the liquid phase at room temperature is $G(H_2) = 4.7$–4.8 [221, 234]. The mechanism of hydrogen formation has been studied by Shida et al. [234] using the deuterium tracer technique, N_2O and SF_6 as electron scavengers and C_2H_4 as additive to capture thermal H-atoms with high efficiency. Table 2.14 contains data on the important intermediates of hydrogen formation in the liquid phase, together with the nature and yields of precursors. These data are in good agreement with those determined for propane radiolysis by similar methods [180], indicating that liquid-phase radiolysis shows several close similarities for different alkanes. For example, the ratio of primary to secondary C—H bond rupture in n-butane radiolysis was measured, using iodine as radical scavengers, as 1 : 3.2 [240, 241], which is close to the value found for n-hexane (1 : 2.8).

The primary products of radiation chemical decomposition of n-butane are excited and/or ionized molecules; further fragmentation results in the rupture of both C—H and C—C bonds. The G-value of all C—C bond-rupture

Table 2.13

Yields from n-butane (radiation: ^{60}Co-γ)

Dose rate, eV g^{-1} s^{-1}			$1.5 \cdot 10^{15}$				$1.2 \cdot 10^{16}$	$1.3 \cdot 10^{16}$
Dose, 10^{19} eV g^{-1}			12.9				12.9	16.7
Temperature, °C			150				0	room
References			[220]				see below	
Phase			Gas				Liquid	
Density, g cm^{-3}	0.125	0.083	0.042	0.022	0.010	0.0018		
Product						G, molecule/100 eV		
Hydrogen							4.80 [234]	4.73^a [221]
Methane	2.31	2.38	2.63	3.72	4.18	4.21		0.52 [239]
Ethane	3.03	2.80	2.91	3.59	3.66	2.88	1.03 [217]	0.99 [239]
Ethylene	0.72	0.76	1.94	1.35	1.52	1.56	0.86 [217]	0.78 [239]
Propane	0.73	0.83	1.14	1.60	1.89	1.97	0.16 [217]	0.16 [239]
Propene	0.35	0.36	0.36	0.40	0.41	0.46	0.20 [217]	0.20 [239]
But-1-ene								0.88^a [221]
trans-But-2-ene							} 2.63 [217]	0.50^a [221]
cis-But-2-ene								0.48^a [221]

^{a} Dose rate: $0.9 \cdot 10^{16}$ eV g^{-1}s^{-1}; extrapolated to zero dose

Table 2.14

Sources of hydrogen formation from liquid n-butane
(radiation: ^{60}Co-γ). After Shida et al. [234]

Sources of hydrogen	Direct excitation and ionization	Neutralization
	G, atoms (or molecules)/100 eV	
Thermal H	0.7 ± 0.2	0.5 ± 0.2
Hot H	1.4 ± 0.2	0.6 ± 0.2
Molecular H$_2$	0.6 ± 0.2	1.0 ± 0.2

processes can be determined on the basis of the yields of stable C$_1$-C$_7$ products as follows [235]:

$$\Sigma G(C-C) = \frac{G(C_1-C_3) + G(C_5-C_7)}{2} = 3.25$$

This method of calculation gives an approximately correct value for ΣG(C—C) if the secondary decomposition of fragment ions and/or radicals is negligible.

Thus, the total bond-rupture yield can be calculated from the yields of end-products; however, this method gives no information about the relative frequency of rupture of individual C—C bonds, or about the ionic or radical character of the intermediate of any given product.

The important intermediates in n-butane radiolysis and their yields have been studied in both the gas and the liquid phase. How these yields vary as a function of the density in the presence of various scavengers is a good indication of the mechanism involved (see below) [219, 220].

Miyazaki [220] has investigated the gas-phase radiolysis of n-butane using ^{137}Cs γ-radiation in the density range 0.0018–0.125 g cm^{-3} (max. 40 bar), with a dose rate of $1.5 \cdot 10^{15}$ eV g^{-1} s^{-1} and a dose of $12.9 \cdot 10^{19}$ eV g^{-1}, at 150°C. It has been reported [220] that the G-values of ionic intermediates participating in propane formation decrease rapidly with increasing density (or pressure). Presumably no propane is produced via neutralization of $C_4H_{10}^+$ ions [220, 221]. The decreasing yield of $C_3H_7^+$ ions at higher densities indicates that, at higher pressures, the rate of collision deactivation of $C_4H_{10}^+$ excited ions increases rapidly. The yield of propyl ion extrapolated to zero density $G_0(C_3H_7^+)$ is 1.3. The yield of propane produced in ion-molecule reaction also drops with increasing density: under such conditions the neutralization of $C_3H_7^+$ takes place.

As for the mechanism of propane formation in gas-phase n-butane radiolysis, it has been reported that at lower pressures, in the presence of ammonia, which reacts rapidly with positive ions, the predominant process is the decomposition of excited butane molecule ions to give methyl radicals and propyl radical ions. Hydride ion transfer to propyl ions then gives propane:

$$C_4H_{10}^{+*} \longrightarrow CH_3 + C_3H_7^+ \qquad (2.32)$$

$$C_3H_7^+ + C_4H_{10} \longrightarrow C_3H_8 + C_4H_9^+$$

With increasing density (and pressure), the neutralization of $C_3H_7^+$ ions and the collision deactivation of excited $C_4H_{10}^{+*}$ ions followed by their radical decomposition, become more important:

$$C_3H_7^+ + e^- \longrightarrow H, CH_3, C_2H_4, C_3H_6$$

$$C_4H_{10}^{+*} + C_4H_{10} \longrightarrow C_4H_{10}^+ + C_4H_{10}$$

$$C_4H_{10}^+ + e^- \longrightarrow CH_3, C_2H_5, C_3H_7, \text{ etc.}$$

Figure 2.2 illustrates that propane formation at lower pressures is mainly ionic, and at higher pressures a radical process. The theoretical calculation of the density dependence of the G-value of the $C_3H_7^+$ ion has been attempted by Miyazaki [220], on the basis of the quasi-equilibrium reaction rate theory of mass spectra [222, 223]. He assumed that the unimolecular decomposition of the excited parent ion is competing with collisional deactivation. The use of an oversimplified model gave, however, no satisfactory agreement with the experimental data [220].

The investigation of C_4H_{10}–C_4D_{10} mixtures in the presence of NH_3 and O_2 scavengers indicated the importance of reaction (2.32) in propane formation [224, 225]. Ausloos et al. found a ratio of 4 : 1 between the yields of

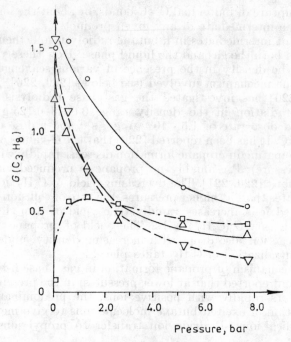

Fig. 2.2. Effect of pressure on the yield of propane in the radiolysis of n-butane, ○ From ionic intermediates; △ from $C_3H_7^+$ ions; ▽ by ion–molecule reactions; □ from C_3H_7 radicals (radiation: ^{60}Co-γ). After Miyazaki [220]

CD_3CHDCD_3 and $CD_3CD_2CHD_2$; with higher concentrations of ammonia they observed higher values for G(cyclopropane). However, they were unable to decide whether $C_3H_7^+$ possesses a primary, secondary or cyclic structure, so they subjected n-butane to noble-gas-sensitized radiolysis. The predominant process was shown to be the decomposition of excited n-butane ion to give n-propyl ions and methyl radicals ($\sim 90\%$) [223]. The n-propyl ion isomerizes within 10^{-10} s to cyclo-$C_3H_7^+$ or sec-$C_3H_7^+$ ion, as indicated by the isotopic composition of deuterated propane fraction. The lower the excess internal energy of the excited n-butane ion, the higher the probability of a rearrangement prior to decomposition. Thus, if krypton is used as sensitizer, the internal energy of $C_4H_{10}^+$ is higher by about 1–2 eV than if xenon is used. Whereas the proportions of $CD_3CDCD_3^+$ and $CD_3CD_2CD_2^+$ ions are 63.3 and 36.7%, respectively, with xenon, the corresponding amounts in the presence of krypton are 93.4 and 6.4%, respectively [229]. $C_3H_7^+$ ions give propane in a hydride transfer reaction. Experiments have indicated that the hydride transfer of n-propyl ions is about five times more rapid than that of sec-propyl ions [229]. The hydride ion transfer of n-propyl ions is always exothermal, but with sec-propyl ions this is true only when secondary or tertiary hydrogens are transferred [229]. The possibility that

$C_3H_7^+$ has a cyclic structure cannot be excluded, as in the presence of NH_3 the yield of cyclo-C_3H_6 increases.

There are contradictory suggestions in the literature concerning the transition state of decomposition of excited n-butane ions. Some authors, such as Vestal [226], have assumed that prior to decomposition the energetically favourable rearrangement

$$n\text{-}C_4H_{10}^+ \longrightarrow iso\text{-}C_4H_{10}^+$$

takes place to an extent depending on the energy content: Liardon and Gäumann [227, 228] have put forward a hypothesis based on the so-called 'internal methyl radical elimination':

$$\left[\begin{array}{c} D_3C—CH\cdots\cdots\cdots CD_3 \\ H \quad\quad CH_2 \end{array} \right]^+$$

At low pressures, the hydride ion transfer between $C_3H_7^+$ and a parent n-butane molecule is a probable process leading to propane and sec-butyl ions. The structure of sec-$C_4H_9^+$ ion has not yet been elucidated. Lias et al. have shown that it isomerizes within 10^{-9}–10^{-8} s to the thermodynamically more stable tert-$C_4H_9^+$ ion [225]. In the pressure range between 30 and 900 mbar, the total yield of butyl ions increases, with a decrease of tertiary ion yield within the total value. The rate of transformation into tertiary structure is enhanced by an increase in the internal energy of the butyl ions. These studies are supported by the experiments that have shown the energy of activation of the transformation to be about 75 kJ mol^{-1} [225]. It is supposed that at least three varieties of $C_4H_9^+$ ion participate in radiation chemical processes: primary, secondary and a species of unknown — possibly cyclic — structure [225]. Miyazaki and Shida have determined the yield of $C_4H_9^+$ ions: $G(C_4H_9^+) = 2.4$ [235].

Fujisaki et al. have observed a drop in the hydrogen yield from $G(H_2) = 6.6$ to 2.4 under the effect of 6 mole% of C_2H_4 as radical scavenger at a pressure of 1.2 bar: the yield of ethane increased simultaneously from $G(C_2H_6) = 1.7$ to 3.3, and that of propane from $G(C_3H_8) = 1.7$ to 2.0 [230]. The admixing of 0.1 mole% of SF_6 caused a sharp decrease in the hydrogen yield, with other G-values practically unchanged. Addition of 1% of propene to the system n-C_4H_{10}-O_2 increased the propane yield by $G=1.5$. These results indicate that the alkene radical scavengers participate in reactions other than capture of H atoms and alkyl radicals. The increase of the propane yield in the presence of propene or O_2 was attributed to H-transfer from $C_4H_{10}^+$ ion to propene, as is the increase of the ethane yield in the presence of ethylene [230–233]. These experiments suggested a yield of butane ions, $G(C_4H_{10}^+)$ of about 1.7, whereas another experimental technique [235] gave $G(C_4H_{10}^+) = 2.6$. It is doubtful whether alkenes mixed with alkanes as radical scavengers can influence the distribution of products by H- and H_2-transfer processes, and if so, to what extent [234]. The charge-transfer process from alkane to alkene is not regarded as likely although its energetics are favourable [236]. Ionic interactions between alkanes and alkenes may contribute to dose dependence phenomena.

Several papers have been published on the radiolysis of liquid n-butane, mainly in the presence of NH_3 positive ion, SF_6 electron and O_2 radical scavengers. The radiolysis of the liquid n-C_4H_{10}–n-C_4D_{10} system has also been investigated. The comparison of yields measured in the gas and the liquid phase led to conclusions on the ratio and importance of decomposition pathways involving electronically excited or ionized states. Tanno and Shida [187] measured the following values for the yields of methane, C_2 and C_3: 0.96, 2.48 and 2.30 in the gas phase and 0.50, 1.87 and 0.45 in the liquid phase, respectively, i.e. the methane yield was lower by about 50%, the C_2 yield by 20% and the C_3 yield by about 70% in the liquid phase. Miyazaki et al. used iodine as radical scavenger, and reported that about twice as much methyl iodide as ethyl iodide is produced in the gas phase, whereas in the liquid phase the yield of ethyl iodide exceeded that of methyl iodide [221]. The decomposition of excited and ionized butane molecules may involve the following reactions:

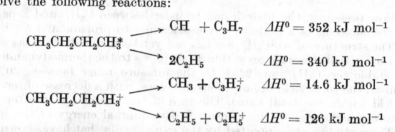

$$CH_3CH_2CH_2CH_3^* \rightarrow CH + C_3H_7 \qquad \Delta H^0 = 352 \text{ kJ mol}^{-1}$$
$$\rightarrow 2C_2H_5 \qquad \Delta H^0 = 340 \text{ kJ mol}^{-1}$$
$$CH_3CH_2CH_2CH_3^+ \rightarrow CH_3 + C_3H_7^+ \qquad \Delta H^0 = 14.6 \text{ kJ mol}^{-1}$$
$$\rightarrow C_2H_5 + C_2H_5^+ \qquad \Delta H^0 = 126 \text{ kJ mol}^{-1}$$

On the basis of a comparison of heats of reaction, more ethyl radicals can be expected from excited molecules, whereas the decomposition of butane ions leads predominantly to methyl radicals [237]. Yields in the presence of radical scavengers and heats of reaction [221] suggest that ionic decomposition is more important in the gas phase, whereas the most favourable process in the liquid phase is a free radical type decomposition of excited molecules.

A significant fraction of products in liquid-phase radiolysis can be attributed to radical combination, disproportionation and abstraction reactions. McCauley and Schuler have determined the yields of important alkyl radicals [243]: $G(CH_3) = 0.4$; $G(C_2H_5) = 0.94$; $G(1\text{-}C_3H_7) = 0.09$.

The yields of individual ionic intermediates and their main reactions in liquid-phase radiolysis have been determined by scavenger studies. Thus, Fujisaki et al. have estimated the yield of ethyl ions to be $G(C_2H_5^+) = 0.25$ [217]. Ethyl ions boost the yield of ethane by hydride ion transfer. This is supported by the liquid-phase radiolysis of an 1 : 1 mixture of C_4H_{10}–C_4D_{10}, where 93.5% of the ethane produced consisted of C_2H_6, C_2D_6, C_2H_5D and C_2HD_5, the remaining 6.5% being $C_2H_4D_2$ and $C_2H_2D_4$ in the presence of oxygen radical scavenger. Similarly, the $C_3H_7^+$ ion is important in the formation of propane [217]. The yield of propyl ions was given as $G(C_3H_7^+) = 0.15$, and scavenger studies have shown $G(C_4H_9^+)$ to be 0.6 [217].

Matsuoka et al. [260] studied the radiation-sensitized thermal cracking of n-butane in the temperature range 17–548°C. The pressure was 0.5 bar at 0°C. At 548°C, the radiation chemical yields of isobutane and isobutene were 31.2 and 12.1, respectively. These products were not observed in the thermal cracking of butane. The authors postulated the following mechanism for the radiation-thermal cracking:

$$sec\text{-}C_4H_9^+ + C_3H_6 \longrightarrow C_7H_{15}^+$$

$$C_7H_{15}^+ \longrightarrow C_7H_{15}^{+'} \text{ (isomer)}$$

$$C_7H_{15}^{+'} \longrightarrow C_3H_7^+ + iso\text{-}C_4H_{10}$$

$$C_3H_7^+ + n\text{-}C_4H_{10} \longrightarrow sec\text{-}C_4H_9^+ + C_3H_8$$

The results described so far have been supported and additional data have been provided by the studies of Munson, which were carried out in a high-pressure mass spectrometer (0.6–2 mbar, 200±10°C) in order to investigate ionic processes [207]. The author showed that 800 V electrons cause total ionization, with 76% of the ions being present as $C_4H_9^+$ (mainly sec-butyl) [209]. The concentration of $C_4H_9^+$ is independent of pressure: its constancy indicates that it is a rather unreactive species. The rate of hydride transfer between $C_2H_5^+$ and $C_3H_7^+$ and n-butane is very great. The relative intensity of $C_3H_7^+$ ions increases with pressure: this suggests the occurrence of reactions other than primary decomposition in $C_3H_7^+$ production. Among these, so-called methyde transfer [207, 211]:

$$R^+ + C_4H_{10} \longrightarrow C_3H_7^+ + RCH_3$$

and H_2 transfer

$$C_3H_5^+ + C_4H_{10} \longrightarrow C_3H_7^+ + C_4H_8$$

can be mentioned, together with dissociative proton transfer processes in which the energy of activation necessary for $C_3H_7^+$ production is supplied by the translation energy of the colliding ion [211, 218]. The overall reaction rate constants of ion-molecule processes between important ionic species and butane molecules are given in Table 2.15 [161, 207, 212].

Table 2.15

Rate constants for reactions of ions with n-C_4H_{10} molecules

Ion	k, 10^{10} $M^{-1}s^{-1}$
$C_4H_9^+$	0.04 [207]
$C_4H_8^+$	0.06 [207]
$C_3H_7^+$	23 ± 18 [207]
$C_3H_6^+$	31 ± 7 [207]; 24.6–27.0 [212]
$C_3H_5^+$	30 ± 4 [207]; 51 [161]
$C_2H_5^+$	45 ± 5 [207]; 40 ± 17 [161]
$C_2H_4^+$	68 ± 8 [207]; 72 ± 24 [161]
$C_2H_3^+$	51 ± 6 [207]; 36 ± 11 [161]

The structure of the ions is unknown, although considering the relatively high energy (800 V) of the irradiating electrons, it can be assumed that re-arrangement to the thermodynamically most stable structure takes place within a very short time ($\sim 10^{-11}$ s) [207, 213].

Novoseltsev et al. [259] studied the temperature dependence of the decomposition of n-butane ions produced by 12.08 eV photons. In the range studied (20–650°C) the yields of molecular ions decreased and that of the unstable radical ions increased with the temperature.

In the presence of alkenes (e.g. propene), H_2 and H_2^- transfer reactions are observed, as well as addition reactions [161, 207, 212, 214, 215]. These processes may provide an explanation of the strong dose dependence of several products.

2.2.5. Isobutane

Compared with the n-alkanes discussed so far, very few papers deal with the radiolysis of isobutane in the gas or liquid phase. Radiolytic yields obtained in liquid-phase irradiation are given in Table 2.16.

Table 2.16

Yields from liquid isobutane
(radiation: ^{60}Co-γ). After Tanno et al. [246]

Dose, 10^{19} eV g^{-1}	5.8	91.5
Product	G, molecule/100 eV	
Hydrogen	4.16	
Methane	1.98	
Propane	0.68	
Propene	1.62	
Isobutene		1.67

Gas-phase radiolysis of isobutane has been studied by Miyazaki, who used NH_3 positive ion and NO radical scavengers over a wide range of density in order to determine the yields and possible reactions of intermediates — mainly ions — as a function of gas density (or pressure) [220]. The experiments were carried out at 153 ± 0.5°C, in the density range 0.0032–0.15 g cm^{-3} (1.9–40 bar). At the lower pressure limit, in the presence of constant amounts of a radical acceptor and varying amounts of NH_3 positive ion scavenger, a sharp decrease in propane yield and an increase in propene yield was observed as a function of the concentration of the latter additive. The changes in the yields were attributed to ionic reactions, as follows:

$$\text{iso-}C_4H_{10} \xrightarrow{\quad \text{-\!\!\!\!\!\wedge\!\!\!\wedge\!\!\!\wedge-} \quad} C_3H_7^+ + CH_3 \qquad (2.33)$$

$$C_3H_7^+ + \text{iso-}C_4H_{10} \longrightarrow C_3H_8 + \text{iso-}C_4H_9^+ \qquad (2.34)$$

$$C_3H_7^+ + NH_3 \longrightarrow C_3H_6 + NH_4^+ \qquad (2.35)$$

The $C_3H_7^+$ ion has a secondary structure: its very rapid H^- transfer reaction leading to C_3H_8 has been verified in several investigations [246, 247].

At higher pressures, Miyazaki found that the yield of propane decreased by about 60% in the presence of NO. The proportion of propane at lower densities produced in hydrogen abstraction reactions of C_3H was about 13%, but this increased to about 58% at 40 bar. The propane yield originating from C_3H_7 radicals decreases on admixture of NH_3; consequently its precursor may be $C_4H_{10}^+$ and/or $C_3H_7^+$ ions. The yield of propane produced in non-radical processes is practically unaffected by NH_3, i.e. the H^- transfer reaction of $C_3H_7^+$ may not be important at higher pressures [220]. These results indicate that the most important processes of C_3H_7 formation at high pressures involve collisional deactivation of excited primary iso-$C_4H_{10}^{+*}$ ions followed by neutralization.

On the basis of extensive scavenger studies, the following mechanism has been suggested for high pressure radiolysis [220, 258]:

$$iso\text{-}C_4H_{10} \longrightarrow\!\!\!\backslash\!\!\backslash\!\!\backslash\!\!\!\longrightarrow iso\text{-}C_4H_{10}^{+*} + e \qquad (2.36)$$

$$iso\text{-}C_4H_{10}^{+*} + iso\text{-}C_4H_{10} \longrightarrow iso\text{-}C_4H_{10} + iso\text{-}C_4H_{10} \qquad (2.37)$$

$$iso\text{-}C_4H_{10}^+ + e^- < \begin{array}{l} C_3H_7 + CH_3 \qquad (2.38a) \\ \\ C_3H_6 + CH_4 \qquad (2.38b) \end{array}$$

$$iso\text{-}C_4H_{10}^{+*} \longrightarrow C_3H_7^+ + CH_3 \qquad (2.39)$$

$$C_3H_7^+ + iso\text{-}C_4H_{10} \longrightarrow iso\text{-}C_3H_8 + C_4H_9^+ \qquad (2.34)$$

$$C_3H_7^+ + e^- < \begin{array}{l} C_3H_6 + H \qquad (2.40a) \\ \\ C_3H_7 \qquad (2.40b) \end{array}$$

$$C_3H_7 + iso\text{-}C_4H_{10} \longrightarrow C_3H_8 + C_4H_9 \qquad (2.41)$$

In the presence of ammonia, the following reactions may have an influence on the product yields:

$$iso\text{-}C_4H_{10}^+ + NH_3 \longrightarrow C_4H_9 + NH_4^+ \qquad \textbf{(2.42)}$$

$$C_3H_7^+ + NH_3 \longrightarrow C_3H_6 + NH_4^+ \qquad \textbf{(2.35)}$$

On the basis of the gas-phase radiolysis of iso-C_4H_{10}—NO, iso-C_4H_{10}— —NO—NH_3 and iso-C_4H_{10}—NH_3 systems, it has been shown that the yield of $C_3H_7^+$ decreases, whereas that of C_3H_7 increases with increasing pressure: at the same time deactivation of iso-$C_4H_{10}^{+*}$ and neutralization of $C_3H_7^+$ and $C_4H_{10}^+$ ions become more important in the formation of products involving radical and molecular decomposition. As a result, G-values vary as a function of the density, as shown in Figure 2.3 [220].

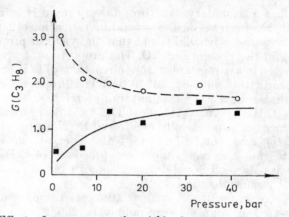

Fig. 2.3. Effect of pressure on the yield of propane in the radiolysis of
isobutane.
○ From ionic intermediates; ■ from radical intermediates (radiation: ^{60}Co-γ).
After Miyazaki [220]

The results obtained by Miyazaki et al. on the variation of the ratio between the G-values of individual products in the gas and liquid phases indicate that the reaction mechanisms are different [220, 246]. A much higher
yield of propene than of propane is observed in the liquid phase but the
reverse is true in the gas phase: $G(C_3H_8) = 3.92$, $G(C_3H_6) = 1.18$. The yield
of methane is also much larger in the gas phase, $G(CH_4) = 6.63$, than in the
liquid phase. Studies with SF_6 and NH_3 scavengers revealed that (2.38b)
and (2.40a) neutralization reactions are the main processes of propene formation in the liquid phase [246].

On the basis of scavenging curves and the reaction mechanism represented
by reactions (2.33)–(2.42), the yield of sec-$C_3H_7^+$ ions in the gas and liquid
phases have been calculated: $G(C_3H_7^+) = 2.3$ and 1.15, respectively. The
difference between these two values indicates greater deactivation of the
$C_4H_{10}^{+*}$ excited ion in the liquid phase: it is also apparent that, even in the
liquid phase, a considerable proportion of the parent ions formed in primary
processes undergoes decomposition. Ausloos et al. elucidated the structure
of $C_3H_7^+$ ions on the basis of the isotope analysis of propane formed in the
radiolysis of the iso-C_4H_{10}–C_4D_{10}–O_2 gaseous system [229]. The composition
100% CD_3CHDCD_3 + 0% $CD_3CD_2CHD_2$ demonstrated that $C_3H_7^+$ ions
produced in the decomposition of iso-$C_4H_{10}^{+*}$ keep their secondary structure
and no isomerization occurs to the thermodynamically less stable n-propyl
ion [229].

The importance of the H_2^- transfer reaction of the $C_3H_6^+$ fragment ion
had earlier been underestimated as a potential source of propane [247];
however the 6% of $C_3H_2D_6$ and $C_3H_6D_2$ formed during radiolysis of the gaseous iso-C_4H_{10}–iso-C_4D_{10}–I_2 system convinced Lias and Ausloos that this
reaction is not negligible [264]:

$$C_3H_6^+ + \text{iso-}C_4H_{10} \longrightarrow C_3H_8 + C_4H_8^+$$

A significant amount of C_3HD_7 has been found in the radiolysis of an iso-C_4D_{10}–H_2S gas mixture [264]; its formation was explained by the following reactions:

$$C_3D_6^+ + \text{iso-}C_4D_{10} \longrightarrow \text{sec-}C_3D_7 + C_4D_9^+$$

$$\text{sec-}C_3D_7 + H_2S \longrightarrow CD_3CHDCD_3 + HS$$

Methane produced in the presence of I_2 radical scavenger contained about 90–91% of CD_4 and CH_4, indicating that the fraction of methane produced in non-radical processes was formed mainly in unimolecular decomposition processes [264]. The relatively weak tertiary C—H bond of the isobutane molecule is inactive for molecular elimination of methane, as indicated by the predominance of CH_4 and the small amount of CH_3D in the radiolysis of $(CH_3)_3CD$. A large proportion of the propane had the composition $C_3H_6D_2$, which is evidence for a bimolecular process [264]. About 85% of propane formed in the decomposition of isobutane irradiated with oxygen radical scavenger is formed by H^- transfer, the remaining 15% by H_2^- transfer [225].

Using the results from stable isotope tracer experiments, Ausloos et al. have calculated the number of excited molecules that decomposes for each ion pair that is formed: the value of $M_{ex}/N = 0.46$ is close to those estimated for propane [182] and n-butane [266] by other methods: propane 0.4 ± 0.1 and butane, 0.27.

The following schematic reaction mechanism has been suggested by Miyazaki et al. [220, 246] for the radiolysis of isobutane, on the basis of the yields as a function of the state and pressure and taking into account the results of tracer and scavenger studies:

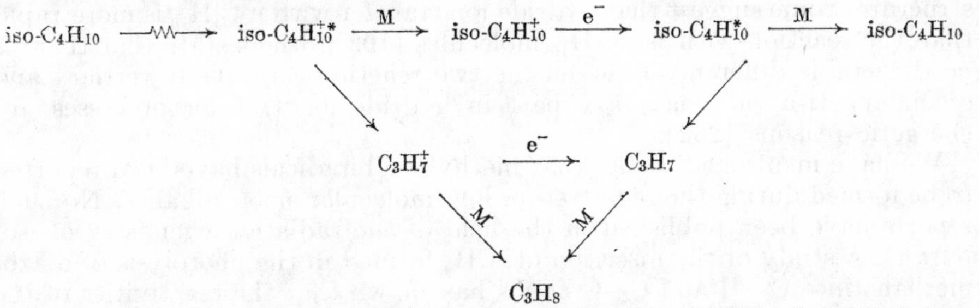

$$\text{iso-}C_4H_{10} \xrightarrow{\text{\raisebox{0pt}{$\sim\!\!\sim$}}} \text{iso-}C_4H_{10}^{+*} \xrightarrow{M} \text{iso-}C_4H_{10}^{+} \xrightarrow{e^-} \text{iso-}C_4H_{10}^{*} \xrightarrow{M} \text{iso-}C_4H_{10}$$

$$C_3H_7^+ \xrightarrow{e^-} C_3H_7$$

$$C_3H_8$$

It can thus be understood that whereas decomposition of excited ion and H^- and H_2^- transfer reactions of ionic fragments are responsible for products such as propane in low pressure gas, the importance of deactivation and neutralization processes without decomposition increases with increasing pressure. The decomposition of excited molecules formed during charge neutralization becomes predominant in the liquid phase. This hypothesis is supported by the values of free ion yields measured by Nishikawa et al. in the gas and liquid phases using N_2O as electron scavenger. The free ion yield at 170°C decreased from $G = 2.4$ to $G = 0.91$ with increasing density between 0.12 and 0.42 g cm^{-3}. The value in the liquid phase was $G = 0.37$ at 23°C [52].

7

Mass spectrometric studies carried out at 1.3–1.8 mbar revealed the properties of several ions with important roles in isobutane radiolysis [209]. With 800 V electrons and a source temperature of $200\pm10°C$ 92% of the ions belonged to $C_4H_9^+$ species (predominantly tertiary isobutyl ion). On the basis of the pressure dependence of relative ion currents corresponding to individual ions, Munson calculated the rate constants of ion-molecule reactions with the participation of isobutane (Table 2.17). The rate of hydride

Table 2.17

Rate constants for reactions of fragment ions
of iso-C_4H_{10} with parent molecules

Ion	k, 10^{10} $M^{-1}s^{-1}$
$C_4H_9^+$	0.006 [207]
$C_4H_8^+$	12.6 $\pm$ 1.8 [207]
$C_3H_7^+$	24 $\pm$ 2 [207]
$C_3H_6^+$	40 $\pm$ 6 [207]; 24–28 [212]
$C_3H_5^+$	35 $\pm$ 2 [207]; 17 [161]
$C_2H_5^+$	61 $\pm$ 8 [207]; 23 [161]
$C_2H_3^+$	68 $\pm$ 8 [207]; 59 [161]

ion transfer depends mainly on the structure of the ion (primary, secondary or tertiary): the n-$C_3H_7^+$ (or n-$C_3D_7^+$) ion is about five times as active towards n-butane molecules as sec-$C_3H_7^+$. There are several contradictory statements in the literature concerning the correlation of reaction rates and molecular structure. Some suggest that hydride ion transfer with n-C_4H_{10} is more rapid than the reaction with iso-C_4H_{10} molecules [103]; others state that there is no detectable difference between the two reaction rates. Only tertiary and secondary H-atoms may take part in hydride ion transfer processes, for energetic reasons [262].

We have mentioned earlier that methylene biradicals have been reported to be formed during the radiolysis of low molecular mass alkanes. No such reports have been published in the field of the radiation chemistry of iso-butane. A study of the insertion of $^1CH_2$ formed in the photolysis of diazomethane into C—H and C—C bonds has shown that the reactivities of the alkanes studied are about the same, with the exception of methane [244]. The reactivities of n-butane and isobutane relative to methane were reported to be 3.1 and 2.6, respectively [244].

Reactions in irradiated solid isobutane have been studied [248–253] using end-product analysis and ESR spectroscopy. The nature of both intermediates and end products depends on the structure of the solid state. ESR studies indicated that iso-C_4H_9 radicals are formed in γ-irradiated crystalline isobutane whereas in the glassy state, tert-C_4H_9 radicals were detected. The yields of methane and propene are considerably higher in the polycrystalline state than in glassy isobutane. The yields of C_4H_9 radicals and hydrogen were not affected by electron scavengers but dropped severely in the presence of CCl_4 acting also as excitation acceptor. These results indicate that C_4H_9 radicals and hydrogen are formed by a direct excitation mechanism:

$$C_4H_{10} \xrightarrow{\quad I \quad} C_4H_{10}^*$$

$$C_4H_{10}^* \xrightarrow{\quad k_d \quad} C_4H_9 + H \qquad\qquad (2.43)$$

$$H + H \longrightarrow H_2$$

$$H + C_4H_{10} \longrightarrow H_2 + C_4H_9$$

$$C_4H_{10}^* + CCl_4 \xrightarrow{\quad k_s \quad} C_4H_{10} + CCl_4^*$$

According to the general equation for the rates of energy transfer processes, the overall order of the reaction depends on the mechanism of the transfer process [254]:

$$R_S = k_S [C_4H_{10}^*][CCl_4]^{n/3} \qquad\qquad (2.44)$$

where $n = 6$ indicates a vibration–relaxation resonance transfer and $n = 3$ exciton transfer. Assuming a steady state concentration for intermediates, on the basis of the scavenged hydrogen yield and reaction (2.44), the relationship between the rate constants of transfer and decomposition processes can be calculated by the following expression:

$$\frac{1}{G'(H_2)} = \frac{1}{I} + \frac{k_S}{I k_d} [CCl_4]^{n/3}$$

where $G'(H_2) = G(H_2) - G(H_2)_S$ and $G(H_2)_S$ the hydrogen yield in the presence of 5 mole % of CCl_4.

The experimental data can be fitted to a curve calculated with $n = 3$, indicating that exciton transfer occurs in solid-state radiolysis. Excitation transfer is a very rapid process, as indicated by its efficient competition with reaction (2.43) proceeding within 10^{-13}–10^{-12} s. These results and conclusions are supported also by yields measured in solid and liquid states: $G(e^-)_{glass} = 1.5$–2.0 and $G(e^-)_{liqu.} = 4.0$. Thus the experimental data support the hypothesis that ionic decomposition processes are more important in the liquid phase than in the solid state. Thus, in turn, in the radiolysis of glassy or crystalline solid iso-C_4H_{10}, the decomposition of 'activated' molecules produced in excitation processes determines the direction of the radiation chemical processes [248–253].

The reaction mechanisms occurring in the glassy state are different from those followed in the polycrystalline state. Saitake et al. studied the radiolysis of solid $(CH_3)_3CD$ at 77 K by ESR spectroscopy: they reported that reaction (2.45) takes place in polycrystalline material and reaction (2.46) in the glassy state [253]:

$$(CH_3)_3CD \xrightarrow{\text{polycryst.}} (CH_3)_2CDCH_2 + {}_{..}H \qquad\qquad (2.45)$$

$$(CH_3)_3CD \xrightarrow{\text{glass}} (CH_3)_3C + D \qquad\qquad (2.46)$$

7*

C—H bond rupture processes can, therefore, be influenced not only by the molecular structure of hydrocarbons but also by the packing of the molecules and interactions between them [248–253, 265, 267].

2.2.6. n-Pentane

Several papers dealing with the radiolysis of gaseous, liquid or solid n-pentane have been published. Radiation chemical yields obtained in liquid-phase radiolysis are summarized in Table 2.18 [271–273].

The comparison of data reported from different laboratories [271–274, 282] shows a good agreement between G-values obtained with low doses. The reliability of the product analysis is demonstrated by the fact that the C/H ratio estimated from the product distribution is nearly the same as that of the starting C_5H_{12} value: ref. [271] gives $C_{4.9}H_{12}$, ref. [273] $C_{4.83}H_{12}$.

The main products of n-pentane radiolysis are hydrogen, methane, ethane, propane, pentenes and 4,5-dimethyloctane [272]. The yields of C_6-C_{10} alkenes are extremely low. Apart from cis- and trans-pent-2-ene, only alk-1-enes with 2–5 carbon atoms could be detected [272]. The overall yield of dimers is significant, but G-values from different papers show great deviations. Radiation chemical yields of C_2-C_3 products show smaller deviations; this, however, is not justified by the difference in dose values, as the dose dependence of their yields is rather small (Bishop and Firestone [272]).

In the dose range 1–$100 \cdot 10^{19}$ eV g^{-1}, no significant dose dependence was observed for the yields of individual products, with the exception of hydrogen and ethylene [272]. Deviations in the reported yields of hydrogen $[G(H_2) = 5.0$–$6.3]$ must be due to dose dependence [271, 274, 276]. Bishop and Firestone think that dose dependence is caused by the capture of H-atoms by unsaturated hydrocarbon products [272]. Further, Rzad and Schuler, as well as Scala et al. [45, 278], have pointed out the possible reactions of ionic intermediates with unsaturated products, such as H_2 and H-transfer processes. Deviations in the yields of higher hydrocarbon products were attributed by Pancini et al. to analytical errors rather than secondary reactions [273].

With increasing temperature, between -196 and $+100°C$, the G-values of hydrogen, C_1-C_4 alkanes and branched C_{10} alkane dimers increase [272], the yield of n-$C_{10}H_{22}$ decreases and those of C_2H_4 and C_3H_6 alkenes show maxima at about $0°C$ [271]. The yield of so-called intermediate C_6-C_9 products is almost independent of temperature, whereas the ratios of yields of individual isomers, particularly C_{10} dimers, exhibit marked temperature dependence [271]. According to the experiments of Koch et al. [271] and Bishop and Firestone [272], using radical scavengers, the fraction of products involving combination reactions of 1-pentyl radical (C_7-C_{10} n-alkanes, 3-ethyloctane, 4-methylnonane) decreases with increasing temperature within the given product groups. Some papers indicate that the temperature dependence of yields is due to the increased role of reactions with higher activation energy (e.g. H-atom abstraction) at higher temperatures, together with the simultaneous decrease of the role of processes requiring nearly zero

Table 2.18

Yields from liquid n-pentane[a] (radiation: ^{60}Co-γ)

Dose, 10^{19} eV g^{-1}	5	14.2	140
References	271	272	273
Product	G, molecule/100 eV		
Hydrogen	5.1 (5.3)[b]	4.20	4.22
Methane	0.26	0.261	0.28
Ethane	0.61	0.417	0.65
Ethylene	0.37	0.219	0.22
Acetylene			0.026
Propane	0.58	0.442	0.67
Propene	0.35	0.24	0.20
Other C$_3$ products			0.19
n-Butane	0.10	0.072	0.14
But-1-ene	0.08	0.051	0.049
trans-But-2-ene	} 0.02		0.012
cis-But-2-ene			0.008
Other C$_4$ products		0.02	0.007
Pent-1-ene	1.00	0.738	0.55
trans-Pent-2-ene	1.23	1.22	0.98
cis-Pent-2-ene	0.53	0.54	0.41
2-Methylbutane		0.0615	
n-Hexane	} 0.05	0.025	0.039
2-Methylpentane		0.016	0.027
3-Methylpentane		0.0072	0.013
n-Heptane		0.0505	0.065
3-Methylhexane	} 0.20	0.119	0.15
3-Ethylpentane		0.0573	0.071
3-Ethylpent-1-ene		0.0047	
4-Methylhex-1-ene		0.0042	
n-Octane		0.041	0.046
4-Methylheptane		0.0917	0.095
3-Ethylhexane	} 0.22	0.040	0.041
2,3-Dimethylhexane		0.011	
3-Ethyl-2-methylpentane		0.014	
Octenes		0.003	
n-Nonane		0.0047	0.007
4-Methyloctane+3-ethylheptane		0.029	0.026
3-Methyloctane	} 0.06	0.010	
3,4-Dimethylheptane + + 4-Ethyl-3-methylhexane		0.003	
n-Decane		0.0344	0.049
3,4-Diethylhexane	} 1.44	} 0.183	0.067
3-Ethyl-4-methylheptane			0.37
4,5-Dimethyloctane		0.751	0.46
3-Ethyloctane		} 0.0734	0.089
4-Methylnonane			0.22
Decenes		0.06	

[a] Room temperature
[b] Dose: $0.37 \cdot 10^{19}$ eV g^{-1}

energy of activation (e.g. radical combination and disproportionation) [271, 272]. In addition, as pointed out by Cramer et al., at higher temperatures the importance of the so-called 'cage effect' decreases [271].

In the presence of radical scavengers the yields of all hydrocarbon products decrease and this, according to several authors, proves the radical character of product formation. Koch et al. have observed a decrease of about 70% in the G-value of C_6-C_9 products [271]. The radical character of these products is also supported by the fact that their G-values increase sharply with increasing dose rates. In connection with the C_6-C_9 products, it is widely accepted that they are formed via combination of pentyl radicals — being present in highest steady state concentration — with fragment radicals. Pancini et al. [273] showed, however, that remarkable deviations from random distribution can occur: namely, more n-hexane is formed than expected from random radical combination. At the same time, the percentage of linear C_7-C_9 products formed by the reactions of 1-pentyl radicals is as low as 20–25%, compared with the 75–80% of products involving 2-pentyl or 3-pentyl radicals. The percentages of dimers corresponding to 1-pentyl, 2-pentyl and 3-pentyl precursors are 15, 46 and 39%, respectively.

Any of the C_6-C_{10} products can — formally — be derived by combination reactions of alkyl radicals formed in the single radical type decomposition of the n-pentane molecule, indicating the relative unimportance of alkyl radical isomerization in radiolytic processes [241, 271], while the proportions of 'parent' 1-pentyl, 2-pentyl and 3-pentyl radicals are the same in reactions leading to C_7-C_9 products (21–25, 52–58 and 21–23%, respectively), the corresponding proportions in hexane formation are rather different (50, 34 and 16%, respectively) [273, 296]. As these latter percentages are the same as the percentages of hydrogen atoms bonded to various carbon atoms in the molecule (50, 33.3 and 16.7% on carbon atoms in positions 1, 2, and 3, respectively), it is assumed that a considerable proportion of C_6 products is formed via random insertion of excited singlet methylene biradicals into the C—H bonds of the parent alkane.

Unsaturated hydrocarbons constitute about 5% of the C_6-C_{10} γ-radiolysis products of n-pentane. These alkenes are formed in secondary reactions of unsaturated fragment end-products accumulating during irradiation with C_n-C_{2n} products, e.g. in the following processes [302]:

$$C_xH_{2x} + R \xrightarrow{\text{H-abstraction}} C_xH_{2x-1} + RH$$

$$C_xH_{2x-1} + C_nH_{2n+1} \xrightarrow{\text{combination}} C_{n+x}H_{2(n+x)} \quad \text{(aliphatic)}$$

$$H + C_{n+x}H_{2(n+x)+2} \xrightarrow{\text{H-abstraction}} H_2 + C_{n+x}H_{-2(n+x)+1}$$

$$2C_{n+x}H_{2(n+x)+1} \xrightarrow{\text{disproportionation}} C_{n+x}H_{2(n+x)+2} + C_{n+x}H_{2(n+x)} \quad \text{(aliphatic)}$$

Bishop and Firestone as well as Holroyd and Klein have calculated the yields of free radicals formed in pentane radiolysis on the basis of the decrease of yields in the presence of iodine [272] and $^{14}C_2H_4$ [241] radical scavengers. Some authors have assumed or shown that iodine may inhibit molecular processes [279] as well as ionic reactions [44]. The G-values of all products reached a saturation value at an iodine concentration of about 0.5 mM; this means that the radical scavenging effect of iodine is of great importance. G-values for radical intermediates are summarized in Table 2.19.

Table 2.19
Yields of radicals formed from liquid n-pentane
(radiation: ^{60}Co-γ)

Radical acceptor	Iodine		$^{14}C_2H_4$
Temperature, °C	25	—78	10
Reference	272		241
Product radicals	G, radical/100 eV		
Methyl	0.12	0.15	0.10
Ethyl	0.50	0.34	0.51
1-Propyl	} 0.44	} 0.42	0.36
2-Propyl			0.04
1-Butyl	0.06	} 0.09	0.07
2-Butyl	0.03		
1-Pentyl	0.25	0.53	0.98
2-Pentyl	2.3	1.5	2.21
3-Pentyl	1.1	0.73	1.10
Pentenyl	0.06		0.05

With the exception of data for the 1-pentyl radical, the decrease of hydrocarbon yield and the radical yields as measured by $^{14}C_2H_5$ radical sampling technique are in good agreement. The ratios of yields of different radicals as determined by scavengers and as calculated on the basis of product distribution also show a good agreement: $CH_3 : C_2H_5 : C_3H_7 : C_4H_9 = 1 : 5.1 :$: 4.0 : 0.7 and 1 : 5 : 5 : 1, respectively [241, 271]. These data unambiguously indicate that the terminal C—C bonds of n-pentane are less likely to decompose than the internal C—C bonds. On the basis of yields of 1-pentyl, 2-pentyl and 3-pentyl radicals, the probability of C—H bond rupture in different positions can be estimated. On the basis of radical yields given in various papers, the relative probability of the rupture of C—H bonds on the first, second and third carbon atom in the carbon chain is 1 : 4 : 3 [271] and 1 : 3.3 : 3.3 [241] per bond. From the ESR experiments of Claes et al. [265] the following yields were obtained: G(1-pentyl) = 0.25, G(2-pentyl) = 0.45 and G(3-pentyl) = 0.10. These values give a slightly different ratio than those mentioned earlier, probably due to radical isomerization processes. Although data from different sources are not quite identical, they all indicate that C—H bond rupture of the terminal methyl group, giving 1-pentyl radical, is less likely than the cleavage of secondary C—H bonds. The data for C—C and C—H bonds indicate that small differences (5–10 kJ mol^{-1}) in the bond dissociation energy influence the yields considerably.

The distribution of pentyl radicals is not an accurate reflection of the primary bond rupture ratios. Koch et al. have shown that the G-values of combination products of 2-pentyl and 3-pentyl radicals decrease to a larger extent in the presence of oxygen radical scavenger than those of products involving 1-pentyl radicals [271]. Most of the secondary pentyl radicals originate from H-atom abstraction reactions by fragment alkyl radicals and hydrogen, this type of reaction being more favourable than the analogous processes leading to 1-pentyl radicals. It can be assumed that primary C—H bond ruptures are less selective than is apparent from a comparison of radical yields.

The overall net yield of C—C bond rupture processes can be calculated by the following formula [271]:

$$\Sigma G(\mathrm{C-C}) = \frac{1}{2} \left[\Sigma G(\mathrm{C_1-C_4}) + \Sigma G(\mathrm{C_6-C_9}) \right]$$

where $\Sigma G(\mathrm{C_1-C_4})$ and $\Sigma G(\mathrm{C_6-C_9})$ are the overall yields of $\mathrm{C_1-C_4}$ and $\mathrm{C_6-C_9}$ hydrocarbons, respectively. On the basis of yields determined at $2_65°$, with a dose of $5 \cdot 10^{19}$ eV g^{-1}, $\Sigma G(\mathrm{C-C}) = 1.5$ [271]. The G-values of decomposition of individual C—C bonds have been calculated similarly, based on product yields: $G(\mathrm{CH_3-CH_2-} \ldots) = 0.3$ and $G(\ldots -\mathrm{CH_2-CH_2-} \ldots) = 1.2$. The value of the overall $G(\mathrm{C-C})$ is strongly temperature dependent: its values at -116, $+25$ and $+95°C$ are $\Sigma G(\mathrm{C-C}) = 1.0$, 1.6 and 1.7, respectively [271]. This phenomenon was attributed by Koch et al. to the negative temperature coefficient of the cage effect [271] although, owing to the rather complex reaction mechanism, this explanation should be treated with caution.

The ESR spectroscopic results obtained by Thyrion et al. also indicate that 2-pentyl radicals were formed with highest yields in n-pentane radiolysis [297].

The method of pulse radiolysis combined with spectrophotometry has been successfully applied to the study of a number of alkanes [290–292]. At 24°C, the overall reaction rate constant of combination and disproportionation reactions of the pentyl radicals produced has been determined: $k = 4.7 \pm 0.5 \cdot 10^9$ M^{-1} s^{-1}. The apparent energy of activation calculated from the temperature dependence is 3 ± 1.3 kJ mol^{-1}. Calculations based on diffusion kinetics [293] have shown that the rate constant is much lower than that of diffusion-controlled processes, i.e. the rate is determined by steric factors.

Several papers have reported that terminal C—C bonds are less likely to rupture than other C—C bonds in n-alkanes. Avdonina et al. [294] have concluded, on the basis of a comparison of product yields obtained from n-pentane, n-hexane and n-heptane as well as scavenger studies, that terminal C—C bonds decompose mainly in substitution reactions of 'hot' hydrogen atoms:

$$\mathrm{RH + H^* \longrightarrow R_1H + R_2}$$

whereas other C—C bonds decompose in unimolecular breakdown reactions of electronically excited or ionized molecules. (In the above equation, H* denotes a 'hot' hydrogen atom and $\mathrm{R_1}$ and $\mathrm{R_2}$ fragment hydrocarbon radicals). Urch and Welch's tritium studies seem to support this hypothesis [298].

In the presence of oxygen as radical scavenger, the yield of C_5 alkene products decreases by about 40 to 60% [272], indicating the role of radical processes in their formation. It was shown that π-bond formation was more likely to occur between two secondary carbon atoms. The G-value of the thermodynamically more stable *trans*-isomers is greater than that of the *cis*-isomers [272].

The product distribution measured by Bienfait and Claes in solid-state radiolysis of n-pentane, at 77 K, is not markedly different from that observed in the liquid phase [283]. G-values extrapolated to zero dose are as follows for the main groups of products: $G_0(H_2) = 5.00$, $G_0(C_2 + C_3) = 1.22$, $G_0(C_5H_{10}) = 3.45$ and $G_0(C_{10}H_{22}) = 0.92$.

There are considerable differences within the dimer product group. In the liquid phase the main products are formed via combination of 2-pentyl and 3-pentyl radicals, but the 1-pentyl radical is more important in the solid state. The rate of H-atom abstraction by alkyl radicals is inhibited because of hindered diffusion, so the yield of secondary pentyl radicals is decreased. The dose dependence observed in the solid state can be explained by the reactions of pentyl radicals trapped in the matrix in addition to the secondary reactions of alkene products [283]. Dimer formation is partly due, and trimer formation ($G = 0.03$) entirely due to ionic condensation reactions [283, 284]. According to Dorfman and Matheson [303], alkene production involves disproportionation as well as monomolecular hydrogen elimination, as in liquid-phase radiolysis [271–273]. The yield of radicals produced from solid n-pentane irradiated at 122 K can be measured by introducing the alkane after irradiation into a solution containing radical scavenger (e.g. oxygen, alkenes). The overall G-value of dimers decreased from $G = 0.54$ to $G = 0.34$; this supports the radical type mechanism of formation [295].

Relatively few reports can be found in the literature either about the main pathways of decomposition of the primary products of n-pentane radiolysis, e.g. electronically excited and/or ionized molecules, or about the ion–molecule reactions. Some results reported in several papers suggest that the radiolysis cannot be treated as a simple radical mechanism, and the study of partial processes cannot yet be regarded as complete.

According to Flesch and Svec, the following main decomposition reactions of the excited n-pentane ion can occur [301]:

$$
\begin{array}{lll}
\text{sec-}C_5H_{11}^+ + H & \Delta H^\circ = & 59 \text{ kJ mol}^{-1} \\
\text{sec-}C_4H_9^+ + CH_3 & \Delta H^\circ = & 33 \text{ kJ mol}^{-1} \\
\text{n-}C_5H_{12}^+ \longrightarrow \quad \text{sec-}C_3H_7^+ + C_2H_5 & \Delta H^\circ = & 25 \text{ kJ mol}^{-1} \\
C_2H_5^+ + 1\text{-}C_3H_7 & \Delta H^\circ = & 160 \text{ kJ mol}^{-1} \\
CH_3^+ + 1\text{-}C_4H_9 & \Delta H^\circ = & 305 \text{ kJ mol}^{-1}
\end{array}
$$

N_2O electron scavenger influences the yields by introducing the following competition:

$$C_5H_{12}^+ + e^- \longrightarrow C_5H_{12}^* \longrightarrow \text{products}$$
$$N_2O + e^- \longrightarrow N_2 + O^-$$

In the presence of electron scavenger, the yield of C_1-C_4 alkane products slightly decreased, and that of C_2-C_4 alkenes showed a considerable drop [271, 272]. The results show that whereas fragment alkanes are formed mainly by direct decomposition processes of excited molecules, alkenes are the products of fragmentation of electronically excited molecules which, in turn, are formed in neutralization processes.

Recent studies on the mechanism of the radiolysis of n-pentane in the liquid state indicate that ionic decomposition processes contribute to the fragment product yields [277] through H^-- and H_2^--transfer reactions. Mass spectrometric works suggest that the most abundant ion formed from n-pentane molecules is the $C_3H_7^+$ ion whose yield is ten times that of the parent ion [280].

The literature dealing with the mechanism of hydrogen formation is much more abundant, and our understanding about it is much more complete, than about the formation of hydrocarbon products. Horowitz and Rajbenbach have developed the following mechanism [275]:

$$
\begin{aligned}
& && C_5H_{12}^+ + e^- \\
C_5H_{12} & \longrightarrow && C_5H_{12}' \\
& && C_5H_{12}''
\end{aligned}
$$

$$C_5H_{12}^+ + e^- \longrightarrow C_5H_{12}^*$$

$$C_5H_{12}^* \longrightarrow H^* + \text{products} \qquad G = 2.3$$

$$C_5H_{12}' \longrightarrow H + \text{products} \qquad G = 1.0$$

$$C_5H_{12}'' \longrightarrow H_2 + \text{products} \qquad G = 1.6$$

where C_5H_{12}', C_5H_{12}'' and $C_5H_{12}^*$ denote three different states of electronically excited pentane molecules, H^* and H 'hot' and thermal hydrogen atoms, respectively. The yields of H-atoms and H_2-molecules have been determined by using hex-1-ene as radical scavenger and CCl_4 as electron scavenger [275]. Yields reported for elementary processes of hydrogen formation are in agreement with yields obtained by different methods for other alkanes [46, 286].

In order to get a deeper insight into the mechanism of decomposition reactions of electronically excited molecules generated in irradiated n-pentane, Holroyd [300] studied the vacuum-UV photolysis of n-pentane at a wavelength of 147 nm (8.4 eV). Although the main products of photolysis and radiolysis are identical, their distributions are different. The main photolytic processes are the elimination of molecular H_2 and decomposition

into H-atom and pentyl radical. The following decomposition reactions and quantum yields have been reported [300]:
Radical reactions:

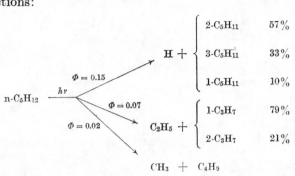

$$n\text{-}C_5H_{12} \xrightarrow[h\nu]{}$$

$\Phi = 0.15$

$\Phi = 0.07$

$\Phi = 0.02$

$$H + \begin{cases} 2\text{-}C_5H_{11} & 57\% \\ 3\text{-}C_5H_{11} & 33\% \\ 1\text{-}C_5H_{11} & 10\% \end{cases}$$

$$C_2H_5 + \begin{cases} 1\text{-}C_3H_7 & 79\% \\ 2\text{-}C_3H_7 & 21\% \end{cases}$$

$$CH_3 + C_4H_9$$

Molecular reactions:

$$n\text{-}C_5H_{12} \xrightarrow[h\nu]{}$$

$\Phi = 0.82$

$\Phi = 0.06$

$\Phi = 0.05$

$\Phi = 0.01$

$$H_2 + C_5H_{10}$$
$$C_2H_4 + C_3H_8$$
$$C_2H_6 + C_3H_6$$
$$CH_4 + C_4H_8$$

The relative importance of molecular elimination processes vis-à-vis radical reactions is greater in photolysis than in radiolysis. Excited molecules produced by the absorption of photons have a lower average energy content than those generated by high-energy radiation, and thus their lifetime is longer and the possibility of the partial rearrangement necessary for molecular elimination increases. Though the hydrogen yield in photolysis is about twice as high as in radiolysis, the total radical yield in radiolysis is 1.5–2.0 times greater than the value measured in photolysis.

2.2.7. Branched-chain pentane isomers

The radiolysis of neopentane has been reported in a relatively large, and that of isopentane in a relatively small number of papers.

Earlier papers on isopentane [304–306] discussed mainly radical reactions out of the spur; recent workers have paid more attention to processes within the spur, together with ionic reactions [246, 277, 297, 301, 307]. Products of liquid-phase and solid-state γ-radiolysis of 2-methylbutane, together with their yields, are summarized in Table 2.20.

Extensive studies have been carried out on the mechanism of radiolysis of isopentane. Investigations have used stable and radioactive isotopic tracers, scavengers and ESR spectroscopy. Scala and Ausloos used deuterium tracers [307] to show that only a small fraction of hydrogen originates from 'molecular' processes. The decrease of $G(H_2)$ in the presence of CCl_4 electron scavenger indicates that most of the hydrogen is produced by

Table 2.20
Yields from isopentane (radiation: ^{60}Co-γ)

Dose, 10^{19} eV g^{-1}	5.8	7–21	7–21
Temperature, K	300	195	77
References	246	307	307
Phase	Liquid		Solid
Products	G, molecule/100 eV		
Hydrogen	3.75	2.85	2.39
Methane	1.40	0.48	0.38
Ethane	1.11	1.00	0.76
Ethylene	0.43	0.45	0.32
Propane	0.57	0.53	0.34
Propene	1.12	0.86	0.51
n-Butane	0.30	0.17	0.08
But-1-ene + isobutene		0.16	0.11
trans-But-2-ene	0.24	0.15	0.08
cis-But-2-ene		0.05	0.02
Isobutane		0.04	
2-Methyl-but-1-ene		0.33	0.50
3-Methyl-but-1-ene		0.37	0.50
2-Methyl-but-2-ene		0.13	0.16

H-atom abstraction by 'hot' hydrogen atoms formed in decomposition reactions subsequent to charge neutralization [307].

In the solid state, at 77 K, about 60% of methane is produced in molecular decomposition processes involving electronically excited isopentane molecules as intermediates. These latter species may be formed either by charge neutralization or by direct excitation [307].

As the G-value of C_2-C_3 products decreases on addition of electron scavengers, it has been concluded [307] that a considerable proportion of these products can be attributed to radical or molecular decompositions of excited molecules. The yields of alkyl radicals produced as a result of fragmentation of electronically excited molecules have been determined by more than one method (Table 2.21) such as direct measurement by means of iodine or $^{14}C_2H_4$ radical scavengers or by calculating the decrease in the yield in the

Table 2.21
Yields of radicals formed from liquid isopentane
(method: $^{14}C_2H_5$-radical sampling). After Holroyd and Klein [306]

Product radical	G, radical/100 eV
Methyl	0.39
Ethyl	0.81
2-Propyl	0.58
2-Butyl	0.18
Isobutyl	0.03

presence of radical scavengers [304–306]. It was verified by ESR spectroscopy that the tertiary C—H bond is cleaved most easily [392].

Scala and Ausloos have studied the gas-phase radiolysis [307] and photoionization decomposition [311] of isopentane. They reported that the main fragmentation reactions of the isopentane ion, and the corresponding appearance potentials, are as follows:

$$C_5H_{12}^+ \begin{cases} \longrightarrow C_3H_7^+ + C_2H_5 & 10.84 \text{ eV} \\ \longrightarrow C_3H_6^+ + C_2H_6 & 10.24 \text{ eV} \\ \longrightarrow C_2H_5^+ + C_3H_7 & 13.2 \text{ eV} \end{cases}$$

Analogously to the radiolysis of other alkanes, it was assumed that the H^- transfer reaction of $C_3H_7^+$ and $C_2H_5^+$ ions as well as the H_2^- transfer reaction of $C_3H_6^+$ ion contributed significantly to the formation of propane and ethane or propane, respectively [277, 307]. The study of an equimolar mixture of iso-C_5H_{12} and iso-C_5D_{12}, irradiated in the liquid phase, revealed [307] that the yields of partially deuterated propane and ethane (C_3HD_7, $C_3H_2D_6$, etc.) drop significantly following the admixing of O_2 (radical scavenger) or CCl_4 (electron scavenger); this indicates that the C_2-C_3 products are formed in radical reactions and not in ion-molecule processes. The considerable increase of the yields with increasing temperature also supports the predominance of radical processes, namely H-abstraction reactions of ethyl and propyl radicals; disproportionation reactions in the 'cage' may also be important [307].

Tanno et al. have observed a decrease of the yield of molecular fragments (except for n-butane) in the presence of electron and positive ion scavengers (such as SF_6, CCl_4, and NH_3, respectively) [246]. When SF_6 and NH_3 were both present, the yield of propene increased and that of ethane decreased. These results have been interpreted in terms of the following reaction mechanism [246]:

$$\text{iso-}C_5H_{12} \xrightarrow{\rotatebox{0}{\leadsto}} \text{sec-}C_3H_7^+ + C_2H_5^{\cdot} + e^- \tag{2.47}$$

$$\text{sec-}C_3H_7^+ + e^- \longrightarrow C_3H_6 + H \tag{2.48}$$

$$\text{iso-}C_5H_{12}^+ + e^- \begin{cases} \longrightarrow C_3H_6 + C_2H_6 & \tag{2.49a} \\ \longrightarrow C_3H_8 + C_2H_4 & \tag{2.49b} \end{cases}$$

$$\text{sec-}C_3H_7^+ + \text{iso-}C_5H_{12} \longrightarrow C_3H_8 + \text{iso-}C_5H_{11}^+ \tag{2.50}$$

Reactions (2.48) and (2.49) are inhibited by electron scavengers, but reaction (2.50) is promoted as they increase the lifetime of positive ions (e.g. $C_3H_7^+$). This is why the yield of propene decreases drastically in the presence of electron scavengers, but that of propane only slightly. On the basis of scavenger studies, the yield of propene formed by charge neutralization has been estimated to be $G = 0.72$ [within this value, that of propene formed in reaction (2.48) to be $G = 0.13$], whereas the yields of reactions (2.49a) and (2.49b) were estimated to be $G = 0.61$ and 0.31, respectively [246].

The yield of n-butane increases in the presence of electron scavengers, whereas those of n-butenes decrease; on this basis the following reactions have been proposed [246].

$$\text{iso-}C_5H_{12} \xrightarrow{\quad\wedge\wedge\wedge\quad} C_4H_9^+ + CH_3 + e^- \qquad\qquad (2.51)$$

$$C_4H_9^+ + e^- \longrightarrow C_4H_8 + H \qquad\qquad (2.52)$$

$$SF_6 + e^- \longrightarrow SF_6^- \qquad\qquad (2.53)$$

$$C_4H_9^+ + \text{iso-}C_5H_{12} \longrightarrow \text{n-}C_4H_{10} + C_5H_{11}^+ \qquad\qquad (2.54)$$

Very small amounts of C_4H_9D and C_4HD_9 have been found as a result of irradiation of the iso-C_5H_{12}–iso-C_5D_{12} system: at the same time, the yields of $C_4H_8D_2$ and $C_4H_2D_8$ were significant: therefore it can be assumed that reactions (2.55) and (2.56) are much more important than reactions (2.51)– (2.54) [307]:

$$\text{iso-}C_5H_{12} \xrightarrow{\quad\wedge\wedge\wedge\quad} C_4H_8^+ + CH_4 + e^- \qquad\qquad (2.55)$$

$$C_4H_8^+ + \text{iso-}C_5H_{12} \longrightarrow C_4H_{10} + C_5H_{10}^+ \qquad\qquad (2.56)$$

The following reaction mechanism has been suggested by Scala and Ausloos [307] as well as by Tanno et al. [246], for the radiolysis of isopentane:

$$
\begin{array}{ll}
 & C_2H_5^+ + \text{sec-}C_3H_7 + e^- \qquad G \approx 0.1 \\
 & \text{sec-}C_3H_7^+ + C_2H_5 + e^- \qquad G \approx 0.23 \\
\text{iso-}C_5H_{12} & C_4H_9^+ + CH_3 + e^- \qquad G \approx 0.12 \\
 & C_4H_8^+ + CH_4 + e^-
\end{array}
$$

$$\text{sec-}C_3H_7^+ + e^- \longrightarrow C_3H_6 + H \qquad\qquad G \approx 0.13$$

$$C_4H_9^+ + e^- \longrightarrow C_4H_8 + H \qquad\qquad G \approx 0.12$$

$$C_4H_8^+ + \text{iso-}C_5H_{12} \longrightarrow C_4H_{10} + \text{iso-}C_5H_{10}^+$$

$$
\begin{array}{ll}
 & C_3H_6 + C_2H_6 \qquad G \approx 0.61 \\
 & C_3H_8 + C_2H_4 \qquad G \approx 0.31 \\
\text{iso-}C_5H_{12}^+ + e^- & \left\{\begin{array}{l} H_2 \\ 2H \end{array}\right\} + \text{products} \qquad G \approx 1.89 \\
 & \left\{\begin{array}{l} CH_4 \\ CH_3 + H \end{array}\right\} + \text{products} \qquad G \approx 0.50
\end{array}
$$

The simple structure of the neopentane molecule makes it a favourite choice as a model compound for the study of ion-molecule reactions [334] as well as of ionic [301], thermal [332, 333] and 'hot'-atom-initiated [298] decomposition. Owing to the high symmetry of the molecule, relatively few types of radical and ionic intermediate are produced in its radiolysis, resulting in a simple product spectrum and higher radiation chemical yields, which can be measured with greater accuracy.

The main products of the liquid-phase radiolysis of neopentane, and their yields as a function of the experimental conditions, have been determined [192, 312, 314]. The details of the reaction mechanism have been clarified only recently, with modern techniques such as pulse radiolysis.

The G-values of the main products are tabulated in Table 2.22 [314]. On

Table 2.22

Yields from liquid neopentane (radiation: ^{60}Co-γ; dose: $1.96 \cdot 10^{19}$ eV g^{-1}). After Holroyd [314]

Product	G, molecule/100 eV
Hydrogen	2.02
Methane	3.76
Ethane	0.43
Propene	0.20
Isobutane	0.71
Isobutene	1.84
2-Methylbutane	0.06
2-Methylbut-2-ene	0.07
2,2-Dimethylbutane	0.35
2,2,4-Trimethylpentane	0.05
2,4,4-Trimethylpent-1-ene	0.16
2,2,3,3-Tetramethylbutane	0.05
2,2,4,4-Tetramethylpentane	0.48
2,2,5,5-Tetramethylhexane	1.22

the basis of the product spectrum and of the dependence of yields on the dose and dose rate, it had been assumed that most products are formed via intermediate H-atoms, or methyl, tert-butyl and neopentyl radicals produced by decomposition of electronically excited neopentane molecules [192, 312, 314].

The yields of hydrogen, 2,2,5,5-tetramethylhexane and isobutene decrease with increasing dose, whereas those of isopentane, 2,2,4-trimethylpentane and 2,2,5-trimethylhexane increase [314]. The extreme dose dependence of the hydrogen yield [$G(H_2) = 2.57$ at $0.4 \cdot 10^{19}$ eV g^{-1}, $G(H_2) = 1.33$ at $24.1 \cdot 10^{19}$ eV g^{-1}] was attributed to the capture of H-atoms by the isobutene produced in high yield ($G = 2.4$) at lower doses [192, 312, 314]. This hypothesis was supported by the experimental facts that the yields of hydrogen, 2,2,5,5-tetramethylhexane and isobutene decreased rapidly in the presence of isobutene additive at concentrations between $6 \cdot 10^{-4}$ and 0.2 M; at the same time the yields of 2,2,4-trimethylpentane and 2,2,4-

trimethylpent-1-ene increased by more than an order of magnitude. Above about 2 mole% of isobutene, the yields of all products decrease except for hydrogen, which exhibits a value of $G(H_2) = 1.28$ that is independent of the concentration of the additive. As an explanation for the dose dependence and the isobutene effect, the following reactions have been proposed [314]:

$$\text{tert-C}_4\text{H}_9 \qquad\qquad (2.57a)$$

$$H + \text{iso-C}_4\text{H}_8 \longrightarrow \text{iso-C}_4\text{H}_9 \qquad\qquad (2.57b)$$

$$H_2 + CH_2C(CH_3)CH_2 \qquad (2.57c)$$

$$H + C(CH_3)_4 \longrightarrow H_2 + (CH_3)_3CCH_2 \qquad\qquad (2.58)$$

$$\text{tert-C}_4\text{H}_9 + \text{iso-C}_4\text{H}_9 \longrightarrow (CH_3)_3CCH_2CH(CH_3)_2 \qquad\qquad (2.59)$$

$$\text{tert-C}_4\text{H}_9 + CH_2C(CH_3)CH_2 \longrightarrow (CH_3)_3CCH_2CCH_2(CH_3) \qquad\qquad (2.60)$$

$$2(CH_3)_3CCH_2 \longrightarrow (CH_3)_3CCH_2CH_2C(CH_3)_3 \qquad\qquad (2.61)$$

The competition between reactions (2.57) and (2.58) followed by the radical combination processes (2.59)–(2.61) explains the experimentally determined dose dependence. On the basis of the plots of iso-C_4H_8 concentration against yields, and taking into account reactions (2.57)–(2.61), the ratio of the reaction rate constants of (2.57) addition and (2.58) H-atom abstraction processes has been calculated: $k_{add}/k_{abs} = 3.3 \cdot 10^4$; the analogous value for methyl radical is $k_{add}/k_{abs} = 100$ [314], in agreement with data reported for other alkanes in another paper [315].

Holroyd explained the variation of the yields of other products as a function of the dose rate in terms of competing radical reactions (see Fig. 1.5) [314]. The H-atom abstraction reaction by methyl radicals is suppressed at higher dose rates in comparison with radical combination reactions; consequently, the G-values of methane and 2,2,5,5-tetramethylhexane decrease, and that of ethane increases [314]:

$$CH_3 + C_5H_{12} \longrightarrow CH_4 + C_5H_{11}$$

$$CH_3 + CH_3 \longrightarrow C_2H_6$$

$$C_5H_{11} + C_5H_{11} \longrightarrow (CH_3)_3CCH_2CH_2C(CH_3)_3$$

Methyl, isobutyl and tert-butyl radicals have been identified in liquid-phase irradiation by the $^{14}C_2H_5$ radical sampling technique, their yields being 2.3, 0.16 and 0.55, respectively [306]. The yield of methyl radicals is much higher than that of the complementary tert-butyl radical: this was explained by the secondary decomposition of excited tert-butyl radicals into methyl radicals and propylene. The yields of several products do not approach zero even at high concentrations of radical scavenger (e.g. 30 mM iodine), indicating the importance in neopentane radiolysis of ionic and molecular decomposition processes, such as

$$C_5H_{12}^* \longrightarrow CH_4 + \text{iso-C}_4\text{H}_8$$

When styrene was used as radical scavenger in liquid-phase radiolysis $G(H_2)$ decreased from 2.2 to 0.6: $G(CH_4)$ dropped simultaneously from 3.5 to 0.9. As styrene had no influence on the yield of N_2 formation from N_2O electron scavenger, Shida et al. [322] reasoned that the yield of methyl radicals was $G(CH_3) \approx 2.6$. Holroyd and Klein used iodine as radical scavenger to obtain the value $G(CH_3) = 2.3$ [306]. $G(H_{therm})$ was found to be 1.6, as indicated by the decrease of $G(H_2)$ in the presence of radical scavenger.

The reactions of H atoms produced in neopentane have recently been discussed by Miyazaki and Hirayama [308]. Electron scavenger (N_2O) studies were carried out on liquid neopentane by Hatano et al. [309].

Field and Lampe have reported mass spectrometric studies of the ions formed in neopentane under the effect of electron impact, as well as of several other ion-molecule reactions of neopentane [316]. Lias and Ausloos have studied the direct and noble-gas-sensitized radiolysis of neo-C_5H_{12}– –neo-C_5D_{12} gas mixtures [224], and reported that $C_3H_3^+$, $C_2H_5^+$, $C_3H_5^+$ and $C_3H_7^+$ fragment ions are formed, giving as products C_2H_4, C_2H_6, C_3H_6 and C_3H_8 in ion–molecule reactions. The neopentane parent ion decomposes in an exothermal reaction giving tert-butyl ion and methyl radical:

$$(CH_3)_4C^+ \begin{cases} \longrightarrow (CH_3)_3C^+ + CH_3 & \Delta H^0 = -4 \, kJ \, mol^{-1} & (2.62a) \\ \longrightarrow (CH_3)_2CC_2H_5^+ + H & \Delta H^0 = 25 \, kJ \, mol^{-1} & (2.62b) \\ \longrightarrow CH_3^+ + (CH_3)_3C & \Delta H^0 = 293 \, kJ \, mol^{-1} & (2.62c) \end{cases}$$

C_3 fragment ions are probably derived from the secondary decomposition of $C_5H_{11}^+$ ions formed in hydride ion transfer processes. Whereas C_3 ions induce mainly hydride ion transfer processes ($k \approx 6 \cdot 10^{11} \, M^{-1} \, s^{-1}$), $C_4H_9^+$ leads to the formation of mainly iso-C_4H_8, in a process not elucidated yet. The yield of several of fragment ions decreases with increasing gas pressure, presumably owing to the collisional deactivation of the parent ion [317].

The gas-phase pulse radiolysis investigations carried out by Rebbert and Ausloos [318] during recent years have provided a lot of new data about the decomposition mechanism of the excited molecules and ions produced in neopentane with very short lifetime (10^{-13}–10^{-12} s):

$$C_5H_{12}^+ \longrightarrow tert-C_4H_9^+ + CH_3$$
$$\longrightarrow iso-C_4H_8^+ + CH_4 \qquad (2.63a)$$
$$\longrightarrow C_3H_5^+ + CH_3 + CH_4 \qquad (2.63b)$$
$$\longrightarrow C_2H_5^+ + CH_3 + C_2H_4 \qquad (2.63c)$$
$$\longrightarrow C_2H_3^+ + H_2 + CH_3 + C_2H_4 \qquad (2.63d)$$
$$\longrightarrow C_3H_3^+ + H_2 + CH_3 + CH_4 \qquad (2.63e)$$
$$C_5H_{12}^* \longrightarrow iso-C_4H_8 + CH_3 + H \qquad (2.64a)$$
$$\longrightarrow C_3H_6 + CH_3 + CH_3 \qquad (2.64b)$$
$$\longrightarrow iso-C_4H_8 + CH_4 \qquad (2.64c)$$

The number of electronically excited molecules per ion pair in gas-phase neopentane radiolysis (0.2 bar) is $N_{ex}/N_i = 0.28$, in good agreement with data reported for other alkanes [320, 321]. The parent ion $(CH_3)_4C^+$, which has excess energy, decomposes even at relatively high pressures, so the rate of collision stabilization is negligible [321]. Its most probable decomposition products are tert-butyl ion and methyl radical (2.62a); Ausloos and Lias suggested that about 2–3% of the neopentyl ions participate in a molecular methane elimination reaction (2.63a) involving a four-centre transition complex [321].

The main decomposition pathway of the $C_5H_{12}^+$ ion is the exothermal reaction (2.62a), but the relative rates of various decomposition processes of excited $C_5H_{12}^+$ are unknown, although Rebbert and Ausloos have concluded on the basis of UV photolysis experiments [318] that (2.64a) is more important than (2.64b) and (2.64c). In the UV photolysis (123.6 nm 10.0 eV) of neo-C_5H_{12}–neo-C_5D_{12} gas mixtures carried out in the presence of NO and H_2S, the relative probabilities of reactions (2.64c), (2.64a) and (2.64b) have been found to be 1.0, 4.5, and 1.1, respectively [224].

Although the high concentration of intermediates in pulse radiolysis diminishes the efficiency of radical scavengers, it was shown that the yield of every fragment decreased. Methane seemed to be produced mainly by H-abstraction by methyl radicals. About 85% of the methane produced during irradiation of the neo-C_5H_{12}–neo-C_5D_{12}–O_2 system consists of CD_4 and CH_4. According to Collin and Ausloos, this indicates that at least some of methane is produced in unimolecular decomposition and/or disproportionation reactions, the latter process taking place within the 'cage' [327]; the unimolecular methane yield was estimated to be $G(CH_{4,mol}) \approx 0.7$–$0.8$ [327]. Other reports suggest that about 10% of the total methane yield corresponds to methane formed in unimolecular decomposition of excited neopentane ion [306].

For energetic reasons, the tert-butyl ions formed in reaction (2.62a) react in hydride ion transfer processes only with molecules containing loose tertiary hydrogen atom(s) [321]. Ausloos and Lias have verified directly the formation of tert-butyl ions by radiolysing the neopentane–ethanol system in the liquid phase, which resulted in the formation of tert-butyl-ethyl ether:

$$(CH_3)_3C^+ + C_2H_5OH \longrightarrow (CH_3)_3COC_2H_5 + H^+$$

Various hydrocarbon mixtures have been studied in order to clarify the reactivities of tert-butyl and isobutene ions. Propene produced as a result of irradiation of neo-C_5H_{12}–neo-C_5D_{12}–CCl_4 system (1 : 1 : 0.1) in the presence of O_2 as radical scavenger contained 88% of $C_3H_6 + C_3D_6$ and 12% of $C_3H_5D + C_3D_5H$ [327] indicating a unimolecular decomposition of tert-butyl ions.

$C_4H_9^+$, $C_4H_8^+$, $C_3H_3^+$ and $C_5H_{11}^+$ ions suffer (homogeneous) neutralization under the high dose rates of pulse radiolysis [224, 328]:

$$tert\text{-}C_4H_9^+ + e^- \longrightarrow tert\text{-}C_4H_9^* \qquad \Delta H^0 = -5.17 \text{ eV}$$

$$\text{tert-C}_4\text{H}_9^* \begin{cases} \longrightarrow \text{iso-C}_4\text{H}_8 + \text{H} & (2.65a) \\ \\ \longrightarrow \text{C}_3\text{H}_6 + \text{CH}_3 & (2.65b \end{cases}$$

$$\text{iso-C}_4\text{H}_8^+ + \text{e}^- \longrightarrow \text{iso-C}_4\text{H}_8^* \qquad \Delta H^0 = -9.23 \text{ eV} \qquad (2.66)$$

$$\text{iso-C}_4\text{H}_8^* \begin{cases} \longrightarrow \text{CH}_2{=}\text{C}{=}\text{CH}_2 + \text{H} + \text{CH} & (2.67a) \\ \\ \longrightarrow \text{CH}_3\text{C}{\equiv}\text{CH} + \text{H} + \text{CH}_3 & (2.67b) \end{cases}$$

The energy released in the neutralization process is sufficient to activate decomposition reactions. Tracer and scavenger studies show that reaction (2.65a) is about six times faster than (2.65b). As iso-C_4H_9 radicals have been shown to form in liquid phase radiolysis, it is assumed that tert-C_4H_9 undergoes isomerization prior to decomposition (2.65b). No direct evidence is available for reactions (2.65)–(2.67), although their products have been detected, however in low yields [192, 314].

For energetic reasons it is probable that $C_5H_{11}^+$ ions formed mainly in hydride ion transfer processes possess a tertiary structure, which is more stable thermodynamically. Its neutralization produces 2-methylbut-1-ene and 2-methylbut-2-ene: the yields are increased by about 30–40% in the presence of electron scavengers.

In a system containing 33% perdeuteromethylcyclopentane and 12% CCl_4 mixed with neopentane, the G-value of the iso-$C_4H_8^+$ ion has been estimated as $G(\text{iso-}C_4H_8^+) \approx 0.9$, on the basis of the yield of partially deuterated isobutane produced in the reaction [327]:

$$\text{iso-C}_4\text{H}_8^+ + \text{cyclo-C}_5\text{D}_9\text{CD}_3 \xrightarrow{\text{H}_2^- \text{ transfer}} (\text{CH}_3)_2\text{CDCH}_2\text{D} + \text{C}_5\text{D}_{10}^+$$

The relative rate constants of the H_2^- transfer reaction of iso-$C_4H_8^+$ with various C_5 and C_6 alkanes have also been determined. The reaction rate constant increases with increasing number of tertiary H-atoms in the alkane molecule, i.e. with the increasing exothermicity of the reaction.

The reaction rate constant of the hydride ion transfer process of tert-butyl ion is smaller by at least an order of magnitude than that of H_2^- transfer of iso-$C_4H_8^+$ [328]. The probability that the tert-butyl ion, with an estimated yield of $G \approx 1.9$, undergoes neutralization, is high. Reaction (2.65b) follows neutralization, as evidenced by the decreasing propene yield in the presence of CCl_4 electron scavenger, and further, by the isotopic composition of propene pointing to a unimolecular type of formation. Reaction (2.65a) competing with (2.65b) gives about 14% of the total yield of tert-butyl ion [321].

The rate constants of reactions of the tert-butyl ion with several alkane molecules have been determined both in the gas and the liquid phases [321, 330]. Their value depends to a great extent on the enthalpy change involved as well as on steric factors (e.g. on the extent of van der Waals interaction between its hydrogen atoms and the hydrogen atoms in β position to the tertiary carbon atom in the alkane molecule). The energy of activation of these processes is not higher than 0.15 eV, i.e. about 15 kJ mol^{-1}.

There are some indications that the yield of tert-$C_4H_9^+$ formed in neopentane radiolysis depends on the phase: G(tert-$C_4H_9^+$) = 1.7–1.8 in the gas phase [320] and 1.0 in the liquid phase [323]. This would mean that collision stabilization of excited parent ions is more important in the liquid phase than in the gas phase [323]. Differences between literature data on the ion yields can be attributed partly to the different experimental conditions, but they also reflect the complex nature of the reaction mechanism and the inaccuracy of current experimental techniques. Similarly, G-values given for the following reactions on the basis of scavenger studies should be treated with caution [306, 322, 323]:

$$neo\text{-}C_5H_{12} \longrightarrow\!\!\!\curvearrowright\!\!\!\longrightarrow (CH_3)_3C^+ + CH_3 + e^- \qquad G \approx 1.0$$

$$\longrightarrow\!\!\!\curvearrowright\!\!\!\longrightarrow (CH_3)_3CCH_3^+ + e^- \qquad G \approx 2.2$$

$$\longrightarrow\!\!\!\curvearrowright\!\!\!\longrightarrow CH_4 + (CH_3)_2C\!\!=\!\!CH_2 \qquad G \approx 0.9$$

$$\longrightarrow\!\!\!\curvearrowright\!\!\!\longrightarrow H_2 + products \qquad G \approx 0.6$$

$$(CH_3)_3CCH_3^+ + e^- \longrightarrow (CH_3)_3C + CH_3 \qquad G \approx 1.6$$

$$\longrightarrow (CH_3)_3CCH_2 + H \qquad G \approx 0.6$$

$$(CH_3)_3C^+ + e^- \longrightarrow H + (CH_3)_2C\!\!=\!\!CH_2 \qquad G \approx 1.0$$

No reliable experimental method exists for the direct detection of ionic intermediates, but the existence of some radical transient species of neopentane radiolysis (such as neopentyl, tert-butyl radicals) in the solid state has been proved by ESR spectroscopy [297, 324]. By changing the energy of the electron beam, Marx et al. found [324] that neopentyl radicals are produced only with energies higher than the ionization potential of neopentane, but small amounts of tert-butyl radicals are formed with energies lower than the ionization potential. Although radical formation was attributed to the decomposition of excited molecules [324], the decomposition of excited ions leading to the above species cannot be excluded. Some authors have suggested that a considerable proportion of neopentane ions (~60%), if not all, would decompose prior to neutralization because the dissociation energy of the C—C bond in the ion is very low: 0.31 eV [323].

The decomposition processes of excited neopentane molecules are less well-known than those of the ionic species. Recent works in this field discovered some features of the excited states of the neopentane molecule [317] in the 7–30 eV energy range. The most promising works, e.g. those of Derai et al., used low and variable energy electron beams for studying the decomposition pathways of the excited neopentane molecule [319].

The decomposition of neopentane ions with a high yield into tert-butyl ons and methyl radicals was used by Stone and Matsushita as a reaction suitable for verifying positive charge-transfer processes in binary alkane mixtures [326]. In the presence of components having a higher ionization potential than neopentane, an abrupt increase of the methane yield has been observed. The methane yield and the rate of positive charge-transfer processes increased sharply in the presence of electron scavengers. The efficiency of the transfer process increased rapidly with increasing difference between the ionization potentials of the two alkanes [326].

2.2.8. n-Hexane

Papers dealing with the radiation chemistry of n-hexane have been well reviewed in a number of works. Therefore, we shall discuss here only the most important conclusions of earlier research, together with the results published since 1970. The reader particularly interested in the radiation chemistry of n-hexane may find useful consulting the comprehensive works of Topchiev [1], Ausloos [2] and Gäumann [5].

Gäumann and co-workers have been studying n-hexane radiolysis for several years; they have determined the yields under varying experimental conditions and described in detail the character and role of individual intermediate species; their results are summarized in ref. [5], in addition to the numerous original papers cited there [336, 367, 369, 372, 373].

The data of several other authors have been critically reviewed by Avdonina and used for building up the reaction mechanism of n-hexane radiolysis [379, 453, 454]. Radiolysis yields of liquid n-hexane are collected in Table 2.23, on the basis of the study made by Shinsaka and Shida [361]. The main product is hydrogen ($G = 5.01$), together with hexenes ($G = 3.06$) and dimers ($G = 1.29$); n-alkanes predominate among the fragments. Of unsaturated products, more trans-alkenes are produced than cis-isomers (the latter being thermodynamically less stable). Only saturated C_7-C_{12} hydrocarbon products could be observed.

Although earlier studies of the vapour-phase radiolysis of n-hexane were not based on a very sophisticated experimental technique, they established the occurrence of several types of reaction, such as hydride ion and H_2^- ion transfer, radical recombination and disproportionation [193, 346, 347].

The data in Table 2.24 indicate that the radiation chemical yields given by Dewhurst for the gas and the liquid phases are rather different [193]. The yield of hydrogen and fragments is higher in the gas phase, whereas hexenes and C_7-C_{12} products are formed in greater yield in the liquid state. This phenomenon is usually attributed to collision deactivation, but the fact that even the ratios between various products are different in the two phases indicates that other factors, such as those connected with the structure of spurs and/or diffusion kinetics, may also be important. For example, the formation of trans-hex-2-ene is attributed to disproportionation within the spur. It can be stated, generally speaking, that the product spectrum of gas phase radiolysis, with more fragments suggesting more C—C bond ruptures, is closer to the mass spectrum than that observed in radiolysis

Table 2.23

Yields from liquid n-hexane
(radiation: ^{60}Co-γ). After Shinsaka and Shida [361]

Dose, 10^{19} eV g^{-1}	5.85	Dose, 10^{19} eV g^{-1}	80
Product	G, molecule/100 eV	Product	G, molecule/100 eV
Hydrogen	5.01	n-Heptane	0.02
Methane	0.18	3-Ethylhexane +	
Ethane	0.42	+ 3-methylheptane	0.13
Ethylene	0.25	n-Octane	0.03
Propane	0.41	4-Ethylheptane +	
Propene	0.19	+ 4-methyloctane	0.10
n-Butane	0.35	n-Nonane	0.02
But-1-ene	0.14	4-Ethyloctane +	
trans-But-2-ene	0.01	+ 5-methylnonane	0.09
n-Pentane	0.10	n-Decane	0.02
Pent-1-ene	0.04	4,5-Diethyloctane	0.22
trans-Pent-2-ene	0.02	4-Ethyl-5-methylnonane	0.51
Hex-1-ene	0.86	5,6-Dimethyldecane	0.31
trans-Hex-3-ene	0.42	4-Ethyldecane	0.10
trans-Hex-2-ene +		5-Methylundecane	0.13
+ cis-hex-3-ene	1.23	n-Dodecane	0.02
cis-Hex-2-ene	0.55	Total dodecanes	1.29

Table 2.24

Effect of phase and radical scavenging on the radiolysis
of n-hexane (radiation: ^{60}Co-γ). After Dewhurst [193]

Additive	10% propene		
Phase	Gas		Liquid
Products	G, molecule/100 eV		
Hydrogen	5.0	2.1	5.0
Methane	0.5	0.48	0.12
Ethane	1.0	0.80	0.30
Ethylene	1.1	1.3	0.30
Acetylene	0.3	0.20	
Propane	2.3	3.0	0.42
Propene	0.3		0.13
n-Butane	2.2	1.2	0.50
Butenes	0.06	0.12	0.03
Isobutane	0.50	0.70	0
Pentanes	0.60	0.60	0.30
Isohexanes	0.30	0.40	0
Isohexenes	0.10		1.2
C$_7$ products	0.50	0.40	0.15
C$_8$ products	1.10	0.48	0.53
C$_9$ products	0.47	1.30	0.45
C$_{10}$ products	0.14	0.10	0.43
C$_{11}$ products	0.10		0.02
C$_{12}$ products	0.40	0.18	2.0

in condensed phases. The latter process leads to the formation of more products involving C—H bond rupture.

Kevan and Libby [340] have examined in detail the radiolysis of solid n-hexane. They observed marked differences between the reactions of liquid and solid n-hexane. The yield of fragments is two times, and that of C_7-C_{11} products six times higher in the liquid than in the solid state. The ratio of *trans* over *cis*-hexenes is ≤ 3 in liquid and ≈ 10 in solid n-hexane. The composition of dimers is entirely different, too. The exact explanation of these facts is not known yet but it seems certain that they are not caused simply by differences in diffusion phenomena.

In order to describe the dose rate dependence of the yields, several diffusion kinetic models have been developed [341, 342]. Burns and Barker [342] calculated the dose rate dependent fraction of the yields of hydrogen, of fragments, of 'intermediate' products and of dimers [$G(H_2)$, $G(C_f)$, $G(C_i)$ and $G(C_{2n})$, respectively]. These values as functions of the dose rate, are plotted in Fig. 2.4. As radiation sources capable of producing the highest dose rates plotted do not exist, experimental data are available only for the low dose-rate range [344, 345]. (The yields independent of dose rate, e.g. from molecular decompositions, should be added to the values plotted.) The theory suggests that at low dose rates ($< 10^{16}$ eV g^{-1} s^{-1}), H-atoms and fragment radicals generated by irradiation abstract H atoms, and hexyl radicals combine and disproportionate; at medium dose rates (between 10^{16} and 10^{25} eV g^{-1} s^{-1}) hydrogen atoms abstract and alkyl radicals combine and disproportionate; at high dose rates ($>10^{25}$ eV g^{-1} s^{-1}) both hydrogen atoms and alkyl radicals combine and disproportionate. The yields of C_7-C_{11} products have been studied extensively as a function of the dose rate [341]. It was shown that the low dose rate limiting value corre-

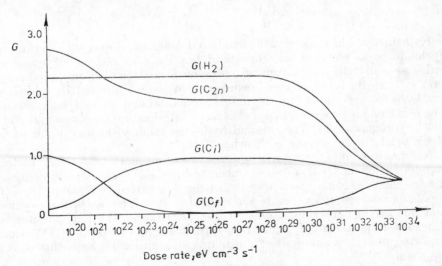

Fig. 2.4. Effect of dose rate on the theoretical yields in the radiolysis of n-hexane. After Burns et al. [342]

sponds to 27% of the total C_7-C_{11} yield at high dose rates. This 'remaining' yield was interpreted in terms of recombination processes within the spur. This is in contradiction with Dewhurst's experiments that indicated that the yields obtained in n-hexane irradiated by electrons, deuterons and helions did not differ significantly [357]; thus the LET effect on the yields is negligible.

The G-values of hydrogen and hexenes show a strong dose dependence whereas those of C_7-C_{11} products are dose independent. Shinsaka and Shida [361] have explained the dose dependence in the following way: hexenes accumulating in irradiated systems produce hexyl radicals by H-atom capture. The dose dependence is caused by the competition between addition and abstraction reactions:

$$H + C_6H_{12} \longrightarrow C_6H_{13}$$

$$H + C_6H_{14} \longrightarrow C_6H_{13} + H_2$$

$$2 C_6H_{13} \longrightarrow C_{12}H_{26}$$

The yield of hydrogen in the liquid phase extrapolated to zero dose is $G_0(H_2) = 5.28$ (Hardwick [358]). Additives have been used in several studies to separate the components of hydrogen yield. The yield of thermal H atoms was found to be $G \approx 1.5$, as shown by the changes in the yield in the presence of various radical scavengers [290, 349, 350, 372, 373]. Thermal hydrogen atoms are stabilized with highest probability in hydrogen abstraction reactions as indicated by the data of Perner and Schuler [351] and Hardwick [352]:

$$H + C_6H_{14} \longrightarrow H_2 + C_6H_{13}$$

Rajbenbach and Kaldor [286] irradiated n-hexane doped with perfluorocyclobutane as electron scavenger and hex-1-ene as radical scavenger in order to study the mechanism of hydrogen formation. The proposed reaction mechanism involves two precursors. One is not affected by additives and leads to the formation of thermal H atoms and molecular H_2 elimination. The other one is produced in ion-electron neutralization processes and gives 'hot' hydrogen atoms. The thermal hydrogen atom yield was found to be $G(H) \approx 1.4$, in agreement with other data.

Gäumann et al. [5] studied the processes of primary 'molecular' hydrogen formation using deuterium tracer techniques. Their hexane–deuterohexane mixture contained about 1% of the deuterated compound; the isotopic composition of hydrogen was measured as a function of the position and number of deuterium atoms in the admixture. With a mixture of n-hexane–n-hexane-2,2-d_2, more D_2 is formed per C—H bond (1.3%) than expected (0.8%), assuming random hydrogen elimination according to the equation:

$$CH_3CD_2CH_2CH_2CH_2CH_3 \longrightarrow D_2 + CH_3\ddot{C}CH_2CH_2CH_2CH_3$$

Using mixtures of n-hexane with n-hexane-2,5-d_2, the value of $G(D_2)$ was lower, and that of $G(HD)$ somewhat higher than the calculated values (2.0 and 24%, respectively), although the difference was close to the limits of experimental error (1.0 and 25%). It was concluded, therefore, that processes leading to monomolecular hydrogen formation also involve reactions giving biradicals: $R-\ddot{C}-R'$.

The yield of fragments decreased slightly upon addition of radical scavengers. This indicates that non-radical processes, e.g. molecular alkane elimination, may be important in fragmentation [357]. Other studies e.g. that of Futrell [347], based primarily on mass spectrometric results, indicate that ion-molecule reactions of ionic fragments are the most probable pathway e.g.:

$$C_2H_3^+ + C_6H_{14} \longrightarrow C_2H_4^* + C_6H_{13}^+$$

$$C_2H_4^* \begin{array}{c} \nearrow^M \ C_2H_4 \\ \searrow \ C_2H_2 + H_2 \end{array}$$

$$C_6H_{13}^+ + e^- \longrightarrow C_6H_{13}^* \longrightarrow C_6H_{12} + H$$

Isildar and Schuler [360] used HI for influencing the neutralization of ions formed in the liquid-phase radiolysis of n-hexane. Their work indicates that charge recombination processes play a minor role in the $C-C$ bond rupture, as compared to the fragmentation of superexcited molecules and excited cations. Energetics suggest the occurrence of ionic decomposition processes followed by hydride ion transfer [347, 348, 360, 365, 376]:

$$n\text{-}C_6H_{14}^+ \begin{array}{c} \nearrow \ \text{sec-}C_4H_9^+ + C_2H_5 \qquad \Delta H^0 = 29 \quad kJ\ mol^{-1} \\ \longrightarrow \ \text{sec-}C_3H_7^+ + n\text{-}C_3H_7 \qquad \Delta H^0 = 42 \quad kJ\ mol^{-1} \\ \searrow \ C_2H_5^+ + n\text{-}C_4H_9 \qquad \Delta H^0 = 170 \quad kJ\ mol^{-1} \end{array}$$

$$C_3H_7^+ + C_6H_{14} \begin{array}{c} \nearrow \ C_3H_7 + C_6H_{14}^+ \qquad \Delta H^0 = 197 \quad kJ\ mol^{-1} \\ \searrow \ C_3H_8 + C_6H_{13}^+ \qquad \Delta H^0 = -25 \quad kJ\ mol^{-1} \end{array}$$

Wexler and Pobo studied the ionic decomposition and ion-molecule reactions of n-hexane in a proton beam mass spectrometer [377]. They elucidated the reactions of the $C_6H_{14}^+$, $C_6H_{13}^+$, $C_6H_{12}^+$, $C_5H_{11}^+$, $C_4H_9^+$ and $C_4H_8^+$ ions. The ion intensities decrease with growing pressure, except those of $C_6H_{13}^+$, $C_6H_{12}^+$ and $C_5H_{11}^+$ which have a maximum intensity as a function of the pressure. These facts have been attributed to H^- and H_2^- transfer processes as well as to the reaction

$$C_6H_{13}^+ + C_6H_{14} \longrightarrow C_{10}H_{21} + C_2H_6$$

Houriet et al. [365] used ion-cyclotron resonance spectroscopy for studying the ionic decomposition processes of C_6 ions. They established that the hexyl ion loses either ethylene to form the butyl ion, or propene to form the propyl ion. The hexyl ion fragments statistically, i.e., all hydrogens are equivalent during the fragmentation process. The protonated hexane (hexonium, $C_6H_{15}^+$) ion was shown to decompose through different reaction pathways depending on the position (or localization) of the proton. Primary species give alkane products and secondary ions lead mainly to molecular H_2 detachment or alkene elimination.

Shinsaka and Shida [361] attempted to elucidate the mechanism of fragmentation by extensive scavenger studies. They have shown that in the presence of SO_2, which can be regarded as a good radical scavenger, the G-value of saturated hydrocarbon fragments decreases by 50–60% but that of the unsaturated fragments changes little if at all. They concluded that both radical and molecular decomposition processes participate in product formation. The simultaneous use of NO as radical scavenger and N_2O as electron scavenger has helped to clarify the formation pathways of C_1-C_4 fragments. The results are shown in Table 2.25.

Table 2.25

Sources of fragment products from liquid hexane
(radiation: ^{60}Co-γ). After Shinsaka and Shida [361]

Primary process	Neutralization		Other than neutralization		Total yield
Type of reaction	radical	molecular	radical	molecular	
Products	G, molecule/100 eV				
Methane	0.05	0.01	0.06	0.10	0.22
Ethane	0.08	0.03	0.10	0.15	0.36
Ethylene	0.03	0.02	0.00	0.08	0.12
Propane	0.09	0.03	0.10	0.14	0.36
Propene	0.03	0.03	0.01	0.06	0.13
n-Butane	0.07	0.03	0.09	0.13	0.32
But-1-ene	0.09	0.03	0.01	0.05	0.18

Avdonina, on the other hand, ascribed fragmentation to three main reactions [364]:

$$RH^* \nearrow \quad R_1 + R_2 \qquad\qquad (2.68a)$$
$$\searrow \quad R_1H + R_2(-H) \qquad\qquad (2.68b)$$

$$RH + H^* \longrightarrow R_1H + R_2 \qquad\qquad (2.69)$$

where R_1H is an alkane and $R_2(-H)$ an alkene.

Analogously to CH_5^+ observed in the mass spectrometry of methane, Kevan and Libby [340] attributed dimer formation to the reaction

$$RH^+ + RH \longrightarrow RH_2^+ + R$$

followed, presumably, by the loss of a proton upon neutralization. Later studies by Wagner and Geymer did not support this hypothesis [338]: in the presence of radical scavengers the yields of both dimers and C_7-C_{11} products decreased to zero, suggesting a radical pathway for these compounds [357, 361]. Similar conclusion can be drawn from the above-mentioned dependence of the yields of C_7-C_{12} products on the dose rate.

Avdonina and Makarov have shown that the fraction of products in the C_8-C_{10} group involving combination of 1-hexyl, 2-hexyl and 3-hexyl radicals with fragment C_2-C_4 radicals are the same (21–22, 42–45 and 34–38%, respectively). A dissimilar distribution has been found however, for heptane products: the fractions involving 1-hexyl, 2-hexyl and 3-hexyl precursors are 52, 26 and 22%, respectively. These latter values are almost equal to the relative numbers of hydrogen atoms on the corresponding carbon atom of the hexane molecule (43, 29 and 29%, respectively). The authors have concluded on this basis that most of the heptane products have been formed via insertion of methylene biradicals into the C—H bonds of the parent hexane molecule. The CH_2 species could be detected by means of iodine radical scavenger.

Radical yields as determined by iodine radical scavenger increase with increasing temperature. The G-value of n-alkyl iodides is constant between -80 and $+25°C$, and that of sec-hexyl iodide increases from 1.4 to 2.3. At 120°C, the total iodide yield exhibits a two-fold increase over the low-temperature yield [335, 336]. This can be attributed partly to HI addition to unsaturated products, and partly to the increasing rate of diffusion of radicals out of the spurs.

The total radical yield of n-hexane radiolysis, expressed as an average of data obtained by different experimental techniques, is $G \approx 5$ [335]; 80% of this yield consists of hexyl radicals and 20% of fragments (Table 2.26). Fragment radicals having a secondary structure have been detected by iodine radical scavenger. They are formed mainly in HI addition reactions of fragment alkenes because the activation energy of decomposition and radical isomerization reactions of alkyl radicals is high (~ 1 and ~ 0.5 eV, respectively) and therefore their role is probably negligible at room temperature in liquids.

Isildar [405] used tritium iodide to scavenge primary radicals produced by γ-rays. He determined the total radical yield to be 6.2 for n-

Table 2.26
Radical yields from liquid n-hexane (radiation: ^{60}Co-γ)

Additive	8 mM I_2	0.1μM $^{14}C_2H_4$	10 mM I_2
Reference	338	241	375
Product	G, radical/100 eV		
Methyl	0.08	0.06	0.08
Ethyl	0.27	0.33	0.32
1-Propyl	0.25	0.32	0.34
2-Propyl	0.04		0.03
1-Butyl	0.23	0.28	0.28
2-Butyl	0.05		0.04
1-Pentyl	0.03	0.06	0.03
1-Hexyl	0.5	0.99	0.64
2- and 3-Hexyl	2.3	3.58	2.44
Total	3.8	5.6	4.20

hexane, i.e. about 20% greater than the average value given by Holroyd [335]. Isildar's results suggest that the great majority of the radicals (~77%) are parent alkyl radicals produced exclusively via geminate neutralization. Direct excitation was shown to yield only fragment free radicals (23%).

The radical yields measured at $-70°C$ in solid and at $+10°C$ in liquid n-hexane lead to some interesting conclusions [241]. According to Holroyd and Klein, the yields of C_1-C_5 radical fragments as well as those of primary hexyl radicals remain unchanged in this temperature range, but the G-value of secondary hexyl radicals increases from 2.35 to 3.58. The yields measured at the lower temperature must be closer to the primary decomposition yields because there is no hydrogen abstraction. The increasing importance of hydrogen abstraction with increasing temperature favours the formation of radicals with secondary structure, and this is in agreement with the experimental data. The results cited suggest that the probability of decomposition of a primary $C-H$ bond at $-70°C$ is about two times lower than that of a secondary one. Other authors, such as Pichuzhkin et al., suggest a random breakage of all the $C-H$ bonds involved [359].

Recently, the importance of disproportionation reactions in processes leading to fragments and C_n unsaturated species has been emphasized. The ratio of the rate constants of disproportionation and combination reactions (k_d/k_c) has been determined for several pairs of aliphatic and cyclic alkyl radicals by measuring the G-values of the products in pure state as well as — in order to separate inhomogeneous processes occurring in the spur — in the presence of radical scavengers. The following expression can be used to determine the k_d/k_c ratio of hexyl radicals:

$$\frac{k_d}{k_c} = \frac{G(\text{scavengable hexenes})}{G(\text{scavengable dimers})}$$

The value of k_d/k_c was estimated to be 1.75, 1.08 and 0.69 at -78, $+25$ and $+150°C$, respectively [296]. However, as the formation of hexyl radicals during radiolysis is not selective (i.e. 1-hexyl, 2-hexyl and 3-hexyl radicals are formed simultaneously), the k_d/k_c value thus calculated is related to the mixture of hexyl radicals. The k_d/k_c value determined by mercury-sensitized photolysis at 25°C is 1.0, in good agreement with the previous data.

An ESR spectroscopic study of a single crystal of n-hexane irradiated at 77 K showed that 80–90% of the radicals is 2-hexyl and 10–20% 1-hexyl [374, 376]. The predominance of 2-hexyl radicals was correlated with the distribution of positive charge in the hexane cation [374], the ratio of positive charges on the first, second and third carbon atoms being 1 : 2.65 : : 2.25, respectively. Neutralization is most likely to occur in the position of greatest positive charge density, consequently the energy liberated causes bond breakage mainly in its vicinity [374].

Phosphorescence phenomena observed in n-hexane–benzene systems, as well as low-energy electron impact studies, permit the examination of excited states of n-hexane [83, 84, 366, 368]. Some excited triplet states

could be related to given decomposition reactions of n-hexane [84]. Intersystem crossing processes take place with high cross-section [370, 371].

One of the most convenient experimental techniques for the study of the
yield and reactivity of excited molecules is the use of so-called quencher
substances with low excitation levels (e.g. benzene) as additives. Makarov
et al. showed that the yield of every radiolysis product of n-hexane decreases
in the presence of benzene, e.g. that of hydrogen by 80% [353]. Although
other processes such as positive ion transfer cannot be excluded, it is probably true that this phenomenon indicates the presence of excited molecules.
The yield of energy transfer in photolysis as determined by the use of additives proved to be much lower [354] than in radiolysis [355]. No precise
experimental results are available with respect to the character of the energy
transferring species [356].

The vacuum UV photolysis of n-hexane results in H_2 elimination having
the highest yield among the decomposition reactions of excited molecules
[378]. C—C bond rupture processes were also observed, with lower quantum
yields.

Shinsaka and Shida have proposed the following decomposition pattern
for the radiolysis of n-hexane, on the basis of scavenger studies [361]:

$$n\text{-}C_6H_{14} \rightsquigarrow \begin{cases} n\text{-}C_6H_{14}^+ + e^- \\ n\text{-}C_6H_{14}^* \end{cases}$$

$$n\text{-}C_6H_{14}^+ + e^- \rightarrow \begin{cases} C_6H_{13} + H \\ C_6H_{12} + H_2 \\ R_1H + R_2H \end{cases}$$

$$n\text{-}C_6H_{14}^* \rightarrow \begin{cases} C_6H_{13} + H \\ C_6H_{12} + H_2 \\ R_1H + R_2H \end{cases}$$

$$H + n\text{-}C_6H_{14} \longrightarrow H_2 + C_6H_{13}$$

$$R_1H + R_2H \rightarrow \begin{cases} R_1H_2 + R_2 \\ R_1 + R_2H_2 \\ R_1H_2 + R_2 \end{cases}$$

$$R_1H + C_6H_{13} \rightarrow \begin{cases} R_1H_2 + C_6H_{12} \\ R_1 + n\text{-}C_6H_{14} \\ R_1H + C_6H_{13} \end{cases}$$

$$C_6H_{13} + C_6H_{13} \left\langle \begin{array}{l} C_6H_{12} + \text{n-}C_6H_{14} \\ C_{12}H_{26} \end{array} \right.$$

$$\text{n-}C_6H_{14} \xrightarrow{\quad\text{\tiny\raisebox{0.5ex}{$\sim\!\!\sim$}}\quad} C_6H_{13}^+ + H + e^-$$

where R_1H and R_2H denote fragment alkyl radicals, by the authors' notation. The primary production of $C_6H_{13}^+$ ion was assumed from mass spectrometric data and was supported by the scavenger studies: in the simultaneous presence of SF_6 as electron scavenger and NH_3 as positive ion scavenger the yield of hexenes increased, and that of dodecanes decreased. This was attributed to the following reactions:

$$\text{n-}C_6H_{14} \longrightarrow C_6H_{13}^+ + H + e^-$$

$$e^- + SF_6 \longrightarrow SF_6^-$$

$$\text{n-}C_6H_{14}^+ + NH_3 \longrightarrow C_6H_{13} + NH_4^+$$

$$NH_4^+ + e^- \longrightarrow NH_3 + H$$

$$C_6H_{13}^+ + SF_6^- \longrightarrow C_6H_{13} + SF_6$$

$$C_6H_{13}^+ + NH_3 \longrightarrow C_6H_{12} + NH_4^+$$

The reactivity of different C—C bonds of n-hexane in fragmentation processes will be discussed later, together with the radiolysis of C_7-C_{12} n-alkanes (see Sections 2.2.10 and 2.4).

2.2.9. Branched-chain hexane isomers

Far fewer papers deal with the radiation chemistry of branched-chain hexane isomers than with that of n-hexane. Radiation chemical yields for 3-methylpentane irradiated in the solid, liquid and gas phases have been reported; for 2,2- and 2,3-dimethylbutanes only liquid-phase data are available; nothing but approximate G-values for product groups have been determined for 2-methylpentane (therefore this compound will not be included here). Several papers have been published on the physical and physico-chemical aspects of 3-methylpentane radiolysis, e.g. the correlation of electric conductivity and lattice structure: this subject is beyond the scope of this book.

The more detailed study of this field would result in a deeper insight into the correlations between molecular structure and reactivity in radiation chemical processes, such as selective decomposition of bonds in the vicinity of tertiary and quaternary carbon atoms, special reactions of branched radicals and ions, etc.

Almost complete product spectra of the radiolysis of 3-methylpentane in the gas, liquid and solid phases have been determined [384, 385, 391]. Table 2.27 contains data on the G-values of the main products determined with

Table 2.27

Yields from 3-methylpentane (radiation: ^{60}Co-γ). After Mainwaring and Willard [384]

Phase	Gas	Liquid	Solid
Product			

Phase	Gas	Liquid	Solid
Product	*G*, molecule/100 eV		
Hydrogen	3.6	3.4	3.2
Methane	0.80	0.20	0.08
Ethane	1.98	1.90	0.45
Ethylene	0.86	1.00	0.20
Propane	0.40	0.09	0.02
Isobutane	0.14	0.02	—
n-Butane	1.45	0.45	0.31
But-1-ene	0.60	0.37	0.19
But-2-enes	0.30	0.22	0.14
2-Methylbutane	0.31	0.06	0.01
2-Methylbut-1-ene	trace	0.03	0.03
n-Pentane	0.03	0.09	0.05
2-Pentenes	0.08	0.07	0.04
2-Methylpentane	0.10	0.65	0.80
3-Methylpent-1-ene	0.11	0.27	0.40
3-Methylpent-2-ene	—	0.55	0.80
3-Methylpenta-1,3-diene	0.16	0.26	0.43
Dimethylpentane	0.19	0.28	0.42
3-Methylhexane + 3-ethylpentane	1.50	0.26	0.18
2-Ethyl-3-methylpentane + + 3-methylheptane + + 3-ethyl-3-methylpentane	1.26	0.51	0.21
3,5-Dimethylheptane	0.07	0.03	—
4-Ethyl-3-methylhexane	0.17	0.04	0.02
5-Ethyl-3-methylheptane	0.05	—	—
3,6-Dimethyloctane + + 3,4-diethylhexane	0.08	—	0.04
3,4,5-Trimethylheptane	—	—	0.07
3,4-Dimethyl-3-ethylhexane	0.20	0.29	0.08
2,4-Dimethyl-5-ethylheptane	0.09	0.05	0.02
3,4-Diethyl-3-methylhexane	—	0.03	—
C_{12} isomer products (identified)	1.37	0.98	0.68

Table 2.28

Effect of phase on the yields from 3-methylpentane (radiation: ^{60}Co-γ). After Mainwaring and Willard [384]

Phase	Gas	Liquid	Solid
Product	*G*, molecule/100 eV		
Hydrogen	3.60	3.40	3.20
C_1-C_5 products	7.09	4.72	1.59
C_6 + dimethylpentane	0.56	2.01	2.85
C_7-C_{11} products	3.87	1.41	0.70
C_{12} products	1.51	1.11	0.68
Total	16.63	12.65	9.02

γ-irradiation with a dose of $400 \cdot 10^{19}$ eV g^{-1} and a dose rate of $6.7 \cdot 10^{16}$ eV g^{-1} s^{-1}. Table 2.28 contains results on the yields obtained in different states of aggregation, divided into product groups [384].

The comparison of G-values obtained in different phases by Mainwaring and Willard (see Table 2.27) reveals that the yields of ethylene, n-pentane, the C$_6$ (so-called congruent) alkenes, dimethylpentane, and 3,4-dimethyl-3-ethylhexane increase in the order: gas < liquid < solid, but the reverse order holds for most other products. The total product G-values for the solid, liquid and gas phases are 9.02, 11.65 and 16.63, respectively. This effect may have two causes: the inhibition of the diffusion of reactive radicals and ionic intermediates, and the phase effect in collisional deactivation of activated species.

The yields and nature of intermediates are also influenced by the phase of irradiation. Weber et al. have found $G = 7$ radicals per 100 eV in the liquid phase using iodine as radical scavenger [305], and Willard et al. have observed $G = 3.1$ in a glassy solid state at 77 K by ESR spectroscopy [383, 384, 387, 388]. Mainly 3-methyl-2-pentyl and 3-methyl-3-pentyl radicals were detected in the solid state [297, 306, 387], but more types of fragment and parent alkyl radical were observed in the fluid phases. Electron spin resonance studies suggest that only 3-methyl-2-pentyl radicals are formed in solid-phase 3-methylpentane at 77 K [392].

The radical yields in liquid-phase radiolysis determined by using iodine as radical scavenger (Table 2.29) show that the decomposition into ethyl and

Table 2.29

Radical yields from liquid 3-methylpentane
(radiation: ^{60}Co-γ). After Dewhurst [385]

Product radical	Radical/100 eV
Ethyl	1.6
Propyl	< 0.1
Butyl	0.7
Pentyl	< 0.1

n-butyl radicals is favoured over other processes leading to pentyl or propyl radicals.

There have been suggestions that thermal and 'hot' hydrogen atoms are the most important intermediates of hydrogen formation in the liquid phase [290, 389]. No H-atoms have yet been detected in the glassy solid state, except in solid methane [390].

Very few papers have been published on the ionic intermediates of 3-methylpentane radiolysis. The dissociation energy of the C—C bond of the parent ion in the vicinity of its tertiary carbon atoms is as low as 95–100 kJ mol^{-1}, leading to the suggestion that — analogously to other alkanes — the role of ionic decomposition reactions may be important even in the liquid phase.

The results of Kimura et al. [394], obtained in the presence of chlorinated methane additives, have clarified the mechanism of hydrogen formation. The hydrogen yield decreased to a different extent in the presence of CH_2Cl_2, $CHCl_3$ and CCl_4. CH_2Cl_2 caused a decrease of about 20% in the $G(H_2)$ value, as it has similar electron capturing properties to N_2O; $G(H_2)$ in the presence of $CHCl_3$ and CCl_4 was about 50% of its original value. It was concluded that CH_2Cl_2 exhibits electron scavenger properties but $CHCl_3$ and CCl_4 are able also to accept excitation energy. Thus, about 20% of hydrogen may have originated from charge neutralization, if we leave the difference in efficiency of electron scavenging out of consideration.

Relatively few data are available on 2,3-dimethylbutane. Although some papers discovered certain features of the mechanism of radiolysis, obtained chiefly by scavenger studies, not even an approximately total product spectrum has yet been determined (Table 2.30). The results of Dewhurst [385],

Table 2.30
Yields from liquid 2,3-dimethylbutane

Radiation	800 keV electrons	^{60}Co-γ	^{60}Co-γ
Dose, 10^{19} eV g^{-1}	290	22.2	5.20
Reference	385	391	246
Products	G, molecule/100 eV		
Hydrogen	2.9	2.90	3.51
Methane	0.5	0.51	0.99
Ethane		0.03	
Ethylene		0.07	
Propane	1.2	1.98	2.79
Propene		2.26	2.88
Isobutane	0.1	0.35	
n-Butane		0.01	
But-1-ene		0.04	
But-2-enes		0.05	
Isobutene		0.05	
C_5 products	0.6	0.65	
C_7 products		0.20	
C_8 products		0.12	
C_9 products		0.35	
C_{10} products		0.05	
C_{11} products		0.10	
C_{12} products		0.05	

Tanno et al. [246], and Földiák and György [391] show that the main products are hydrogen, methane, propane, propene and C_5 hydrocarbons. The G-values indicate that both C—H and C—C bond rupture processes are important. C_7-C_{12} hydrocarbon yields are rather low as compared with alkanes mentioned so far.

The radiation chemical yields of two types of C—C bond rupture have been calculated by Földiák and György [391] on the basis of the product spectrum:

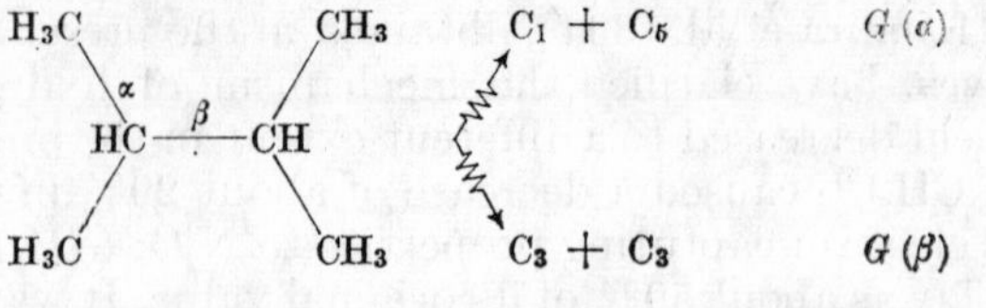

$$\begin{array}{ccc}
H_3C & CH_3 & C_1 + C_5 \qquad G(\alpha) \\
 & \alpha \quad \beta \\
HC & CH \\
H_3C & CH_3 & C_3 + C_3 \qquad G(\beta)
\end{array}$$

Simplifications had to be made for these calculations, such as that radicals produced in the decomposition of excited or ionized 2,3-dimethylbutane molecules are stabilized in H-atom abstraction or radical combination and disproportionation reactions, whereas fragment ions undergo hydride ion transfer processes. C_7-C_{11} and C_{12} products are supposedly formed in combination reactions of parent alkyl radicals [399] with fragments and with each other, respectively. The yields per bond of the two types of ruptures are: $G(\alpha) = 0.3$ and $G(\beta) = 2.6$.

In addition to the very high yields of propane and propene, the low standard dissociation energy (70–100 kJ mol^{-1}) of the central C—C bond of the parent 2,3-dimethylbutane ion with one positive charge also suggests that the C—C bond in the β-position is highly reactive [398].

Tanno et al. [246] have studied the mechanism of radiolysis of 2,3-dimethylbutane using SF_6 as electron scavenger and NH_3 as positive ion scavenger. The G-value of propene decreases to a greater extent than that of propane in the presence of SF_6. With 5% SF_6 and increasing amounts of NH_3, the yield of propene increased, while that of propane diminished. The results were interpreted in terms of competition between the following reactions:

$$C_6H_{14} \longrightarrow sec\text{-}C_3H_7^+ + sec\text{-}C_3H_7 + e^-$$
$$C_6H_{14} \longrightarrow C_6H_{14}^+ + e^-$$

$$sec\text{-}C_3H_7^+ + e^- \longrightarrow C_3H_6 + H \qquad (2.70)$$

$$C_6H_{14}^+ + e^- \longrightarrow \cdot C_3H_8 + C_3H_6 \qquad (2.71)$$

$$e^- + SF_6 \longrightarrow SF_6^- \qquad (2.72)$$

$$sec\text{-}C_3H_7^+ + C_6H_{14} \longrightarrow C_3H_8 + C_6H_{13}^+ \qquad (2.73)$$

$$sec\text{-}C_3H_7^+ + NH_3 \longrightarrow C_3H_6 + NH_4^+ \qquad (2.74)$$

According to this sequence, the electron-capture reaction (2.72) competes with the charge neutralization processes (2.70) and (2.71), and the yields of propane and propene decrease. At the same time, the acceleration of reactions (2.73) and (2.74) leads to increased yields. The experimental results do not contradict this hypothesis. The considerable decrease of propane and propene yields in the presence of an electron scavenger indicates that neutralization of parent ions followed by fragmentation of these electronically excited molecules is very important in liquid-phase radiolysis. The existence of excited alkane molecules has been indirectly verified by Miyazaki et al. by solid-state radiolysis of the 2,3-dimethylbutane–toluene system [399]. By varying the concentration of toluene, they measured the emission of the irradiated system in the wavelength range between 250 and 500 nm. They showed that transfer of electron excitation energy from 2,3-dimethylbutane to toluene takes place during radiolysis. As the intensity of luminescence was not correlated with the concentration of toluene anions produced, the light emission could not be caused by a neutralization process between positive alkane ions and toluene anions. The production of 2,3-dimethyl-1-butyl radicals was verified by ESR spectroscopy [399].

It is regrettable that the radiolysis of 2,2-dimethylbutane (called also neohexane) has been so little studied, because its quaternary carbon atom makes it nearly as interesting a model substance as neopentane (see Section 2.2.7). Although individual radical, ionic etc. processes occurring in neohexane radiolysis have been studied by electron radiation [395, 401], X-rays [276, 352] and γ-rays [274, 291, 409], the product analyses published do not include the full product spectrum. Radiation chemical yields obtained with γ-rays and accelerated electrons, respectively, by Castello et al. [409] and Dewhurst [395], are summarized in Table 2.31. The G-values show that hydrogen, methane, ethane, ethylene, isobutene and 2-methylbut-1-ene are the main radiolysis products.

Castello et al. studied the effect of dose in the range $5.5–115 \cdot 10^{19}$ eV g^{-1}; they found that the G(total) decreased from 16.98 to 8.81 within this, the decreased yields of hydrogen, ethylene, 3,3-dimethylbut-1-ene and isopentene were most marked [409].

Radical reactions have been studied using iodine as radical scavenger. Addition of iodine decreased the value of G(total) from 16.98 to 9.32, with an overall yield of iodides $G = 5.13$ [409]. The deficit is accounted for by the formation of hydrogen iodide and alkyl di-iodides (e.g. di-iodomethane, di-iodoethane). The hydrogen yield drops from $G(H_2) = 2.50$ to 1.63 (with a dose of $5.5 \cdot 10^{19}$ eV g^{-1}) [409] or from $G(H_2) = 2.58$ to 1.64 ($0.9 \cdot 10^{19}$ eV g^{-1}) [274] in the presence of iodine, i.e. about 30% of the hydrogen yield can be scavenged; the rest is due to 'hot' hydrogen atoms, molecular eliminations, etc.

From the high G-values of C_2, isobutene and isobutane yields, it can be concluded that the main decomposition pathway of the neohexane molecule involves $C—C$ bond rupture in the vicinity of the quaternary carbon atom; $C—C$ bond rupture leading to methyl elimination is also significant. An increase in the dose of an order of magnitude does not significantly influence the ratio of the total yield of C_1-C_5 fragments within the total yield, yet the distribution of the fragment products changes greatly with the dose. The ratio of methane increases with increasing dose, indicating the importance of molecular elimination [409]:

$$C_6H_{14} \xrightarrow{\;\;\text{\tiny W}\;\;} CH_4 + C_5H_{10}$$

The yield of iso-C_4H_{10} is lower than that of iso-C_4H_8 [409]. As the hydride ion transfer reaction of tert-$C_4H_9^+$ ion is energetically unfavourable owing

9*

Table 2.31
Yields from liquid 2,2-dimethylbutane

Radiation	^{60}Co-γ	800 keV electrons
Dose rate, eV g^{-1} s^{-1}	$2.2 \cdot 10^{15}$	$1.3 \cdot 10^{19}$
Dose, 10^{19} eV g^{-1}	5.5	290
Reference	409	385
Products	\multicolumn{2}{c}{G, molecule/100 eV}	
Hydrogen	2.50	2.0
Methane	1.76	1.2
Ethane	1.25	
Ethylene	1.87	~1.5
Acetylene	0.012	
Propane	0.21	
Propene	0.15	
Isobutane	0.56	
Isobutene	2.07	~2.4
n-Butane	0.05	
n-Butenes		
Neopentane	0.06	
Isopentane	0.46	
2-Methylbut-1-ene	1.01	1.2
2-Methylbut-2-ene	0.64	
3-Methylbut-1-ene	0.04	
C$_6$ isomers	0.56	
3,3-Dimethylbut-1-ene	1.18	
C$_7$ products	0.70	0.7
C$_8$ products	0.39	0.75
C$_9$ products	0.14	0.2
C$_{10}$ products	0.37	0.4
C$_{11}$ products	0.18	0.6
C$_{12}$ products	0.80	
C$_{12}<$ products	0.02	

to the large number of primary C–H bonds in neohexane, the higher yield of butene can be attributed to reactions (2.75) and (2.76):

$$C_6H_{14} \longrightarrow C_2H_5 + C_4H_9^+ + e^- \qquad (2.75)$$

$$C_4H_9^+ + e^- \longrightarrow C_4H_8 + H \qquad (2.76)$$

The formation of C$_3$ products has been explained by the reaction sequence

$$C_6H_{14} \longrightarrow C_2H_5 + C_4H_9$$

$$C_4H_9 \longrightarrow CH_3 + C_3H_6$$

$$CH_3 + C_2H_5 \longrightarrow C_3H_8$$

and, in addition, by consecutive multiple ionic decomposition processes. Castello et al. proposed the simultaneous ionic decomposition of 1–2 and 2–3 C–C bonds, on the basis of mass-spectrometric results [409]. The same conclusion was arrived at by Rucinska–Starosta et al. [410], who measured

C_1-C_4 fragment product yields in the liquid phase and used N_2O and SF_6 scavengers.

Alkenes with six carbon atoms were detected in very low yields. Their formation was interpreted in terms of ionic rearrangement within the spur, as their yields were not influenced by radical scavengers. The following processes have been proposed as responsible for congruent alkene formation [409]:

$$C_6H_{14}^* \begin{cases} \longrightarrow C_6H_{12} + H_2 \\ \longrightarrow C_6H_{13} + H \end{cases}$$

$$C_6H_{13} + R \longrightarrow C_6H_{12} + RH$$

The yield decreases when iodine radical scavenger is added to the system [409].

Most of the C_7-C_{11} alkane products are formed in radical combination reactions. As only about 20% of the radicals formed in neohexane radiolysis are the parent alkyl radicals, recombination processes of fragment radicals must be important in the formation of these products. This contrasts with the radiolysis of n-alkanes, wherein the concentration of parent radicals is much higher than that of fragments [194, 243, 401, 409]. The G-values of C_7-C_{11} products and C_{12} dimers also decrease with increasing dose and are similarly influenced by radical scavengers [409].

The importance of radical processes has been supported by the ESR studies of Fessenden and Schuler [407], and Thyrion et al. [297], who detected methyl, ethyl, tert-butyl, tert-pentyl, $(CH_3)_2\dot{C}(CH_2)C_2H_5$ and $(CH_3)_3C\dot{C}HCH_3$ radicals in irradiated solid neohexane. The yields of radical intermediates have been determined by scavenger studies: $G(CH_3) = 0.99$ [274], $G(C_2H_5) = 1.2$ [401], $G(\text{iso-}C_4H_9) = 0.07$ [401].

The decomposition yields of individual C—C bonds in neohexane have been calculated on the basis of the G-values of the stable end-products of radiolysis. Castello et al. and Földiák and György found the G-value of decomposition into C_2 and tert-C_4 fragments to be 3.12 [409] and 2.9 [391], and the value for decomposition into C_1 and C_5 fragments to be 3.28 [409] and 2.25 [391]. Considering the number of C—C bonds of neohexane in different positions, these values mean a 3–4 times higher probability for the former type of bond rupture than for the latter [391]. These calculations give no information on whether the decomposition of excited molecules or excited ions is more important. This question cannot be answered from direct experimental evidence, but it can be assumed from the radiolysis of other branched-chain alkanes that both types of reaction take place in liquid-phase radiolysis. Ionic bond rupture is promoted by the favourable thermodynamic conditions embodied in the very low bond dissociation energy ($E_d = 45$–60 kJ mol^{-1}) [376]:

$$(CH_3)_3C^+CH_2CH_3 \begin{cases} \longrightarrow (CH_3)_3C^+ + C_2H_5 & \Delta H^0 = -4 \text{ kJ mol}^{-1} \\ \longrightarrow (CH_3)_2\dot{C}^+CH_2CH_3 + CH_3 & \Delta H^0 = -33 \text{ kJ mol}^{-1} \\ \longrightarrow (CH_3)_3C^+\dot{C}HCH_3 + H & \Delta H^0 = 110 \text{ kJ mol}^{-1} \end{cases}$$

2.2.10. C_7-C_{12} n-alkanes

This section attempts to summarize the experimental results obtained in the radiolysis of C_7-C_{12} n-alkanes. The purpose of this review is to point out the several common features of these compounds in radiolysis, indicated by the closely similar radiation chemical yields. Differences have also been found in the radiolysis of C_7-C_{12} n-alkanes; these reveal, mainly, the variation of radiation chemical behaviour with molecular size.

Radiolysis of n-heptane will be discussed in somewhat more detail because it has been studied most widely. The review works of Rappoport and Gäumann [414, 435] were a useful source of data in compiling this section.

Earlier studies revealed that the gaseous product of n-alkane radiolysis consists mainly of hydrogen [415–417]. Dewhurst studied the radiolysis of liquid C_5-C_{10} n-alkanes with 800 kV electrons and found that the hydrogen yield was independent of the length of the carbon chain and its value was $G \approx 4.9$ [240]. The decreasing yield of hydrogen and unsaturated hydrocarbon products with increasing dose was attributed to the 'internal scavanger effect' of the latter species [240, 414, 434]. According to IR spectroscopic data [240], most of the unsaturated products have the *trans*-vinylene structure. The methane yield decreases from $G = 0.4$ to $G = 0.04$ as the length of the carbon chain in the parent molecule increases from C_5 to C_{16}. Processes involving both C—H and C—C bond rupture are important: $G(C—C)/G(C—H) \approx 0.5$.

More detailed studies have been carried out during recent years. The main products of n-heptane radiolysis are hydrogen and C_7 (so-called congruent) alkenes (Table 2.32) [414, 419, 420]. It can be seen from the G-values that *trans*-alkenes, which are more stable thermodynamically, are formed with higher probability than *cis*-alkenes. Recent studies by Melekhonova et al. [418, 458] report data obtained with aromatic compounds as additives. They attributed the decrease of $G(H_2)$ in the presence of the additives to electron capture and energy transfer processes occurring at higher concentrations of the solutes. Similar studies were carried out by Baba and Fueki [423] who used bromobenzene and 1-hexene additives.

Of the fragments produced in relatively low yields, n-alkanes predominate; this indicates that fragment isomerization is a process of minor importance.

The yield of C_8-C_{13} 'intermediate' products is very small, but that of C_{14} dimers is significant. Of the latter compounds, 5-ethyl-6-methylundecane and 6,7-dimethyldodecane are produced in the greatest amounts: their precursors are 2-heptyl and 3-heptyl radicals. As C_8-C_{14} hydrocarbons are the products of the random combination of alkyl radicals with each other, it seems that the steady state concentration of 2-heptyl and 3-heptyl radicals is highest in the system [435].

The almost complete product spectra of C_8-C_{12} n-alkanes indicate [414, 435] that the above conclusions are also valid for these compounds. The G-values of various product groups are nearly the same within the limits of experimental error, therefore these data have been included in a single summarizing table (Table 2.33) [414].

Table 2.32

Yields from liquid n-heptane

Radiation	^{60}Co-γ	^{60}Co-γ	4.5 MeV electrons	^{60}Co-γ
Dose rate, eV g^{-1} s^{-1}	$1.3 \cdot 10^{16}$	$1 \cdot 10^{16}$	10^{16}	
Dose, 10^{19} eV g^{-1}	140	2–10	0.01–0.04	6–60
Temperature, °C	25	0	25	25
Reference	419	420	421	414
Product	G, molecule/100 eV			
Hydrogen	3.43	4.9		5.25
Methane	0.11		0.16	0.03
Ethane	0.28		0.29	0.43
Ethylene	0.092		0.16	
Acetylene	0.11			
Propane	0.30	0.25		0.46
Propene	0.058	0.12		0.12
Other C$_3$ products	0.006			
Isobutane	0.005	} 0.27		} 0.32
n-Butane	0.32			
But-1-ene	0.052			0.08
trans-But-2-ene	0.006	} 0.12		0.01
cis-But-2-ene	0.004			
Isopentane	0.003	} 0.27		} 0.32
n-Pentane	0.35			
Pent-1-ene	0.051			0.07
trans-Pent-2-ene	0.010	0.10		0.02
cis-Pent-2-ene	0.005			0.02
Hex-1-ene	0.020			0.02
trans-Hex-2-ene	0.006			
cis-Hex-2-ene	0.005		0.04	
trans-Hex-3-ene	0.003			
cis-Hex-3-ene	0.003			
Hept-1-ene	0.36			0.49
trans-Hept-2-ene	0.72			0.69
cis-Hept-2-ene	0.34			0.24
trans-Hept-3-ene	0.62			0.57
cis-3-Heptene	0.23			0.13
2-Methylheptane	0.010			
3-Methylheptane	0.010			
4-Methylheptane	0.005			
n-Octane	0.021			
3-Ethylheptane	0.029		0.15	0.027
4-Ethylheptane	0.017			0.013
3-Methyloctane	0.052			0.036
n-Nonane	0.026			0.021
4-Propylheptane	0.013			0.13
4-Ethyloctane	0.028		0.15	0.026
4-Methylnonane	0.029			0.029
n-Decane	0.018			0.015
3-Propyloctane	—		0.021	0.012
4-Propyloctane	0.012			
5-Ethylnonane	0.024		0.044	0.026

(Cont'd on next page)

Table 2.32 (Cont'd)

Radiation	$^{60}Co-\gamma$	$^{60}Co-\gamma$	4.5 MeV electrons	$^{60}Co-\gamma$
Dose rate, eV g^{-1} s^{-1}	$1.3 \cdot 10^{16}$	$1 \cdot 10^{16}$	10^{16}	
Dose, 10^{19} eV g^{-1}	140	2–10	0.01–0.04	6–60
Temperature, °C	25	0	2b	25
Reference	419	420	421	414
Product	G, molecule/100 eV			
5-Methyldecane	0.028		0.053	0.028
n-Undecane	0.016		0.027	0.011
4-Propylnonane	0.011		0.019	0.009
5-Ethyldecane	0.022		0.049	0.022
6-Methylundecane	0.024		0.053	0.025
n-Dodecane	0.019		0.027	0.011
(4-4)4,5-Dipropyloctane	0.036		0.06	0.025
5-Ethyl-4-propylnonane	0.113		0.18	0.118
5-Methyl-4-propyldecane	0.125		0.18	0.131
4-Propylundecane	0.028		0.24^a	0.029
5,6-Diethyldecane	0.120		0.17	0.122
5-Ethyl-6-methylundecane	0.243		0.36	0.267
5-Ethyldodecane	0.059		0.08	0.060
6,7-Dimethyldodecane	0.145		0.24^a	0.148
7-Methyltridecane	0.073		0.08	0.067
n-Tetradecane	0.018		0.03	0.016

^{a} 4-Propylundecane + 6,7-dimethydodecane

Table 2.33

Average yields from C$_7$-C$_{12}$ n-alkanes and the apparent activation
energies (E$_A$). After Rappoport and Gäumann [414]

Product	Temperature, °C	G, molecule/100 eV	E$_A$, kJ mol^{-1}
Hydrogen	25	5.2	1.5
Fragment alkanes (C$_2$-C$_{n+1}$)	−50	1.1	2.1
Fragment alkenes (C$_2$-C$_{n-1}$)	−50	0.38	0.8
Congruent olefins (C$_n$H$_{2n}$)		2.2	0
Intermediate products (C$_{n+1}$-C$_{2n-1}$)	−25	0.56	0, if $t < −25$°C −6.3, if $t > −25$°C
Dimers (C$_{2n}$)	−25	0.7–1.4	2.5–2.8

The temperature dependence of product forming reactions reveals some
details of their mechanisms [414, 434, 437]. The yields of hydrogen, fragments and dimers increase and those of intermediate products decrease with
increasing temperature: the G-values of congruent alkenes are practically
temperature invariant. This result can be interpreted in terms of increasing
rates of hydrogen atom abstraction by hydrogen atoms and alkyl radicals,

which have considerable energies of activation, and therefore their importance relative to combination and disproportionation reactions increases at higher temperatures. Hydrogen abstraction becomes significant above about −25°C [435].

Several papers deal with the mechanisms of hydrogen [429] and fragment [359, 420, 421] formation. Muratbekov and Saraeva have shown that the presence of N_2O as electron scavenger decreases the G-values of these products; NH_3 positive ion scavenger increases the hydrogen yield and decreases the amount of fragments [420]. Thus, charge neutralization processes may lead to both hydrogen formation and C−C bond rupture. Directly produced excited molecules are stabilized predominantly via molecular hydrogen formation, although the carbon chain cleavage of these species as well as their deactivation without decomposition could also be indicated [420]. About 25% of the fragments were produced in radical processes, and 75% in molecular reactions [435].

Avdonina has proposed an interesting hypothesis concerning the fragmentation pathway of C_4-C_7 n-alkanes, on the basis of scavenger studies: she assumed that these products were formed in radical decomposition processes of excited alkyl radicals [430]:

$$RH^+ + e^- \longrightarrow RH^*$$

$$RH^* \longrightarrow R^* + H$$

$$R^* \longrightarrow alkene + R_1$$

where RH^* denotes excited molecules produced predominantly in charge neutralization (and eventually also in direct excitation) processes [432, 433]; R_1 is the fragment radical formed by C−C bond rupture of the R^* excited parent alkyl radical in the β-position with respect to the position of the electron of the original radical. The energy of activation of radical decomposition is about 1 eV (100 kJ mol⁻¹). The product distribution calculated on the basis of this hypothesis is in fairly good agreement with experimental results.

The yields and reactions of radicals produced in n-alkane radiolysis have been reported in some papers. It has been shown that the overall yield of radicals increase with increasing number of carbon atoms in the irradiated compound: at −25°C, for n-heptane, G(radical) = 4.5, for n-decane, G(radical) = 6.8 [435]. The increase of overall radical yield is connected with the increased probability of C−H bond rupture; the yield of C−C bond rupture remains unchanged with varying carbon number. The observation that the yield of dimers formed via combination of parent alkyl radicals increases with the length of the carbon chain can also be related to this phenomenon [414, 434, 435]. Table 2.32 and the original papers indicate that the dimer product group is the only one that exhibits such a dependence on molecular structure.

Pichoozhkin and Arshakuni [397] studied the reactions of heptyl radicals in the liquid and solid phases. They observed an increase of k_d/k_c with temperature between −196 and 85°C. The k_d/k_c values were found to obey the relationship $\log(k_d/k_c) = a\Delta S + b$, where ΔS is the difference between the entropies of products of the disproportionation and combination reactions; a and b are constants.

A better insight into the production and consumption of alkyl radicals is offered by pulse radiolysis and ESR methods. These investigations not only reveal the kinetics of radical reactions but also give information about the structure and stereochemistry of alkyl radicals. Lund et al. have shown by an ESR study of C_4-C_{16} n-alkane single crystals, irradiated at 77 K, that both $CH_3\dot{C}HCH_2R$ and $R'CH_2\dot{C}HCH_2R''$ radicals are formed [374, 422, 424, 425]. Willard and Henderson [393] proposed that mostly 2-alkyl radicals are produced at 77 K. There is evidence, however, that at 4 K, the $C-H$ bond rupture processes occur randomly [395], and at room temperature, in 3-methylpentane, only tertiary alkyl radicals were found [396].

If all the $C-H$ bonds of n-alkanes are equally likely to suffer primary rupture [359], than it was supposed that 1-alkyl radicals isomerize rapidly into 2-alkyl radicals, which explains why 2-alkyl radicals predominate in irradiated systems. Similar results have been obtained by comparative studies with high dose rate pulse radiolysis and low dose rate steady-state irradiation [431]. It has been shown that an increase in the dose rate of seven orders of magnitude does not influence the overall yield of heptyl radicals produced by irradiation ($G \approx 3.4 \pm 0.5$), but does change slightly the ratio of radical isomers and the product distribution of the dimers. The steady-state concentration of heptyl radicals is high at high dose rates ($\sim 10^{23}$ eV g^{-1} s^{-1}), and consequently the ratio of primary to secondary heptyl radicals in dimer products of radical combination corresponds to the random distribution. At low dose rates ($\sim 10^{16}$ eV g^{-1} s^{-1}) primary heptyl radicals have sufficient time to isomerize into the thermodynamically more stable secondary radicals in intra- or inter-molecular processes, and the products of combination of 2-heptyl and 3-heptyl radicals predominate. For the same reason, the ratio of 1-heptyl to 2-heptyl radicals in dimers decreases with increasing temperature [359].

Gillbro and Lund have shown by ESR spectroscopy that radical pairs are formed in irradiated n-heptane [426]. They proposed the spur reaction (2.77) and (2.78):

$$RH^+ + RH \longrightarrow R + RH_2^+ \qquad (2.77)$$

$$RH_2^+ + e^- \longrightarrow R + H_2 \qquad (2.78)$$

Diamagnetic RH_2^+ species have been detected in the gas phase [107]. This hypothesis suggests that radical pairs contribute to dimer formation [426].

2.2.11. Branched-chain C_7-C_{12} alkane isomers

2,2,4-Trimethylpentane (iso-octane) is one of the branched-chain alkanes most frequently studied in the radiation chemistry of hydrocarbons. Owing to the high number of carbon atoms contained in it, a great variety of products is formed. The G-values of higher products are generally unknown, but the

Table 2.34

Yields from liquid 2,2,4-trimethylpentane (radiation: ^{60}Co-γ)

Dose rate, eV g^{-1} s^{-1}	$2.2 \cdot 10^{15}$	$1.4 \cdot 10^{16}$
Dose, 10^{19} eV g^{-1}	5	1.2–12
Temperature, °C	25	0
Reference	438	436
Product	G, molecule/100 eV	
Hydrogen	3.53	2.0
Methane	1.17	
Ethane	0.13	
Ethylene	0.05	
Acetylene	0.01	
Propane	0.11	
Propene	0.26	
n-Butane	0.02	
Isobutane	2.49	2.40
n-Butenes	0.04	
Isobutene	2.07	1.90
2-Methylbutane	0.08	
Neopentane	0.31	
2-Methylbut-1-ene	0.02	
2,2-Dimethylbutane	0.02	
2-Methylpentane	0.04	
4-Methylpent-1-ene	0.02	
4-Methylpent-2-ene	0.01	
2,2-Dimethylpentane	0.05	
2,4-Dimethylpentane	0.22	
2,4-Dimethylpent-1-ene	0.53	
2,4-Dimethylpent-2-ene	0.26	
2,4,4-Trimethylpent-1-ene	0.66 ⎫	1.10
2,4,4-Trimethylpent-2-ene	0.04 ⎭	
C_9 products	0.07	0.20
C_{10} products	0.008	
C_{11} products	0.07	0.16
C_{12} products	0.55	0.48
C_{13} products	0.04	0.09
C_{14} products	0.03	
C_{15} products	0.16	0.09
C_{16} products	0.85	0.36
C_{16} < products	0.10	
Total	14.12	

yields of fragments obtained under various experimental conditions have been reported by several authors. The mechanism of fragmentation is fairly well established, because the presence of many methyl groups in a molecule containing both tertiary and quaternary carbon atoms attracted the attention of researchers.

Table 2.34 lists the important products of isooctane radiolysis, together with the corresponding yield values (after Castello et al. [438] and Murat-bekov et al. [436]). Apart from the relatively low hydrogen yield, the

G-values of methane, isobutane, isobutene and C_8 alkenes should be emphasized [436, 438]. About 85% of the total product consists of hydrogen and C_1-C_8 hydrocarbons. The yields show that the preferred positions of decomposition are the C—C bonds in the vicinity of the quaternary carbon atom.

The reaction mechanism has been studied by investigating the product distribution as a function of dose and dose rate as well as the effect of various additives. The G-value of hydrogen decreases markedly in the dose range 5–$90 \cdot 10^{19}$ eV g^{-1}, however, above the upper limit, only slight changes are observed. Other products show hardly any dose dependence [438]. $G(H_2)$ values reported in the literature have been determined with rather large doses, so most values are between 2.0 and 2.5 [274, 322, 386, 436, 446, 447]. Individual components of H_2 production — such as the G-value of thermal H-atoms — have been determined by means of radical scavengers. Bansal and Rzad [386] proposed the following reaction scheme in the presence of ethylene:

$$H + C_2H_4 \xrightarrow{\ k_1\ } C_2H_5 \tag{2.79}$$

$$H + \text{iso-}C_8H_{18} \xrightarrow{\ k_2\ } H_2 + C_8H_{17} \tag{2.80}$$

$$C_2H_5 + {}^{131}I_2 \longrightarrow C_2H_5{}^{131}I + {}^{131}I \tag{2.81}$$

Based on reactions (2.79)–(2.81) the kinetic equation is:

$$G(C_2H_5I) = G_0(H) \cfrac{1}{1 + \cfrac{k_2[\text{iso-}C_8H_{18}]}{k_1[C_2H_4]}} \tag{2.82}$$

where $G_0(H)$ denotes the yield of thermal H-atoms in pure solvent. By fitting the parameters to the experimental values, $G_0(H) = 1.00$ and $k_2/k_1 = 0.0043$ are obtained, in satisfactory agreement with other results [43, 350]. Ionic and non-ionic components of hydrogen and methane yield have been determined using electron, positive ion and radical scavengers, respectively [274, 436, 450 – 452]. Experimental results as reported by different laboratories are in fairly good agreement, in spite of the rather different experimental techniques.

The yields of hydrogen and iso-C_4 fragments decrease in the presence of electron scavengers such as N_2O, those of C_8 alkenes and C_{16} dimers, on the other hand, increase [432, 436, 439, 440]. Two hypotheses have been suggested to explain this phenomenon; the first one is as follows [436]:

$$N_2O + e^- \longrightarrow N_2 + O^-$$

$$\text{iso-}C_8H_{18}^+ + O^- \Big\langle \begin{array}{l} C_8H_{17}OH \\[4pt] C_8H_{17} + OH \end{array}$$

$$OH + \text{iso-}C_8H_{18} \longrightarrow H_2O + C_8H_{17}$$

$$C_8H_{17} + C_8H_{17} \diagdown \begin{array}{l} \nearrow \ C_8H_{16} + C_8H_{18} \\ \searrow \ C_{16}H_{34} \ \text{(dimer)} \end{array}$$

and the second one is:

$$C_8H_{17} + R_i \diagdown \begin{array}{l} \nearrow \ \text{iso-}C_8H_{16} + R_iH \\ \searrow \ C_8H_{18} + R_i(-H) \ \text{(alkene)} \end{array}$$

$$C_8H_{17} + CH_3 \longrightarrow \text{iso-}C_8H_{16} + CH_4$$

where R_i denotes a fragment radical, such as C_3H_7 or C_4H_9. The yield of dimers decreases, and those of hydrogen, octene and dimers increase in the presence of ammonia, presumably as a result of the following reaction sequence [436]:

$$\text{iso-}C_8H_{18}^+ + NH_3 \longrightarrow NH_4^+ + C_8H_{17}$$

$$NH_4^+ + e^- \longrightarrow NH_3 + H$$

$$H + \text{iso-}C_8H_{18} \longrightarrow H_2 + C_8H_{17}$$

Schuler and Kuntz [274] have studied the methane yield and the decay kinetics of methyl radicals using radioactive iodine as tracer: $G(CH_4) = 1.19$ and $G(CH_3) = 0.686$. They have shown that the fate of methyl radicals and the relative rates of their reactions depend strongly on the dose rate (Fig. 2.5), presumably owing to competition between reactions (2.83) and (2.84a):

$$CH_3 + \text{iso-}C_8H_{18} \longrightarrow CH_4 + C_8H_{17} \tag{2.83}$$

$$CH_3 + R \diagdown \begin{array}{l} \nearrow \ R(-H) + CH_4 \quad\quad\quad\quad\quad (2.84a) \\ \searrow \ RCH_3 \quad\quad\quad\quad\quad\quad\quad\quad\; (2.84b) \end{array}$$

If the dose rate is very low, abstraction (2.83) predominates; at very high dose rates, radical combination becomes the rule. Thus, in the former case, $G(CH_4) = G[\text{Eq.}(2.83)] + G(CH_4)_m = 1.19$; in the latter case, $G(CH_4) = [\text{Eq.}(2.84a)] + G(CH_4)_m$; where $G[\text{Eq.}(2.83)]$ denotes the yield of abstraction, $G[\text{Eq.}(2.84a)]$ that of disproportionation, and $G(CH_4)_m$ that of molecular methane (0.5) [274]. On the basis of these two expressions, as well as Fig. 2.5, the values obtained are $G[\text{Eq.}(2.84a)] = 0.28$ and $G[\text{Eq.}(2.84b)] = 0.41$, and the mean k_d/k_c ratio of methyl radicals will be $0.28/0.41 = 0.68$. Considering that secondary and tertiary radicals will react with methyl radicals, this is a realistic value.

On the basis of the temperature dependence of G-values as well as oxygen scavenger studies, it has been reported that dimers are the products of combination reactions between secondary and tertiary radicals; these, in turn, are formed mainly in secondary reactions of primary radicals. The product spectrum indicated an overall yield for octyl radicals of $G \approx 7.2$ [436]. The importance of radical combination processes is supported by calorimetric results [457]. A considerable portion of fragments (e.g. C_4 molecules) originate from unimolecular decomposition of excited molecules [300, 454–456].

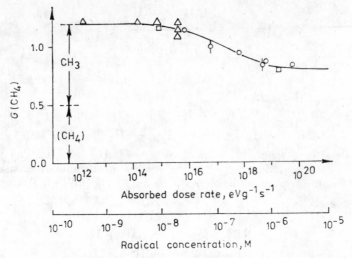

Fig. 2.5. Yields of methane in the radiolysis of 2,2,4-trimethylpentane.
○ 2.38 MeV, electrons; ⊘ same, but sample stirred during irradiation; □ 36.8 MeV
He⁺-ions; △ ⁶⁰Co-γ After Schuler and Kuntz [274]

The following generalized conclusions can be drawn on the basis of studies
with additives [436]:

1. Species formed as a result of direct excitation are stabilized by molecular H_2 elimination rather than by C—H bond rupture;
2. C_{11}-C_{15} products result from combination reactions between parent
alkyl and fragment radicals;
3. Excited molecules produced in charge neutralization reactions react
predominantly via C—C bond rupture to give fragments.

Shida et al. and Tanno et al. have proposed the following decomposition
pattern and yields for the radiolysis of isooctane [322, 444]

$$iso\text{-}C_8\dot{H}_{18}$$

$$H_2 + products \qquad\qquad G = 0.7$$
$$C_8H_{18}^+ + e^- \qquad\qquad G = 3.1$$
$$CH_4 + products \qquad\qquad G = 0.2$$
$$C_4H_9^+ + C_4H_9 + e^- \qquad G = 0.4$$

$$iso\text{-}C_8H_{18}^+ + e^-$$

$$CH_3 + C_7H_{15} \qquad\qquad G = 0.9$$
$$2\,C_4H_9 \qquad\qquad\qquad G = 1.1$$
$$H + C_8H_{17}$$

$$C_4H_9^+ + e^- \longrightarrow H + C_4H_8 \qquad\qquad G = 1.5$$

$G = 3.8$–4.0 has been reported for the total ionization yield of (based on scavenger studies [436]); this value is lower by about 10–15% than those obtained by other methods, e.g. conductometry.

Castello et al. [438] estimated the G-values of decomposition of individual C–C bonds:

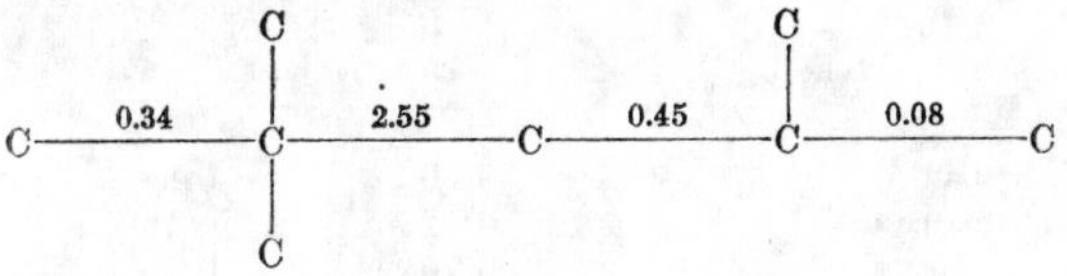

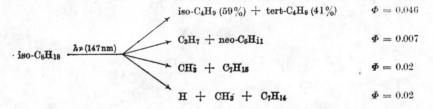

The overall yield of C–C bond rupture is $\Sigma\, G(\text{C–C}) \approx 4.2$–$4.4$ [438]: 80% of these cleavages occur at C–C bonds adjacent to the quaternary carbon atom.

Low pressure (1–75 mbar) gas-phase photolysis of isooctane was studied by Gowenlock et al. using 123.6 and 147 nm (10.0 and 8.4 eV) photons [448]. Holroyd investigated the reactions of excited isooctane molecules in liquid-phase photolysis experiments at a wavelength of 147 nm (8.4 eV) [300]. It has been shown that mainly unimolecular elimination processes are responsible for product formation, although radical-type C–C bond rupture also occurs. The following reactions and quantum yields have been determined by means of iodine and ethylene radical scavengers [300]:

Radical processes:

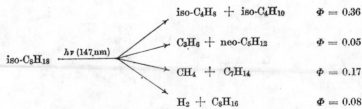

Molecular processes:

The quantum yields were found to increase with the temperature in the ranges 173–293 K (for $\lambda = 147$ nm; 8.4 eV) and 293–373 K (for $\lambda = 160$–180 nm, 6.9–7.7 eV) [310]. The mechanism that has been discussed above was confirmed in recent works which applied several additives [313, 325, 425, 428].

Table 2.35

Relative yields of radicals in the radiolysis and
photolysis of liquid
2,2,4-trimethylpentane. After Holroyd [300]

Source	Photolysis	Radiolysis
Radical	Relative yields	
Methyl	0.69	0.86
Isopropyl	0.12	0.16
tert-Butyl	0.70	0.68
Isobutyl	1.00	1.00
Neopentyl	0.14	0.21
Heptyl	0.36	
Octyl	0.45	2.80

Table 2.35 compares radical yields determined in photolysis and radiolysis [300]. The two techniques give very similar results, except for octyl
radicals produced in a process of C—H bond rupture predominating in radiolysis only. Another difference is the approximately twofold yield of molecular decomposition in photolysis (as compared with radiolysis where the
rates of the two types of reaction are comparable).

The yields of the important fragments obtained by liquid-phase radiolysis
of 2,3,4-trimethylpentane are listed in Table 2.36 (after Kudo [446]). The

Table 2.36

Overall (G_{total}) and molecular (G_M) yields from liquid
2,3,4-trimethylpentane (radiation: ^{60}Co-γ). After Kudo [446]

Product	G_{total}	G_M
	molecule/100 eV	
Hydrogen	3.32	2.06
Methane	0.49	0.23
Propane	1.44	0.67
Propene	1.68	0.89
Isopentane	2.70	1.13
Isopentene	0.42	0.27

G-values indicate that the most abundant fragment products are C_1, C_3 and
C_5 hydrocarbons. The nearly equal yields of C_3 and C_5 suggests that secondary decompositions are not important. The Table also contains molecular
yields obtained as 'residual yields' in experiments using methylmethacry-
late and iodine radical scavengers. On this basis the following decomposition pattern and G-values were suggested by Kudo [446]:

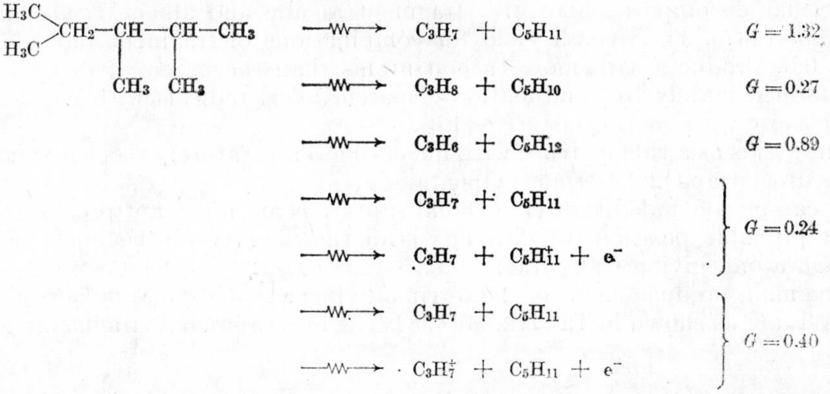

$$\begin{array}{c} H_3C \\ \diagdown \\ CH_2-CH-CH-CH_3 \\ H_3C \diagup \\ CH_3CH_3 \end{array}$$

$\xrightarrow{}\quad C_3H_7 \;+\; C_5H_{11}\qquad\qquad G = 1.32$

$\xrightarrow{}\quad C_3H_8 \;+\; C_5H_{10}\qquad\qquad G = 0.27$

$\xrightarrow{}\quad C_3H_6 \;+\; C_5H_{12}\qquad\qquad G = 0.89$

$\xrightarrow{}\quad C_3H_7 \;+\; C_5H_{11}$

$\xrightarrow{}\quad C_3H_7 \;+\; C_5H_{11}^+ + \text{e}^-$ $\qquad\Big\} \; G = 0.24$

$\xrightarrow{}\quad C_3H_7 \;+\; C_5H_{11}$

$\xrightarrow{}\quad \cdot C_3H_7^+ \;+\; C_5H_{11} + \text{e}^-$ $\qquad\Big\} \; G = 0.40$

The vacuum-UV photolysis of 2,2,3,3-tetramethylbutane was studied by Bondjouk and Koob [443], and Gowenlock et al. [448]. To our knowledge, the radiolysis of this compound has not been studied yet.

2,2,4,4-Tetramethylpentane has been studied by Knight and Lewis using a ^{137}Cs γ-radiation source [437], at -196, -78 and $+35°C$: Table 2.37 lists the main product yields ($G > 0.3$). G-values for the rupture of individual bonds are as follows (estimated from product yields):

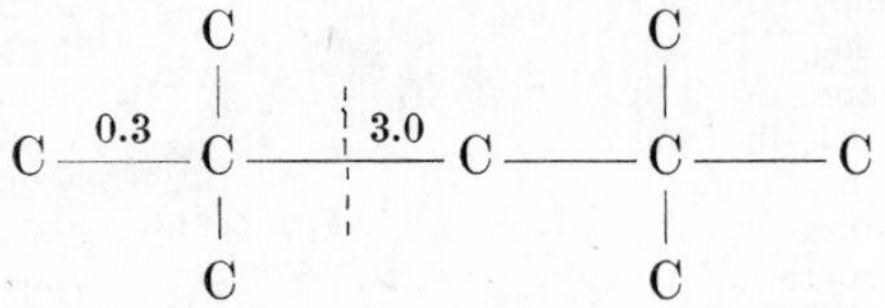

$$\begin{array}{ccccccc} & & C & & & C & \\ & & | & & & | & \\ C & \overset{0.3}{\rule{1cm}{0.4pt}} & C & \overset{3.0}{\vrule}\ \rule{1cm}{0.4pt} & C \rule{1cm}{0.4pt} & C & \rule{1cm}{0.4pt}\ C \\ & & | & & & | & \\ & & C & & & C & \end{array}$$

The almost complete product spectrum indicated that the principal pathway leading to stable products is hydrogen abstraction of fragment radicals;

Table 2.37

Yields from 2,2,4,4-tetramethylpentane
(radiation: ^{137}Cs-γ). After Knight and Lewis [437]

Temperature, °C	35	—78	—196
Phase	Liquid	Solid	Solid
Product	G, molecule/100 eV		
Hydrogen	2.21	2.95	3.53
Methane	1.20	0.76	0.64
Isobutane	1.01	0.26	0.22
Isobutene	3.82	1.67	1.27
Neopentane	4.89	2.38	1.57
2,2,4-Trimethylpentane	0.32	0.80	1.12
2,4,4-Trimethylpent-1-ene	1.24	0.60	0.44
C_{13} products	0.40		
C_{14} products	0.74		

10

molecular decomposition to give fragment alkane and alkene is also important [437]. The overall yield for combinations of fragment radicals is $G \approx 0.8$. Products with more carbon atoms than the parent hydrocarbon are formed mainly by combination of parent alkyl radicals with fragment radicals or with each other ($G \approx 1.8$).

The yields as a rule increase with increasing temperature, except for those of hydrogen and 2,2,4-trimethylpentane.

It can be concluded that C—C bond rupture is an important process: the most probable position for cleavage is in the vicinity of the quaternary carbon atom, giving C_4-C_5 fragments.

The main product yields of 2,2,5-trimethylhexane ($G \geq 0.3$) in ^{60}Co γ-radiolysis are as shown in Table 2.38 [438]. The most abundant fragments are

Table 2.38

Yields from liquid 2,2,5-trimethylhexane
(radiation: ^{60}Co-γ). After Castello et al. [438]

Product	G, molecule/100 eV
Hydrogen	2.85
Methane	0.58
Isobutane	0.60
Isobutene	1.31
Isopentane	1.36
2,4-Dimethylhexane	0.86
2,2,5-Trimethylhexenes	2.25
C_{14} products	0.53
C_{18} products	0.54

C_4 and C_5 hydrocarbons. G-values for the rupture of individual C—C bonds are as follows [438]:

$$C \xrightarrow{0.28} \underset{\overset{|}{C}}{\overset{\overset{C}{|}}{C}} \xrightarrow{1.95} C \xrightarrow{0.11} C \xrightarrow{0.37} \underset{}{\overset{\overset{C}{|}}{C}} \xrightarrow{0.07} C$$

The increased likelihood of rupture of C—C bonds in the vicinity of the quaternary carbon atom is well demonstrated and can be attributed to the relatively low bond dissociation energies in that site of the molecule.

Yields of hydrogen and C_1-C_5 fragments of 2,6-dimethyloctane extrapolated to zero dose have been reported by Shelberg and Pestaner [460]: $G(H_2) = 3.32$, $G(CH_4) = 0.28$, $G(C_2H_4) = 0.14$, $G(C_3H_6) = 0.10$, others $G < 0.1$. It can be seen that the methane yield is higher than that obtained from the n-alkane with the same carbon number (n-decane), but as the branchings are some way from each other, the tendency to fragment is much lower than with the branched-chain alkanes discussed so far. Tertiary carbon atoms give the molecule a much lower 'fragility' than quaternary ones, and this can be interpreted in terms of molecular energetics.

The yields of main products and groups of products of 2,2,4,6,6-penta-methylheptane ($G \geq 0.2$) are listed in Table 2.39 (after Knight and Sicilio

Table 2.39

Yields from liquid 2,2,4,6,6-pentamethylheptane
(radiation: X-ray). After Knight and Sicilio [461]

Product	G, molecule/100 eV
Hydrogen	1.3
Methane	0.74
Isobutane	0.34
Isobutene	1.2
Neopentane	0.25
2,2,4-Trimethylpentane	1.05
2,4,4-Trimethylpent-1-ene	0.2
Σ C_1-C_{11} products	4.93
Σ $C_{12} \leq$ products	2.14

[461]). It can be seen that the most probable fragmentation of the molecule is as shown below:

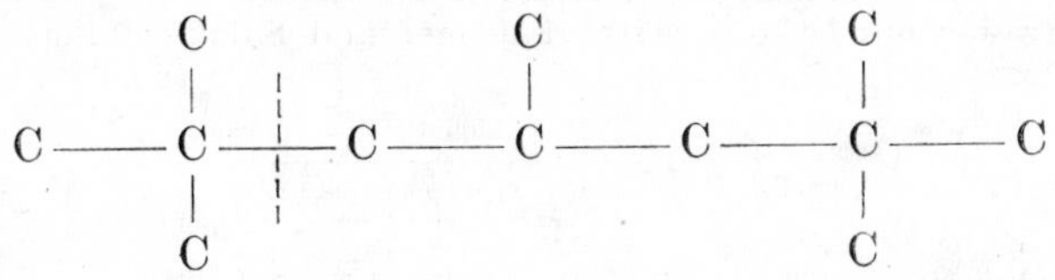

resulting in C_4 and C_8 products as the main fragments. The preference shown is supported by the product distribution, and is in agreement with the facts discussed in connection with other branched-chain alkanes pointing to the enhanced reactivity of C—C bonds near the quaternary carbon atom. A good deal of methane is formed, which must be due to the large number of methyl groups in the molecule. Individual products, especially those containing more than 12 carbon atoms, can be identified only with great difficulty, therefore the yields of individual C—C bond ruptures could not be estimated.

2.2.12. n-Hexadecane and n-heptadecane

Some unexpected phenomena observed in radiation chemical polymerization, as well as the necessity for a deeper insight into crosslinking processes, have promoted the study of the radiolysis of long-chain alkanes.

n-Hexadecane which has a melting point of 18°C, proved to be the most suitable model substance for such studies. Its radiation chemistry was studied by Falconer and Salovey [462–466] in a wide temperature range (be-

10*

tween −196 and +19°C) using different types of radiation (1 MeV electrons, ^{60}Co-γ). Radiation chemical yields are shown in Tables 2.40–2.42. More sophisticated details of the mechanism, the temperature dependence of fragmentation and phase effects have also been studied in the presence of various radical scavengers [463]. Irradiation with various doses was carried

Table 2.40

Dependence of yields from n-hexadecane on experimental conditions.
After Salovey and Falconer [462]

Radiation		1 MeV electrons		^{60}Co-γ
Dose, 10^{19} eV g^{-1}		6.9-125		15.6
Phase	Temperature, °C	$G(H_2)$	G(dimers)	G(dimers)
		molecule/100 eV		
Liquid	19	4.1	2.60	1.59
Solid	4	3.0	2.30	
Solid	−80		2.01	1.50
Solid	−196		1.23	0.99

Table 2.41

Yields from liquid n-hexadecane
(radiation: ^{60}Co-γ). After Falconer and Salovey [463]

Phase	Liquid	Solid
Products	G, molecule/100 eV	
CH_4	0.038	0.028
C_2 products	0.126	0.064
C_3 products	0.060	0.032
C_4 products	0.025	0.020
C_9 products	0.075	0.033
C_{10} products	0.069	0.032
C_{11} products	0.072	0.032
C_{12} products	0.072	0.033
C_{13} products	0.073	0.041
C_{14} products	0.075	0.042
C_{15} products	0.012	0.005

Table 2.42

Phase dependence of yields from n-hexadecane
(radiation: ^{60}Co-γ). After Falconer and Salovey [463]

Phase	$G(H_2)$	G(fragments)	$G(C_{n+1}-C_{2n-1})$	G(dimers)
	molecule/100 eV			
Liquid	4.5	1.00	0.31	1.7
Solid	4.0	0.50	0.05	1.6

out using 1 MeV (van de Graaff) electrons, with a dose rate of $7 \cdot 10^{15}$ eV $g^{-1} s^{-1}$. It was reported that the dose rate and the LET value do not affect G-values within the limits of experimental error [462–465].

In liquid-phase radiolysis, the yields of methane, hydrogen and fragments show hardly any change due to varying dose over two orders of magnitude, whereas in the solid phase, where lattice defects created by irradiation play an important role, the yields show a marked dose dependence: it has been shown that the yield of hydrogen decreases considerably, but that of dimers less markedly, when the dose changes in the range $2.5–200 \cdot 10^{19}$ eV g^{-1}. The dissimilar dose dependences of $G(H_2)$ in the liquid and solid states indicate different mechanisms of hydrogen formation [463].

The yield of C_9-C_{14} products is $G = 0.036$ and 0.073 in the solid and liquid state, respectively, and the total yield of fragments is $G(C_1$-$C_{15}) = 0.5$ and 1.00. Overall yields for various groups of products are shown in Table 2.42 [463]. The approximately twofold amount of fragments in the liquid phase can be attributed to the recombination of alkyl radicals within the 'cage' [462].

The yields of congruent C_{16} alkenes extrapolated to zero dose are $G = 1.6$ and 2.8 in the solid and liquid phase, respectively, whereas the corresponding values for total alkene yields are $G = 2.2$ and 4.2; those for fragment alkenes were estimated to be $G = 0.5$ and 1.0, respectively [466]. Both *cis* and *trans* isomers were observed among the congruent alkenes. The phase dependence of G-values was interpreted in terms of the inhibition of the disproportionation reactions of hexadecyl radicals with fragment radicals in the solid state [466].

The formation of C_{17}-C_{31} products was attributed to combination of fragment and parent alkyl radicals [462, 463]. The effect of the state of aggregation on this group of products is much more pronounced than in the case of other products. Their yield is higher in the liquid phase: this may be due to both diffusion and the cage effect.

Dimers are the main products of radiolysis. Although the total dimer yields are similar in solid- and liquid-phase radiolysis, the relative abundance of individual isomers is considerably different. The relative abundance of n-dotriacontane which is obtained via combination of two 1-hexadecyl radicals, is only about 0.2% of the total dimer yield. In the presence of about 5% n-octadecane, however, the abundance of n-dotriacontane increases by about one order of magnitude. Falconer et al. have explained this phenomenon by the transformation of the triclinic crystal structure into an orthorhombic one under the effect of admixing. This result supports the suggestion about the importance of the structure of the solid state in determining the direction of reactions, presumably by influencing the orientation of radical intermediates. Dimers are formed mainly via combination of sec-hexadecyl radicals. Crosslinking among various secondary carbon atoms has nearly a random probability.

Földiák has found that the rate of decomposition of n-hexadecane is nearly constant up to about 250°C; however, it then increases rapidly with increasing temperature, owing to the transformation of the process into a chain reaction (radiation–thermal cracking) with an apparent energy of activation of about 60 kJ mol^{-1} [469, 470]. At about 310–320°C, thermal chain initiation also commences, and the relative importance of radiation chemical initiation and its influence on the product composition dwindles [470].

The yields of hydrogen and dimer products decrease sharply, those of other products — mainly C_1–C_4 fragments — slightly, with increasing temperature.

In the presence of iodine and 2-methylpent-1-ene radical scavengers admixed into solid state, only the yields of dimers decrease (by about 40%), whereas admixture into the liquid phase decreases the G-values of every product.

The temperature dependence of the yields and the results of scavenger studies indicate that mainly H-atoms, fragment and parent alkyl radicals are responsible for product formation [462, 463], although the existence of 'limiting yields' means that 'hot' atom, ion-molecule, combination and disproportionation processes within the 'cage' cannot be neglected entirely.

On the basis of qualitative and quantitative determination of alkyl iodide products obtained in iodine scavenger studies, it was shown [463] that the abundance of 1-iodohexadecane is as low as 6.7%, as opposed to the 17.6% of primary H atoms among the hydrogens of the molecule: i.e. the probability of secondary C—H bond rupture is about 2.5 times greater than that of the primary C—H bond rupture. As the yield of n-dotriacontane is very low, it was assumed that 1-hexadecyl radicals formed in primary processes isomerize into secondary radicals, which are thermodynamically more favoured.

Fragment radicals formed in C—C bond rupture processes combine with hexadecyl radicals with high probability. The yield of C_1 and C_{15} products indicates that terminal C—C bonds are more stable than the others: fragmentation of the latter is obviously a random process. The probability of the rupture of the CH_3–$C_{15}H_{31}$ bond is about seven times lower than that of any of the RCH_2–CH_2R bonds. This result contradicts the data of Miller et al. on the radiolysis of solid octacosane [467], which indicate that the terminal C—C bond is the least stable. It must be mentioned, however, that the conclusion reached by Falconer et al. [462–466] is in agreement with most experimental results obtained in the field of alkane radiolysis.

Molecular processes, which are insensitive to the presence of radical scavengers, give nearly all the products in the solid phase and about half in the liquid phase. About 80% of C_{17}-C_{31} products are formed in radical combination processes: the combination of hexadecyl radicals is responsible for 40% and 70% of the dimers in solid and liquid phase, respectively.

The above results indicate that the yields of radiolysis products of solid n-hexadecane also depend on the crystal structure and the physical state of the solid substance [462–467]. Whereas C_8-C_{24} n-alkanes with even carbon number crystallize in the triclinic symmetry group, C_9-C_{35} n-alkanes with odd carbon number give orthorhombic crystals in which molecules may rotate freely around their longitudinal axis [462].

The radiolysis of n-heptadecane was compared with that of n-hexadecane in order to decide how the different crystal structure would affect the product formation processes [471].

G-values of C_4-C_{16} products formed under the effect of 1 MeV electrons with a dose of $25 \cdot 10^{19}$ eV g^{-1} were similar to those measured in n-hexadecane radiolysis. The mean yield of C_9-C_{15} products is $G = 0.034$, and the overall G-value of C_1-C_{16} products is $G \approx 0.51$ in solid state; the corresponding values in the liquid phase are $G = 0.071$ and 1.03, respectively [471]. The overall G-values of tetratriacontane dimers are $G = 1.13$ and 1.63, respectively, in the solid and liquid states; among these, linear dimers were formed with the following yields: $G = 0.043$ and 0.0045, respectively [471].

The yields of fragments form n-heptadecane showed a clear minimum at C_{10} ($G = 0.024$), although the mean yields of fragments of n-hexadecane and n-heptadecane in the solid state were nearly equal. This phenomenon was interpreted in terms of secondary decomposition reactions of fragment radicals [471]. The results indicate that the excess energy which otherwise leads to fragmentation can be dissipated more easily in the triclinic crystals of n-C_{16}.

The G-values for dimer formation are equal for both n-hexadecane and n-heptadecane in the liquid phase ($G \approx 1.7$). On the other hand, in the solid state, G(dimer) for n-heptadecane is 1.6 and (at the same temperature) G(dimer) for n-hexadecane is about 30% lower: 1.13. The abundance of linear dimers is about 4% for heptadecane, but only 1% for n-hexadecane. These results were interpreted in terms of less close packing of the orthorhombic crystals of odd n-alkanes compared with the triclinic crystals of even n-alkanes. Thus, crystal structure influences both the extent of energy dissipation and the spatial orientation of intermediate radicals [471]. As the radiolysis of alkanes having more than 17 carbon atoms are not discussed in this work, we refer at this point to the recent studies of Ahmad and Charlesby [483] on the changes occurring in the crystal structure of n-$C_{36}H_{74}$ upon irradiation.

2.3. MIXTURES OF ALIPHATIC ALKANES

The study of the radiolysis of mixtures can lead to conclusions about the character of basic radiation chemical processes, their relative rates and the precursors of given products.

In principle, if no interactions occurred during radiolysis between the components, e.g. if all processes were unimolecular, the G-values ought to be linear functions of the composition. As will be illustrated later, this argument may not necessarily be true: the linearity found in the yield of a product as a function of the composition does not mean that there is no interaction between the components. Since the energy absorbed by the medium upon irradiation is, mainly, due to the interaction of radiation with the electrons contained in the matter (e.g. photoelectric effect, Compton scattering), the G-value of a product formed in a mixture where no interaction takes place between the components should obey the 'additivity rule' in the form

$$G(P) = (1 - e_B)G_A(P) + e_B G_B(P)$$

where $G_A(P)$ and $G_B(P)$ denote the yields of P products from pure A and B hydrocarbons, respectively, $e_B = E_B/(E_A + E_A)$, the electron fraction of B in the mixture, and E_A and E_B the number of electrons in the corresponding components.

It is well known, however, that yields of products formed from irradiated binary mixtures of A and B hydrocarbons generally do not correspond to the theoretical values calculated on the basis of the mixing rule [472–477].

There are at least three possible reasons for this phenomenon:

1. The energy absorbed by individual components in the system is not proportional to their electron fractions, as would be expected on the basis of additivity. According to the optical approximation, the preferential primary energy absorption by one of the components can be assumed in the radiolysis of mixtures. Bednar suggested that energy absorption proportional to the electron fraction occurs only for isoelectronic components having similar excitation spectra [406]. The structural selectivity of electrons with respect to radiation absorption may also be important (e.g. valence electrons or n-electrons in alkenes or aromatics) [481]. The existence of subexcitation or collective excitation, which is generally assumed for condensed-phase systems (mainly aromatics) [481], may also cause preferential energy absorption.

Although the possibility cannot be excluded, experimental and theoretical studies to date do not support the suggestion that preferential primary energy absorption is the main reason why the mixing rule is not valid in hydrocarbon mixtures: on the contrary, it is generally assumed that the energy absorption is proportional to the electron fractions.

2. Some of the energy absorbed proportionally to the electron fraction, is transferred in secondary processes — such as electron excitation transfer or charge transfer — from the particles of one of the components to those of the other (physical interactions, 'physical protection' [408]). Activated alkane molecules produced in transfer processes can behave similarly to those generated directly by irradiation: they may lose their excess energy in physical or physicochemical processes, e.g. by collision, polarization interactions, light emission ('sponge type protection'), or in chemical processes such as decomposition or ion-molecule reactions ('sacrificial-type protection'). The extent of radiolytic decomposition of the transferring component simultaneously decreases, thus the other (accepting) component exerts a 'protecting effect' on the former [408].

It is generally assumed that charge transfer may occur when the ionization potential of the accepting component is lower than that of the donor component: for excitation transfer, the energy of the lowest excitation level of the acceptor should be lower than the corresponding level of the donor.

3. An intermediate of the radiolysis of one of the components reacts chemically with the other component or with one of its intermediate species (chemical interaction, 'chemical protection').

n-Alkane–isoalkane mixtures

Few data are available on the radiolysis of mixtures of n-alkanes with isoalkanes. Wojnárovits and Földiák have reported results on n-hexane––2,2-dimethylbutane and n-heptane–2,2,4-trimethylpentane mixtures [497]; György et al. identified some hydrocarbon products from mixtures of n-hexane and branched-chain hexanes [496]; Horváth and Földiák did the same with n-butane–isobutane systems [478].

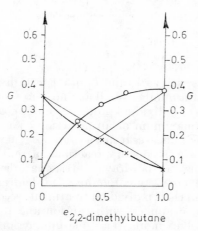

$Fig.$ $2.6.$ Yields of butane isomers in the radiolysis of n-hexane–2,2-dimethylbutane mixtures.
○ iso-C_4H_{10};   × n-C_4H_{20} (radiation: ^{60}Co-γ). After György and Földiák [473]

Hardwick and then Kudo and Shida have shown that yields of hydrogen and some fragments do not correspond to the mixing rule in the radiolysis of n-hexane–2,2-dimethylbutane [476], n-hexane–2,2,4-trimethylpentane [474] and n-hexane–2,3-dimethylbutane [476] mixtures. These results were interpreted in terms of positive charge transfer and/or excitation transfer from n-hexane, which has the higher ionization potential, to the branched-chain component [474, 476]. Studies of binary mixtures containing other hexane isomers [473, 496] support this hypothesis (Fig. 2.6). Since the formation of n-butane is largely due to the decomposition of n-hexane, and the formation of isobutane is characteristic of the radiolysis of neohexane, the explanation that a transfer process is responsible for the nonlinearity of the curves in Fig. 2.6 seems reasonable. György and Földiák have concluded on the basis of G-values as a function of the composition of various mixtures containing branched-chain alkanes that the extent of 'protection' and the resulting direction of the transfer process are also influenced by the radiation chemical reactivity and the structure of the component molecules [473]. Consequently, less stable alkanes with more branching 'protect' the other component more efficiently. This hypothesis suggests a competition between primary decomposition, deactivation and transfer processes of activated molecules as an explanation for the interactions: this concept assumes the inseparability of 'physical' and 'chemical' 'protection'.

$$A \xrightarrow{\wedge\wedge} A^*$$

$$B \xrightarrow{\wedge\wedge} B^*$$

$$A^* \xrightarrow{k_1} products$$

$$B^* \xrightarrow{k_2} products$$

$$A^* + B \rightleftharpoons A + B^*$$

where the resultant direction of the transfer processes is influenced by photo-ionization potentials as well as k_1 and k_2 rate constants [473]. Horváth and Földiák [478] found for a number of alkane mixtures (such as n-butane–iso-butane) that the yields of individual products can be characterized by curves having a minimum, a maximum or an inflection as a function of the com-position (Fig. 2.7). This indicates a more complex reaction mechanism than that discussed above, e.g. secondary reactions of radical or ionic interme-diates. Two conclusions can be drawn: first, that the characterization of the behaviour of a given system requires the determination of as many products as possible and their individual treatment; second, that the concept of 'protection' should be applied with caution and it must be always re-membered that it involves more, than one interaction.

When mixtures of n-alkanes and neopentane were irradiated in the pres-ence of electron scavengers, the rate of transfer processes and the yields of their products both increase. Stone [479] thus deduced that ionic transfer takes place. A study of the methane–propane–oxygen ternary system show-ed that the oxidation products of methane react with ionic intermediates,

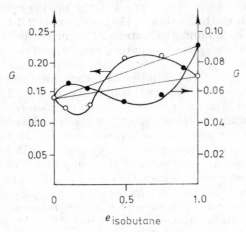

Fig. 2.7. Yields of pentane-isomers in the radiolysis of the n-butane–isobutane mixtures. ○ iso-C_5H_{12}; ● n-C_5H_{12} (radiation: ^{60}Co-γ). After Horváth and Földiák [478]

thus inhibiting ethane formation [480]. Studies of the reaction mechanisms of transfer processes have been reviewed by Kroh and Karolczak [481].

Energy transfer has been observed not only in the fluid phases but also in the solid state [437, 482, 483]. It has been shown by ESR spectroscopy of a γ-irradiated n-$C_{10}D_{22}$ single crystal containing n-$C_{10}H_{22}$ [482] that in addition to $-CD_2\dot{C}DCD_2-$ radicals, $CH_3\dot{C}HCH_2-$ radicals are also produced, in about double the yield expected on the basis of the composition. Kinetic studies indicated electron excitation transfer.

Electron mobility in various liquid alkanes may differ by several orders of magnitude: e.g. the values in n-hexane and neopentane at room temperature are $\mu = 0.076$ and 70 cm^2 V^{-1} s^{-1}, respectively [59]. The mobility measured in n-hexane–neopentane mixtures is not a linear function of the composition. Minday et al. have found the following linear correlation between the logarithm of the mobility and the molar fraction of n-hexane [59]:

$$\mu_{mix} = \mu_{np} \exp(Ax_h) \qquad (2.85)$$

where μ_{mix} and μ_{np} denote electron mobilities in the mixture and pure neopentane, respectively, x_h is the molar fraction of n-hexane, $A = -6.8$, an empirical constant. It was also shown that the mobility in low pressure gas mixtures follows the rule of mixing [59]. The relation in Eq. (2.85) is supported by several other reports (e.g. [484–486]). Although the value of the electron mobility in alkane mixtures has been discussed in a number of papers, e.g. [484–487], a correlation between electron mobility and product yields is not yet established.

Ausloos et al. have used the radiolysis of binary alkane mixtures to determine the reaction rate constants of some ion-molecule reactions and the ion-electron recombination process [321, 330]. The rate of hydride ion transfer of tert-$C_4H_9^+$ and that of H_2^- transfer of iso-$C_4H_8^+$, both produced from neopentane, were measured in irradiated mixtures of neopentane with C_5-C_8 alkanes containing tertiary hydrogen atoms [133]:

$$neo\text{-}C_5H_{12}^+ \begin{cases} \nearrow tert\text{-}C_4H_9^+ + CH_3 \\ \searrow iso\text{-}C_4H_8^+ + CH_4 \end{cases}$$

The rate constant of hydride ion transfer is very small ($6 \cdot 10^9$–$6 \cdot 10^{10}$ M^{-1} s^{-1}) and depends on the heat of reaction as well as on the stereochemistry of the molecule [321]:

$$tert\text{-}C_4H_9^+ + RH \longrightarrow iso\text{-}C_4H_{10} + R^+ \qquad (2.86)$$

The energy of activation of reaction (2.86) has been found to be $E_a \leq 0.15$ eV [321].

The rate of H_2^- transfer of iso-$C_4H_8^+$ is higher than that of hydride ion transfer of tert-$\dot{C}_4H_9^+$ [327], and it increases with the exothermicity of the reaction. The dependence on the heat of reaction is more pronounced in the liquid than in the gas phase.

$$iso\text{-}C_4H_8^+ + RH \longrightarrow iso\text{-}C_4H_{10} + R(-H)^+$$

Ausloos et al. have irradiated gaseous mixtures of $CH_4\text{-}C_3H_8$, $CH_4\text{-}n\text{-}C_4H_{10}$, $CH_4\text{-}iso\text{-}C_4H_{10}$ and $CH_4\text{-}n\text{-}C_5H_{12}$ in order to investigate the chemical properties of the CH_5^+ ion [132]. On the basis of end product analysis, they concluded that the following ion-molecule reactions take place:

$$CH_4 \quad \longrightarrow\!\!\wedge\!\!\wedge\!\!\wedge\!\!\longrightarrow \quad CH_4^+$$

$$CH_4^+ + CH_4 \quad \longrightarrow \quad CH_5^+ + CH_3$$

$$CH_5^+ + C_3H_8 \quad \longrightarrow \quad C_2H_5^+ + products$$

$$CH_5^+ + n\text{-}C_4H_{10} \quad \longrightarrow \quad C_2H_5^+, \; C_3H_7^+, \text{ etc.}$$

Isotopic tracer experiments have shown that the hydrogen atoms of CH_5^+ do not appear in the ionic products, e.g.:

$$CH_5^+ + C_3D_8 \longrightarrow CH_4 + CD_3H + C_2D_5^+$$

Similar reactions were observed in $CH_4\text{-}C_6$-alkane systems [133].

Harrison and co-workers have studied the ion-molecule reactions occurring in the $CH_4\text{-}C_3H_8$ system in a mass spectrometer [130]. CH_5^+ and $C_2H_5^+$ ions formed from methane react with propane: practically every collision of CH_5^+ and about 40% of the collisions of $C_2H_5^+$ are effective. For the process:

$$C_2H_5^+ + C_3H_8 \xrightarrow{\ k\ } C_2H_6 + C_3H_7^+$$

the value of k is $3.4 \cdot 10^{11}$ M^{-1} s^{-1}. The formation of both $iso\text{-}C_3H_7^+$ and $n\text{-}C_3H_7^+$ is exothermal: $\Delta H^0 = -92$ and -42 kJ mol^{-1}, respectively. The reaction between propane and CH_5^+ leads to three products: 71–75% of ethyl ion ($\Delta H^\circ = -54$ kJ mol^{-1}), 4–5% of n-propyl ion ($\Delta H^\circ = 25$ kJ mol^{-1}) and about 25% of sec-propyl ion ($\Delta H^\circ = -92$ kJ mol^{-1}). As there is no simple correlation between the abundance of the products and the exothermicity of the reactions, it was concluded [130, 132] that no $C_3H_9^+$ intermediate is formed and that the relative rates of simultaneous reactions are influenced by the geometry of the collision. From a stereochemical point of view, attack at the terminal methyl group is most favourable:

$$CH_5^+ + CH_3 \text{---}\vdots\text{---} CH_2CH_3 \longrightarrow C_2H_5^+ + 2\,CH_4$$

Lias et al. [139] used an ion-cyclotron resonance spectrometer to study electron transfer phenomena in various binary mixtures, e.g. in n-octane–3-methylpentane mixtures irradiated by pulses of 3 ms duration and 13 ± 1 eV energy [132A]. They found that the rate constant of the transfer process is

significantly lower than the rate constant of the collisions when the exothermicity of charge transfer between aliphatic alkane components is lower
than 0.5 eV per molecule: the probability of the process in such cases was
determined mainly by thermochemical factors. It has also been shown that
if the heat of reaction is lower than 0.15 eV per molecule, charge transfer
takes place in both directions: the data are characteristic of a dynamic equilibrium. This process occurs in both radiolysis and photolysis, both in the
liquid and in the gas phase.

Isoalkane–isoalkane mixtures

Hardly any work has been done so far in this field, although the study of
the effect of C—C bonds with different dissociation energies or carbon atoms
of different order on the transfer processes could produce interesting results.

György and Földiák [473] have studied the radiolysis of binary mixtures
of branched-chain hexane isomers in the liquid phase. The photoionization
potentials of these compounds are very similar. The plots of G-values as a
function of the composition indicate hardly any interaction between the
components, in agreement with Hardwick's hypothesis; for example, the
yields are an almost linear function of the concentration for the mixture of
3-methylpentane (10.06 eV) and 2,2-dimethylbutane (10.04 eV), and in the
case of the mixtures 3-methylpentane–2,3-dimethylbutane (10.01 eV) and
2,2-dimethylbutane–2,3-dimethylbutane, the yield vs. composition curves
indicate no detectable transfer processes. The nature of the reactions in such
systems, and the possible interactions between components and intermediates, can probably be elucidated only by means of tracer and/or scavenger
studies.

In recent years a number of works discussed the hydrogen formation in
solid-phase isoalkane–isoalkane mixtures [459, 488, 490, 493]. These studies
showed that hydrogen atoms abstract randomly at 4 K, whereas they abstract selectively secondary hydrogen atoms already at 77 K.

2.4. CONCLUSIONS

The radiolysis of individual alkanes, product yields and the possible mechanisms of radiation chemical reactions are discussed in detail in Section
2.2. The present section is devoted to the analysis of structural factors (such
as bond energies, number of branchings and chain length) governing the
radiolytic G-values. This will be done by comparing some of the data detailed
above and some contained in review papers.

Processes related to C—H bond ruptures

When alkanes are irradiated, the product with highest G-value is generally
hydrogen. This fact, together with the relatively simple analysis, explains
why most publications have presented $G(H_2)$ values only. As can be seen

from Fig. 2.8, $G(H_2)$ values of n-alkanes are nearly constant as a function of the carbon number [414]. Small deviations have to be accepted with reservations. For example, in principle, only values extrapolated to zero conversion can be strictly compared because alkenes produced during irradiation decrease hydrogen yields already in very low concentrations (see Section 4.3). Similar distortions can be caused by differences in the type of radiation, the LET values, the state of aggregation of hydrocarbons, the temperature of irradiation, etc.

It has been verified by several experiments that the mechanism of hydrogen formation is more or less the same for every alkane: direct excitation and ion-electron recombination excitation lead to molecular H_2 elimination as well as to the formation of 'hot' and thermal hydrogen atoms. It has been found that the hydrogen yield involving a molecular mechanism and 'hot' atoms (i.e. the yield unaffected by radical scavengers) is about 33–43% of the total hydrogen yield, independently of the structure of the carbon chain and the overall hydrogen yield [352].

Isildar and Schuler [360, 405] have studied the radiation chemistry of n-alkanes, ranging from n-hexane to n-hexadecane, in order to establish

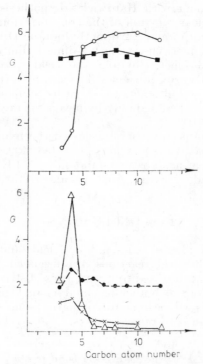

Fig. 2.8. Yields of products in the radiolysis of aliphatic and cyclic alkanes, as a function of the carbon atom number of the molecule [489].
■ Hydrogen from n-alkanes; ○ hydrogen from cycloalkanes;
● fragments from n-alkanes; △ fragments from cycloalkanes; × open-chain C_n-products from cycloalkanes (radiation: ^{60}Co-γ)

general rules which control the C—H bond rupture processes. They have found that the main source of H atom formation is geminate neutralization, whereas direct excitation of the n-alkane molecule leads, primarily, to C—C bond rupture. Further, they observed a slight increase of the parent alkyl radical yield with the carbon chain length between C_6 and C_{16}. The increase of the parent alkyl radical yields parallels the electron density in the parent hydrocarbon indicating that the source of the parent radicals is ionic.

The $G(H_2)$ value of branched-chain alkanes is lower than that of the corresponding n-alkane: the larger the number of methyl groups in the molecule, the lower the hydrogen yield (Table 2.43).

Table 2.43

Yields of hydrogen and fragment products from liquid alkanes
(radiation: ^{60}Co-γ)

Products	Ethane	Propane	n-Butane	Isobutane	n-Pentane	Isopentane	Neopentane	n-Hexane	3-Methyl-pentane	2,3-Dimethyl-butane	2,2-Dimethyl-butene
H_2	3.68	4.90	4.80	4.16	5.10	3.75	2.2	5.00	3.40	2.90	2.00
CH_4	0.44	0.75	0.52	1.98	0.26	1.40	3.76	0.14	0.24	0.51	1.14
C_2 products		1.25	1.77	—	0.98	1.54	0.43	0.55	1.35	0.10	1.48
C_3 products			0.36	2.3	0.93	1.69	0.20	0.52	0.12	4.24	0.21
C_4 products					0.20	0.54	2.55	0.70	1.60	0.50	1.82
C_5 products								0.15	0.30	0.65	0.80
Total hydrocarbon product	0.44	2.00	2.65	4.28	2.37	5.17	6.94	2.06	3.61	6.00	5.45

It has been suggested in several papers that the differences in $G(H_2)$ values from aliphatic alkane isomers with different structures can be attributed to the different rates of original C—H bond rupture of activated molecules, which, in turn, depend on the number of primary, secondary and tertiary C—H bonds in the molecule. Although vacuum-UV photolysis and ESR spectroscopic studies seem to support this view, it has not yet been verified beyond doubt, because these methods measure the yields of parent alkyl radicals remaining after C—H bond dissociation; but such radicals can also be formed in hydrogen abstraction processes. According to these investigations, the 'overall reactivity' of C—H bonds decreases with increasing bond strength; on the basis of results reported in several papers, the relative reactivities correspond on average to the ratio [240, 271, 281, 422]:

$$C_{prim}-H : C_{sec}-H : C_{tert}-H = 1 : 3 : 9.$$

According to experiments carried out at −70°C, the ratio $C_{prim}-H : C_{sec}-H$ is closer to 2 than to 3. This figure may be a better approximation to the real value because, at such a low temperature, the extent of hydrogen atom abstraction must be negligible. For the sake of comparison it should be noted that the ratio of ruptures of primary, secondary and tertiary C—H bonds in mercury-sensitized photolysis has been found to be 1 : 65 : 350 [181]. Thus photolysis is clearly a more selective process, owing to the fact that the amount of energy transferred to the alkane molecules by excited mercury atoms is lower than the average energy content of molecules activated by irradiation.

Congruent alkenes are among the main products of radiolysis ($G = 1$–3). Their yields are strongly dependent on dose and temperature. The two main pathways of their formation are molecular hydrogen elimination and disproportionation of parent alkyl radicals. There is a general tendency for π-bonds to be formed in an internal rather than in a terminal position. The thermodynamically more stable *trans*-isomers are produced in higher yields than *cis*-isomers.

C_{n+1}—C_{2n} products are formed mainly in radical combination reactions. Some C_{n+1} products may also be formed via CH_2 biradical insertion. Whereas the formation of C_{n+1}—C_{2n-1} products in n-alkane radiolysis involve mainly the combination of fragment radicals with parent alkyl radicals, in the case of branched chain alkanes the combination of fragments with each other may also be considerable.

The yield of dimers from branched chain alkanes is much lower than from n-alkanes, because the stereochemistry of the parent alkyl radicals hinders radical combination.

Processes related to C—C *bond ruptures*

In order to illustrate the correlation between molecular structure and the reactivity of C—C bonds in radiolysis, the yields of fragments (summarized according to their carbon number) from C_2-C_6 aliphatic alkanes are listed in Table 2.43.

Of isomers with identical numbers of methyl groups, the one with its methane-producing C—C bond linked to a carbon atom of higher order gives more methane (and less hydrogen) [391]; consequently, the methane-forming tendency of isoalkanes containing a quaternary ('central') carbon atom exceeds that of alkanes with tertiary carbon atom(s).

From a comparison of the yields of other C_2—C_{n-1} fragments from iso-alkanes, similar conclusions can be drawn. With regard to complementary products of C—C bond fission reactions, the yields of those having lower molecular mass are generally higher. For example, for neopentane: $G(CH_4) = 3.76$ whereas $G(\text{iso-}C_4H_{10}) = 0.71$ and $G(\text{iso-}C_4H_8) = 1.84$; or for 2,2-dimethylbutane: $G(CH_4) = 1.14$ and $G(C_5) = 0.8$. This fact can be interpreted in terms of secondary decomposition processes of higher molecular mass species (radicals or ions).

The data in the Table also indicate that for C_2—C_8 hydrocarbons, bonds in the vicinity of a branching point are more likely to be cleaved. Thus, tertiary and especially quaternary alkanes produce much higher amounts of fragments than the corresponding n-alkanes. Radical scavenger studies support this conclusion, and empirical formulae have been derived for approximate calculation of radical yields from e.g. n-alkanes [274]:

$$G(CH_3) = \frac{2.04}{(n-1)^2}$$

$$G(C_2H_5) = \frac{7.0}{(n-1)^2}$$

where n denotes the number of carbon atoms in the molecule. The comparison of these equations reveals that the probability of the rupture of the terminal (primary-secondary) C—C bond is about 3.5 times lower than the decomposition of other, secondary–secondary C—C bonds. The following formula [302]:

$$G(R) = \frac{1}{(n-1)^2}(1.0\,C_1 + 2.8\,C_2 + 8.6\,C_3 + 29\,C_4)$$

holds for branched alkanes: C_i ($i = 1, 2, 3$ or 4) denotes the number of i-th order carbon atoms in the bonds whose decomposition gives radical R.

Denoting the order of carbon atoms at both ends of the bond by Roman numbers, the following $G(C-C)$ yields can be calculated for C_6 alkanes as parent hydrocarbons [391]:

	$G(C-C)$
I–II	0.1–0.15
I–III	0.3
II–II	0.4–0.45
I–IV	0.7
II–III	0.9
III–III	2.6
II–IV	2.9

C—C bond dissociation energies of aliphatic alkanes can be expressed as follows [401]:

$$E_D = 353.5 - 8.8\,i$$

where E_D is the dissociation energy of the C—C bond, in kJ mol^{-1}, i is the number of bonds in the vicinity of the given bond. Figure 2.9 demonstrates that $G(C-C)$ values decrease nearly exponentially as a function of the bond dissociation energy of the bond to be dissociated. The shift of the curves with different carbon numbers has been explained — at least qualitatively — in terms of a changing number of vibrational degrees of freedom. With increasing carbon number and thus more vibrational degrees of freedom, the probability of concentration of the 'excess' energy in the activated molecule on any single C—C bond or a group of degrees of freedom important for a given bond fission becomes lower: at the same time, deactivation by energy dissipation without decomposition is more likely.

11

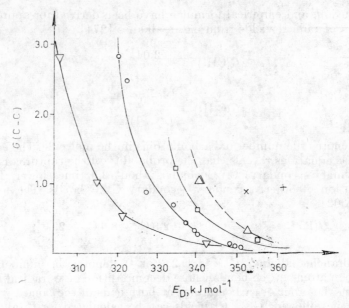

Fig. 2.9. G-values of C—C bond fission in the radiolysis of aliphatic alkanes, as a
function of the standard dissociation energy of the bond in question.
▽ C_8; ○ C_6; □ C_5; △ C_4; × C_3; + C_2 (radiation: ^{60}Co-γ). After Földiák et al. [489, 491]

A similar explanation has been given by other authors for the experimental difference between the yields of irradiated isobutane giving methyl radical and sec-propyl ions with $G = 1.15$ and iso-octane giving isobutyl radicals and tert-butyl ions with a G-value of ~ 0.4, in spite of the nearly
identical reaction enthalpies: $\Delta H^\circ = 54$ and 59 kJ mol^{-1}, respectively
[411–413].

Isildar and Schuler [360, 405] attempted to find general rules that govern
radiation-induced bond rupture processes in C_6-C_{16} alkanes. As it was
found, the electron density, the yield of parent alkyl radicals, the yields
of fragment radicals and C—C bond ruptures decrease with the number of
C—C bonds in the molecules. Since the parent alkyl radicals have been shown
to form through geminate neutralization of ion pairs, the increase of their
yield with the number of C—C bonds can be explained by the increase of
the ion pair yields, i.e. the electron density of the hydrocarbons. On the
other hand, if the chain is longer, the number of C—C bonds increases so
that the excitation energy is distributed among more C—C bonds and, consequently, the efficiency of rupture of any bond decreases. Since the fragmentation of the chain in n-alkanes was shown to occur upon direct excitation, these arguments can account for the observed decrease of G(total
fragment) with the number of C—C bonds.

Isildar [405] also studied the decomposition of C_6-C_{16} n-alkanes into radicals
using radical acceptors. He obtained a relationship describing the yield of

radicals as a function of the carbon atom number of the radical produced:

$$G(R) = \frac{9.7}{(n-1)^{1.8}} \left[1 - \frac{(x - n/2)^2}{(n/2 - 0.8)^2} \right]$$

where n is the carbon atom number of the parent alkane and x is that of the radical, R. The equation suggests a maximum in $G(R)$ that is shifting towards the central C—C bond as the chain length of the parent alkane is increased.

Several hypotheses have been proposed to explain the greater tendency of branched carbon chains to decompose. Some authors assumed that it was due to ionic reactions. Quantum chemical calculations show that in alkane ions with one positive charge, the dissociation energy of bonds in the vicinity of the branching is very low, thus decomposition may even be exothermal. The G-value for the decomposition of ions decreases with increasing carbon number, owing to collision deactivation without decomposition promoting random energy dissipation. The most favourable positions of alkane ions from the point of view of scission are the C—C bonds in the vicinity of the branching, because the density of positive charge is highest at these sites (Fig. 2.10). The charge distribution within the molecule ion has been shown to be in direct correlation with the mass spectrum.

An investigation of the radiation-induced fluorescence of straight and branched-chain alkanes has shown that an increase of the extent of ramification leads to a decrease in the quantum yield of fluorescence and to an increase in the yields of processes other than light emission, e.g. chemical decomposition.

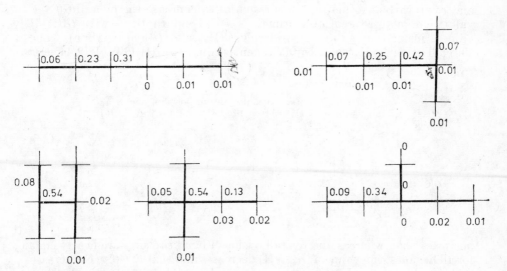

Fig. 2.10. Net positive charge distribution in the ground state of the ions of hexane isomers after Lorquet [494]. (Values above the lines are those of the C—C bonds, those below are of the C—H bonds)

11*

The molecular exciton theory of σ-bonds has also been applied to the interpretation of radiolysis [87]. It has been shown that energy propagates very rapidly along the chain of an electronically excited alkanes. Intramolecular energy transfer is more rapid than the movement of atomic nuclei, therefore long straight chains are less likely to break. The elimination of side-groups is the result of primary energy deposition. A high $G(C-C)$ can be expected when more than one side-group is linked to the same carbon atom.

The relation between C—H and C—C bond ruptures

Radiolysis of alkanes leads to the formation of several hydrocarbons, some in very small amounts. Most fragments are straight-chain saturated or unsaturated compounds. Of products with an identical number of carbon atoms, the yields of the saturated ones, generally exceed those of alkenes [391].

The yields of C_1-C_{n-1} fragments generally increase with increasing branching in isoalkanes. The yields of hydrogen and dimers are usually 25–60% less than the corresponding values with n-alkanes, but the proportion of fragments is usually higher. For example, $G(H_2)$ values for hexane isomers are as follows: 5.0 for n-hexane, 4.0 for 2-methylpentane, 3.4 for 3-methylpentane, 2.9 for 2,3-dimethylbutane and 2.0 for 2,2-dimethylbutane: the corresponding $G(CH_4)$ values are: 0.14, $\sim$ 0.5, 0.24, 0.51 and 1.14, respectively. Thus as the number of methyl groups in the parent molecule increases, $G(H_2)$ values decrease and $G(CH_4)$ values increase, thereby indicating a close antibathic correlation [391, 489, 491].

In order to obtain a deeper insight into the correlations between hydrogen and hydrocarbon formation, it is essential to compare the overall $G(C-C)$ values characteristic of the primary C—C bond ruptures with $G(C-H)$ values, instead of simply considering $G(H_2)$ and G(hydrocarbon) values [391]. The $G(C-H)$ value can be taken as equal to $G(H_2)$. Table 2.44 dem-

Table 2.44

Comparison of unimolecular decomposition yields
(bond-fission/100 eV). After Földiák and György [391]

	n-Hexane	3-Methylpentane	2,3-Dimethyl-butane	2,2-Dimethyl-butane
$G(C-H)$	5.0	3.4	2.9	2.0
$G(C-C)$	1.5	2.4	3.9	5.1
$j \equiv G(C-C)/G(C-H)$	0.3	0.7	1.4	2.5

onstrates that whereas the overall yields of bond ruptures are nearly equal for all hexanes, the value of $G(C-H)$ decreases and that of $G(C-C)$ increases with increasing number of methyl groups. It can be concluded that processes leading to H_2 evolution and those involving C—C bond fission compete with each other during the radiolysis of aliphatic alkanes.

According to the exciton theory, the energy localized on $C-H$ bonds does not participate in intramolecular transfer processes: this may explain the experimental result that hydrogen is the main product of alkane radiolysis in spite of the dissociation energy of the $C-H$ bond being by about 35–50 kJ mol^{-1} higher than that of the $C-C$ bond [79, 87].

REFERENCES

1. TOPCHIEV, A. V., *Radiolysis of Hydrocarbons*, Elsevier, Amsterdam, 1964
2. AUSLOOS, P. (Ed.) *Fundamental Processes in Radiation Chemistry*, Wiley, New York, 1968
3. ALLEN, A. O., in *Current Topics in Radiation Chemistry* (Eds EBERT, M. and HOWARD, A.), North-Holland, Amsterdam, 1969, Vol. 4, p. 1
4. HARDWICK, T. J., in *Actions Chimiques et Biologiques des Radiations*, Vol. 10 (Ed. HAISSINSKY, M.), Masson, Paris, 1966
5. GÄUMANN, T., in *Aspects of Hydrocarbon Radiolysis* (Eds GÄUMANN, T. and HOIGNÉ, J.), Academic Press, London, 1968
6. COFFEY, S. (Ed.) *Rodd's Chemistry of Carbon Compounds*, Elsevier, Amsterdam, 1964, Vol. 1
7. *Handbook of Chemistry and Physics*, 51st Edition, The Chemical Rubber Co., Cleveland, 1970—1971
8. PAULING, L., *The Nature of the Chemical Bond*, Cornell University Press, New York, 1960
9. MISLOV, K., *Introduction to Stereochemistry*, Benjamin, New York, 1965
10. ELIEL, E. L., ALLINGER, N. L., ANGYAL, S. J. and MORRISON, G. A., *Conformational Analysis*, Interscience, New York, 1965
11. SCHLEYER, P. v. R., WILLIAMS, J. E. and BLANCHARD, K. R., *J. Am. Chem. Soc.*, **92**, 2377 (1970)
12. ALLINGER, N. L., TRIBBLE, M. T., MILLER, M. A. and WERTZ, D. H., *J.Aom. Chem. Soc.*, **93**, 1637 (1971)
13. STULL, D. R., WESTRUM, E. F. JR. and SINKE, G. C., *The Chemical Thermodynamics of Organic Compounds*, Wiley, New York, 1969
14. COX, J. D. and PILCHER, G., *Thermochemistry of Organic and Organometallic Compounds*, Academic Press, London, 1970
15. STEACIE, E. W. R., *Atomic and Free Radical Reactions*, Reinhold, New York, 1954
16. BENSON, S. W., *Thermochemical Kinetics*, Wiley, New York, 1968
17. ROBINSON, P. I. and HOLBROOK, K. A., *Unimolecular Reactions*, Wiley, London, 1972
18. LAIDLER, K. J. and LOUCKS, L. F., in *Comprehensive Chemical Kinetics* (Eds BAMFORD, C. H. and TIPPER, C. F. H.), Elsevier, Amsterdam, 1972, Vol. 5.
19. BOOZ, J. and EBERT, H. G., *Z. Angew. Phys.*, **13**, 376 (1961)
20. BOOZ, J. and EBERT, H. G., *Z. Angew. Phys.*, **14**, 385 (1962)
21. WHITE, G. N., *Radiat. Res.*, **18**, 265 (1963)
22. MEISELS, G. G., *J. Chem. Phys.*, 41, 51 (1964)
23. ADLER, P. and BOTHE, H.-K., *Z. Naturforsch.*, *A*, **20**, 1700 (1965)
24. LE BLANC, R. M. and HERMAN, J. A., *J. Chim. Phys.*, **63**, 1055 (1966)
25. FREEMAN, G. R. and STOVER, E. D., *Can. J. Chem.*, **46**, 3235 (1968)
26. STONEHAM, T. A., ETHRIDGE, D. R. and MEISELS, G. G., *J. Chem. Phys.*, **54**, 4054 (1971)
27. KLOTS, C. E., *J. Chem. Phys.*, 44, 2715 (1966); **46**, 3468 (1967)
28. COOPER, R. and MOORING, R. M., *Australian J. Chem.*, **21**, 2417 (1968)
29. FREEMAN, G. R. and DODELET, J. P., *Int. J. Radiat. Phys. Chem.*, **5**, 371 (1973)
30. SHIDA, T. *J. Phys. Chem.*, **74**, 3055 (1970)
31. TERLECKI, J. and FIUTAK, J., *Int. J. Radiat. Phys. Chem.*, **4**, 469 (1972)
32. HO, H. W. and MOZUMDER, A. *J. Chem. Phys.*, **57**, 4724 (1972)

33. WADA, T. and HATANO, Y., *J. Phys. Chem.*, **79**, 2210 (1975)
34. WADA, T. and HATANO, Y., *J. Phys. Chem.*, **81**, 1057 (1977)
35. FREEMAN, G. R. and SAMBROOK, T. E. M., *J. Phys. Chem.*, **78**, 102 (1974)
36. HOLROYD, R. A., *Advan. Chem. Ser.*, **82**, 488 (1968)
37. SHERMAN, W. V., *J. Chem. Soc. A*, 599 (1966)
38. STOVER, E. D. and FREEMAN, G. R., *J. Chem. Phys.*, **48**, 3902 (1968)
39. RZAD, S. J. and WARMAN, J. M., *J. Chem. Phys.*, **49**, 2861 (1968)
40. INFELTA, P. P. and SCHULER, R. H., *Int. J. Radiat. Phys. Chem.*, **5**, 41 (1973)
41. ROBINSON, M. G. and FREEMAN, G. R., *Can. J. Chem.*, **52**, 440 (1974)
42. WARMAN, J. M., ASMUS, K.-D. and SCHULER, R. H., *J. Phys. Chem.*, **3**, 931 (1969)
43. ASMUS, K.-D., WARMAN, J. M. and SCHULER R. H., *J. Phys. Chem.*, **74**, 246 (1970)
44. BANSAL, K. M. and SCHULER, R. H., *J. Phys. Chem.*, **74**, 3924 (1970)
45. RZAD, S. J. and SCHULER, R. H., *J. Phys. Chem.*, **72**, 228 (1968)
46. HUMMEL, A. and ALLEN, A. O., *J. Chem. Phys.*, **44**, 3426 (1966)
47. SCHMIDT, W. F. and ALLEN, A. O., *J. Chem. Phys.*, **52**, 2345 (1970)
48. VANNIKOV, A. V., KOVALEV, V. O. and ZOLOTAREVSKIJ, V. U., *Khim. Vys. Energ.*, **5**, 49 (1971)
49A. TEWARI, P. H. and FREEMAN, G. R., *J. Chem. Phys.*, **49**, 4394 (1968)
49B. TEWARI, P. H. and FREEMAN, G. R., *J. Chem. Phys.*, **51**, 1276 (1969)
50. FUOCHI, P. G. and FREEMAN, G. R., *J. Chem. Phys.*, **56**, 2333 (1972)
51. DODELET, J. P. and FREEMAN, G. R., *Can. J. Chem.*, **50**, 2667 (1972)
52. NISHIKAWA, M., YAMAGUCHI, Y. and FUJITA, K., *J. Chem. Phys.*, **61**, 2356 (1974)
53. KROH, J., PIEKARSKA, J. and KAROLCZAK, S., *Bull. Acad. Pol. Sci., Ser. Sci. Chim.*, **21**, 279 (1973)
54. HOLROYD, R. A., DIETRICH, B. K. and SCHWARZ, H. A., *J. Phys. Chem.*, **76**, 3794 (1972)
55. SCHMIDT, W. F. and ALLEN, A. O., *J. Phys. Chem.*, **72**, 3730 (1968)
56. FREEMAN, G. R. and FAYADH, J. M., *J. Chem. Phys.*, **43**, 86 (1965)
57. SCHMIDT, W. F. and ALLEN, A. O., *J. Chem. Phys.*, **52**, 4788 (1970)
58. ALLEN, A. O. and HOLROYD, R. A., *J. Phys. Chem.*, **78**, 796 (1974)
59. MINDAY, R. M., SCHMIDT, L. D. and DAVIS, H. T., *J. Phys. Chem.*, **76**, 442 (1972)
60. MINDAY, R. M., SCHMIDT, L. D. and DAVIS, H. T., *J. Phys Chem.*, **54**, 3112 (1971)
61. ALLEN, A. O., GANGWER, T. E. and HOLROYD, R. A., *J. Phys. Chem.*, **79**, 25 (1975)
62. HAMILL, W. H., in *Radical Ions* (Eds KAISER, E. T. and KEVAN, L. K.), Wiley, NEW YORK, 1969
63A. BAXENDALE, J. H., BELL, C. and WARDMAN, P., *Chem. Phys. Letters.*, **12**, 347 (1971)
63B. BAXENDALE, J. H., BELL, C. and WARDMAN, P., *J. Chem. Soc., Faraday Trans. I.*, **4**, 776 (1973)
64. a. RICHARDS, J. T. and THOMAS, J. K., *Chem. Phys. Lett.*, **10**, 317 (1971)
    b. RICHARDS, J. T. and THOMAS, J. K., *J. Chem. Phys.*, **53**, 218 (1970)
65. ROBINSON, M. G., FUOCHI, P. G. and FREEMAN, G. R., *Can. J. Chem.*, **49**, 984 (1971)
66. BALOG, J., TÓTH, L. and PINTÉR, K., *Izotóptechnika*, **14**, 346 (1971)
67. SCHILLER, R., *J. Chem. Phys.*, **57**, 2222 (1972)
68. SCHILLER, R. and VASS, Sz., *Int. J. Radiat. Phys. Chem.*, **7**, 193 (1975)
69. AUSLOOS, P. and LIAS, S. G., *Radiat. Res. Rev.*, **1**, 75 (1968)
70. AUSLOOS, P., *Mol. Photochem.*, **4**, 39 (1972)
71. AUSLOOS, P., *Proc. 4th Intern. Conf. Rad. Res.* Vol. **2**. *Phys. and Chem.* (Eds DUPLAND, J. F. and CHAPIRO, A.) Gordon and Breach Science Publishers, London, 1973
72. AUSLOOS, P. and LIAS, S. G., *Chemical Spectroscopy and Photochemistry in the Vacuum Ultraviolet* (Eds SANDORFY, C., AUSLOOS, P. and ROBIN, M. B.), Reidel Publ. Co, Dordrecht, 1974
73. THOMAS, J. K., *Rec. Chem. Progr.*, **32**, 145 (1972)
74. MAGEE, J. L. and HUANG, J.-T., *J. Phys. Chem.*, **76**, 3801 (1972)
75. MAGEE, J. L., *Radiat. Res.*, **20**, 71 (1963)
76. SINGH, A., *Radiat. Res. Rev.*, **4**, 1 (1972)

77. PLATZMAN, R. L., in *Radiation Research* (Ed. SILINI, G.), North-Holland, Amsterdam, 1967, p. 20
78. SIECK, L. W., in *Fundamental Processes in Radiation Chemistry* (Ed. AUSLOOS, P.), Wiley, New York, 1968
79. RAYMODA, J. W. and SIMPSON, W. T., *J. Chem. Phys.*, **47**, 430 (1967)
80. KATAGIRI, S. and SANDORFY, C., *Theoret. Chim. Acta.*, **4**, 203 (1966)
81. LOMBOS, B. A., SAUVAGEAU, P. and SANDORFY, C., *Mol. Spectrosc.*, **24**, 253 (1967)
82A. HIRAYAMA, F., ROTHMAN, W. and LIPSKY, S., *Chem. Phys. Lett.*, **5**, 296 (1970)
82B. ROTHMAN, W., HIRAYAMA, F. and LIPSKY, S., *J. Chem. Phys.*, **58**, 1300 (1973)
83. HIRAOKA, K. and HAMILL, W. H., *J. Chem. Phys.*, **57**, 3870 (1972)
84. MATSUSHIGE, T. and HAMILL, W. H., *J. Phys. Chem.*, **76**, 1255 (1972)
85. HUANG, T. and HAMILL, W. H., *J. Phys. Chem.*, **78**, 2077 (1974)
86. HIRAOKA, K. and HAMILL, W. H., *J. Chem. Phys.*, **59**, 5749 (1973)   ←
87. PARTRIDGE, R. H., *J. Chem. Phys.*, **52**, 2485 (1970)
88. SIECK, L. W. and JOHNSEN, R. H., *J. Phys. Chem.*, **67**, 2281 (1963)
89. FREEMAN, G. R., *Radiat. Res. Rev.*, **1**, 1 (1968)
90. HAYNES, R. M. and KEBARLE, P., *J. Chem Phys.*, **45**, 3899 (1966)
91. MEISELS, G. G., HAMILL, W. H. and WILLIAMS, W. H., *J. Phys. Chem.*, **61**, 1456 (1957)
92. HUNTRESS, W. T. JR., PINIZZOTTO, R. F. JR. and LANDENSLAGER, J. B., *J. Am. Chem. Soc.*, **95**, 4107 (1973)
93. WEINER, J., SMITH, G. P. K., SAUNDERS, M. and CROSS, R. J., *J. Am. Chem. Soc.*, **95**, 4115 (1973)
94. FIELD, F. H. and BEGGS, D. P., *J. Am. Chem. Soc.*, **93**, 1585 (1971)
95. AUSLOOS, P. and LIAS, S. G., *J. Chem. Phys.*, **38**, 2207 (1963)
96. HUMMEL, R. W., *Disc. Faraday Soc.*, **36**, 246 (1963)
97. HUMMEL, R. W., *Int. J. Radiat. Phys. Chem.*, **2**, 31 (1970)
98. HUMMEL, R. W., *Disc. Faraday Soc.*, **36**, 75 (1963)
99. GORDEN, R. and AUSLOOS, P., *J. Chem. Phys.*, **46**, 4823 (1967)
100. HUMMEL, R. W., *Int. J. Radiat. Phys. Chem.*, **2**, 119 (1970)
101. AQUILANTI, V., GIARDINI-GUIDONI, A., VOLPI, G. G. and ZOCCHI, F., *Int. J. Radiat. Phys. Chem.*, **2**, 217 (1970)
102. FUTRELL, J. H. and TIERNAN, T. O., in *Fundamental Processes in Radiation Chemistry* (Ed. AUSLOOS, P.), Wiley-Interscience, New York, 1968
103. FIELD, F. H. and FRANKLIN, J. L., *Electron Impact Phenomena*, Academic Press, New York, 1957
104. BONE, L. I. and FIRESTONE, R. F., *J. Phys. Chem.*, **69**, 3652 (1965)
105. BOSNALI, M. W. and PERNER, D., *Z. Naturforsch. A.*, **26**, 1768 (1971)
106. COLLIN, G. J. and HERMAN, J. A., *J. Chem. Phys.*, **66**, 496 (1969)
107. AUSLOOS, P. and LIAS, S. G., *Disc. Faraday Soc.*, **39**, 36 (1965)
108. FINNEY, C. D. and MOSER, H. C., *J. Phys. Chem.*, **75**, 2405 (1971)
109. MELTON, C. E. and RUDOLF, P. S., *J. Chem. Phys.*, **47**, 1771 (1967)
110. MELTON, C. E. and RUDOLF, P. S., *J. Phys. Chem.*, **71**, 4572 (1967)
111. BOWMAN, C. R. and MILLER, W. D., *J. Chem. Phys.*, **42**, 681 (1965)
112. HAMLET, P., MOSS, J., MITTAL, J. P. and LIBBY, W. F., *J. Am. Chem. Soc.*, **91**, 258 (1969)
113. HUMMEL, R. W., *J. Phys. Chem.*, **79**, 2685 (1966)
114. CAHILL, R. W., SEELER, A. K. and GLASS, R. A., *J. Phys. Chem.*, **71**, 4564 (1967)
115. HUMMEL, R. W. and HEARNE, J. A., *J. Phys. Chem.*, **75**, 1164 (1971)
116. HUNTRESS, W. T. Jr., *J. Chem. Phys.*, **56**, 5111 (1972)
117. KLASSEN, N. V., *J. Phys. Chem.*, **72**, 1076 (1968)
118. REBBERT, R. E. and AUSLOOS, P., *J. Photochem.*, **1**, 171 (1972/73)
119. BRAUN, W., MCNESBY, J. R. and BASS, A. M., *J. Chem. Phys.*, **46**, 2071 (1967)
120. BRAUN, W., BASS, A. M., DAVIS, D. D. and SIMMONS, J. D., *Proc. Roy. Soc.*, **A312**, 417 (1969)
121. HELLNER, L., MASANET, J. and VARMEIL, C., *J. Chem. Phys.*, **55**, 1022 (1971)
122. REBBERT, R. E. and AUSLOOS, P., *J. Res. Natl. Bur. Stand.*, **77A**, 101 (1973)
123. MEABURN, G. M. and PERNER, D., *Nature*, **212**, 1042 (1966)
124. HUSAIN, D. and KIRSCH, L. J., *Trans. Faraday Soc.*, **67**, 2125 (1971)

125. OGATA, Y., OBI, K., AKIMOTO, H. and TANAKA, I., *Bull. Chem. Soc. Jpn.*, **44**, 2671 (1971)
126. REBBERT, R. E. and AUSLOOS, P., *J. Res. Natl. Bur. Stand.*, **77A**, 109 (1973)
127. SHERIDAN, M. E., GREER, E. and LIBBY, W. F., *J. Am. Chem. Soc.*, **94**, 2614 (1972)
128. DAVIS, D. R., LIBBY, W. F. and MEINSEKEIN, W. G., *J. Chem. Phys.*, **45**, 4481 (1966)
129. SCOTT, W. M. and WILSON, H. W., *Org. Mass Spectrom.*, **1**, 519 (1968)
130. SOLKA, B. H., LAU, A. Y. K. and HARRISON, A. G., *Can. J. Chem.*, **52**, 1798 (1974)
131. BLAIR, A. S., HESLIN, E. J. and HARRISON, A. G., *J. Am. Chem. Soc.*, **94**, 2935 (1972)
132. AUSLOOS, P., LIAS, S. G. and GORDEN, R. JR., *J. Chem. Phys.*, **39**, 3341 (1963)
132A. LIAS, S. G., REYLER, and AUSLOOS, P., *Mass Spectrometry Ion Phys.* **19**, 219 (1976)
133. FIELD, F. H., MUNSON, M. S. B. and BECKER, D. A., *Adv. Chem. Ser.*, **58**, 167 (1966)
134. STEVENS, G. C., CLARKE, R. M. and HART, E. J., *J. Phys. Chem.*, **76**, 3863 (1972)
135. BRAUN, W., BASS, A. M. and PILLING, M., *J. Chem. Phys.*, **52**, 5131 (1970)
136. REBBERT, R. E., LIAS, S. G. and AUSLOOS, P., *Chem. Phys. Lett.*, **12**, 323 (1971)
137. GILLIS, H. A., *J. Phys. Chem.*, **67**, 1399 (1963)
138. FESSENDEN, R. W. and SCHULER, R. H., *J. Chem. Phys.*, **33**, 935 (1960)
139. LIAS, S. G., COLLIN, G. J., REBBERT, R. E. and AUSLOOS, P., *J. Chem. Phys.*, **52**, 1841 (1970)
140. SHAULOV, A. YU., TITOV, V. B. and BRODSKII, A. M., *Khim. Vys. Energ.*, **6**, 192 (1972)
141. BAKALE, G., *Thesis.* Cleveland, Case Western Reserve Univ., 1968
142. GAWLOWSKI, J., NIEDZIELSKI, J. and SOBKOWSKI, J., *Proc. 4th Tihany Symp. Rad. Chem.* (Eds HEDVIG, P. and SCHILLER, R.), Akadémiai Kiadó, Budapest, 1977
143. BACK, R. A., *J. Phys. Chem.*, **64**, 124 (1960)
144. YANG, K. and GANT, P. C., *J. Phys. Chem.*, **65**, 1861 (1961)
145. INOUE, M., UNO, T., SATO, S. and SHIDA, S., *Bull. Chem. Soc. Jpn.*, **41**, 2005 (1968)
146. CARMICHAEL, H. H., GORDEN, R. JR. and AUSLOOS, P., *J. Chem. Phys.*, **42**, 343 (1965)
147. HATANO, Y., SHIDA, S. and SATO, S. *Bull. Chem. Soc. Jpn.*, **41**, 1120 (1968)
148. PEARSON, E. F. and INNES, K. K., *J. Mol. Spectr.*, **30**, 232 (1969)
149. SROKA, W. Z., *Naturforsch.*, **A24**, 1724 (1969)
150. AUSLOOS, P. and LIAS, S. G., *Ann. Rev. Phys. Chem.*, **22**, 2524 (1971)
151. BEENAKKER, D. I. M. and DE HEER, F. J., *J. Chem. Phys.*, **7**, 130 (1975)
152. DE HEER, F. J., *Int. J. Radiat. Phys. Chem.*, **7**, 137 (1975)
153. AUSLOOS, P., REBBERT, R. E. and SIECK, L. W., *J. Chem. Phys.*, **54**, 2612 (1971)
154. BENNET, S. L., LIAS, S. G. and FIELD, F. H., *J. Phys. Chem.*, **76**, 3919 (1972)
155. FINN, T. G., CARNAHAN, B. L., WELLS, W. C. and ZIPF, E. C., *J. Chem. Phys.*, **63**, 1469 (1975)
156. TSUJI, M., OGAWA, T., NISHIMURA, Y., and ISHIBASHI, N., *Bull. Chem. Soc. Jpn.*, **49**, 2913 (1976)
157. BAKALE, G. and GILLIS, H. A., *J. Phys. Chem.*, **73**, 2178 (1969)
158. WEXLER, S. and POBO, C. G., *J. Am. Chem. Soc.*, **93**, 1327 (1971)
159. URCH, D. S. and WELCH, M. J., *Radiochim. Acta.*, **5**, 202 (1966)
160. REBBERT, R. E., LIAS, S. G. and AUSLOOS, P., *J. Res. Natl. Bur. Stand.*, 607 (1971)
161. MUNSON, M. S. B., FRANKLIN, J. L. and FIELD, F. H., *J. Phys. Chem.*, **68**, 3098 (1964)
162. GIOUMOUSIS, G. and STEVENSON, D. P., *J. Chem. Phys.*, **29**, 294 (1958)
163. DERWISH, G. A. W., GALLI, A., GIARDINI-GUIDONI, A. and VOLPI, G. G., *J. Chem. Phys.*, **5**, 40 (1964)
164. STONE, J. A. and HOLLAND, P. T., *Can. J. Chem.*, **48**, 3282 (1970)
165. HOLLAND, P. T. and STONE, J. A., *Can. J. Chem.*, **48**, 1078 (1970)
166. HOLLAND, P. T. and STONE, J. A., *Can. J. Chem.*, **48**, 3277 (1970)

167. SIDEBOTTOM, H. W., TEDDER, J. M. and WALTON, J. L., *Trans. Faraday Soc.*, **65**, 2103 (1969)
168. HOLLAND, P. T. and STONE, J. A., *Can. J. Chem.*, **52**, 221 (1974)
169. MCNESBY, J. R. and OKABE, H., *Adv. Photochem.*, **3**, 157 (1964)
170. WODETZKI, C. M., MCCUSKER, P. A. and PETERSON, D. B., *J. Phys. Chem.*, **69**, 1056 (1965)
171. AUSLOOS, P., REBBERT, R. E., SIECK, L. W. and TIERNAN, T. V., *J. Am. Chem. Soc.*, **94**, 8939 (1972)
172. SEARLES, S. K., SIECK, L. W. and AUSLOOS, P., *J. Chem. Phys.*, **53**, 849 (1970)
173. SHAW, H., MENCZEL, J. H. and TOBY, S., *J. Phys. Chem.*, **71**, 4180 (1967)
174. AUSLOOS, P. and LIAS, S. G., *J. Chem. Phys.*, **44**, 521 (1966)
175. AKIMOTO, H. and TANAKA, I., *J. Phys. Chem.*, **71**, 4135 (1967)
176. FESSENDEN, R. W., *J. Phys. Chem.*, **68**, 1508 (1964)
177. OKABE, H. and MCNESBY, R., *J. Chem. Phys.*, **37**, 1340 (1962)
178. BIRDWELL, B. F. and CRAWFORD, G. W., *J. Chem. Phys.*, **33**, 928 (1960)
179. YANG, K. P. and MANNO, P. JR., *J. Am. Chem. Soc.*, **81**, 3507 (1959)
180. FUJISAKI, N., SHIDA, S. and HATANO, Y., *J. Chem. Phys.*, **52**, 556 (1970)
181. HOLROYD, R. A. and KLEIN, G. W., *J. Phys. Chem.*, **67**, 2273 (1963)
182. AUSLOOS, P. and LIAS, S. G., *J. Chem. Phys.*, **44**, 521 (1966)
183. AUSLOOS, P. and LIAS, S. G., *Ber. Bunsenges. Phys. Chem.*, **72**, 187 (1968)
184. ZHITNEVA, G. P., KOZHEMYAKINA, L. F. and PSEZHECKII, S. Ya., *Khim. Vys. Energ.*, **8**, 181 (1974)
185. SIECK, L. W. and FUTRELL, J. H., *J. Phys. Chem.*, **69**, 900 (1965)
186. WEI, L. Y. and BONE, L. I., *J. Phys. Chem.*, **75**, 2272 (1971)
187. TANNO, K. and SHIDA, S., *Bull. Chem. Soc. Jpn.*, **42**, 2128 (1969)
188. KOOB, R. D. and KEVAN, L., *Trans. Faraday Soc.*, **64**, 706 (1968)
189. KOOB, R. D. and KEVAN, L., *Trans. Faraday Soc.*, **64**, 422 (1968)
190. See Ref. 12 in Ref. 139
191. LAMPE, P. W., *J. Phys. Chem.*, **61**, 1015 (1957)
192. TAYLOR, W. H., MORI, S. and BURTON, M., *J. Am. Chem. Soc.*, **82**, 5817 (1960)
193. DEWHURST, H. A., *J. Am. Chem. Soc.*, **83**, 1050 (1961)
194. WIDMER, H. and GÄUMANN, T., *Helv. Chim. Acta*, **46**, 2766 (1963)
195. YANG, K., *J. Am. Chem. Soc.*, **84**, 719 (1962)
196. DIXON, P. S., STEFANI, A. P. and SZWARCZ, M., *J. Am. Chem. Soc.*, **85**, 2551 (1963)
197. SIECK, L. W., BLOCHER, N. K. and FUTRELL, J. H., *J. Phys. Chem.*, **69**, 888 (1965)
198. see DODELET, J. P., *Can. J. Chem.*, **55**, 2050 (1977) and the references cited therein
199. BONE, L. I., SIECK, L. W. and FUTRELL, J. H., *J. Chem. Phys.*, **44**, 3667 (1966)
200. YAMAGUCHI, Y. and NISHIKAWA, M., *J. Chem. Phys.*, **59**, 1298 (1973)
201. ROBINSON, M. G., FUOCHI, P. G. and FREEMAN, G. R., *Can. J. Chem.*, **49**, 3657 (1971)
202. ROBINSON, M. G. and FREEMAN, G. R., *J. Chem. Phys.*, **55**, 5644 (1971)
203. LUCCHESI, P. J., TARMY, B. L., LONG, R. B., BAEDER, D. L. and LINGWELL, J. P. *Ind. Eng. Chem.*, **50**, 879 (1958)
204. POLAK, L. S., ENDUSKIN, P. N., UGLEV, V. N. and VOLODIN, N. L., *Khim. Vys. Energ.*, **3**, 184 (1969)
205. BONE, L. I., SIECK, L. W. and FUTRELL, J. H., in *The Chemistry of Ionisation and Excitation* (Eds JOHNSON, G. R. A. and SCHOLES, G.), Taylor and Francis, London, 1967
206. LIAS, S. G., REBBERT, R. E. and AUSLOOS, P., *J. Chem. Phys.*, **52**, 773 (1970)
207. MUNSON, M. S. B., *J. Phys. Chem.*, **71**, 3966 (1967)
208. GAWLOWSKI, J., NIEDZIELSKI, J. and HERMAN, J. A., *Can. J. Chem.*, **55**, 463 (1977)
209. MUNSON, M. S. B., *J. Am. Chem. Soc.*, **89**, 1771 (1967)
210. RYAN, K. R. and FUTRELL, J. H., *J. Chem. Phys.*, **42**, 819 (1965)
211. BONE, L. I. and FUTRELL, J. H., *J. Chem. Phys.*, **46**, 4084 (1967)
212. SIECK, L. W. and FUTRELL, J. H., *J. Chem. Phys.*, **45**, 560 (1966)
213. SKELL, P. S. and MAXWELL, R. J., *J. Am. Chem. Soc.*, **84**, 3963 (1962)
214. AUSLOOS, P. and LIAS, S. G., *J. Chem. Phys.*, **43**, 127 (1965)
215. ABRAMSON, F. P. and FUTRELL, J. H., *J. Phys. Chem.*, **71**, 1233 (1967)
216. KISSIAKOWSKY, G. B. and MICHAEL, J. V., *J. Chem. Phys.*, **40**, 1447 (1964)

217. FUJISAKI, N., WADA, T., SHIDA, S. and HATANO, Y., *J. Phys. Chem.*, **77**, 755 (1973)
218. GAWLOWSKI, J., HERMAN, J. A. and GAGNON, P., *Can. J. Chem.*, **53**, 1348 (1975)
219. MIYAZAKI, T. and SHIDA, S., *Bull. Chem. Soc. Jpn.*, **38**, 716 (1965)
220. MIYAZAKI, T., *J. Phys. Chem.*, **71**, 4282 (1967)
221. MIYAZAKI, T., ARAI, S., SHIDA, S. and SUNOHARA, S., *Bull. Chem. Soc. Jpn.*, **37**, 1352 (1964)
222. ROSENSTOCK, H. M., WALLENSTEIN, M. B., WAHRHAPTIG, A. C. and EYRING, H., *Proc. Natl. Acad. Sci. U. S.*, **38**, 667 (1952)
223. STEVENSON, D. P., *Radiat. Res.*, **10**, 610 (1959)
224. LIAS, S. G. and AUSLOOS, P., *J. Chem. Phys.*, **43**, 2748 (1965)
225. LIAS, S. G., REBBERT, R. E. and AUSLOOS, P., *J. Am. Chem. Soc.*, **92**, 6430 (1970)
226. VESTAL, M. L., in *Fundamental Processes in Radiation Chemistry* (Ed. AUSLOOS, P.) Wiley, New York, 1968
227. LIARDON, R. and GÄUMANN, T., *Helv. Chim. Acta*, **52**, 528 (1969)
228. LIARDON, R. and GÄUMANN, T., *Helv. Chim. Acta*, **52**, 1042 (1969)
229. AUSLOOS, P., REBBERT, R. E. and LIAS, S. G., *J. Am. Chem. Soc.*, **90**, 5031 (1968)
230. FUJISAKI, N., FUJIMOTO, I. and HATANO, Y., *J. Phys. Chem.*, **76**, 1260 (1972)
231. AUSLOOS, P., *Progress in Reaction Kinetics*, **5**, 113 (1969)
232. AUSLOOS, P. and LIAS, S. G., *J. Chem. Phys.*, **45**, 524 (1966)
233. AUSLOOS, P., SCALA, A. A. and LIAS, S. G., *J. Am. Chem. Soc.*, **89**, 3677 (1967)
234. SHIDA, S., FUJISAKI, N. and HATANO, Y., *J. Chem. Phys.*, **49**, 4571 (1968)
235. MIYAZAKI, T. and SHIDA, S., *Bull. Chem. Soc. Jpn.*, **38**, 2114 (1965)
236. STONE, A., QUIRT, A. R. and MILLER, O. A., *Can. J. Chem.*, **44**, 1175 (1966)
237. HIRAOKA, K. and KEBARLE, P., *J. Am. Chem. Soc.*, **97**, 4179 (1975)
238. AUSLOOS, P. and LIAS, S. G., *Ber. Bunsenges.*, **72**, 187 (1968)
239. FUJISAKI, N., SHIDA, S., HATANO, Y. and TANNO, K., *J. Phys. Chem.*, **75**, 2854 (1971)
240. DEWHURST, H. A., *J. Phys. Chem.*, **61**, 1466 (1957)
241. HOLROYD, R. A. and KLEIN, G. W., *J. Am. Chem. Soc.*, **84**, 4000 (1962)
242. HIRAOKA, K. and KEBARLE, P., *Can. J. Chem.*, **53**, 970 (1975)
243. McCAULEY, C. E. and SCHULER, R. H., *J. Am. Chem. Soc.*, **79**, 4008 (1957)
244. HASE, W. C. and SIMONS, J. W., *J. Chem. Phys.*, **54**, 1277 (1971)
245. WEXLER, S., LIFSHITZ, A. and QUATROCHI, A., in *Ion Molecule Reactions in the Gas Phase*, *Adv. Chem. Ser.*, **58**, Am. Chem. Soc., Washington, 1966
246. TANNO, K., SHIDA, S. and MIYAZAKI, T., *J. Phys. Chem.*, **72**, 3496 (1968)
247. BORKOWSKI, R. P. and AUSLOOS, P., *J. Chem. Phys.*, **38**, 36 (1962)
248. MIYAZAKI, T., YAMADA, T., WAKAYAMA, T., FUEKI, K. and KURI, Z., *Bull. Chem. Soc. Jpn.*, **44**, 934 (1971)
249. WAKAYAMA, T., MIYAZAKI, T., FUEKI, K. and KURI, Z., *Bull. Chem. Soc. Jpn.*, **42**, 1164 (1969)
250. WAKAYAMA, T., KIMURA, T., MIYAZAKI, T., FUEKI, K. and KURI, Z., *Bull. Chem. Soc. Jpn.*, **43**, 1017 (1970)
251. MIYAZAKI, T., WAKAYAMA, T., FUEKI, K. and KURI, Z., *Bull. Chem. Soc. Jpn.*, **42**, 2086 (1969)
252. WAKAYAMA, T., MIYAZAKI, T., FUEKI, K. and KURI, Z., *J. Phys. Chem.*, **74**, 3854 (1970)
253. SAITAKE, Y., WAKAYAMA, T., KIMURA, T., FUEKI, K. and KURI, Z., *Bull. Chem. Soc. Jpn.*, **44**, 301 (1971)
254. KOLANOVSKY, U. A. and POLAK, L. S., *Dokl.*, *Akad. Nauk SSSR, Ser. Khim.* **135**, 361 (1960)
255. FIELD, F. H., in *Ion Molecule Reactions* (Ed. FRANKLIN J. L.), Plenum Press, N. Y. 1972
256. HIRAOKA, K. and KEBARLE, P., *Can. J. Chem.*, **54**, 1739 (1976)
257. LAWSON, G., BONNER, R. F., MATHER, R. E., TODD, J. F. J. and MARCH, R. E., *J. Chem. Soc. Faraday Trans. I*, **72**, 545 (1976), and the refs cited
258. WADA, T., SHIDA, S. and HATANO, Y., *J. Phys. Chem.*, **79**, 561 (1975)
259. NOVOSELTSEV, A. M., KOMAROV, V. W., POTAPOV, V. K. and KUPRIYANOV, S. E., *Khym. Vys. Energ.*, **9**, 558 (1975)
260. MATSUOKA, S., TAMURA, T., OSHIMA, K. and OSHIMA, A., *Can. J. Chem.*, **92**, 92 (1975)

261. BABA, M. and FUEKI, K., *Bull. Chem. Soc. Jpn.*, **48**, 2240 (1975)
262. BORKOWSKI, R. P. and AUSLOOS, P., *J. Chem. Phys.*, **40**, 1138 (1964)
263. DODELET, J.-P. and FREEMAN, G. R., *Can. J. Chem.*, **50**, 2667 (1972)
264. LIAS, S. G. and AUSLOOS, P., *J. Chem. Phys.*, **48**, 392 (1968)
265. CLAES, P. and TILQUIN, B., *Proc. 4th Tihany Symp. Rad. Chem.* (Eds HEDVIG, P. and SCHILLER, R.), Akadémiai Kiadó, Budapest 1977
266. AUSLOOS, P. and LIAS, S. G., *Actions Chim. Biol. Radiations*, **11**, 1 (1967)
267. WAKAYAMA, T., MIYAZAKI, T., FUEKI, K. and KURI, Z., *Bull. Chem. Soc. Jpn.*, **43**, 3761 (1970)
268. HARDWICK, T. J., *J. Phys. Chem.*, **64**, 1632 (1960)
269. WAGNER, C. D., *J. Phys. Chem.*, **64**, 231 (1960)
270. DeVRIES, A. E. and ALLEN, A. O., *J. Phys. Chem.*, **63**, 879 (1959)
271. KOCH, R. O., HOUTMAN, J. P. W. and CRAMER, W. A., *J. Am. Chem. Soc.*, **90**, 3326 (1968)
272. BISHOP, W. P. and FIRESTONE, R. F., *J. Phys. Chem.*, **74**, 2274 (1970)
273. PANCINI, E., SANTORO, V. and SPADACCINI, G., *Int. J. Radiat. Phys. Chem.*, **2**, 147 (1970)
274. SCHULER, R. H. and KUNTZ, R. R., *J. Phys. Chem.*, **67**, 1004 (1963)
275. HOROWITZ, A. and RAJBENBACH, L. A., *J. Chem. Phys.*, **48**, 4278 (1968)
276. HARDWICK, T. J., *J. Phys. Chem.*, **66**, 1611 (1962)
277. BRYL-SANDELEWSKA, T. and BROSZKIEWICZ, R. K., *Radiat. Phys. Chem.*, **10**, 309 (1977)
278. SCALA, A., LIAS, S. G. and AUSLOOS, P., *J. Am. Chem. Soc.*, **88**, 5701 (1966)
279. DYNE, P. J. and JENKINSON, W. M., *Can. J. Chem.*, **39**, 2163 (1961)
280. WINCEL, H., *INR Report No. 1338 (XVII) C*, Warsaw, 1971
281. ROBINSON, M. G. and FREEMAN, G. R., *Can. J. Chem.*, **52**, 440 (1974)
282. CLAES, P. and RZAD, S., *Bull. Soc. Sci. Belges.*, **73**, 689 (1964)
283. BIENFAIT, C. and CLAES, P., *Int. J. Radiat. Phys. Chem.*, **2**, 101 (1970)
284. BIENFAIT, C., CEULEMANS, J. and CLAES, P., *Advan. Chem. Ser.*, **82**, 300 (1968)
285. DYNE, P. J., *Can. J. Chem.*, **43**, 1080 (1965)
286. RAJBENBACH, L. A. and KALDOR, U., *J. Chem. Phys.*, **47**, 242 (1967)
287. TOMA, S. Z. and HAMILL, W. H., *J. Am. Chem. Soc.*, **86**, 1477 (1964)
288. SHINSAKA, K. and FREEMAN, G. R., *Can. J. Chem.*, **52**, 3495 (1974)
289. SCHILLER, R., in *Radiation Research, Biomedical, Chemical, and Physical Perspectives* (Eds NYGAARD, O. F., ADLER, H. I. and SINCLAIR, W. K.), Academic Press, New York, 1975
290. SAUER, M. C. and MANI, I., *J. Phys. Chem.*, **72**, 3856 (1968)
291. REITBERGER, T., *Radiochem. Radioanal. Lett.*, **6**, 349 (1970)
292. BURGGRAF, L. W. and FIRESTONE, R. F., *J. Phys. Chem.*, **78**, 508 (1974)
293. NOYES, R. M., *Progr. Reaction Kinet.*, **1**, 126 (1961)
294. AVDONINA, E. N., *Proc. 4th 'Tihany' Symp. Rad. Chem.* (Eds HEDVIG, P. and SCHILLER, R.), Akadémiai Kiadó, Budapest, 1977, p. 83
295. ALLAERT, J., CLAES, P. and TILQUIN, B., *Int. J. Radiat. Phys. Chem.*, **8**, 387 (1976)
296. WIDMER, H. and GÄUMANN, T., *Helv. Chim. Acta*, **46**, 944, 2766, 2780 (1964
297. THYRION, F., DODELET, J. P., FAUQUENOIT, C. and CLAES, P., *J. Chem. Phys.*, **65**, 227 (1968)
298. URCH, D. S. and WELCH, M. J., *Trans. Faraday Soc.*, **64**, 1547 (1968)
299. TILQUIN, B., ALLAERT, J. and CLAES, P., *Symposium on Mechanisms of Hydrocarbon Reactions* (Eds MÁRTA, F. and KALLÓ, D.), Akadémiai Kiadó, Budapest 1975, p. 639
300. HOLROYD, R. A., *J. Am. Chem. Soc.*, **91**, 2208 (1969)
301. FLESCH, G. D. and SVEC, H. J., *J. Chem. Soc., Faraday Trans. 2.*, **69**, 1187 (1973)
302. HOLROYD, R. A., in *Fundamental Processes in Radiation Chemistry* (Ed. AUSLOOS, P.), Wiley, New York, 1968
303. DORFMAN, L. M., MATHESON, M. S., *Progress in Reaction Kinetics* (Ed. PORTER, G.), Pergamon Press, Oxford, 1965, Vol. **3**.
304. DAUGOKIN, J., *J. Chem. Phys.*, **59**, 1207 (1962)
305. WEBER, E. N., FORSYTH, P. F. and SCHULER, R. H., *Radiat. Res.*, **3**, 68 (1955
306. HOLROYD, R. A. and KLEIN, G. W., *J. Am. Chem. Soc.*, **87**, 4983 (1965)

307. SCALA, A. A. and AUSLOOS, P., *J. Chem. Phys.*, **47**, 5129 (1967)
308. MIYAZAKI, T. and HIRAYAMA, T., *J. Phys. Chem.*, **79**, 566 (1975)
309. HATANO, Y., ITO, K. and TAKAR, S., *Int. J. Radiat. Phys. Chem.*, **7**, 39 (1975)
310. PICHOOZHKIN, V. I. and ANTONOVA, E. A., *Khim. Vys. Energ.*, **9**, 546 (1975)
311. SCALA, A. A. and AUSLOOS, P., *J. Chem. Phys.*, **45**, 847 (1966)
312. HAMAZHIMA, M., REDDY, M. P. and BURTON, M., *J. Phys. Chem.*, **62**, 246 (1958)
313. PICHOOZHKIN, V. I. and ANTONOVA, E. A., *Vest., Mosk. Univ., Khim.*, **16**, 302 (1977)
314. HOLROYD, R. A., *J. Phys. Chem.*, **65**, 1352 (1961)
315. FELD, M. and SZWARC, M., *J. Am. Chem. Soc.*, **82**, 3791 (1960)
316. FIELD, E. H. and LAMPE, F. W., *J. Am. Chem. Soc.*, **80**, 5587 (1958)
317. KOCHE, E., SAILE, V. and SCHWARTNER, N., *Chem. Phys. Lett.*, **33**, 322 (1975)
318. REBBERT, R. E. and AUSLOOS, P., *J. Res. Natl. Bur. Standards*, **76A**, 329 (1972)
319. DERAI, R., NECTOUX, P. and DANON, J., *J. Phys. Chem.*, **80**, 1664 (1976)
320. MIYAZAKI, T. and SHIDA, S., *Bull. Chem. Soc. Jpn.*, **39**, 2344 (1966)
321. AUSLOOS, P. and LIAS, S. G., *J. Am. Chem. Soc.*, **92**, 5037 (1970)
322. SHIDA, S., YUGETA, R. and SATO, S., *Bull. Chem. Soc. Jpn.*, **43**, 2758 (1970)
323. TANNO, K., MIYAZAKI, T., SHINSAKA, K. and SHIDA, S., *Bull. Chem. Soc. Jpn.*, **71**, 4290 (1967)
324. MARX, R., McCLAIR, G. and WALLART, M., *J. Chim. Phys. Physicochim. Biol.*, **68**, 1680 (1971)
325. ANTONOVA, E. A. and PICHOOZHKIN, V. I., *Khim. Vys. Energ.*, **11**, 242 (1977)
326. STONE, J. A. and MATSUSHITA, G., *Can. J. Chem.*, **49**, 3287 (1971)
327. COLLIN, G. J. and AUSLOOS, P., *J. Am. Chem. Soc.*, **93**, 1336 (1971)
328. HELLNER, L. and SIECK, L. W., *J. Res. Natl. Bur. Standards*, **75A**, 487 (1971)
329. ALLEN, A. O., GANGWER, T. E. and HOLROYD, R. A., *J. Phys. Chem.*, **79**, 25 (1975)
330. LIAS, S. G., REBBERT, R. E. and AUSLOOS, P., *J. Chem. Phys.*, **57**, 2080 (1972)
331. BALAKIN, A. A. and YAKOVLEV, B. S., *Khim. Vys. Energ.*, **9**, 29, 69 (1975)
332. TAYLOR, J. E., HUTCHING, D. A. and FRECH, K. J., *J. Am. Chem. Soc.*, **91**, 2215 (1969)
333. BRADLEY, J. N. and WEST, K. O., *J. Chem. Soc., Faraday Trans I*, **72**, 1 (1976)
334. HOGEVEEN, H. and GAASBEEK, C. J., *Recl. Trav. Chim. Pays-Bas.*, **89**, 857 (1970)
335. HOLROYD, R. A., in *Aspects of Hydrocarbon Radiolysis* (Eds GÄUMANN, T. and HOIGNÉ, J.), Academic Press, London, 1968, p. 1
336. WIDMER, H. and GÄUMANN, T., *Helv. Chim. Acta*, **46**, 944 (1963)
337. GOWENLOCK, B. G., JOHNSON, C. A. F., *J. Chem. Soc., Perkin Trans. II.*, 1150 (1972)
338. WAGNER, C. D. and GEYMER, D. O., *J. Phys. Chem.*, **71**, 1551 (1967)
339. TILQUIN, B., LOUVEAUX, M., BOMBAERT, C. and CLAES, P., *Radiat. Eff.*, **32**, 37 (1977)
340. KEVAN, L. and LIBBY, W. F., *J. Chem. Phys.*, **39**, 1288 (1963)
341. BURNS, W. G., and BARKER, R., *Progr. Reaction Kinetics*, **3**, 303 (1965)
342. BURNS, W. G., BARKER, R., LEWIS, B. A. and YORK, E. J., *Trans. Faraday Soc.*, **61**, 2691 (1965)
343. BURNS, W. G. and BARKER, R., in *Aspects of Hydrocarbon Radiolysis* (Eds GÄUMANN, T. and HOIGNÉ, J.), Academic Press, London, 1968
344. DEWHURST, H. A. and SCHULER, R. H., *J. Am. Chem. Soc.*, **81**, 3210 (1959)
345. DEWHURST, H. A. and WINSLOW, E. H., *J. Chem. Phys.*, **26**, 959 (1957)
346. BACK, R. A. and MILLER, N., *Trans. Faraday Soc.*, **55**, 911 (1959)
347. FUTRELL, J. H., *J. Am. Chem. Soc.*, **81**, 5921 (1959)
348. LOSSING, F. P. and MACCOLL, A., *Can. J. Chem*, **54**, 990 (1976)
349. MEISSNER, G. and HENGLEIN, A., *Ber. Bunsenges. Physic. Chem.*, **69**, 264 (1965)
350. HOLROYD, R. A., *J. Phys. Chem.*, **70**, 1341 (1966)
351. PERNER, D. and SCHULER, R. H., *J. Phys. Chem.*, **70**, 317 (1966)
352. HARDWICK, T. J., *J. Phys. Chem.*, **65**, 101 (1961)
353. MAKAROV, V. I., POLAK, L. S., CHERNIAK, N. YA. and SCHERBAKOVÁ, A. S., *Neftekhim.*, **6**, 58 (1966)
354. KILIN, S. F. and ROZMAN, I. M., *Opt. Spectroscopy.*, **17**, 380 (1964)
355. LAOR, U. and WEINREB, A., *J. Chem. Phys.*, **43**, 1565 (1965)
356. BURTON, M., *Strahlentherapie.*, **51**, Suppl., 1, (1962)
357. DEWHURST, H. A., *J. Phys. Chem.*, **62**, 15 (1958)

358. HARDWICK, T. J., *J. Phys. Chem.*, **64**, 1623 (1960)
359. ARSHAKUNI, A. A., PICHUZHKIN, V. I. and SARAEVA, V. V., *Khim. Vys. Energ.*, **7**, 465 (1973)
360. ISILDAR, M. and SCHULER, R. H., *Radiat. Phys. Chem.*, **11**, 11 (1978)
361. SHINSAKA, K. and SHIDA, S., *Bull. Chem. Soc. Jpn.*, **43**, 3728 (1970)
362. SHINSAKA, K. and SHIDA, S., *Bull. Chem. Soc. Jpn.*, **40**, 2796 (1967)
363. AVDONINA, E. N. and MAKAROV, V. I., *Int. J. Radiat. Phys. Chem.*, **4**, 259 (1972)
364. AVDONINA, E. N., *Khim. Vys. Energ.*, **5**, 236 (1971)
365. HOURIET, R., PARISOD, G. and GÄUMANN, T., *J. Am. Chem. Soc.*, **99**, 3599 (1977)
366. BRONGERSMA, H. H. and OOSTERHOFF, L. J., *Chem. Phys. Lett.*, **3**, 437 (1969)
367. WIDMER, H. and GÄUMANN, T., *Helv. Chim. Acta*, 2766 (1964)
368. HUANG, T. and HAMILL, W. H., *J. Phys. Chem.*, **78**, 2081 (1974)
369. WIDMER, H. and GÄUMANN, T., *Helv. Chim. Acta*, 2780 (1964)
370. SAUER, M. C. JR. and MULAC, W. A., *J. Phys. Chem.*, **78**, 22 (1974)
371. HENRY, M. S. and HELMAN, W. P., *J. Chem. Phys.*, **56**, 5734 (1972)
372. MENGER, A. and GÄUMANN, T., *Helv. Chim. Acta*, **52**, 2129 (1969)
373. MENGER, A. and GÄUMANN, T., *Helv. Chim. Acta*, **52**, 2477 (1969)
374. LUND, A., *J. Phys. Chem.*, **76**, 1411 (1972)
375. ZAJTSEV, V. M., SEROVA, V. A. and TIKHONOV, V. I., *Khim. Vys. Energ.*, **7**, 174 (1973)
376. WILLIAMS, T. Ff., *Trans. Faraday Soc.*, **57**, 755 (1961)
377. WEXLER, S. and POBO, L. G., *J. Am. Chem. Soc.*, **91**, 7233 (1969)
378. PITCHOOZHKIN, V. I., YAMAZAKI, H. and SHIDA, S., *Bull. Chem. Soc. Jpn.*, **46**, 67 (1973)
379. AVDONINA, E. N., *Khym. Vys. Energ.*, **11**, 327 (1977)
380. POLYAKOVA, G. N., FIZGEER, B. M. and ERKO, V. F., *Khim. Vys. Energ.*, **9**, 367 (1975)
381. POLYAKOVA, G. N., FIZGEER, B. M. and RANYUK, A. I., *Khim. Vys. Energ.*, **9**, 369 (1975)
382. HORIE, T., NAGURA, T. and OTSUKA, M., *J. Phys. Soc. Jpn.*, **15**, 641 (1960)
383. POLYAKOVA, G. N. and FIZGEER, B. M., *Khim. Vys. Energ.*, **8**, 301 (1974)
384. MAINWARING, D. D. and WILLARD, J. E., *J. Phys. Chem.*, **77**, 2864 (1973)
385. DEWHURST, H. A., *J. Am. Chem. Soc.*, **80**, 5607 (1958)
386. POLYAKOVA, G. N. and FIZGEER, B. M., *Khim. Vys. Energ.*, **8**, 387 (1974)
387. HENDERSON, D. and WILLARD, J. E., *J. Am. Chem. Soc.*, **91**, 3014 (1969)
388. APARINA, E. V., KIR'YAKOV, N. V., MARKIN, M. I. and TAL'ROZE, V. L., *Khim. Vys. Energ.*, **9**, 349 (1975)
389. BANSAL, K. M. and RZAD, S. J., *J. Phys. Chem.*, **74**, 3486 (1970)
390. TIMM, D. and WILLARD, J. E., *J. Phys. Chem.*, **73**, 2403 (1969)
391. FÖLDIÁK, G. and GYÖRGY, I., *Acta Chim. Acad. Sci. Hung.*, **80**, 385 (1974)
392. ICHIKAWA, T. and OHTA, N., *J. Phys. Chem.*, **81**, 560 (1977)
393. HENDERSON, D. J. and WILLARD, J. E., *J. Am. Chem. Soc.*, **91**, 3014 (1969)
394. KIMURA, T., FUKAYA, M., HADA, M., WAKAYAMA, T., FUEKI, K. and KURI, Z., *Bull. Chem. Soc. Jpn.*, **43**, 3400 (1970)
395. IWASAKI, M., TORIYAMA, K., MUTO, H. and NUNONE, K., *J. Chem. Phys.*, **65**, 596 (1976)
396. KANIC, S. W., LINDER, R. E. and LING, A. C., *J. Chem. Soc. A.*, 297 (1971)
397. PICHOOZHKIN, V. I. and ARSHAKUNI, A. A., *Khim. Vys. Energ.*, **9**, 207 (1975)
398. FRANKLIN, J. L. and LAMPE, F. W., *Trans. Faraday Soc.*, **57**, 1449 (1961)
399. MIYAZAKI, T., SAITAKE, Y., KURI, Z. and SAKAI, S., *Bull. Chem. Soc. Jpn.*, **46**, 70 (1973)
400. SIMONS, J. W. and CURRY, R., *Chem. Phys. Lett.*, **38**, 171 (1976); ZABRONSKY, V. P. and CARR, R. W. Jr., *J. Am. Chem. Soc.*, **98**, 1130 (1976)
401. HOLROYD, R. A., *J. Am. Chem. Soc.*, **88**, 5381 (1966)
402. SIMONS, J. W. and CURRY, R., *Chem. Phys. Lett.*, **38**, 171 (1976)
403. ZABRONSKY, V. P. and CARR, R. W. Jr., *J. Am. Chem. Soc.*, **98**, 1130 (1976)
404. KUNTZ, R. R. and MAINS, G., *J. Am. Chem. Soc.*, **85**, 2219 (1963)
405. ISILDAR, M., *Studies of Electron Scavenging and Fragmentation Processes in Radiolysis of Hydrocarbons*, Ph. D. Thesis, Radiation Research Laboratories, Mellon Institute of Sciences, Carnegie-Mellon University, Pittsburgh, 1976

406. BEDNAR, I., PRASIL, Z. and VACEK, K., *Coll. Czech. Chem. Commun.*, **30**, 2693 (1965)
407. FESSENDEN, R. W. and SCHULER, R. H., *J. Chem. Phys.*, **39**, 2147 (1963)
408. MANION, J. P. and BURTON, M., *J. Phys. Chem.*, **56**, 560 (1952)
409. CASTELLO, G., GRANDI, F. and MUNARI, S., *Radiat. Res.*, **45**, 399 (1971)
410. RUCINSKA-STAROSTA, A., KAROLCZAK, S. and KROH, J., *Radiochem. Radioanal. Lett.*, **23**, 205 (1975)
411. TANNO, K., MIYAZAKI, T., SHINSAKA, K. and SHIDA, S., *J. Phys. Chem.*, **71**, 4290 (1967)
412. TANNO, K. and SHIDA, S., *Bull. Chem. Soc. Jpn.*, **42**, 2128 (1969)
413. TANNO, K., SHIDA, S. and MIYAZAKI, T., *J. Phys. Chem.*, **72**, 3496 (1968)
414. RAPPOPORT, S. and GÄUMANN, T., *Helv. Chim. Acta*, **56**, 531 (1973)
415. HONIG, R. E. and SHEPPARD, C.-W., *J. Phys. Chem.*, **50**, 119 (1946)
416. SCHOEPFLE, C .S. and FELLOWS, C .H., *Ind. Eng. Chem.*, **23**, 1396 (1931)
417. BREGER, I. A., *J. Phys. Chem.*, **52**, 551 (1948)
418. MELEKHONOVA, I. I., ROMANTSEV, M. F. and SARAEVA, V. V., *Khim. Vys. Energ.*, **10**, 179 (1976)
419. ARMENANTE, M., CALICCHIO, C. F., SANTORO, V. and SPINELLI, N., *Int. J. Radiat. Phys. Chem.*, **6**, 177 (1974)
420. MURATBEKOV, M. B. and SARAEVA, V. V., *Vest. Mosk. Univ.*, *Khim.*, **6**, 68 (1969)
421. PICHUZKHIN, V. I., SARAEVA, V. V. and BAH, N. A., *Khim. Vys. Energ.*, **2**, 151 (1968)
422. GILLBRO, T., KINELL, P.-D. and LUND, A., *J. Phys. Chem.*, **73**, 4167 (1969)
423. BABA, M. and FUEKI, K., *Bull. Chem. Soc. Jpn.*, **48**, 3039 (1975)
424. LUND, A. and KEVAN, L., *J. Phys. Chem.*, **77**, 2180 (1973)
425. GILLBRO, T. and LUND, A., *J. Chem. Phys.*, **5**, 283 (1974)
426. GILLBRO, T. and LUND, A., *J. Chem. Phys.*, **61**, 1469 (1974)
427. ANTONOVA, E. A., PICHUZHKIN, V. I. and BAH, N. A., *Vestn. Mosk. Univ. Khim.*, **19**, 45 (1978)
428. ANTONOVA, E. A., PICHUZHKIN, V. I. and BAH, N. A., *Khim. Vys. Energ.*, **11**, 242 (1977)
429. GÄUMANN, T. and RUF, A., *Symp. Mech. Hydrocarbon Reactions* (Eds MÁRTA, F. and KALLÓ, D.), Akadémiai Kiadó, Budapest, 1975
430. AVDONINA, E. N., *Radiat. Eff.*, **20**, 245 (1973)
431. ZIMINA, T. I., PICHUZHKIN, V. I., CHUDAKOVA, T. A. and BAH, N. A. *Khim. Vys. Energ.*, **7**, 154 (1973)
432. MURATBEKOV, M. B. and SARAEVA, V. V., *Vestn. Mosk. Univ.*, *Khim.*, **6**, 73 (1969)
433. FRANKEVICH, E. L. and YAKOVLEV, B. S., *Izv. Akad. Nakl. SSSR*, *Khim.*, **12**, 226 (1965)
434. GABSATAROVA, S. A. and KOBAKCHI, A. U., *Khim. Vys. Energ.*, **3**, 126 (1969)
435. RAPPOPORT, S. and GÄUMANN, T., *Helv. Chim. Acta*, **57**, 2861 (1973)
436. MURATBEKOV, M. B., SARAEVA, V. V. and ZATONSKII, S. V., *Khim. Vys. Energ.* **5**, 134 (1971)
437. KNIGHT, J. A. and LEWIS, C. T., *Radiat. Res.*, **23**, 319 (1964)
438. CASTELLO, G., GRANDI, F. and MUNARI, S., *Radiat. Res.*, **62**, 323 (1975)
439. POLAK, L. C., CERNIAK, N. Yo. and SERBAKOV, A. S., *Khim. Vys. Energ.*, **2**, 317 (1968)
440. SAGERT, N. H. and BLAIR, A. S., *Can. J. Chem.*, **45**, 1351 (1967)
441. RIGGIN, M. and DUNBAR, R.C., *Chem. Phys. Lett.*, **31**, 539 (1975)
442. CARRINGTON, A., MILVERTON, D. R. J. and SARRE, P. J., *Mol. Phys.*, **32**, 297 (1976)
443. BONDJOUK, P. and KOOB, R. D., *J. Am. Chem. Soc.*, **97**, 6595 (1975)
444. TANNO, K., MIYAZAKI, T., SHINSAKA, S. and SHIDA, S., *J. Phys. Chem.*, **71**, 4290 (1967)
445. CHUPKA, W. A ., *J. Chem. Phys.* **48**, 3227 (1968)
446. KUDO, T., *J. Phys. Chem.*, **71**, 3681 (1967)
447. KNIGHT, J. A., McDANIEL, R. L. and SICILIO, F., *J. Phys. Chem.*, **67**, 2273 (1963)
448. GOWENLOCK, B. G., JOHNSON, C. A. F. and SENOGLES, E., *J. Chem. Soc. Perkin Trans. II.*, 1386 (1972)
449. McGILVERY, D.C., MORRISON, J.D. and SMITH, D.L., *J. Chem. Phys.*, **68**, 3949 (1978)

450. RZAD, S. J. and BANSAL, K. M., *J. Phys. Chem.*, **76**, 2374 (1972)
451. BANSAL, K. M. and RZAD, S. J., *J. Phys. Chem.*, **76**, 2381 (1972)
452. KUDO, T. and SHIDA, S., *J. Phys. Chem.*, **67**, 2871 (1963)
453. AVDONINA, E. N., *Khim. Vys. Energ.*, **11**, 327 (1977)
454. AVDONINA, E. N. and LEBEDIEV, V. A., *Khim. Vys. Energ.*, **11**, 376 (1977)
455. PICHUZHKIN, V. I., MURATBEKOV, M. B. and BAH, N. A., *Khim. Vys. Energ.*, **4**, 317 (1970)
456. PICHUZHKIN, V. I., MURATBEKOV, M. B. and SARAEVA, V. V., *Vestn. Mosk. Univ., Khim.*, **30**, 13 (1970)
457. KIM, I. P., KAPLAN, A. M., MIKHAILOV, A. I., BARKALOV, I. M. and GOLDANSKII, B. I., *Dokl. Akad. Nauk, SSSR, Ser. Khim.*, **194**, 359 (1970)
458. MELENKHONOVA, I. I., ROMANTSEVA, M. F. and SARAEVA, V. V., *Vestn. Mosk. Univ., Khim.*, 2, **18**, 489 (1977)
459. MIYAZAKI, T. and HIRAYAMA, T., *J. Phys. Chem.*, **79**, 566 (1975)
460. SHELBERG, W. E. and PESTANER, J. F., *Nature*, **200**, 754 (1963)
461. KNIGHT, J. A. and SICILIO, F., *Radiat. Res.*, **19**, 359 (1963)
462. SALOVEY, R. and FALCONER, W. E., *J. Phys. Chem.*, **69**, 2345 (1965)
463. FALCONER, W. E. and SALOVEY, R., *J. Chem. Phys.*, **49**, 3151 (1966)
464. FALCONER, W. E. and SALOVEY, R., *J. Chem. Phys.*, **46**, 387 (1967)
465. FALCONER, W. E., SALOVEY, R., DUNDER, W. A. and WALKER, L. G., *Radiat. Res.*, **47**, 41 (1971)
466. FALCONER, W. E., SUNDER, W. A. and WALKER, L. G., *Can. J. Chem.*, **49**, 3892 (1971)
467. MILLER, A. W., LAWTON, E. J. and BALEVIT, J. S., *J. Phys. Chem.*, **60**, 599 (1956)
468. IWASAKI, M., TORIYAMA, K., NUNONE, K., KUKAYA, M. and MUTO, H., *J. Phys. Chem.*, **81**, 1410 (1977)
469. FÖLDIÁK, G. and PAAL, Z., *Acta Chim. Acad. Sci. Hung.*, **42**, 421 (1964)
470. FÖLDIÁK, G., D. Sc. Thesis, Hungarian Academy of Sciences, Budapest, 1965
471. SALOVEY, R. and FALCONER, W. E., *J. Phys. Chem.*, **70**, 3203 (1966)
472. DYNE, P. J. and DENHARTOG, J., *Can. J. Chem.*, **40**, 1616 (1962)
473. GYÖRGY, I. and FÖLDIÁK, G., *Acta Chim. Acad. Sci. Hung.*, **81**, 455 (1974)
474. KUDO, T. and SHIDA, S., *J. Phys. Chem.*, **71**, 1971 (1967)
475. MANION, J. P. and BURTON, M., *J. Phys. Chem.*, **56**, 560 (1952)
476. HARDWICK, T. J., *J. Phys. Chem.*, **66**, 3132 (1962)
477. EKSTROM, A., and GARNETT, J. L., *J. Labelled Comp.*, **3**, 167 (1968)
478. HORVÁTH, Zs. and FÖLDIÁK, G., *Radiochem. Radioanal. Lett.*, **19**, 67 (1974)
479. STONE, J. A., *Chem. Commun.*, **24**, 1677 (1968)
480. CLAY, J. G. and SIDDIQI, A. A., *J. Chem. Soc. D.*, **21**, 1271 (1969)
481. KROH, J. and KAROLCZAK, S., *Radiat. Res. Rev.*, **1**, 411 (1969)
482. GILLBRO, T. and LUND, A., *Chem. Phys. Lett.*, **27**, 300 (1974)
483. AHMAD, S. R. and CHARLESBY, A., *Radiat. Phys. Chem.*, **11**, 29 (1978)
484. HOLROYD, R. A. and TAUCHERT, W., *J. Chem. Phys.*, **60**, 3715 (1974)
485. MINDAY, R. M., SCHMIDT, L. D. and DAVIS, H. T., *Phys. Rev. Lett.*, **26**, 360 (1971)
486. BECK, G. and THOMAS, J. K., *J. Chem. Phys.*, **57**, 3649 (1972)
487. NYIKOS, L. and SCHILLER, R., *Chem. Phys. Lett.*, **34**, 128 (1975)
488. WAKAYAMA, T., MIYAZAKI, T., FUEKI, K. and KURI, Z., *J. Phys. Chem.*, **77**, 2365 (1973)
489. FÖLDIÁK, G., CSERÉP, GY., GYÖRGY, I., HORVÁTH, Zs. and WOJNÁROVITS, L., *Hung. J. Ind. Chem.*, Suppl. 2, 277 (1974)
490. MIYAZAKI, T., KINUGAWA, K. and KASUGAI, J., *Radiat. Phys. Chem.*, **10**, 155 (1977)
491. FÖLDIÁK, G., GYÖRGY, I. and WOJNÁROVITS, L., *Int. J. Radiat. Phys. Chem.*, **8**, 575 (1976)
492. CLOW, R. P. and FUTRELL, J. H., *J. Am. Chem. Soc.*, **94**, 3748 (1972)
493. KINUGAWA, K. and MIYAZAKI, T., *Radiat. Phys. Chem.*, **10**, 341 (1977)
494. LORQUET, J. C., *Mol. Phys.*, **9**, 101 (1965)
495. HIRAYAMA, F. and LIPSKY, S., *J. Chem. Phys.*, **51**, 3616 (1969)
496. GYÖRGY, I., HORVÁTH, Zs., WOJNÁROVITS, L. and FÖLDIÁK, G., *Proc. 3rd Tihany Symp. Rad. Chem.* (Eds DOBÓ, J. and HEDVIG, P.), Akadémiai Kiadó, Budapest, 1972, p. 311
497. WOJNÁROVITS, L. and FÖLDIÁK, G., *Acta Chim. Acad. Sci. Hung.*, **67**, 221 (1971)

3. CYCLOALKANES

The literature dealing with the radiolysis of cycloalkanes shows a great variety. Some 500 publications have been devoted to investigations of cyclohexane, a figure approached only by the reports concerning hexane and benzene. Cyclopentane has received attention in fewer than 50 papers, and other cycloalkanes have been studied even less. The diversity and inhomogeneity of the literature are especially apparent when publications concerning primary processes, i.e. the physical stage of radiolysis, are considered. The predominance of studies on cyclohexane and to a much lesser extent on cyclopentane is most striking.

3.1. INTRODUCTORY REMARKS

Compared with the aliphatic alkanes, the cycloalkane series shows considerable changes in the physical and chemical properties of its members as a function of the carbon atom number of the ring [1–4]. For example, during thermal decomposition of cyclopropane and cyclobutane, $C-C$ bond rupture proceeds with a comparatively low activation energy (260–280 kJ mol^{-1}), whereas the activation energy of similar processes in the case of larger rings (e.g. methylcyclopentane and cyclohexane) is some 60 kJ mol^{-1} greater [5–7].

The reactivity of cyclic molecules is influenced chiefly by the ring strain, which depends largely on the carbon atom number of the ring, i.e. on the mass and size of the molecule [7, 8]. The magnitude of the ring strain is generally characterized by the so-called strain energy, or rather by its value normalized to one CH_2 unit. The strain energy, which has a varying magnitude and can be traced back to different sources, depending on the carbon atom number of the ring, is usually quoted in the literature as the excess energy content of the cyclic molecule compared with the analogous aliphatic molecule (Table 3.1) [2, 3].

Practically all the characteristic features of the molecule are, to a greater or lesser extent, influenced by the fact that the molecule has a cyclic structure. The excess energy of formation is usually considered to arise from the following interrelated effects [2, 4]:

1. The excess energy ensuing from the increase in $C-C$ bond distances;
2. The repulsion energy arising between the groups 1 and 4 of the $-\overset{1}{C}-\overset{2}{C}-\overset{3}{C}-\overset{4}{C}-$ carbon chain, being the function of the torsional angle between the 1–2 and 3–4 $C-C$ bonds (it has a minimum at 180°);

Table 3.1

Some thermodynamic data of cycloalkanes taken from refs [4, 7, 8]

Hydrocarbon	Strain energy	Strain energy referred to one CH_2-unit	Heats of dehydrogenation
	kJ mol^{-1}		
Cyclopropane	115.7	38.6	223.3
Cyclobutane	109.8	27.4	134.5
Cyclopentane	26.4	5.0	110.2
Cyclohexane	0.0	0.0	118.2
Cycloheptane	26.8	3.8	110.2
Cyclo-octane	41.5	5.2	97.6
Cyclononane	53.6	5.9	98.9
Cyclodecane	55.7	5.6	86.7
Cyclododecane	37.7	3.1	

3. The excess heat of formation arising from the distortions of C—C—C bonds angles;

4. The van der Waals repulsive interactions between hydrogen atoms bonded to different carbon atoms.

The strain energies of cyclopropane and cyclobutane molecules are chiefly related to the differences between the actual values of C—C—C bond angles and the 'normal' tetrahedral ones. Accordingly, the most favourable chemical reactions of these molecules result in a decrease of the considerable ring strain by ring opening. For example, thermal decomposition proceeds with relatively small activation energies.

The rupture of C—H bonds in cyclopropane and cyclobutane results in an increased strain in the ring of the monoradical formed in hydrogen atom elimination of about 12-25 kJ mol^{-1} [7-9].

Figure 3.1 displays the strain and steric energies of cycloalkanes. In the Figure the steric energies are shown in detail because reliable data are published in the literature concerning the components of this excess energy. The steric energies are defined somewhat differently from the strain energies [4]. The actual values of ring strain and steric energy differ by an additive value depending on the carbon atom number: steric energy $\approx$ strain energy $+4.3\,n$ (kJ mol^{-1}), where n is the number of carbon atoms in the ring.

The energies corresponding to angle distortions, torsions and van der Waals repulsive interactions contribute in about equal proportions to the ring strain energy of cyclopentane (Fig. 3.1) [4, 8]. However, the excess energy of cycloheptane, cyclo-octane, cyclononane and cyclodecane is mainly due to the repulsive interactions of hydrogen atoms resulting also in bond angle distortions and bond elongations [2, 4]. The repulsive interactions in these hydrocarbons weaken the C—H bonds, as demonstrated by the relatively low heats of dehydrogenation in their cycloalkane → cycloalkene transformations (Table 3.1). A change in the state of hybridization of a ring carbon atom from sp^3 to sp^2 leads to a lessening of the intra-

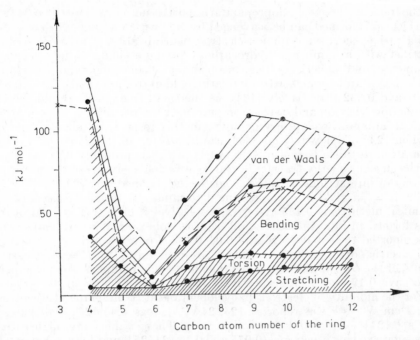

Fig. 3.1. Energetic data of cycloalkanes as a function of the carbon number in the
rings. Data taken from refs [4, 8].
———— Components of steric energy; — · — · — · — total steric energy;
— — — — — total strain energy

molecular repulsive interactions and, consequently, to a greater likelihood
that the molecule will assume an energetically more favourable conformation [1].

During the past decade a great amount of data has been published on the
primary processes of cycloalkane radiolysis, dealing with, for example, the
formation and decay of ion-pairs, the yields and the fate of secondary electrons brought about by irradiation. A value of $G = 4.4$ was reported for
the total yield of ions in the liquid-phase radiolysis of cyclopentane [10, 11].
This value is somewhat higher than $G_{ti} = 3.94$, estimated by taking into
account the energy needed for the formation of one ion-pair in the gas phase,
$W = 25.39$ eV [12]. The gas-phase W value of cyclohexane is 25.05 eV [13],
corresponding to a total ion-pair yield of $G_{ti} = 4.0$. In a number of studies
[14–16] published in 1968–70, Schuler et al. assumed a value of $G_{ti} = 4$ for
the liquid-phase ion-pair yield of cyclohexane; however, some time later a
revised value of $G_{ti} = 4.3$ was reported [17, 18]. When the data proposed
in more recent studies [19] are considered, the total ion-pair yield seems to
be closer to 5 than to 4. This means that the total ion-pair yield in the liquid
phase might exceed somewhat that in the gas phase. Studies on cyclopropane, cyclohexane and methylcyclopentane radiolysis consider this feature
to be generally true [20].

The free ion yield of cyclopropane at $-90°C$ is not more than $G=0.04$ [21]. The electrons formed can be scavenged by N_2O with a low efficiency suggesting a rather short time for ion-electron neutralization. This is in striking contrast with a number of hydrocarbons having a similarly high degree of symmetry and sphericity (e.g. neopentane, neohexane, and isooctane), as the latter are characterized by relatively high free ion yields ($G_{fi} = 0.857$, 0.304 and 0.332 respectively [13]). As the small free ion yield and the low interaction with ion and/or electron scavenger are peculiar to alkenes rather than to alkanes, cyclopropane shows, in this respect, an alkene type behaviour [23]. The low free ion yield can be a result of the short electron–ion separation distance (see also Section 1.1), which is probably brought about by the efficient energy transfer of the penetrating electrons to the molecules by way of interactions with the bonds in the strained ring [22, 23].

The free ion yield of cyclopentane determined by means of physical and chemical methods is 0.155 and 0.12, respectively [11, 13]. The yield of free ions formed in liquid phase radiolysis of cyclohexane at room temperature was reported to be 0.08–0.16 and 0.06–0.156 G-unit according to scavenger and electric conductivity experiments, respectively [13]. Electric measurements of the G_{fi}-values of cycloheptane and cyclo-octane at room temperature gave 0.19 and 0.17, respectively [22].

For the mobility of electrons in cyclopentane at room temperature a value of 1.1 $cm^2\ V^{-1}\ s^{-1}$ was reported [24, 24A] whereas 0.45 [25] and 0.38 $cm^2\ V^{-1}\ s^{-1}$ [24] were found for cyclohexane. These values are higher than those observed for n-pentane (0.075 and 0.16 [24, 25]) and for n-hexane (0.09 and 0.118 $cm^2\ V^{-1}\ s^{-1}$ [24–26]). The electron mobilities in cycloheptane and cyclo-octane at room temperature are 0.44 and 0.17 $cm^2\ V^{-1}\ s^{-1}$, respectively [22].

The free electron yields and electron mobilities of polycyclic alkanes (bicyclo[2,2,1]heptane, bicyclo[2,2,2]octane) are much higher than those of the monocyclic alkanes and approach the values measured for highly spherical branched chain molecules (e.g. neopentane). The high values appear to be associated with the high degree of sphericity, relative rigidity, and lack of distorted orbitals in these molecules [22].

According to some reports, the reaction rates of positive charge carriers and, consequently, their mobilities in cyclohexane and its derivates are higher by about one order of magnitude than corresponding to molecular diffusion [19, 27, 28]. The rapidly moving charge carriers relax in a relatively short time to slowly moving positive ions. Brede et al. [28A] supposed a connection between the rapid motion of positive charges and the possible formation of vibrationally or electronically excited radical cations that have an estimated lifetime of 10^{-11}–10^{-10} s. They mention two possible explanations for the mechanism of the rapid charge transport: first, resonance transfer with full conservation of the excitation energy of the cation, and second, a hopping mechanism with a partial loss of energy at each step. As the excited cations have very short relaxation times, they are in favour of the second mechanism [28A].

The detection as well as the determination or even the estimation of yields of excited states produced on the interaction of radiation with matter have presented considerable difficulties [29, 29A]. Excited states formed by nano-second pulses of radiation were detected in cyclohexane–aromatic hydro-carbon systems. On the basis of time-dependence studies, as well as of in-vestigations on changes observed in the presence of charge acceptors, it was concluded that these excited states can be produced as a result of elec-tron-ion recombination [30]. At the same time the UV photochemistry of a number of cycloalkanes studied in all the three phases at wavelengths shorter than 200 nm may reveal the importance of primarily produced ex-cited states that can also transfer their energy to different solutes [31–39].

Pulse radiolysis of cyclohexane in the presence of additives has been studied in the picosecond range [40]. From the results it was assumed that singlet-state cyclohexane molecules, formed via direct excitation or fast ion-electron recombination, have a reaction rate of decomposition of about $3.6 \cdot 10^9$ s^{-1} or a lifetime of $2.8 \cdot 10^{-10}$ s. The excited states were reported to transfer energy to benzene, carbon tetrachloride and 9,10-diphenyl-anthracene with a rate of $k = 2.2 \cdot 10^{11}$, $2.5 \cdot 10^{11}$, and $3.4 \cdot 10^{11}$ M^{-1} s^{-1}, respectively. Several other works also report data on the lifetimes of excited states produced in cyclohexane ($3 \cdot 10^{-10}$ [41], $6.8 \cdot 10^{-10}$ [41A], $9 \cdot 10^{-10}$ s [41B]), methylcyclohexane ($6.5 \cdot 10^{-10}$ s [41A]), bicyclohexane ($1.6 \cdot 10^{-9}$ [41], $1.58 \cdot 10^{-9}$ s [41A]), cis-decalin ($2.18 \cdot 10^{-9}$ s [41A]), trans-decalin ($2.82 \cdot 10^{-9}$ s [41B]) and a mixture of cis- and trans-decalins ($2.3 \cdot 10^{-9}$ s [41]).

Several attempts have been made to estimate the G-value of excited cycloalkane molecules. In 1968 Holroyd published a work in which he inter-preted the radiolysis and photolysis of cyclohexane–N$_2$O solutions [33A]. He supposed that during radiolysis, N$_2$O scavenges both electrons and ex-cited states, whereas in the course of photolysis only the quenching of ex-cited states occurs. Estimating the G-value of electrons scavenged, he cal-culated a G-value of 1.2 for the excited singlet (S$_1$) cyclohexane formation. Later, in 1972, Beck and Thomas [40] obtained a value of $G(S_1) \approx 1$ from pulse radiolysis studies. Baxendale and Mayer [42] reported $G = 0.3$, fol-lowing fluorescence intensity investigations of solutions of toluene in cyclo-hexane in low concentrations. The applicability of this technique has been criticized by Walter and Lipsky [29], who arrived at a value of $G(S_1) = = 1.4$–1.7 by comparing the fluorescence intensities of pure cyclohexane with that of neat benzene. Most recently Wojnárovits et al. estimated $G(S_1) = 1.4$ by comparing the unimolecular H$_2$ formation in radiolysis and bromine lamp photolysis (163 nm, 7.6 eV) experiments [39]. During photo-lysis the hydrogen forms mainly by unimolecular H$_2$ detachment, which is also known to occur in liquid-phase radiolysis. The authors, in estimating the G-value given above, supposed that the same excited state is responsible for unimolecular H$_2$ detachment in both liquid-phase radiolysis and photo-lysis.

The same technique has been used to evaluate the formation of the first excited singlet state in cyclopentane, cycloheptane, cyclo-octane and cyclo-decane: $G(S_1) = 1.8$, 1.6, 2.1 and 2.2, respectively [37, 39]. The yields of S$_1$ states of methylcyclohexane, cis-decalin and bicyclohexane are 1.9–2.2, 3.4 and 3.5, respectively, as reported by Walter and Lipsky [29]. In bicyclohexane radiolysis about 90% of S_1 states are generated via recombination of geminate ion pairs [29A].

In a number of hydrocarbons, including several cycloalkanes, phenomena related to the vacuum-UV light-induced fluorescence have been studied, i.e. the yields of energy dissipation from electronically excited states and the dependence of the wavelength of the emitted light on molecular structure [43, 44]. It was established that cyclopentane, cycloheptane, cyclooctane and cyclodecane show quite different behaviour from cyclohexane and its derivatives. The latter produce considerable fluorescence intensities ($\Phi \approx 0.01$), whereas C_5 and C_7-C_{10} cycloalkanes show practically no fluorescence ($\Phi \leq 10^{-5}$).

A possible explanation of these phenomena may be as follows: the system, being in an excited state, strives to decrease the effects that hinder reorganization into the configuration with least strain. Therefore, the configurations in the ground state and in the excited state are more or less different, depending on the strain in the molecule. As the C_5 and C_7-C_{10} cycloalkanes are considerably strained (Table 3.1), there is a large disparity in the equilibrium configurations of the ground and excited states, which presumably increases the rate of radiationless transitions.

The energy content of a certain fraction of excited molecules may exceed the photoionization potential, i.e. so-called superexcited states can also be produced [45–47]. In a work published in 1968 [47A] the yield of superexcited molecules was estimated for a number of alkanes (including cyclopropane) and alkenes to be about 0.8. Some later photochemical studies [46] have reported that the quantum yield of ionization attains unity at about 16 eV. The chemical decomposition of superexcited molecules does not differ much from that of ordinary excited molecules [31, 45, 46].

It is rather difficult to obtain values for the yields of directly produced excited molecules that do not autoionize. There are no experimental methods available for the satisfactorily accurate determination of their yields. A comparison of the photochemical and radiation chemical product distributions, for the ratio of excitation to ionization brought about in C_2-C_5 aliphatic alkanes and cycloalkanes, led to an estimated value of about 0.15–0.30 [45, 46]. This value appears to be reasonable, if we consider that on irradiation of aliphatic alkanes (see Chapter 2) and cycloalkanes (see Section 3.4), the absorption of every 100 eV of energy leads to the primary decomposition of about 6.5 molecules, whereas the yield of ionization is about 5, corresponding to the results outlined above.

Whereas the C—H bond rupture processes of cycloalkanes are essentially similar to those of the aliphatic alkanes, the C—C bond dissociations of the two classes of compounds show remarkable differences. The rupture of a C—C bond in aliphatic alkanes gives two monoradicals that can diffuse apart and, subsequently, do not affect the reaction possibilities of each other. The analogous decomposition of the ring of a cycloalkane, however, can produce a 'biradical', the two radical sites of which are connected by a carbon chain.

The term 'biradical' has a rather different meaning for scientists working in different fields of physics and chemistry [48]. For example, in a wider sense of the word, certain triplet states can also be considered as biradicals. However, according to a more strict definition [49], the biradical is a hypothetical bifunctional intermediate having radical character, in which the close proximity of the two radical sites acts as an extraordinarily high local concentration of radicals (~ 10 M) and therefore the subsequent reactions of the radicals are strongly interdependent. This explains why biradical processes are not affected by the presence of radical acceptors. At the same time, the 'hypothetical' character of biradicals poses the question: do the reactions attributed to the biradicals exist in reality or does another reaction type, namely concerted processes, lead to the formation of the given products? These questions cannot yet be answered. However, there is no doubt, that the reactions of bifunctional biradicals play an important role in radiation-induced C—C bond rupture processes [50].

3.2. INDIVIDUAL COMPOUNDS

The radiolysis of the individual cycloalkanes will be discussed following the order of their molecular weights. Separate sections are devoted to the radiation chemistry of both alkylcycloalkanes and compounds having a bridged ring.

3.2.1. Cyclopropane and cyclobutane

The radiolysis of cyclopropane and cyclobutane has been a subject of interest for scientists for nearly 20 years [51–53]. Initially, the photochemical ctions in the gas phase were studied, including investigations with scavengers and deuterated compounds. Since the experiments of Ausloos et al. [54], cyclopropane has often been applied as a positive ion scavenger for alkane radiolysis (Section 1.4.1.).

The $G(H_2)$ value of cyclopropane and cyclobutane are conspicuously low compared with those of the corresponding aliphatic alkanes and higher cycloalkanes, being closer to those of propene and butene than to the higher cycloalkanes (Table 3.2 and 3.3).

The mechanism of formation of hydrogen gas during irradiation is not yet quite clear. Various papers [35, 51, 55–57] are contradictory in several respects, but it seems that there is one point which is undisputed: hydrogen formation is a complex process consisting of several sequential and parallel reactions whose character is different from those reactions which produce hydrogen from larger rings. It is probable that the process is still more complicated in condensed phases.

According to Yang's experiments [51], the yield of hydrogen produced in gas-phase irradiations of cyclopropane (Table 3.2) does not decrease detectably even in the presence of 10% NO, whereas von Bünau and Kühnert [55] observed a decrease of 40% in the G-value on addition of 4.5% NO. Scala and Ausloos [35] have studied the gas-phase photolysis and radiolysis of cyclo-C_3D_6–H_2S mixtures; they observed that HD formation was several times greater than that of D_2 even in mixtures containing as little as

Table 3.2

Yields of the photolysis and radiolysis of cyclopropane

Radiation	Photolysis					60Co-γ			
Energy, eV	8.4	11.6–11.8	8.4	8.4	10.0			—	
Temperature, K	298	298	298	77	77	298	298	195	298
Reference	35	35	59	35	35	35	55	64	56A
Phase	Gas (6 mbar)	Gas (13 mbar)	Gas (27 mbar)	Solid	Solid	Gas^a (6 mbar)	Gas (600 mbar)	Liquid	Liquid
Products	Relative yield					G, molecule/100 eV			
Hydrogen			15.0	15.0	26.0		1.38	1.1	0.91
Methane			4.6				0.16	0.2	0.23
Ethane	19.0	7.0	20.0	2.8	0.6	0.65	0.35	0.2	0.11
Ethylene	100.0	100.0	100.0	100.0	100.0	3.45	3.28	1.6	1.5
Acetylene	14.0	15.2	14.0	8.0	1.2	1.04	0.72		0.32
Propane							1.05	0.2	0.27
Propene	7.0	15.6	11.0	172	299	1.0		0.9	0.70
Propyne			21.0^b				0.20		
Propadiene (allene)	19.0	17.9	31	4.0	21	0.38	0.37		0.25
n-Butane			trace				0.39		
Isobutane			trace				0.27		
But-1-ene	10.0		17.0				0.08		0.13
But-2-ene	7.0		18.7						0.53
Methylcyclopropane	7.0		8.4	64	75		0.51		
Pentanes							0.42		0.08

^{a} The authors published M/N-values. The present data were obtained by using $W = 23.7$ eV/ionpair
^{b} Propyne + propane

10% of H_2S. This indicated a relatively high yield of deuterium atoms reacting with H_2S:

$$D + H_2S \longrightarrow HD + HS\cdot \qquad (3.1)$$

It could be supposed therefore that the major part of hydrogen gas would be produced in radical processes. In fact the opposite is the case: the isotopic composition of hydrogen gas produced from a ternary mixture (47.8% cyclo-C_3H_6 – 4.7% cyclo-C_3D_5H – 47.5% cyclo-C_3D_6) was as follows [55]: 53% H_2, 9% HD, 38% D_2 thereby indicating the minor importance of radical processes in hydrogen production. As gas-phase photolysis and radiolysis produce considerable amounts of C_2H_2 and C_3H_4, it can be concluded that hydrogen production is related to the formation of these products [35, 55]; its most important source is ethylene which possesses a considerable amount of excess energy.

von Bünau and Kühnert [55] supposed that the energetic hydrogen atoms can cleave the cyclopropane ring:

$$H + \text{cyclo-}C_3H_6 \longrightarrow C_3H_7 \qquad (3.2)$$

Table 3.3

Yields of photolysis and radiolysis of cyclobutane (Temperature: 298 K)

Radiation	Photolysis				^{60}Co-γ		
Energy, eV	8.4	10.0	10.6–10.8	Hg (^{1}P$_1$) photosensitization	—		
Pressure, mbar	20	20	7	20	26	26	
Reference	65			68	60^a	66	56A
Phase	Gas						Liquid
Products	Relative yield				G, molecule/100 eV		
Hydrogen	14.4	38.0		50.0		1.86	1.70
Methane	0.52	1.1			0.34	0.06	0.06
Ethane	0.88	1.58	1.58	8.6	0.23	0.10	0.44
Ethylene	100.0	100.0	100.0	100.0	3.81	1.41	4.93
Acetylene	12.0	29.0	35.0		1.02	0.45	0.29
Propane				3.5		0.12	0.004
Cyclopropane	0.59						0.15
Propene	2.1	3.4	4.86		0.53		0.05
Allene	1.06	4.5	2.04		0.12	0.10	
n-Butane				10.0			
Methylcyclopropane				1.0			0.24
Methylcyclobutane				1.0		0.10	0.24
Ethylcyclobutane							0.15
But-1-ene			4.68^b		0.26^b	0.07	0.72
trans-2-Butene			2.43			0.096	0.05
Buta-1,3-diene	1.78	5.0				0.12	0.24
Cyclobutene				10.0		0.07	0.72
CH$_3$C≡C—C≡CCH$_3$						0.21	
Heptynes						0.38	
Heptadienes						0.61	
Octenes						0.10	
(cyclo-C$_4$H$_7$)$_2$				16.0			

^{a} The authors published M/N values [60]. The present data obtained by using $W = 25$ eV/ionpair [66]
^{b} But-1-ene + buta-1,3-diene

Their calculations suggest that in the gas phase about 95% of hydrogen atoms formed in radiolysis produce propyl radicals according to reaction (3.2) and about 5% of them form H$_2$ molecules and cyclopropyl radicals:

$$H + \text{cyclo-C}_3H_6 \longrightarrow \text{cyclo-C}_3H_5 + H_2 \qquad (3.3)$$

Thus the phenomenon observed in mixtures of cyclo-C$_3$D$_6$ and H$_2$S can be attributed to the more efficient reaction of D-atoms with hydrogen sulphide (which is present in high concentrations) [reactions (3.1)] than with cyclopropane [reactions (3.2) and (3.3)].

According to the experiments of Horváth and Földiák [56A] radiolysis in the liquid phase also produces C$_2$H$_2$ and C$_3$H$_4$ dehydrogenation products, but their combined G-value (0.57) is definitely lower than that of hydrogen [G(H$_2$) = 0.91]. Thus, other products containing less hydrogen than the starting cyclopropane should also be formed (e.g. cyclopropene or C_6 hydrocarbons) but such compounds have not been found so far in the radiolysis and photolysis experiments. It is interesting to note, however, that cyclopropene formation was observed in the gas phase low energy electron impact investigations [57].

Scala and Ausloos have shown that the relative yields of acetylene and allene produced in radiolysis in the solid state increase with increasing total conversion; these molecules should therefore be produced in secondary processes between accumulating reaction products [35].

In the multiphoton CO_2-infrared laser-induced photochemistry of cyclopropane acetylene was formed in the highest amount, and the distribution of products was completely different from those found in radiolytic, vacuum-UV photolytic and thermal decompositions [58].

Photolysis experiments have supplied a fair amount of indirect information on the decomposition possibilities of excited species formed in irradiated cyclopropane [35, 58, 59]. The contributing excited states are the following: triplet states at 7.4, 9.0 and 9.8 eV, singlet states at 6.7 (or 7.7), 8.55, 9.4 and 9.95 eV, and superexcited states lying around 10.2 eV [57]. The most important reactions would be (according to Currie et al. [59]):

$$
\text{cyclo-}C_3H_6^*
\begin{cases}
C_2H_4 \ + \ CH_2 & (3.4a) \\
C_3H_4 \ + \ H_2 \text{ or} & \\
C_3H_4 \ + \ 2\,H & (3.4b) \\
C_3H_5 \ + \ H & (3.4c)
\end{cases}
$$

In principle, the term 'cyclo' for the decomposing species 'cyclo-$C_3H_6^*$' could well be omitted as it is virtually impossible to tell whether it has a cyclic or an open chain structure [35, 60]. Scala and Ausloos assumed that 'cyclo-$C_3H_6^*$' is mostly an excited trimethylene biradical giving excited propene in addition to the processes shown in reaction 3.4. If deactivation is not rapid enough, propene suffers fragmentation:

$$
CH_2{=}CH{-}CH_3^*
\begin{cases}
CH_2{=}CH \quad\ \ + \ CH_3 \\
CH{\equiv}CH \quad\ \ + \ CH_3 \ + \ H \\
CH_2{=}CH{-}CH_2 \ + \ H
\end{cases}
$$

Data listed in Table 3.2 support this latter assumption indicating that irradiation produces propene in all phases but in a greater proportion in condensed phases because deactivation prevents its decomposition.

Contrary to the suggestion of Scala and Ausloos, Dhingra and Koob proposed [61] that there is no need to postulate the formation of trimethylene in the primary process, at least as a source of methylene and ethylene. The latter authors quoted at least two pieces of evidence to support their proposal. First, they did not observe a singlet to triplet conversion of the hypothetical biradical intermediate in the pressure dependence studies. Second, *cis*- and *trans*-1,2-dimethylcyclopropane yield only *cis*- and *trans*-but-2-ene, respectively. This means, that the rotation about a bond in the proposed biradical is not competitive with decomposition to give methylene biradical and butene. As there is no direct method of detecting the proposed trimethylene chemically, the authors have suggested that, if it exists it must be very short lived.

The methylene biradical formed in reaction (3.4a) is very reactive; insertion into $C-H$ bonds and hydrogen abstraction are among its important reactions:

$$CH_2 + \text{cyclo-}C_3H_6 \longrightarrow CH_3\text{—cyclo-}C_3H_5$$

$$CH_2 + \text{cyclo-}C_3H_6 \longrightarrow \text{cyclo-}C_3H_5 + CH_3$$

Excited methylcyclopropane becomes deactivated partly in collisions, partly in a ring-opening process, leading to butenes [35, 56B, 62]. Another possible pathway for methylcyclopropane formation is the combination of methyl and cyclopropyl radicals, as evidenced by the decrease in the methylcyclopropane yield in gas-phase photolysis with scavengers [35].

The following unimolecular decomposition reactions can be regarded as proved by mass spectrometry:

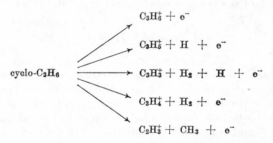

von Bünau et al. have found [55] the ratios of the relative importance of these processes in the gas phase to be: 1 : 0.86 : 0.57 : 0.29 : 0.23. As cyclopropane has a strong tendency to take part in ion-molecule reactions, primary ions in a high-pressure mass spectrometer participate in various ion-molecule reactions [58, 63], such as

$$
\text{cyclo-}C_3H_6 + C_3H_6^+
\begin{cases}
C_3H_7^+ + C_3H_5 \\
C_5H_9^+ + CH_3 \\
C_4H_8^+ + C_2H_4 \\
C_4H_7^+ + C_2H_5
\end{cases}
$$

Decompositions subsequent to neutralization of positive ions, as well as interactions between radicals, bring about additional branching reactions.

Thus the variation of the possible end-products of decomposition increases very sharply, together with the number of parallel ways leading to the same end-product [35, 55].

Scala and Ausloos observed the formation of ethyl and propyl radicals when photons with energies above the ionization potential or high-energy radiation were applied; such species were not observed in decomposition with lower energy photons. Thus these are formed in ionic processes [35].

The formation of ions containing more than six carbon atoms has been observed by mass spectrometry. The importance of these ions may be even greater in radiolysis, leading to considerable amounts of 'polymer'. In fact, the yield of cyclopropane consumption has been found by von Bünau and Kühnert to be $G = 16.3$ in the gas phase, whereas this value would be only 7.4 on the basis of summarized yields of C_1-C_6 products [55].

In the gas-phase radiolysis of cyclobutane (Table 3.3) at room temperature and a pressure of 26 mbar, the G-value of hydrogen gas formation is 1.86. Doepker and Ausloos [65] and Heckel and Hanrahan [66] have suggested that the main reaction of hydrogen production is H_2 elimination from excited intermediates. The contribution of H-atoms to H_2 formation is probably small, partly because a considerable proportion of thermal hydrogen atoms is captured by unsaturated hydrocarbon products formed in high yield (e.g. ethylene), and partly because in collisions of hot hydrogen atoms with cyclobutane there is, according to Heckel and Hanrahan, a tendency to ring opening (similar to cyclopropane radiolysis) [66, 66A].

$$H + \text{cyclo-}C_4H_8 \longrightarrow C_4H_9$$

The hydrogen yield from irradiated liquid cyclobutane is $G = 1.70$, as shown by the studies of Horváth and Földiák [56A]. The hydrogen deficit of hydrocarbon products is $\Delta G \approx 0.5$. Although the work cited contains no data on bicyclobutane formation, it can be concluded from the high yields of cyclobutene and relatively low hydrogen deficiency in the hydrocarbon products that the mechanism of hydrogen formation is more similar to that characteristic of larger rings in the liquid state than in the gas phase.

Ausloos et al. and Tanaka et al. have used vacuum UV photochemistry to investigate the reaction pathways of excited states generated in cyclobutane [35, 65, 67, 67A]. Further important information has been obtained from experiments in which the substance was irradiated in an electric field. Electrons produced in radiolysis and accelerated in the electric field generate additional excited molecules, thus yields of their products increase, whereas those originating from ion decomposition remain unchanged (Fig. 3.2) [60]. Spittler and Klein [68] have reported the $Hg(^1P_1)$ photosensitized decomposition of cyclobutane; these reactions also take place via further transformations of excited molecules. Studies of pyrolysis [6, 7] and recoil hot atom reactions [69, 69A] also assist the understanding of radiation-induced reactions.

The most important steps in the decomposition of excited states are as follows [35, 65, 67]:

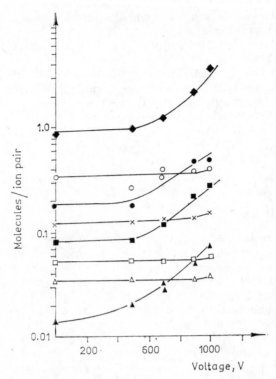

Fig. 3.2. Effect of an applied electric field in the radiolysis of the cyclobutane–NO (100 : 5) mixture (radiation: ^{60}Co-γ).
$\blacklozenge$ C_2H_4; $\bigcirc$ 2-C_4H_8; $\bullet$ C_2H_2; $\times$ C_3H_6; $\blacksquare$ 1,3-C_4H_6; $\square$ CH_4; $\triangle$ C_2H_6; $\blacktriangle$ C_3H_4. After Doepker and Ausloos [60]

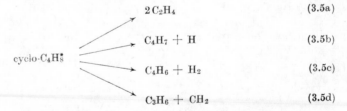

$$\text{cyclo-}C_4H_8^* \quad \begin{cases} 2\,C_2H_4 & (3.5\text{a}) \\ C_4H_7 + H & (3.5\text{b}) \\ C_4H_6 + H_2 & (3.5\text{c}) \\ C_3H_6 + CH_2 & (3.5\text{d}) \end{cases}$$

The term 'cyclo-$C_4H_8^*$' (similarly to 'cyclo-$C_3H_6^*$') may denote either a cyclic or an open-chain species; whereas transition species generated from cyclopropane mostly possess an open-chain structure, analogous ones from cyclobutane may contain a high proportion of cyclic entities, though this is very difficult to verify [66]. Cyclobutene formation can be observed both in photolysis and in radiolysis; its formation can be visualized not only via reaction (3.5c), but also by the disproportionation of cyclobutyl radicals. The latter, in turn, can be produced — apart from reaction (3.5b) — by hydrogen abstraction of radicals (which is probably not a very important process) and in ionic reactions.

It is remarkable that cyclobutane was observed among the products of ethylene photolysis [70, 71].

According to Doepker and Ausloos and Tanaka et al. ethylene molecules produced in reaction (3.5a) possess a considerable excess energy (up to 7.6 eV in the case of 147 nm photon irradiation); consequently they decompose further in reactions (3.6a) and, to a lesser extent (3.6b) and (3.6c) [60, 65, 67]:

$$C_2H_4 \begin{cases} C_2H_2 + H_2 & (3.6a) \\ C_2H_2 + 2\,H & (3.6b) \\ C_2H_3 + H & (3.6c) \end{cases}$$

The occurrence of these reactions is supported by the facts that in gas-phase photolysis, the ratio of acetylene to ethylene decreases with increasing pressure (where deactivation becomes higher) and that in radiolysis in the liquid state, the same ratio is several times less than in the gas phase. The rate constant of ethylene decomposition during the 147 nm photolysis experiments in the gas phase was found to be $9.2 \cdot 10^8 \ s^{-1}$ [67A].

No unambiguous evidence is available for the actual occurrence of reactions (3.5b) and (3.5c) in the gas phase. However, both processes have been rendered probable by Spittler and Klein when they used scavengers and deuterated compounds in $Hg(^1P_1)$ sensitized photolysis [68]. The yields of hydrogen and, especially, cyclobutene and buta-1,3-diene, are low in vacuum UV photolysis and radiolysis indicating that reactions (3.5b) and (3.5c) are of minor importance [60, 65, 67]. Buta-1,3-diene is possibly produced via decomposition of excited cyclobutene molecules; in pyrolysis, it is the only product of decomposition of vibrationally excited cyclobutene [72].

A proportion of the methylene biradicals formed in reaction (3.5d) can be transformed into methyl radical in a hydrogen abstracting process, as reported by Doepker and Ausloos, who also studied the reactions of cyclo-C_4D_8 and H_2S mixtures [65].

Some of the important decomposition processes of $C_4H_8^+$ ions are as follows [60, 65]:

$$C_4H_8^+ \begin{cases} C_2H_4^+ + C_2H_4 & (3.7a) \\ C_3H_5^+ + CH_3 & (3.7b) \\ C_2H_5^+ + C_2H_3 & (3.7c) \\ C_3H_4^+ + CH_4 & (3.7d) \end{cases}$$

In addition to the ions shown in reactions (3.7a)–(3.7d) $C_2H_3^+$, $C_2H_2^+$, $C_3H_3^+$ and $C_4H_7^+$ have been detected by mass spectrometry. Hughes and Tiernan have reported that if the pressure is right, practically all ions participate in transfer reactions with molecules of the starting hydrocarbon. The $C_3H_3^+$ ion is, however, rather inactive towards cyclobutane molecules [73]. Important reactions of radiolysis are [66]:

$$C_2H_4^+ + \text{cyclo-}C_4H_8 \longrightarrow C_4H_8^+ + C_2H_4$$

$$C_3H_5^+ + \text{cyclo-}C_4H_8 \longrightarrow C_4H_7^+ + C_3H_6$$

$$C_2H_3^+ + \text{cyclo-}C_4H_8 \longrightarrow C_4H_7^+ + C_2H_4$$

$$C_2H_2^+ + \text{cyclo-}C_4H_8 \longrightarrow C_4H_8^+ + C_2H_2$$

$$C_2H_5^+ + \text{cyclo-}C_4H_8 \longrightarrow C_4H_7^+ + C_2H_6$$

$$C_3H_4^+ + \text{cyclo-}C_4H_8 \longrightarrow C_4H_8^+ + C_3H_4$$

As a consequence, the neutralization in radiolysis takes place between positive ions $C_4H_8^+$, $C_4H_7^+$, $C_3H_3^+$ and electrons:

$$C_4H_8^+ + e^- \longrightarrow C_4H_8^* \qquad (3.8)$$

$$C_4H_7^+ + e^- \longrightarrow C_4H_7^* \qquad (3.9)$$

$$C_3H_3^+ + e^- \longrightarrow C_3H_3^* \qquad (3.10)$$

The $C_4H_8^*$ species produced in reaction (3.8) probably have an open chain structure and, either prior to neutralization or subsequent to it rearrange partly into butenes, and butadiene [66]:

$$C_4H_8^* \begin{cases} \nearrow CH_3-CH=CH-CH_3 \\ \longrightarrow CH_2=CH-CH_2-CH_3 \\ \searrow CH_2=CH-CH=CH_2 + H_2 \end{cases}$$

It is remarkable that whereas in photolysis with energies below the ionization potential no butenes are formed, in higher energy photolysis and radiolysis their yield is rather high. Their formation in ionic reactions is supported by the increase of their G-values in the gas phase, under the effect of NO addition [65]. The role of NO has been interpreted by Lias and Ausloos in terms of charge transfer which promotes the stabilization of the butene structure [74]:

$$C_4H_8^+ + NO \longrightarrow C_4H_8 + NO^+$$

The yield of butene is increased by any additive with a lower ionization potential than butenes (e.g. toluene, trimethylamine), whereas those with a higher ionization potential have no effect [60].

The species $C_4H_7^*$ formed in reaction (3.9) decomposes further. The $C_3H_3^*$ transition species produced in reaction (3.10) is assumed by Heckel and Hanrahan to have a propenyl radical structure. Hexa-2,4-diyne can be formed by the combination of two such radicals [66]. It can be seen from Table 3.3 that, in addition to this product, other hydrocarbons with more than four carbon atoms have been observed. As these could not be detected in the presence of 10% C_2H_4 or O_2, it is propable that they are products of radical combination.

3.2.2. Cyclopentane

Photolytic and radiolytic studies of cyclopentane (Table 3.4) are hindered by the fact that the commercially available compound contains $\sim 10^{-2}\%$ 2,2-dimethylbutane, which is very tiresome to get rid of. This impurity can considerably affect the yields of certain products, e.g. fragments, as a consequence of charge-transfer reactions due to its gas-phase ionization potential being lower than that of cyclopentane [75] (10.01 and 10.53 eV, respectively).

As the melting point of cyclopentane is relatively low ($-93.9°C$) as compared with that of cyclohexane (6.5°C), the former seems to be more useful for temperature dependence studies in the liquid phase. At the same time, the relatively low boiling point (49.3°C) of cyclopentane may be a serious inconvenience in some experiments, because cyclopentane has a considerable vapour pressure even at room temperature (~ 400 mbar).

During the past few years several experiments have been done on hydrocarbons dissolved in water, mostly using pulse techniques. As a result of hydrogen atom abstraction by H and OH radicals formed from the water, relatively large amounts of hydrocarbon radicals are produced; their decay kinetics can be followed by means of the spectrophotometric method of pulse radiolysis techniques. Cyclopentane dissolves in water relatively easily (1 litre of water dissolves, at a temperature of 23°C, about 2.5 mmol of cyclopentane [76], compared with 0.53 mmol of pentane [77] and 0.68 mmol of cyclohexane [76]), therefore it has been applied as a model compound in these types of experiment [76, 78].

Dehydrogenation products

The G-value of hydrogen formed from cyclopentane at low doses or extrapolated to zero dose, in the liquid phase at room temperature, has been reported in a number of papers as $G(H_2) = 5.0–5.4$ [11, 79–85].

On the basis of experiments performed in the presence of HCl, Busi and Freeman [86] deduced that the real G-value of hydrogen, extrapolated to zero dose, is ~ 5.7. The value determined in dose dependence studies can be lower than this as a result of the presence of some impurities, e.g. CO_2. The G-value of hydrogen shows only a slight, if any, dose dependence in the dose ranges $0.2–2 \cdot 10^{19}$ eV g^{-1} [86] and $0.7–5 \cdot 10^{19}$ eV g^{-1} [82], respectively.

Table 3.4

Yields of the radiolysis and photolysis of cyclopentane

Radiation	Photolysis					⁶⁰Co-γ				
Energy, eV	8.4	10.0	11.6–11.8	7.6	10.0					
Temperature, K	Room temperature	Room temperature	Room temperature	288	77	343	298	195	298ᵇ	77
Pressure, mbar	7	7	7(+1.3 NO)			1150				
References	45	45		39	45	100	85	103	11	88, 100
Phase	Gas	Gas	Gas	Liquid	Solid	Gas	Liquid	Liquid	Liquid	Solid
Product	Relative yield					G, molecule/100 eV				
Hydrogen	298	110	70	2430		0.45	5.35	4.3		4.25
Methane	3.5	4.0	0.14			0.56	0.03	0.025	0.03	0.014
Ethane	5.4	4.4	0.6				0.03	0.04	0.03	0.017
Ethylene	100.0	100.0	100.0	100.0	100.0	4.43	0.45	0.31	0.39	0.15
Acetylene	9.3	10.5	8.8			0.28	0.05	0.03		0.01
Propane	4.7	2.8	8.2			2.23	0.08	0.05	0.041	0.015
Cyclopropane	2.1	1.3	2.4	33	46	0.44	0.12	0.11	} 0.22	0.05
Propene	27.0	27.0	14.6	67	66	0.67	0.28	0.15	}	0.08
Propadiene + propyne	8.4ᵃ	19.9ᵃ	10.1ᵃ			nm	0.05			
n-Pentane				100		< 0.03	0.14	0.11	0.03	0.1
Pent-1-ene			5			~ 0.1	0.74	0.47	0.62	0.47
Pent-2-enes								0.05	0.02	0.02
Cyclopentene	168.0	41.5	17.4	2380	560	0.97	2.97	1.91	1.92	2.75
C₆-C₈ Alkyl- or alkenyl-cyclopentane	7.9	10.6	nm	100	3330		0.18			nm
n-Pentylcyclopentane							0.06			0.01
Bicyclopentane	3.4	5.3	nm			0.68	1.29			0.65
Cyclopentylcyclopentene							0.11		0.46	0.07

ᵃ Propadiene to propyne ratio ~3:1
ᵇ Yields determined in presence of 5 mM I₂

13

The $G(H_2)$ value of solid cyclopentane at 77 K is 4.2–4.4 [87, 88], i.e. 1 G-unit lower than in the liquid phase, at room temperature. Kennedy and Stone [89] consider that the yield of hydrogen is smaller at lower temperatures, because the thermalized hydrogen atoms react more slowly, so their lifetime becomes longer and, consequently, they are more likely to react with product alkenes that accumulate during irradiation. Thus, in fact, at low temperatures one cannot determine initial hydrogen yields.

In samples irradiated at 77 K, frozen cyclopentyl radicals can be observed by ESR spectroscopy. These are transformed into stable end-products only when the sample is warmed to 122 K, the temperature of the solid–solid state transition [90, 90A].

As shown in Table 3.4, during gas-phase photolysis relatively small amounts of hydrogen and cyclopentene are formed compared with those of fragment products. The radiation-chemical yield of cyclopentene is also rather low, and the formation of bicyclopentane becomes important only when a relatively high pressure (1.15 bar) is applied. These observations reflect the partial fragmentation of reactive intermediates in the gas phase.

Hentz and Brazier [91] have reported data on the liquid-phase radiolysis of cyclopentane carried out under 0–6.2 kbar pressure. The yields of hydrogen and bicyclopentane remained unchanged, that of cyclopentene decreased by about 10% under the highest pressure applied. Two possible explanations were given:

1. The high pressure affects the decomposition of activated cyclopentane species in such a manner that the yield of unimolecular H_2 elimination decreases, whereas that of bimolecular hydrogen formation increases.
2. The application of high pressure modifies the k_d/k_c ratio of cyclopentyl radicals, provided that disproportionation and recombination do not have a common transition state. The preferred reaction is that which has the smaller activation volume. The experimental data suggest the formation of bicyclopentane to be the preferred reaction. As noted by the authors, this conclusion is in disagreement with the one drawn by Dixon et al., who studied the reaction of ethyl radicals. They found the activation volume of the recombination of ethyl radicals to be the larger, on the basis of the dependence of the reaction rate on the internal pressures of solvents [92].

In cyclopentane, containing 60 mM N_2O, the yields of hydrogen and nitrogen ($G = 3.7$ and 3.0, respectively), and consequently the yield of electrons also, were found to be independent of pressure. The yield of HD produced from irradiated cyclopentane solutions containing 60 mM ND_3 or 0.32 M C_2H_5OD increased by 35% in the pressure range studied, whereas $G(H_2 + HD + D_2)$ remained constant at $G = 5.2$. The simultaneous decrease of the yields of H_2 and D_2 indicates an enhanced yield of scavenged cations, as a result of a decrease in the diffusion coefficient of electrons and/or an increase in that of positive charge carriers.

Taking $G(H_2)$ as 5.7, Busi and Freeman summarized hydrogen formation as follows [86]:

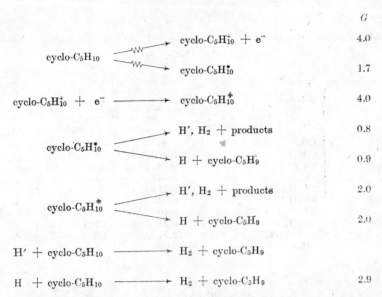

G

cyclo-C$_5$H$_{10}$ → cyclo-C$_5$H$_{10}^+$ + e$^-$ 4.0

→ cyclo-C$_5$H$_{10}^*$ 1.7

cyclo-C$_5$H$_{10}^+$ + e$^-$ ⟶ cyclo-C$_5$H$_{10}^\#$ 4.0

cyclo-C$_5$H$_{10}^*$ → H′, H$_2$ + products   0.8

→ H + cyclo-C$_5$H$_9$ 0.9

cyclo-C$_5$H$_{10}^\#$ → H′, H$_2$ + products   2.0

→ H + cyclo-C$_5$H$_9$ 2.0

H′ + cyclo-C$_5$H$_{10}$ ⟶ H$_2$ + cyclo-C$_5$H$_9$

H + cyclo-C$_5$H$_{10}$ ⟶ H$_2$ + cyclo-C$_5$H$_9$ 2.9

where cyclo-C$_5$H$_{10}^*$ and cyclo-C$_5$H$_{10}^\#$ stand for excited cyclopentane molecules produced directly and after electron-ion recombination, respectively.

Thus, the G-values of hydrogen that can be traced back to ion-electron recombination are 4, 3.2 and 3.4 as reported in refs [10, 16, 86], respectively. In the reaction scheme, H′ denotes hot hydrogen atoms that cannot be scavenged by radical acceptors, while H denotes scavengeable, i.e. thermal, hydrogen atoms. The yield of the latter species, $G = 2.9$, considerably exceeds those determined in other scavenger studies, e.g. 1.6 [11, 87], 1.8 [93] and 1.96 [82].

Thus, according to the proposed reaction mechanism, the hydrogen is produced partly via atomic hydrogen as an intermediate and partly via a unimolecular elimination. Along the former pathway cyclopentyl radicals are also formed, the reactions of which have been studied by means of radical scavengers [11, 79, 82, 88, 94, 95], pulse radiolysis techniques [76, 78] and ESR spectroscopy [76, 96].

The radiation-chemical yields of iodides formed from cyclopentane–I$_2$ solutions are listed in Table 3.5, and the distribution of radicals determined with ethylene as radical scavenger is shown in Table 3.6. The G-value of cyclopentyl iodide was reported to be 1.95, 2.41 and 2.96, in refs [11], [94] and [81], respectively. However, the concentration of I$_2$ in all the three cases, as well as the dose applied in the first work, was too high to give acceptable results, i.e. close to the initial yields. If we take into account secondary and subsidiary reactions and make the necessary corrections as has been done for the radiolysis of cyclohexane–I$_2$ systems [97], the G-value of cyclopentyl iodide and thus the corrected yield of scavengeable cyclopentyl radicals can be estimated as $G = 4$. Holroyd and Klein [95] determined the yield of scavengeable cyclopentyl radicals as $G = 4.1$ by using the ^{14}C$_2$H$_4$-radical sampling technique.

13*

Table 3.5

Yields of iodides formed from cyclopentane–I_2 systems (radiation: ^{60}Co-γ)

Dose, 10^{19} eV g^{-1}	3.12	31.2
I_2, mM	0.65	8.5
References	94	11
Products	G, molecule/100 eV	
CH_3I	0.28	0.026
C_2H_5I	0.066	0.019
2-C_3H_7I	0.046	0.016
1-C_3H_7I	0.048	0.049
2-C_4H_9I	0.014	
1-C_4H_9I	0.027	
2-$C_5H_{11}I$	0.10	0.047
1-$C_5H_{11}I$	0.23	0.147
cyclo-C_5H_9I	2.91	1.95
CH_2I_2	0.096	

Table 3.6

Distribution of alkyl radicals formed from cyclopentane in the presence of ethylene (radiation: 2.8 MeV electrons). After Holroyd and Klein [95]

Radical	%
Methyl	0.506
Cyclopentyl	88.7
Cyclopent-3-enyl	4.7
Pent-4-enyl	2.6
n-Pentyl	3.6

In the absence of radical acceptors, cyclopentyl radicals are consumed chiefly in disproportionation and combination reactions (see also Section 1.2.2.):

$$2 \text{ cyclo-}C_5H_9 \begin{array}{c} \xrightarrow{k_d} \text{cyclo-}C_5H_{10} + \text{cyclo-}C_5H_8 \qquad (3.11a) \\ \xrightarrow{k_c} (\text{cyclo-}C_5H_9)_2 \qquad\qquad\quad (3.11b) \end{array}$$

Scavenger studies gave a k_d/k_c ratio for cyclopentyl radicals of about unity, in both the liquid [11] and gas [98] phases, and $k_d/k_c = 1.4$ for radicals formed from cyclopentane dissolved in water [76]. These latter investigations were performed with pulse radiolysis techniques, by which it was established that at least 90% of the radicals are consumed with second-order kinetics, with a rate constant of $2 k_{3.11} = 2 \cdot 10^9$ M^{-1} s^{-1}, which falls in the range of diffusion-controlled reactions [76, 78]. The k_d/k_c ratio at 195 K in the liquid phase is 3.0 ± 1.1 [100], while $k_d/k_c = 3.05 \pm 1.1$ was reported in ref. 90 for the radical disappearance at 122 K during the solid–solid transition.

The most important products of dehydrogenation, formed on irradiation in the liquid phase at room temperature, are cyclopentene and bicyclopentane, with G-values of 2.9–3.1 and 1.2–1.3, respectively [79, 81, 84, 85, 99]. In the presence of 5 mM I_2 G-value of cyclopentene decreases to 1.92 and that of bicyclopentane to 0.46 [11]. The yield of bicyclopentane formed in the presence of radical acceptors can be accounted for by the fast recombination of cyclopentyl radicals produced near each other. Provided that the k_d/k_c ratio is also about unity for this reaction, the yield of cyclopentyl radicals reacting with each other can be as high as $G = 4.8$–5.2. As to the estimation of the total yield of cyclopentyl radicals, it should be taken into account that cyclopentyl radicals can react with other radicals in the system, with comparatively small yields (see Tables 3.5 and 3.6). These latter reactions lead to the formation of alkylcyclopentanes with a total G-value of ~ 0.24 [11].

As the k_d/k_c ratio for cyclopentyl radicals is close to unity, a fraction of the yield of cyclopentene ($G = 2.9$–3.1), corresponding to the yield of bicyclopentane, i.e. 1.2–1.3. G-unit, can result from a bimolecular reaction [reaction (3.11a)]. The remaining fraction, i.e. about 1.7 G-units, can be formed by unimolecular H_2 elimination:

$$\text{cyclo-}C_5H_{10}^* \longrightarrow \text{cyclo-}C_5H_8' + H_2 \qquad (3.12)$$

where the prime denotes a vibrationally excited species.

As far as the elimination of hydrogen molecules is concerned, it has been suggested [45] to occur chiefly with hydrogen atoms situated on the same carbon atom (this problem was discussed in Section 2.3.8. in relation to the radiolysis of hexane, and will be dealt with again in the treatment of cyclohexane). The vibrationally excited residual cyclo-C_5H_8' species will undergo either fragmentation or stabilization without decomposition:

$$\text{cyclo-}C_5H_8' \begin{cases} \longrightarrow \text{cyclo-}C_5H_8 & (3.13a) \\ \longrightarrow C_2H_4 + C_3H_4 & (3.13b) \end{cases}$$

The energy content of the cyclo-C_5H_8' species must be very large, as the energy of its precursor, cyclo-$C_5H_{10}^*$, certainly falls in the range 5–10 eV, and the energy demand of reaction (3.12) was calculated to be not more than about 1.2 eV. The excess energy of the activated species is likely to be transferred to the environment, if one considers the large number of deactivating encounters in the liquid phase. Thus, the stabilization takes place mostly via cyclo-C_5H_8 formation, as is evidenced by the rather low yield of C_3H_4 observed in liquid-phase experiments (Table 3.4). On the other hand, decomposition into fragments according to reaction (3.13b) proceeds with a considerable yield in the gas phase.

The investigations of Doepker et al. [45] and Wojnárovits et al. [39, 100] support the idea that unimolecular H_2 elimination is mainly a result of the decomposition of low-energy excited states. This reaction has been shown to be the most important primary decomposition process during photolysis brought about by photons of 7.6 or 8.4 eV energy, emitted by a Br_2 or Xe lamp. This is evidenced both by the relatively large yield of hydrogen (Table 3.4) and by the relatively small amounts of HD formed from cyclo-C_5H_{10}–cyclo-C_5D_{10} (1 : 1) mixtures; the isotopic distribution of the hydrogen was: 24.4% D_2, 13.4% HD and 62.4% H_2. However, when the 10.0 eV photons of a Kr lamp were applied, the C—H and C—C bond rupture processes leading to hydrogen atoms and alkyl radicals, respectively, were shown to be of considerable importance and consequently the extent of the unimolecular H_2 elimination was much less than at the lower energy irradiation [45].

Keszei et al. [11] established that in the liquid-phase, about 60–80% of the H_2 produced in unimolecular processes is formed subsequent to ion-electron recombination. This is in fair agreement with the previous observation, as the energy released on charge neutralization is relatively small. Huang and Hamill [101] have reported that the first ionization potential of solid-state cyclopentane is as small as 8.5 eV, and that the liquid-phase ionization potential and the energy of charge neutralization are probably of the same order.

A number of papers [11, 82, 85, 100, 102] have reported approximately complete product distributions of cyclopentane radiolysis; the combined G-value of dehydrogenation products, i.e. cyclopentene and bicyclopentane, is 1–1.5 G-units less than the yield of hydrogen. The hydrogen balance can be improved somewhat, i.e. by a few tenths of a G-unit, if the yields of other products are also taken into account. It has also been reported [85] that a 'polymer' fraction containing mostly hexamers and tetramers is also produced with a yield of $G(-\text{cyclo-}C_5H_{10}) = 0.7$. If this is taken into account, then the hydrogen deficit is reduced to 0.2 G-unit [85].

C—C bond rupture products

As can be seen from Table 3.4, C—C bond rupture products of the liquid-phase room-temperature radiolysis of cyclopentane include fragment products, and straight-chain C_5 hydrocarbons, as well as alkyl- and alkenylcyclopentanes having G-values of 1, 0.9 and ∼0.25, respectively. The isomerization of the cyclopentane ring into ethylcyclopropane was also observed, although with a rather small yield, i.e. $G = 0.05$ [85]. The yield of C—C bond rupture products is lower in the solid than in the liquid phase, and the extent of fragmentation is rather high in the gas phase.

Among the fragment products in the liquid phase, the yield of methane is not more than $G = 0.03$, and only traces of its complementary C_4 products were detected [82]; thus, ring cleavage produces mainly units of two and three carbon atoms. These species form chiefly ethylene and propene or cyclopropane in condensed phases, and ethene and propane and, to a lesser extent, propene in the gas phase. In condensed phases, pent-1-ene is the most important product of C—C bond rupture reactions, whereas only traces of this compound can be detected in the gas phase. Pent-2-enes are formed in all three phases.

Numerous experiments have been carried out by Ausloos et al. [45, 103, 104] to explore the photolytic and radiolytic evolution of fragment products and pentenes. According to their results, the formation of the majority of C_2H_4 and C_3H_6 can be traced back to the decomposition of excited molecules:

$$\text{cyclo-}C_5H_{10}^* \longrightarrow C_2H_4 + C_3H_6 \tag{3.14}$$

In their opinion, this is supported by the fact that photolysis of solid cyclopentane with 123.6 nm photons at 77 K produces C_2H_4 and C_3H_6 with identical yields. The fact that the C_3H_6 fraction contains both propene and cyclopropane points to a trimethylene intermediate. A number of authors (e.g. [105]) consider that the main reaction of trimethylene biradicals is the ring-closure process forming cyclopropane. A subsequent rearrangement can then lead to the formation of propene, provided that the intermediates have enough internal energy to supply the activation energy of the process.

The cyclo-C_3H_6/C_3H_6 ratio is considerably larger in the liquid, than in the gas phase at low pressures (Table 3.4). The excess cyclo-C_3H_6 formed in the liquid phase could be the result of a higher efficiency of deactivation. The energy requirement of reaction (3.14) is not more than $\sim$2 eV, thus the cyclopropane retains enough energy for the rearrangement into propene; the activation energy requirement of the latter reaction is 2.8 eV [45].

Pent-1-ene is an important product of photolysis in the solid state at 123.6 nm (10.0 eV) (Table 3.4). Its yield decreases in condensed-phase radiolysis in the presence of positive ion and electron scavengers, but changes only slightly, if at all, when radical acceptors are added to the cyclopentane. Ausloos et al. [104] thus came to the conclusion that pent-1-ene is formed largely via isomerization subsequent to ion-electron neutralization:

$$\text{cyclo-}C_5H_{10}^+ + e^- \longrightarrow \text{cyclo-}C_5H_{10}^* \longrightarrow 1\text{-}C_5H_{10}$$

The pent-1-ene molecule possesses a considerable amount of energy, about 5–10 eV, so it readily decomposes in the gas phase.

The highest intensity peaks in the mass spectrum of cyclopentane have mass-to-charge ratios of 70, 55, 42 and 41, corresponding to $C_5H_{10}^+$, $C_4H_7^+$, $C_3H_6^+$ and $C_3H_5^+$ ions, respectively. According to the 70 eV mass-spectral fragmentation pattern, the ion currents corresponding to the $C_3H_6^+$, $C_3H_5^+$ and $C_4H_7^+$ species amount to 44%, 12.5% and 12.3% of the total ionization, respectively. The formation of low-molecular mass fragment ions can be formulated as follows [82]:

$$\text{cyclo-}C_5H_{10}^+ \longrightarrow C_3H_6^+ + C_2H_4 \tag{3.15}$$

$$\text{cyclo-}C_5H_{10}^+ \longrightarrow C_4H_7^+ + CH_3 \tag{3.16}$$

$$\text{cyclo-}C_5H_{10}^+ \longrightarrow C_3H_5^+ + C_2H_5 \tag{3.17}$$

Taking into account the relative intensities of the ions detected, the proportion of the fragmentation occurring via reaction (3.15) exceeds by several times those of reactions (3.16) and (3.17). The $C_3H_5^+$ ions can presumably be produced not only by reaction (3.17), but also by the decomposition of $C_3H_6^+$ and $C_4H_7^+$ ions [104]. Although most of cyclo-$C_5H_{10}^+$ ions are deactivated in condensed phases, some undoubtedly suffer fragmentation, albeit in low yield.

On studying the solid-phase radiolysis of cyclo-C_5H_{10}–cyclo-C_5D_{10} systems, Ausloos et al. found that, on the one hand, in the presence of the electron acceptor CCl_4 the G-value of propane did not decrease, and, on the other, in the presence of a radical scavenger the G-value of C_3D_7H decreased sharply, whereas those of C_3D_8, $C_3D_6H_2$, $C_3D_2H_6$ and C_3H_8 decreased to a much lesser extent. The authors suggested that the latter compounds arise from H_2^--transfer processes subsequent to the fragmentation shown in reaction (3.15):

$$C_3H_6^+ + \text{cyclo-}C_5H_{10} \longrightarrow C_3H_8 + \text{cyclo-}C_5H_8^+ \quad \varDelta H_{gas} = -163 \text{ kJ mol}^{-1}$$

$$(3.18)$$

From the yields of the propane fraction, the ions decomposing via reaction (3.15) amount to 1 and 0.2% of the total amount of cyclopentane cations at 195 and 77 K, respectively [103].

The frequency of decomposition of ions can thus be calculated from the propane yield. This is indicated by the result [45, 104] that if cyclopentane is photolysed with photons of lower energy than the ionization potential, propane is formed only in trace amounts, whereas its yield is considerably larger when photons of higher energy than the ionization potential, or high-energy ionizing radiation, are applied. The importance of ionic decomposition is supported by the gas-phase radiolysis experiments performed in the presence of NO as a radical scavenger [104]. The yield of propane decreases, whereas that of propene slightly increases when NO is added to the samples before irradiation. As the ionization potential of propene is higher than that of NO (9.8 and 9.25 eV, respectively), the observations can be explained by an exothermic charge transfer reaction:

$$C_3H_6^+ + NO \longrightarrow C_3H_6 + NO^+$$

Another ion-molecule reaction can take place between $C_3H_6^+$ ions and cyclopentane molecules that can affect the product distribution, chiefly in the gas phase, as it probably has a G-value less than that of reaction (3.18) by a factor of about 3.5:

$$C_3H_6^+ + \text{cyclo-}C_5H_{10} \longrightarrow \text{sec-}C_3H_7 + \text{cyclo-}C_5H_9^+$$

This latter reaction is evidenced by the formation of $CD_3-CDH-CD_3$ in the radiolysis of cyclo-C_5D_{10}–H_2S systems and, moreover, by the fact that in the photolysis of the same system with photons of lower energy than the ionization potential, propyl radicals could not be detected [45, 104]. The yield of sec-propyl radicals in the liquid phase was not more than $G \approx 0.02$ [11].

3.2.3. Cyclohexane

Cyclohexane is undoubtedly the most popular model compound for investigations on the radiolysis of hydrocarbons. The great majority of papers aim at establishing certain functions, such as the dose dependence of G-values, the investigation of some physical phenomena, e.g. the mobilities of ions, or the study of some types of chemical reactions, e.g. the mechanism of hydrogen evolution. Only a small proportion of the reports are concerned with correlations between molecular structure of cyclohexane and radiation chemical yields, by comparing the G-values obtained with those of other hydrocarbons, etc.

Cyclohexane has been studied so frequently because it is relatively easy to handle, cheap and can be obtained in high purity. It has one other advantage over the aliphatic alkanes, namely that it contains only one kind of C—H and C—C bond and, consequently, the product distribution is relatively simple.

At the same time, the use of cyclohexane as a model compound has the disadvantage that it may be a somewhat structured liquid at room temperature, which is not much above the 'abnormally' high melting point of 6.5°C [85]. However, only the very rapid positive charge migration can be mentioned as an unusual property of its radiolysis (see Section 3.1).

A brilliant review of the radiolysis of cyclohexane was published by Cramer [107] in 1968. Therefore, this Section covers mainly the results published since then, as well as the areas not fully covered in Cramer's work.

Hydrogen formation

The $G(\mathrm{H_2})$ values measured in the gas and the liquid phases do not differ significantly. The yield of hydrogen changes only slightly with temperature up to 500 K (Table 3.7). Blachford and Dyne [108] established by investigation of cyclohexane–perdeuterocyclohexane (4.77 mol%) systems in the gas phase that the ratio of components of the hydrogen formed, i.e. $\mathrm{H_2}$, HD and $\mathrm{D_2}$, is independent of the temperature in this range and differs only very slightly from that obtained in the liquid phase. This is rather surprising in view of a number of reports that the mechanisms of hydrogen formation in the two phases differ in many respects. The yield of hydrogen formed in the gas phase via recombination of electrons with cyclohexyl cations is not more than $G = 1.5$ [140], in a sharp contrast to that obtained in the liquid phase, namely $G = 4.0$ [16]. Moreover, a certain fraction of the ionic contribution to the yield of hydrogen originates in the neutralization of fragment ions resulting from the decomposition of positive cyclohexane ions.

The sharp increase of the yield of hydrogen observed above about 250°C, in the opinion of Blachford and Dyne [108] can be traced back to radiation induced pyrolysis. The data given by Jones [110] show a considerable decline in the yield of hydrogen as a function of the pressure of gaseous cyclohexane at 300°C (Table 3.7). This phenomenon was attributed by the author partly to a cage effect and partly to collisional deactivation, both factors becoming of greater importance with increasing pressure.

Table 3.7

Dependence of yields from cyclohexane on experimental conditions
(radiation: ^{60}Co-γ)

Temperature, K	373	398	406	508	573	573	573
Density, g cm^{-3}	0.004	0.0047	0.0025	0.0047	0.0079	0.079	0.42
Reference	141	108	140	108	110		
Phase	Gas						
Product	G, molecule/100 eV						
Hydrogen	4.7	5.3	4.6	5.55	8.7	6.3	5.7
cyclo-C_6H_{10}	·1.0	2.2	2.4	3.1	2.6	2.4	1.7
(cyclo-C_6H_{11})$_2$	0.8	1.0	0.75	1.15	0.5	1.2	1.1

Temperature, K	295	373	433	493	77	178	195
Density, g cm^{-1}							
Reference	133				139		
Phase	Liquid				Solid		
Product	G, molecule/100 eV						
Hydrogen	5.6	5.85	6.24	6.36	4.53	4.4	4.73
cyclo-C_6H_{10}	3.26	2.99	2.24	1.64	2.27	2.11	1.76
(cyclo-C_6H_{11})$_2$	1.81	1.87	1.85	1.68	1.32	1.45	1.34

The relatively small hydrogen yield observed in solid-phase radiolysis of cyclohexane was attributed by van Dusen and Truby [111] and Kimura et al. [87, 112] to the fact that at lower temperatures the thermal hydrogen atoms are unable to abstract hydrogen but, instead, recombine. In Stone's opinion, some other factors contribute to the low yield: the frozen cyclohexyl radicals may act as electron and hydrogen atom acceptors, and the hydrogen atoms can readily add to the unsaturated species produced during irradiation [113]. The important role of these latter processes is also indicated by the strong dose dependence of the hydrogen yield observed when the cyclohexane is irradiated in a xenon matrix [113]. Recent results contradict the idea that thermal hydrogen atoms do not react with hydrocarbons at low temperature; in methane matrix, hydrogen atoms produced at 4 K were shown to abstract a hydrogen atom from the ethane solute upon warming up the sample to 10–30 K [113A].

Sagert discovered that in the presence of electron acceptors, hydrogen formation by electron-ion neutralization has a lower efficiency in the solid than in the liquid phase [114].

The dose dependence of the G-values of some products of cyclohexane radiolysis is shown in Fig. 1.3.

Asmus et al. [16] determined the components of hydrogen formation on basis of comprehensive studies involving a number of scavengers of different kinds. Their results are presented in Table 3.8. The term 'ionic' yield refers to the hydrogen that originates from decomposition subsequent to electron–ion neutralization. Freeman and Stover [10] published a value 3.4 for the same quantity, whereas the former authors reported a value of 4.0 based on the experimental results of Robinson and Freeman [115]. The 'molecular' yield can be understood as the one not influenced by radical acceptors, while the term 'atomic' yield stands for the so-called 'thermal hydrogen' yield, i.e. the yield of hydrogen formed through hydrogen atoms, which react with radical scavengers such as ethylene. In earlier works [116] several values ranging from 0.2 to 3.1 were given for the yield of thermal hydrogen atoms. The reason for this rather wide range has already been discussed in detail in Section 1.4.1. However, a number of papers published more recently have reported data that differ only slightly from the value shown in Table 3.8, i.e. $G = 1.46$, for example 0.8 [117], 1.0 [87], 1.3 [118], 1.4 [119]. Asmus et al. made the necessary corrections for the reactions taking place between the ethylene radical scavenger and the positive ions in the system. The accuracy of their results is supported by studies in which complex scavenging processes have been successfully described using the value $G(H)_{thermal} = 1.46$ [120–122].

Table 3.8

Sources of the hydrogen formation from liquid cyclohexane
(radiation: ^{60}Co-γ). After Asmus et al. [16]

Reaction types	Molecular[a]	Atomic	Total
	G, atom or molecule/100 eV		
Ionic	3.01	0.89	3.90
Nonionic	1.20	0.57	1.77
Total	4.21	1.46	5.67

[a] Includes any H_2 produced by hot hydrogen atoms through abstraction of hydrogen from the solvent

Pulse radiolysis of cyclohexane dissolved in water gave the rate constant of the reaction taking place between hydrogen atoms and cyclohexane molecules as $7 \cdot 10^7 \, M^{-1} \, s^{-1}$. This value hardly differs from that measured for cyclopentane, namely $10^8 \, M^{-1} \, s^{-1}$ [76].

The so-called molecular yield results from a combination of fast bimolecular non-scavengeable reactions and unimolecular H_2 elimination. It is difficult to differentiate between these two processes. Investigations carried out with cyclohexane–perdeuterocyclohexane systems might give some information on the unimolecular hydrogen formation. However, the interpretation of such experimental results presents problems, because

1. There is a very large kinetic isotope effect in the hydrogen abstraction step for thermal hydrogen atoms [123];
2. The isotopic purity of the samples is generally less than 100% by a few tenths of a per cent;

Table 3.9

Hydrogen yields of the photolysis and radiolysis of
cyclo-C_6H_{12}–cyclo-C_6D_{12} mixtures

Radiation	Energy, eV	$\dfrac{\text{cyclo-}C_6H_{12}}{\text{cyclo-}C_6D_{12}}$	Phase	Pressure, mbar	Temperature, K	Additive	H_2, %	HD, %	D_2, %	References
^{60}Co-γ	—	1 : 1	Gas	26	298	5% NO	45.0	29.7	25.3	125
Photolysis	8.4	1 : 1	Gas	13	298	—	56.2	4.7	39.1	106
Photolysis	8.4	1 : 1	Gas	13	298	5% NO	59.3	5.3	35.4	106
Photolysis	10.0	1 : 1	Gas	13	298	—	56.2	7.4	36.4	106
Photolysis	10.0	1 : 1	Gas	13	298	5% NO	56.7	8.7	34.6	106
Photolysis	11.6–11.8	1 : 1	Gas	26	298	5% NO	60.1	12.3	27.5	125
^{60}Co-γ	—	1 : 1	Liquid		283	—	56.8	32.1	11.1	129
Photolysis	8.4	3 : 2	Liquid		283	—	83.8	3.7	12.5	128
Photolysis	10.0	3 : 2	Liquid		283	—	84.2	4.6	11.2	128
^{60}Co-γ	—	1 : 1	Solid		77	—	54.5	32.3	13.2	113, 127

3. A charge-transfer interaction between the light and heavy cyclohexane molecules is possible, preferentially forming cyclo-$C_6H_{12}^+$ ions [103, 124–126]:

$$\text{cyclo-}C_6H_{12}^+ + \text{cyclo-}C_6D_{12}^+ \rightleftharpoons \text{cyclo-}C_6H_{12}^+ + \text{cyclo-}C_6D_{12}$$

The data listed in Table 3.9 show the isotopic distribution of the hydrogen yield produced in different phases by radiolysis and photolysis of mixtures containing light and heavy cyclohexanes in proportions of 1 : 1 and 3 : 2. If the hydrogen were produced exclusively by reactions of hydrogen atoms, the proportion of HD formed in 1 : 1 mixtures would be 50%, provided that no isotope effect arises in the hydrogen-abstraction step. However, the proportion of HD is in each case lower than 40%. Especially small amounts of HD are formed in photolysis, though its yield increases with increasing photon energy. Consequently, a unimolecular H_2 elimination takes place as a result of the decomposition of molecules in lower excited states, whereas the atomic pathway is preferred by molecules being in higher excited states. According to Ausloos et al., the proportion of molecular D_2 elimination amounts to 80, 60 and 55% on photolysis of cyclo-C_6D_{12} in the gas phase with wavelengths of 147.0, 123.6 and 104.8–106.7 nm, respectively [125]. In liquid-phase radiolysis a value of $\sim$7% was reported for the same quantity [129–131]. The proportion of the molecular H_2 elimination from cyclo-C_6H_{12} in the gas-phase amounts to $\sim$95 and 90% with photons of 147.0 and 123.6 nm, respectively [106], whereas it is about 10–20% in liquid-phase radiolysis, as reported in an earlier paper by Dyne and Jenkinson [130]. An investigation of the formation of hydrogen and dehydrogenation products in the presence of scavengers, led the authors [81, 132] to estimate the unimolecular yield in the liquid phase as $G = 1$, whereas taking into account LET effects a value of $G = 0.81$ was calculated [133]. The unimolecular yield is probably higher in the solid phase at 77 K, i.e. $G = 1.8$–2.0 [114].

Unimolecular H_2 elimination can be accomplished from two adjacent carbon atoms as well as from a single carbon atom (see Sections 2.2.8 and 3.2.2). Doepker and Ausloos have published experimental evidence that H_2 elimination from cyclohexane takes place from one carbon atom in gas-phase photolysis [106]. They irradiated 1,1,2,2,3,3-hexadeuterocyclohexane with Xe and Kr resonance lines, and found the distributions of H_2, HD and D_2 to be as follows: 72.8, 10.6, 16.6 and 67.3, 15.9, 16.8, respectively. Thus the yield of HD is lower than those of both H_2 and D_2; the sequence $H_2 >$ $> HD > D_2$ would be expected from consideration of the results obtained in cyclo-C_6H_{12}–cyclo-C_6D_{12} mixtures [103, 106]. Ballenger et al. established the ratio 6 : 3 : 1 for the D_2 yields formed in the solid and liquid phase radiolysis of gem-, cis-1,2- and trans-1,2-dideuterocyclohexanes. The considerable yield of D_2 produced from the gem isomer is further evidence that unimolecular H_2 elimination takes place from one carbon atom [129].

The question remains as to how that part of the hydrogen is produced which cannot be accounted for either by unimolecular elimination or by the reactions of thermal hydrogen atoms. It has been assumed [107, 116] that this fraction arises from hot hydrogen atoms that have enough energy to abstract hydrogen during the first collisions with the molecules of the solvent, and therefore the radical acceptors are unable to interact with them. This latter fact illustrates well the difficulties encountered in studying the reactions of hot hydrogen atoms, and explains why all the information we have on them originates from indirect sources. As will be seen later in this Section, the yield of cyclohexyl radicals acceptable by radical scavengers is 5–5.7, whereas the estimated yield, considering the k_d/k_c ratio of cyclohexyl radicals, is about 8. This means that it is higher than the yield of cyclohexyl radicals corresponding to the yield of thermal hydrogen atoms. If the total yield of cyclohexyl radicals is taken to be $G = 8$, then a value of 2.5 can be derived for the yield of hot hydrogen atoms.

Experiments carried out with recoil tritium atoms can also give some information on the reactions of hot hydrogen atoms, although the influence of the energy content and the mass numbers, which are different in the two cases, cannot be neglected [134]. Having measured the yields of different labelled products, Avdonina summarized the more important reactions as follows [135, 136]:

$$T + \text{cyclo-}C_6H_{12} \longrightarrow HT + \text{cyclo-}C_6H_{11} \quad 50\%$$

$$T + \text{cyclo-}C_6H_{12} \longrightarrow \text{cyclo-}C_6H_{11}T + H \quad 25\% \qquad (3.19)$$

$$T + \text{cyclo-}C_6H_{12} \longrightarrow \text{cyclo-}C_6H_{11}T + H$$
$$\downarrow$$
$$\text{fragments} \quad 1.5\%$$

$$T + \text{cyclo-}C_6H_{12} \longrightarrow C_6H_{12}T^{\cdot} \qquad (3.20)$$

$$C_6H_{12}T \longrightarrow C_nH_{2n} + C_{6-n}H_{12-2n}T$$

$$C_nH_{2n}T + \text{cyclo-}C_6H_{12} \longrightarrow C_nH_{2n+1}T + \text{cyclo-}C_6H_{11} \qquad 5\% \qquad (3.21)$$

$$T_{hot} \longrightarrow T_{thermal}$$

$$T_{thermal} + \text{cyclo-}C_6H_{12} \longrightarrow HT + \text{cyclo-}C_6H_{11} \qquad 11\% \qquad (3.22)$$

$$T + \text{cyclo-}C_6H_{12} \longrightarrow \text{polymer and other products having}$$
$$\text{low yields} \qquad 7\%$$

From all these processes only the reactions (3.21) and (3.22) are sensitive to the presence of oxygen. The important role played by the reaction (3.19) is noteworthy: an analogous 'substitution' may possibly take place when hot hydrogen atoms collide with cyclohexane molecules. In reaction (3.20), the hot tritium atom breaks the cyclohexane ring forming thus aliphatic hexyl radicals. These latter species possess a considerable amount of excess energy resulting, at least in part, in a secondary fragmentation of the radical, or they become stabilized simply by hydrogen abstraction or recombination with other radicals [136–138].

Dehydrogenation products

The earlier and widely cited G-values for cyclohexene and bicyclohexane formed on room-temperature liquid-phase radiolysis of cyclohexane are 3.27 ± 0.06 and 1.95 ± 0.06 [142] and 3.2 ± 0.2 and 1.76 ± 0.05 [143], respectively. The data in more recent papers do not differ considerably from these, e.g., 3.2 and 1.8 [121]; 3.25 and 1.9 [144]; 3.0 and 1.85 [145]. If the temperature is increased, the G-value of cyclohexene decreases somewhat, whereas that of bicyclohexane remains unchanged in the range 20–220°C [133]. The yields of both products are slightly lower in the gas and solid phases than in the liquid phase [108, 139–141].

The yields of cyclohexene and bicyclohexane measured in different phases are shown in Table 3.7. The ratio of yields of cyclohexene and bicyclohexane determined in the photolysis of cyclohexane considerably exceeds that obtained by radiolysis under the same experimental conditions [34, 125, 128].

The formation and decay of cyclohexyl radicals have been studied several times by means of ESR spectroscopy in the gas [146, 147], liquid [96] and solid [148–155] phases. The major objective of some of these studies has been the determination of the chemical nature of radicals produced, as well as collection of characteristic ESR lines; other authors were concerned with ESR spectroscopy as a means of quantitative analysis. Some papers have reported data obtained by studying cyclohexyl radicals formed on pulse radiolysis [119, 157, 158]..

Bansal and Schuler used 0.1 mM I_2 to determine the yield of cyclohexyl radicals and obtained $G = 5.3$–5.4. Taking into account the scavenging of hydrogen atoms and positive ions, they corrected this value to $G = 5.7$ [97]. Stone et al. reported a value of 5.2 for the yield of cyclohexyl and alkyl radicals produced during cyclohexane radiolysis using propanethiol additive [120, 121]. In the opinion of the latter authors, the role of the otherwise disturbing reactions of hydrogen atoms and ions can be neglected in this case (see Section 1.4.1). As the total yield of radicals other than cyclo-hexyl amounts to $G = 0.25$ [156], Stone's overall value is undoubtedly smaller than the one obtained with I_2 radical scavenger.

Ebert et al. [157] and Sauer and Mani [119] investigated the reactions of cyclohexyl radicals formed on pulse radiolysis. They established that the disappearence of radicals takes place according to diffusion-controlled second order kinetics. Applying a careful experimental method for the study of radical reactions, Makarov and Kabakchi came to the conclusion that only about 80% of the radicals take part in second-order reactions, whereas the rest disappear by fast, formally first-order reactions. As the unimolecular decomposition of cyclohexyl radicals looks highly improbable under the given experimental conditions, the apparently first-order kinetics can be the result of 'spur' reactions, i.e. fast recombination between radicals pro-duced in close proximity [158]. Considering these latter data, the total yield of cyclohexyl radicals is surely above $G = 6$, thus exceeding the yield of cyclopentyl radicals formed on the radiolysis of cyclopentane by at least about 10% (see Section 3.2.2). The stabilization of cyclohexyl radicals leads to the formation of cyclohexene and bicyclohexane:

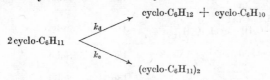

The ratio of rate constants, k_d/k_c, in liquid phase at room temperature is 1.1, according to Ho and Freeman [143], and Cramer [159] (see Table 3.10). The correctness of this result has been checked and confirmed several times in the past few years [120, 160]. A number of experimental results indicate that neither the yield of cyclohexene, nor that of bicyclohexane can be de-creased to zero by adding radical acceptors to the samples. The G-values of cyclohexene and bicyclohexane decrease in the presence of about 10–100 mM propanethiol [120, 121], n-propyl disulphide [120, 160], benzoquinone or oxygen [142], phenyl-N-tert-butylnitrone [161] as radical acceptors to 1.3–1.8 and 0.2–0.4, respectively.

According to the model experiments of Blackburn and Charlesby [150], for an explanation of the formation of non-scavengeable dimers one need not assume a reaction pathway different from that producing the scavenge-able part. The formation of the non-scavengeable dimers can be related to the fast recombination of radicals produced very close to each other, as discussed earlier in connection with pulse radiolysis experiments. For these types of reaction taking place in the 'spur' no k_d/k_c value is available; how-ever, if we suppose that it does not differ considerably from that in the bulk medium, the yield of cyclohexyl radicals forming cyclohexene and bicyclo-hexane products can be estimated as $G = 8$ [133]. To get an approximately correct estimate of the total yield of cyclohexyl radicals, those that are transformed into hexyl-, hexenyl- and ethylcyclohexane, with a combined yield of $G = 0.2$, should also be taken into account. On the radical pathway

Table 3.10

Ratios of rate constants of disproportionation and combination
reactions (k_d/k_c) for cyclohexyl radicals

Source of radicals	Temperature, K	Phase	k_d/k_c	References
Hg-photosensitization	298	Gas	0.455	164
Hg-photosensitization	Room temperature	Liquid	1.31	109
Radiolysis, LET effect	295	Liquid	1.24	133
Radiolysis, scavenging	295	Liquid	1.1 ± 0.3	143
$Hg(C_6H_5)_2 \xrightarrow{UV} Hg + 2 C_6H_5$ $C_6H_5 + cyclo\text{-}C_6H_{12} \to C_6H_6 + cyclo\text{-}C_6H_{11}$	296	Liquid	1.1 ± 0.14	159
Hg-photosensitization	318	Liquid	1.47	165
$cyclo\text{-}C_6H_{10} + T \to cyclo\text{-}C_6H_{10}T$	77 112	Solid Solid	49 ± 2 4.5 ± 0.5	166 166
$C_6H_5I + Na \to C_6H_5 + NaI$ $C_6H_5 + cyclo\text{-}C_6H_{12} \to C_6H_6 + cyclo\text{-}C_6H_{11}$	173	Solid	0.7 ± 0.15	167
$^{60}Co\text{-}\gamma$ dissolution method	186	Solid	0.65	90A
$^{60}Co\text{-}\gamma$	186	Solid	0.3 ± 0.2	114
$^{60}Co\text{-}\gamma$	196	Solid	0.5 ± 0.1	163

of hydrogen formation the production of one H_2 molecule is accompanied
by that of two cyclohexyl radicals. Thus, the unimolecular fraction of the
total hydrogen yield, $G(H_2) = 5.6$, which is not accompanied by a simultaneous formation of cyclohexyl radicals, can be given by the equation $G =$
$= G(H_2) - 1/2G(cyclo\text{-}C_6H_{11}) = 1.5$ molecule/100 eV. This quantity is the
so-called unimolecular H_2 elimination discussed earlier.

If both atoms of the H_2 molecule come from the same carbon atom, the
residual carbene intermediate rearranges rapidly to cyclohexene [106, 125,
129]. Whether the hydrogen formation had taken place from one carbon
atom or from two adjacent ones, the cyclohexene molecule produced possesses an internal energy of up to a few electronvolts. This is why the cyclohexene molecule is excited, and leads to the formation of ethylene and butadiene in about 50% when the cyclohexane is irradiated in the gas phase
at a pressure of about 1 mbar. The ratio of decomposition decreases rapidly with increasing pressure, i.e. with increasing rate of collisional deactivation. In the liquid phase, the extent of fragmentation must be less
than 3%, as the yield of butadiene formation is low both in photolysis
($\Phi = 0.02$ [128]) and radiolysis ($G = 0.004$ [162]).

At 77 K, the cyclohexyl radicals are largely frozen ($G = 3.1 \pm 1.2$ [163];
3.8 ± 0.3 [153, 154]). Transformation into stable end-products occurs when

the samples are warmed to the temperature of the phase change, namely $\sim$186 K [90A]. Sagert's [114] results show that in the solid phase the cyclohexene is produced mainly via unimolecular hydrogen elimination.

As can be seen from Table 3.10, the k_d/k_c ratio for cyclohexyl radicals decreases markedly with increasing temperature. Accordingly, the activation energy of recombination must be greater than that of disproportionation $(E_c - E_d = 5.0–5.5$ kJ mol^{-1}). Similar results have been reported for ethyl radicals in the gas, liquid and solid phases [92] and for isopropyl radicals in the liquid state [168].

The experimental fact that the yield of cyclohexene formed in liquid-phase radiolysis decreases with increasing temperature in the range 22–220°C, has been explained by Burns and Reed mainly in terms of secondary reactions, e.g.

$$\text{cyclo-C}_6\text{H}_{11} + \text{cyclo-C}_6\text{H}_{10} \longrightarrow \text{C}_{12}\text{H}_{21}$$

The increase in the yield of cyclohexylcyclohexene from 0.15 to 0.54 observed in the same temperature range supports the increasingly important role of the latter reaction [133].

Both cyclohexene and bicyclohexane have lower yields in the gas than in the liquid phase. This observation can be explained, in part, by the fragmentation of positive ions prior to neutralization with electrons and, in part, by the decomposition of cyclohexene formed by unimolecular hydrogen elimination, as discussed earlier. On the basis of the considerable deficit in the hydrogen balance, Blachford and Dyne [108] suggested that polymer formation also occurs.

Other experiments conducted in the liquid phase showed a continuous increase in the yield of cyclohexylcyclohexene up to a dose of 15–45 · 10^{19} eV g^{-1} and afterwards it remains constant ($G = 0.3$). As its initial yield is practically zero, this compound is probably formed via secondary reactions involving cyclohexyl radicals. This is supported by the observation that the yield is not more than $\sim$0.01 when radical acceptors are present [143].

C–C bond rupture products

The C–C bond rupture products of cyclohexane radiolysis carried out in gas, liquid and solid phases are listed in Table 3.11. Among the products of liquid-phase room-temperature radiolysis are the following, with G-values shown in parentheses: fragment hydrocarbons (0.2), straight-chain C$_n$ hydrocarbons ($\sim$0.5), methylcyclopentane (0.15–0.2) and alkylcyclohexanes (0.1–0.2).

According to the detailed investigations of Makarov et al. on the formation of fragment, hex-1-ene and hexane products, the yield of these increases with increasing temperature in the range from -196 to $+60$°C. However, whereas in the case of hex-1-ene only a phase effect can be observed, i.e. at the melting point the G-value doubles and the yield is otherwise strictly temperature-independent, for the other products both phase and temperature effects can be recognized [170–172].

14

210   3. CYCLOALKANES

Table 3.11

C—C-bond rupture products from cyclohexane (radiation: ^{60}Co-γ)

Temperature, K	373	296	298	300	300	77	185
Reference	141	143	102, 169	170	143, 170	170, 172	170, 172
Phase	Gas (3,2 mbar)	Liquid			Liquid + 10 mM O$_2$	Solid	
Product	\multicolumn{7}{c}{G, molecule/100 eV}						
Methane	0.36	0.006	0.01	0.01	0.009	0.002	0.002
Ethane	1.4		0.015	0.0057	0.0022	0.002	—
Ethylene	1.7		0.10	0.119	0.13	0.021	—
Acetylene	0.35		0.025	0.0143	0.0155		
Propane	0.44		0.011	0.0043	0.0018	0.005	0.005
Cyclopropane			0.006	0.0045	0.0042	0.004	0.004
Propene	0.55		0.025	0.036	0.038	0.012	0.009
n-Butane			0.008	0.011	0.005	0.006	0.006
But-1-ene	} 0.6		} 0.025	} 0.044	0.0035	0.009	0.009
But-2-ene					} 0.031	0.009	0.009
Buta-1,3-diene			0.004			0.015	0.015
n-Hexane		0.08	0.08	0.071	0.029		
Hex-1-ene		0.40	0.36	0.37	0.256	0.19	0.19
Hex-2- and -3-enes			0.02	0.018			
Methylcyclopentane		0.15	0.20		0.03		
Ethylcyclohexane	0.3	0.04			0.01		
n-Hexylcyclohexane			0.08				
6-Cyclohexylhex-1-ene		0.12	0.03		0.03		

Makarov and Chernyak [173] reported the following G-values for the C$_2$, C$_3$ and C$_4$ fragment hydrocarbons: 0.123, 0.048 and 0.041.

1   2

The fact that the yield of C$_2$ products exceeds that of the C$_4$ products, the latter being the complementary products of the former ones, was explained by the authors by dissociation of the molecule into three C$_2$ units (fission type **3**). Similar results have been reported in a number of other studies [102, 170–172, 174, 175].

Holroyd and co-workers irradiated cyclohexane in the gas and liquid phases with photons of energies lower (8.4 eV, 147 nm) as well as higher (10.0 eV, 124 nm) than the ionization potential (9.88 eV) [33, 128]. They detected only relatively small amounts of C$_2$-C$_4$ products when the photolysis was carried out with 8.4 eV photons, whereas a 20-fold increase in the yields was observed with 10.0 eV photons. Makarov et al. have reported a 30, 25 and 10% reduction in the yields of most important unsaturated fragment products in the presence of N$_2$O, CO$_2$ and NH$_3$ as ion and electron scavengers, respectively [170, 172]. Making use of the results of Holroyd et al. also, they concluded that the major part ($\sim 70\%$) of the monoalkene products is formed by decomposition of molecules in superexcited states [171], whereas the minor part springs from fragmentation of molecular ions. The hypothesis concerning the role of superexcited molecules is supported by the results of Ausloos et al. [125]. Photolysing gas-phase cyclohexane with 8.4 eV photons, they established that the yields of fragments decrease strongly with increasing pressure, probably owing to the increasing efficiency of collisional deactivation. In the liquid phase, where collisional deactivation is more important, the decomposition of low-energy excited states forming fragment products may be considerably less extensive.

Makarov et al. observed a reduction of $\sim 40\%$ in the yield of hex-1-ene in the presence of 0.1–0.4 M electron and positive ion acceptors, such as N_2O, CO_2 and NH_3 [170, 172]. The trend of the yield of hex-1-ene in the radiolysis of cyclohexane–benzene mixtures indicates a strong 'protective effect'. These results lead to the conclusion that the transformation of cyclohexane into straight-chain hex-1-ene is probably due to fission of the excited states produced in part, via ion-electron neutralization and in part directly, giving excited states of lower energy than the ionization potential. On the contrary, it has been directly proved that hex-1-ene or other C—C bond rupture products are not produced in the liquid-phase 163 nm (7.6 eV) photolysis [34, 36, 37]. It means that the lowest excited state of cyclohexane, which may be populated under these circumstances, does not contribute to the ring decomposition.

The yields of straight-chain alkenes are reduced by about 30% on the addition of oxygen. Since the common feature of products formed by disproportionation reactions of radicals is that their yields are greatly reduced in the presence of radical acceptors (i.e. by about 80–90%), some authors have concluded that in the formation of these products disproportionation reactions of monoalkyl radical intermediates (e.g. straight-chain alkyl radicals) are of minor importance [160, 170, 172]. It seems reasonable to assume that among the possible intermediates an especially important role is played by 'biradicals' formed by the fission of C—C bonds and capable of undergoing very fast chemical transformations. Because they decay so rapidly, the biradicals are not directly detectable. However, their existence and importance in radiation chemical processes have been evidenced by several experimental results [102, 105, 137, 138, 162, 173, 176–178], e.g. from studies applying radical scavengers, and the phase effect observed in the yield of hex-1-ene. The solid phase hinders the rearrangement of the biradical necessary for the transformation into hex-1-ene, thus promoting re-formation of the ring [170].

The existence of biradicals is also supported by the fact that hex-1-ene is not produced in the gas-phase radiolysis of cyclohexane, whereas the yields of fragments are considerably higher than those in the liquid phase [106, 110, 170]. At the instant of ring-opening, the biradical still possesses some excess energy amounting to a few electron volts that will rapidly be transferred, in the liquid phase, in part or completely, to the molecules in the environment thus promoting isomerization into aliphatic hexene or the

14*

reclosure of the ring instead of fragmentation of the biradical. The efficiency
of deactivation is considerably lower in the gas phase and thus fragmentation can be the preferred reaction:

cyclo-$C_6H_{12}^*$ $\longrightarrow$ $\cdot CH_2-CH_2-CH_2-CH_2-CH_2-CH_2^\cdot$ energy-rich hexamethylene biradical

$$\cdot CH_2-CH_2-CH_2-CH_2-CH_2-CH_2^\cdot \Longleftarrow$$

cyclo-C_6H_{12} deactivation

1-C_6H_{12} intramolecular rearrangement

$C_2H_4 +$ $\cdot CH_2-CH_2-CH_2-CH_2^\cdot$

2 $\cdot CH_2-CH_2-CH_2^\cdot$

$\cdot CH_2^\cdot + \cdot CH_2-CH_2-CH_2-CH_2-CH_2^\cdot$

Among the biradicals formed on the fragmentation of hexamethylene,
$\cdot CH_2-CH_2-CH_2^\cdot$ can undergo unimolecular stabilization, as its decomposition is endothermic by 1–2 eV, whereas the $\cdot CH_2-CH_2-CH_2-CH_2^\cdot$ and
$\cdot CH_2-CH_2-CH_2-CH_2-CH_2^\cdot$ can readily undergo secondary decomposition
[170], e.g.:

$$\cdot CH_2-CH_2-CH_2-CH_2^\cdot$$

1-C_4H_8 or 2-C_4H_8

2 C_2H_4

The investigations of Makarov et al. suggested that the saturated hydrocarbon fragments of cyclohexane radiolysis are formed via reactions basically different from that producing the unsaturated fragments [170, 172].
This is illustrated by the fact that the yields of the former compounds decreased considerably, i.e. by about 55–70%, in the presence of 10 mM oxygen, whereas those of the monoalkene products decrease by 30% at most.
The yields of saturated fragments formed from the cyclohexane component
of irradiated cyclohexane–benzene mixtures suggest a strong 'protective
effect', whereas in the G-values of monoalkene products only negligibly small
'protective effect' is observed. It has been assumed by Avdonina et al. [135,
136] that, analogously to the reactions of recoil tritium atoms, the hot hydrogen atoms produced during radiolysis are also able to break the ring [re-
action (3.20)]. The n-hexyl radicals thus formed can suffer fragmentation,
producing alkyl radicals that can be the precursors of alkane products, and
also undergo bimolecular reactions without fragmentation:

$$CH_3-CH_2-CH_2-CH_2-CH_2-CH_2 \longrightarrow C_nH_{2n+1} + C_{6-n}H_{12-2n}$$

$$CH_3-CH_2-CH_2-CH_2-CH_2-CH_2 + cyclo\text{-}C_6H_{12} \longrightarrow n\text{-}C_6H_{14} + cyclo\text{-}C_6H_{11}$$

$$CH_3-CH_2-CH_2-CH_2-CH_2-CH_2 + cyclo\text{-}C_6H_{11}$$

$\cdot n\text{-}C_6H_{14} + cyclo\text{-}C_6H_{10}$

1-$C_6H_{12} + cyclo\text{-}C_6H_{12}$

cyclo-$C_6H_{11}-C_6H_{13}$

The formation of C_1-C_6 alkyl radicals was verified by Zaitsev et al. [156] by investigations using I_2 as radical acceptor; the yields were as follows: $G(CH_3I) = 0.004$, $G(C_2H_5I) = 0.005$, $G(1\text{-}C_4H_9I) = 0.014$ and $G(C_6H_{13}I) = 0.20$.

As already mentioned, the fragmentation of cyclohexyl positive ions in the liquid phase leads to the formation of fragment products with a yield of $G = 0.05$. The corresponding value in the gas phase [174, 179] is considerably larger. Consequently, the yields determined in the gas as well as in the liquid phase differ considerably, not only quantitatively but also qualitatively. The production of alkanes is much more extensive in the gas than in the liquid phase (Table 3.11).

The formation of methylcyclopentane from cyclohexane is one of the most interesting phenomena. It has been observed that this product is formed in significant amounts not only in the radiolysis of cyclohexane, but also on Hg or Cd sensitized photolysis [180, 181], when cyclohexyl radicals are produced in the first step. Shida et al. suggested the following reaction sequence for the transformation of cyclohexyl radicals into methylcyclopentane:

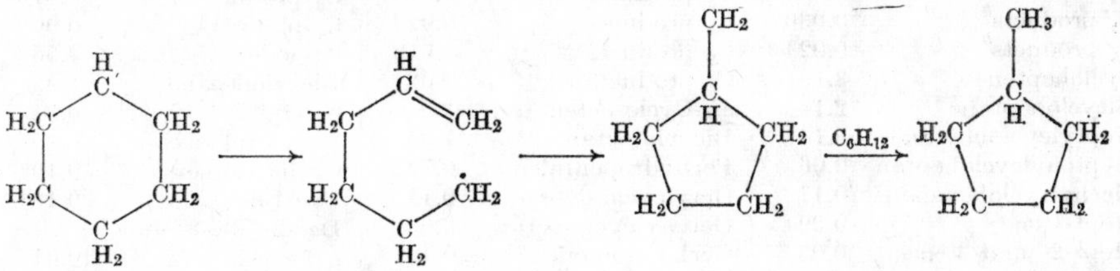

On investigating the reactions of cyclohexyl radicals formed in the gas phase via radical reactions, Gordon [182] assumed that the isomerization resulting in the formation of methylcyclopentyl radicals can take place via concerted reactions, i.e. without any intermediate ring-opening. The formation of methylcyclopentane via methylcyclopentyl radicals is supported by the observation that on the radiolysis of cyclohexane, methylenecyclopentane and cyclopentylcyclohexylmethane products were also detected, though in low yield, $G = 0.005$ and 0.03, respectively. These products may arise as a result of disproportionation and combination of methylcyclopentyl and cyclohexyl radicals [169].

In a number of papers the detection of dimers other than bicyclohexane and cyclohexylcyclohexene has been reported [133, 142, 150, 169]. Dyne and Stone [142] observed 6-cyclohexylhex-1-ene in the highest yield, $G = 0.27$. Ho and Freeman [143] used the identification made by the former authors, and arrived at a value of 0.12. Wojnárovits and Földiák reported G-values that indicated a low efficiency of formation of this product, namely $G = 0.03$. Among these types of product the most important compounds are not the unsaturated but the saturated ones: $G(\text{n-hexylcyclohexane}) = 0.08$ [169]; however, apart from these, a number of other alkyl- and alkenyl-cyclohexanes have also been detected ($G \approx 0.05$).

3.2.4. C_7-C_{10} cycloalkanes

Only a very limited number of data are available on the radiolysis of cycloalkanes with more than six carbon atoms [183, 184] (Table 3.12). Apart from these, some papers have been published concerning the ESR spectra of irradiated cycloheptane [96, 152, 155] and cyclo-octane [152, 155], as well as the UV-light induced fluorescence of these hydrocarbons [43, 44].

Table 3.12

Yields from liquid cycloheptane, cyclo-octane and cyclodecane
(radiation: ^{60}Co-γ). After Földiák and Wojnárovits [184]

Cycloheptane	G, molecule/100 eV	Cyclo-octane	G, molecule/100 eV	Cyclodecane	G, molecule/100 eV
Hydrogen	5.75	Hydrogen	5.9	Hydrogen	5.95
CH_4	0.005	CH_4	0.004	CH_4	0.004
C_2 products[b]	0.075	C_2 products	0.08	C_2 products	0.055
C_3 products[b]	0.048	C_3 products	0.016	C_3 products	0.016
C_4 products[b]	0.030	C_4 products	0.023	C_4 products	0.008
C_5 products[b]	0.024	C_5 products	0.013	Cyclodecene	3.55
Cycloheptene	3.1	C_6 products	0.02	Bicyclodecane	1.9
Bicycloheptane	2.1	*cis*-Cyclo-octene	3.0	$C_{10}H_{18}$[a]	0.3
Heptylcycloheptane	0.14	Bicyclo-octane	1.95	Decylcyclodecane	0.15
Heptenylcycloheptane	0.06	Perhydropentalene	0.72	Cyclic isomers	~0.10
Methylcyclohexane	0.17	Octylcyclo-octane	0.11	Dec-1-ene	0.16
Hept-1-ene	0.29	Octenylcyclo-octane	0.03	Dec-2-,-3-,-4- and	
Hept-2- and-3-enes	0.03	Cyclic isomers	~0.12	5-enes	0.01
n-Heptane	0.10	Oct-1-ene	0.17	n-Decane	0.09
Heptadienes	0.03	Oct-2- and -3-enes	0.02	Decadienes	0.05
		n-Octane	0.08		
		Octadienes	0.04		

[a] Most likely bridged bicyclodecane, but not decalin
[b] Alkene products having the formulas C_nH_{2n} and C_nH_{2n-2} contribute to the fragment products with ~ 70 and ~ 15% respectively, while n-alkanes constitute the remaining part

The major types of hydrocarbon product formed on the radiolysis of cycloheptane, cyclo-octane and cyclodecane, as well as their behaviour in the presence of radical acceptors, are generally the same or similar to those observed for cyclopentane and cyclohexane; however, some differences can be recognized [184].

In the case of C_8 and larger rings, there is a possibility that both *cis*- and *trans*-cycloalkenes can be formed. However, in the radiolysis of cyclo-octane, the *trans* form, having a strain energy 39 kJ mol^{-1} higher than the *cis* form, could not be detected. On the other hand, the corresponding difference in C_{10} and larger rings is much smaller, i.e. 17 kJ mol^{-1} [185–187]. In the radiolysis of cyclodecane and cyclododecane both the *cis* and *trans* isomers were observed, the latter having the higher yield [184].

A study of the radiolysis of cyclo-octane detected a bridged-ring compound, namely perhydropentalene (bicyclo[3.0.3]octane), on the basis of its gas chromatographic retention data, with a yield of $G = 0.72$.

cyclooctane   perhydropentalene

This reveals that partial dehydrogenation of the cyclo-octane molecule can take place leading to a cross bridge in the ring instead of the usual double bond production. According to the gas-phase data, the enthalpy change is more favourable for this reaction (36 kJ mol^{-1}) than that for the reaction cyclo-octane → cyclo-octene + H_2 (97.6 kJ mol^{-1}) [184]. As the yield of perhydropentalene is reduced in the presence of radical scavengers by not more than 20%, one can suppose that its precursor is not the same as that of cyclo-octene and bicyclo-octane, the latter products being to a large extent scavengeable by radical acceptors.

In the course of the 163 nm (7.6 eV) photolysis of liquid cyclo-octane, perhydropentalene forms with a quantum yield of $\Phi = 0.36$, while the quantum yield of the other products are: hydrogen 1.0, cyclo-octene 0.57, bicyclo-octane 0.07. The low yield of bicyclo-octane reflects the little importance of hydrogen formation through an atomic pathway; consequently, the perhydropentalene formation must be connected with the unimolecular H_2 detachment. When supposed that both atoms of the H_2 molecule detached originate from the same carbon atom, as discussed earlier in the case of cyclohexane (see Section 3.2.3), perhydropentalene formation is seen as a result of the rearrangement of the carbene intermediate. The geometry of the molecule does support such a notion [37].

The ratios of the rate constants, k_d/k_c, determined by means of radical scavengers for the cycloheptyl, cyclo-octyl and cyclodecyl radicals are as follows: ~ 0.95, ~ 0.95 and ~ 1.0, respectively. This means that k_d/k_c ratios are very similar for C_5-C_{10} cycloalkyl radicals [90A, 184].

3.2.5. Alkylcycloalkanes

The radiolysis of methylcyclopentane and methylcyclohexane has been studied several times, but much fewer data are available on the high-energy induced decompositions of other alkylcycloalkanes. In the course of alkyl-cycloalkane radiolysis large amounts of C—C bond rupture products are formed, some of them having relatively high G-values. For this reason more attention will be paid here to the C—C bond rupture reactions than to the hydrogen formation.

Dehydrogenation products

The yields of hydrogen for monoalkylcycloalkanes listed in Table 3.13 are lower by 0.5–3.0 G-units than the $G(H_2)$ values of the cycloalkanes. Apart from this, the alkanes with six carbon atoms in the ring show slightly

Table 3.13

Yields of hydrogen from liquid alkylcycloalkanes (radiation: ^{60}Co-γ)

Cycloalkane	$G(H_2)$	Dose 10^{19} eV g^{-1}	Ref.	$G(H_{thermal})$	Ref.
Cyclopentane	5.35	a	85	1.6[b]	87
Methylcyclopentane	4.2	a	188, 189	1.1[b]	87
	4.5	a	10		
Ethylcyclopentane	4.2	18.7	102		
Cyclohexane	5.67	a	16	1.0[b]	87
				1.46[b]	16
Methylcyclohexane	4.8	a	190	1.0[b]	87
Ethylcyclohexane	4.5	1.3	173		
	4.7		102		
Isopropylcyclohexane	3.2	1.2	173		
Bicyclohexane	4.0	a	191		
1,1-Dimethylcyclohexane	3.3	0.5	192		
cis-1,2-Dimethylcyclohexane	3.69	5.0	193	0.65[c]	193
trans-1,2-Dimethylcyclohexane	3.6	5.2	193	1.18[c]	193
cis-1,3,5-Trimethylcyclohexane	4.4	1.3	173		
trans-1,3,5-Trimethylcyclohexane	4.5	1.3	173		
1,2,4,5-Tetramethylcyclohexane	3.6	1.3	173		

[a] Yields extrapolated to zero dose
[b] Yields determined with ethylene scavenger
[c] Yields determined with 2 mM I$_2$

higher $G(H_2)$ values than the corresponding cycloalkane having five carbon atoms in the ring.

The fifth column of the table shows the yields of thermal hydrogen atoms. Each numerical value, except for one, falls in the range 1.0–1.6, the exception being that cis-1,2-dimethylcyclohexane gives a G-value (0.65) little more than half that of the trans form (1.18). Eberhardt [193] suggested that the thermal hydrogen atoms are formed mainly by detachment of tertiary hydrogen atoms in axial positions. This hypothesis seems to be supported by the fact that

cis-1,2-dimethylcyclohexane   trans-1,2-dimethylcyclohexane

the $G(H_{thermal})$ of trans-decalin is also high as compared with that of the cis form: 1.72 and 0.72, respectively. The non-scavengeable yield of hydrogen for the cis-1,2-dimethylcyclohexane is greater than that for the trans form: 3.05 and 2.36, respectively. These results were attributed by the author to the fact that in the case of the trans form (which contains methyl groups at least in 99% in the diequatorial position) the four-centre unimolecular hydrogen elimination is hindered to a greater extent:

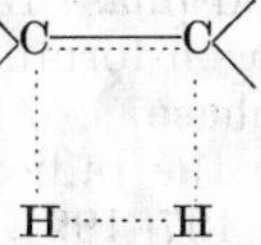

Although this interesting hypothesis is in agreement with the stereochemistry, according to observations made with n-hexane and cyclohexane the unimolecular H_2 elimination takes place to a high extent from a single carbon atom rather than from two adjacent ones [106, 125, 129, 178].

It was shown by Stover and Freeman that the G-value of hydrogen for methylcyclopentane is reduced in the presence of electron acceptors, N_2O and SF_6, from 4.2 to a limiting value of 2.2, whereas it is enhanced under the influence of ND_3. From these results they concluded that in the absence of ND_3 not every positive ion leads to the formation of H_2; a hydrogen yield of not more than $G = 2.0$ originates in recombination [10, 188]. The same figure was estimated by Schuler et al. to be about 2.4 [16], using the results of Robinson and Freeman [115] (Table 3.14). That is, the remaining part of the hydrogen yield, 1.8–2.2 molecules/100 eV, can be traced back to the reactions of directly excited molecules.

Table 3.14

Sources of hydrogen formation from liquid cycloalkanes (radiation: $^{60}Co\text{-}\gamma$)

Sources of hydrogen	Total	Ionic	Non-ionic	References
Hydrocarbon	G, molecule/100 eV			
Cyclopentane	5.4	3.4	2.0	10
Cyclopentane	4.9	3.2	1.7	16, 115
Cyclopentane	5.7	4.0	1.7	86
Methylcyclopentane	4.2	2.0	2.2	10
Methylcyclopentane	4.2	2.4	1.8	16, 188
Cyclohexane	5.6	3.4	2.2	10
Cyclohexane	5.6	4.0	1.6	16, 115
Methylcyclohexane	4.7	3.2	1.5	16, 115
Methylcyclohexane	4.9	2.6	2.3	16, 194
Methylcyclohexane[a]	5.2	2.1	3.1	10
1,1-Dimethylcyclohexane	3.3	2.0	1.3	192

[a] Gas phase (500 mbar, 110°C)

The data in Table 3.14 show that the hydrogen-forming efficiency of ion-electron neutralization is higher for unbranched than for branched hydrocarbons. On the other hand, the contribution of directly produced excited molecules to hydrogen formation seems to be only very slightly dependent on molecular structure [16]. The yield of C—C bond rupture for alkylcycloalkanes is about twice that of unbranched molecules [192, 195], and scission occurs mostly at the branch point (see below).

On irradiation of methylcyclohexane vapour [196] at 110°C, ion-electron neutralization leads to hydrogen formation with a yield of $G = 2.1$, slightly lower than that obtained in liquid-phase radiolysis ($G = 2.6$ or 3.2) (Table 3.14). It appears that the hydrogen-forming efficiency of neutralization is only slightly influenced by the phase.

Kimura et al. have dealt with the radiolysis of methylcyclohexane in a number of papers [87, 112, 194, 197–199]. They established that the yield of scavengeable electrons is smaller at $-196°C$ than at room temperature. The authors concluded that the electrons can be regarded as being 'quasi-free' in the solid phase, but 'solvated' in the liquid state.

The values of G(cycloalkene) extrapolated to zero dose, 3.70 and 3.48, respectively, for methylcyclopentane [10] and methylcyclohexane [190, 200], the most studied alkylcycloalkanes, are somewhat higher than those for cyclopentane [85] and cyclohexane [143], 2.97 and 3.20, respectively. Similar conclusions can be reached concerning the radiolysis of ethylcyclopentane and ethylcyclohexane [102]: at a dose of $19 \cdot 10^{19}$ eV g^{-1} the yields of cycloalkenes were found to be $G = 2.5$ and 2.7, respectively; the equivalent values for cyclopentane [102] and cyclohexane [143] are 2.1 and 2.3, respectively.

The experimental results show that in the radiolysis of methyl- and ethylcyclopentanes and -cyclohexanes every cyclomonoalkene that does not need the reorganization of the carbon structure is formed; however, their distribution differs considerably from the statistical one. The alkenes with π-bonds at the branch point are formed in higher amounts than all the other isomers together (Fig. 3.3; dose: $19 \cdot 10^{19}$ eV g^{-1}). The G-values of bicycloalkane products of both methyl- and ethylcyclopentane are 0.9 [10, 102], those of methyl- and ethylcyclohexane 0.75 [190] and 0.8 [102], and those of cis- and trans-1,2-dimethylcyclohexane 0.4 and 0.63 [193], respectively. All these yields are lower than those of cyclopentane [95] and cyclohexane [143], 1.29 and 1.76, respectively.

These observations can be explained using methylcyclopentane as an example [10, 102]. As a result of the ruptures of C—H bonds of methylcyclopentane four different radical isomers can be formed:

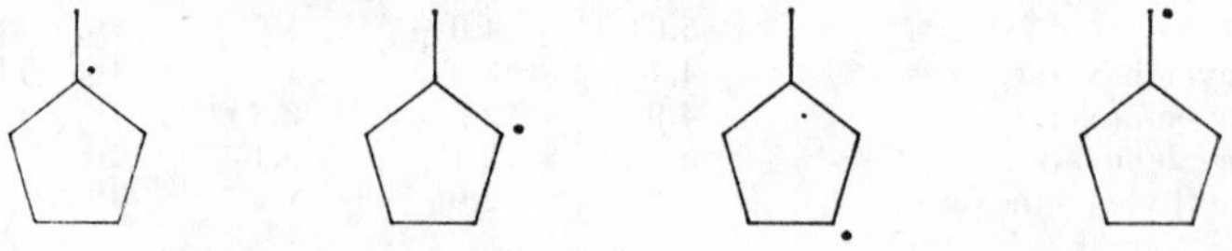

Among them, the radical 1 may have high k_d/k_c ratio ($\sim$4-5) [201], taking into consideration the data available on radicals having analogous stuctures, e.g. for tertiary butyl radicals, $k_d/k_c = 4.38$. This can be explained by the steric hindrance in recombination, and by the high number (7) of so-called 'active' hydrogen atoms situated in the vicinity of the radical site and being more readily attacked in disproportionation reactions [203]. The k_d/k_c ratios for the radicals 2 and 3 may be approximately equal to that of cycloalkyl

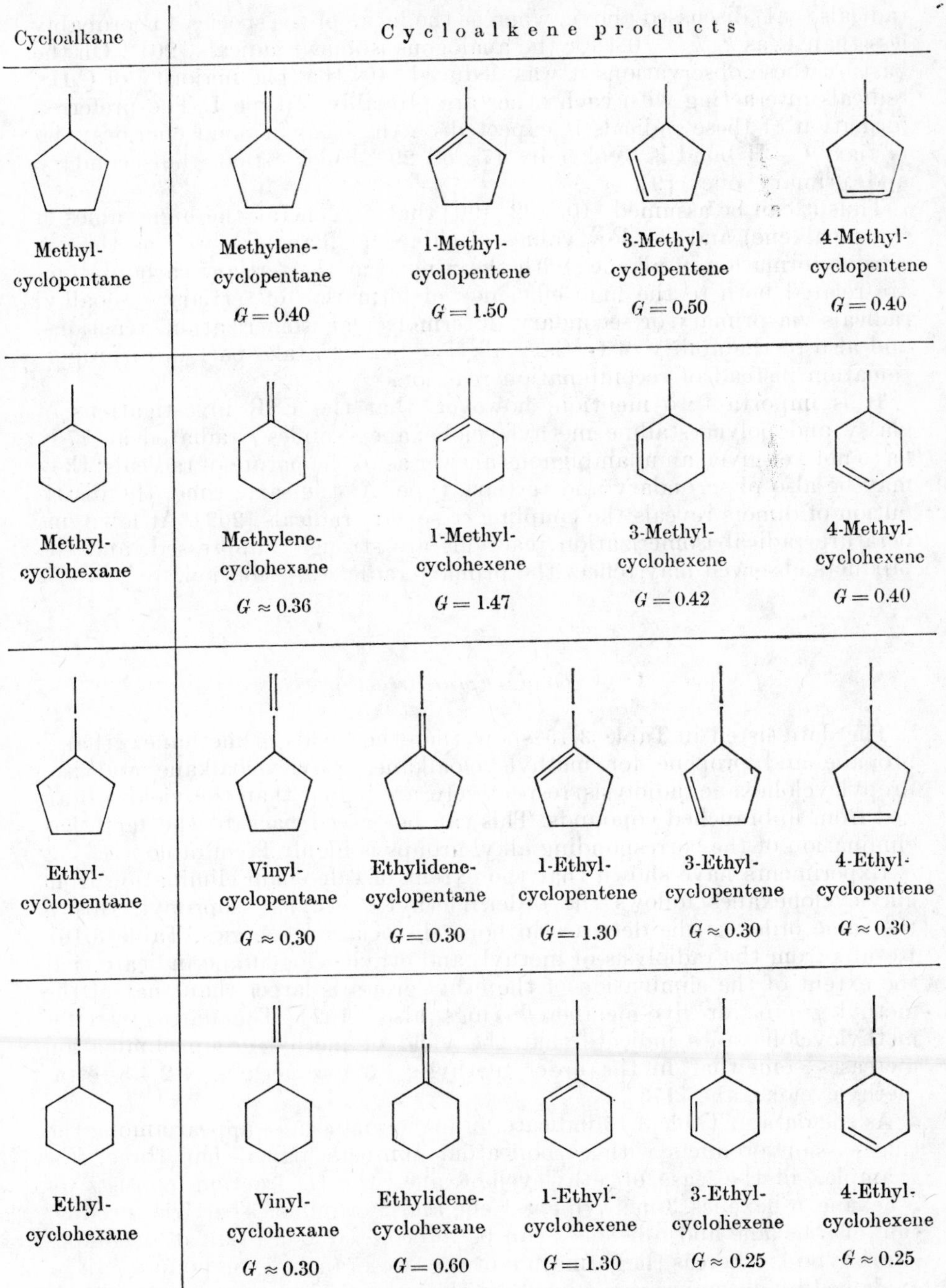

Cycloalkane	Cycloalkene products			
Methyl-cyclopentane	Methylene-cyclopentane $G = 0.40$	1-Methyl-cyclopentene $G = 1.50$	3-Methyl-cyclopentene $G = 0.50$	4-Methyl-cyclopentene $G = 0.40$
Methyl-cyclohexane	Methylene-cyclohexane $G \approx 0.36$	1-Methyl-cyclohexene $G = 1.47$	3-Methyl-cyclohexene $G = 0.42$	4-Methyl-cyclohexene $G = 0.40$
Ethyl-cyclopentane	Vinyl-cyclopentane $G \approx 0.30$	Ethylidene-cyclopentane $G = 0.30$	1-Ethyl-cyclopentene $G = 1.30$	3-Ethyl-cyclopentene $G \approx 0.30$	4-Ethyl-cyclopentene $G \approx 0.30$
Ethyl-cyclohexane	Vinyl-cyclohexane $G \approx 0.30$	Ethylidene-cyclohexane $G = 0.60$	1-Ethyl-cyclohexene $G = 1.30$	3-Ethyl-cyclohexene $G \approx 0.25$	4-Ethyl-cyclohexene $G \approx 0.25$

Fig. 3.3. G-values of alkenes produced during the radiolysis of alkylcycloalkanes (radiation: ^{60}Co-γ). After Wojnárovits and Földiák [102]

radicals ($\sim$1) discussed above, whereas the ratio for the species **4** is probably less than 1, as $k_d/k_c = 0.4$ for the analogous isobutyl radicals [201]. On the basis of these observations it was deduced [10] that the majority of C_6H_{11} radicals interacting with each other are probably of type **1**. The preferred formation of these radicals is expected on the basis of bond energies: the tertiary C—H bond is weaker by 17 and 29 kJ mol $^{-1}$ than the secondary and primary ones [9].

Thus it can be assumed [10, 102, 190] that, in general, the high values of G(cycloalkene) and the low values of G(bicycloalkane), as well as the extensive formation of alkenes with the π-bond at the tertiary carbon atom, are related both to the high efficiency of formation of tertiary cycloalkyl radicals via primary or secondary (intermolecular isomerization) reactions, and also to the ability of tertiary alkyl radicals to take part in disproportionation instead of recombination reactions.

It is important to mention, however, that the ESR investigations of glassy and polycrystalline methylcyclohexane samples irradiated at 77 K have not yet given an unambiguous answer as to the nature of radicals; they may be also of secondary and tertiary type. At the same time, the distribution of dimers reveals the coupling of several radicals [202]. At low temperature radical isomerization reactions are strongly suppressed, and the products observed may reflect the primary radical distribution.

C—C *bond rupture reactions*

The data listed in Table 3.15 show that the yields of methane, ethane, propane and propene for methylcycloalkane, ethylcycloalkane and isopropylcyclohexane radiolysis, respectively, are higher than the yields obtained from unbranched copounds. This can be traced back to the fact that elimination of the corresponding alkyl groups is highly favourable.

Experiments have shown that the extent of side-chain elimination from alkylcyclohexanes follows the order methyl < ethyl < isopropyl; this is the same order as the decrease in bond dissociation energies (Table 3.16). Results from the radiolysis of methyl- and ethylcyclopentane indicate that the extent of the elimination of the ethyl group is larger than that of the methyl group for five-membered rings also [102]. The data given for methylcyclohexanes indicate that the yield of methyl group elimination increases somewhat in the order methyl-, 1,3,5-trimethyl-, 1,2,4,5-tetramethylcyclohexane [173].

As the data in Table 3.15 indicate, branched molecules appear among the ring-scission products with carbon atom numbers higher than three. For example, in the case of ethylcyclohexane, the C_6 fraction consists of n-hexane, n-hexenes, 3-methylpent-1-ene and 2-ethylbut-1-ene. The production of n-hexane and n-hexenes can be interpreted as a result of decomposition type **1**, whereas the formation of the two branched compounds can be explained by decomposition type **2**. Neither 2-methylpentane, nor 2-methylpentenes could be detected among the products. For the production of these

Table 3.15

Yields of fragments from liquid alkylcycloalkanes (radiation: ^{60}Co-γ)

Hydrocarbon	Methyl-cyclo-pentane	Ethyl-cyclo-pentane	Methyl-cyclo-hexane	Ethyl-cyclo-hexane	Iso-propyl-cyclo-hexane	1,4,5-Tri-methyl-cyclo-hexane	1,2,4,5-Tetra-methyl-cyclo-hexane
Reference	162, 204	162	162	162	173	173	173
Product	G, molecule/100 eV						
Methane	0.128	0.05	0.095	0.028	0.045	0.136	0.148
Ethane	0.026	0.16	0.01	0.12	0.001	0.004	0.004
Ethylene	0.40	0.18	0.076	0.10	0.017	0.008	0.008
Acetylene	0.02	0.015	0.01	0.01	0.001	0.012	0.007
Propane	0.038	0.01	0.011	0.005	0.505	0.003	0.002
Cyclopropane	0.058	0.015	0.005	0.003			
Propene	0.312	0.05	0.05	0.015	0.505	0.123	0.096
Allene	0.008						
n-Butane	0.027	0.01	0.006	0.01		0.005	
Isobutane	0.006		0.003				
Methylcyclopropane	0.068						
But-1-ene	0.08	0.06	0.012	0.02		0.06	
trans- and cis-But-2-ene	0.08	0.005	0.008	0.01	0.011	0.011	0.047
Isobutene	0.062		0.003				
Butadiene	0.03	0.01	0.007	0.004			
n-Pentane		0.01	0.002	0.004			
Ethylcyclopropane		0.025		0.004			
n-Pentenes		0.085	0.011	0.01			
Isopentenes		0.026	0.02	0.01			
n-Hexane				0.007			
n-Hexenes				0.01			
Isohexenes				0.01			

Table 3.16

Yields of elimination of alkyl groups from alkylcycloalkanes
and the apparent bond dissociation energies (E_D)
(radiation ^{60}Co-γ). After Wojnárovits and Földiák [50, 192]

Hydrocarbons	Alkyl group	G (alkyl)	E_D, kJ mol^{-1}
Methylcyclopentane	methyl	0.2	347
Ethylcyclopentane	ethyl	0.3	334
Methylcyclohexane	methyl	0.2	351
Ethylcyclohexane	ethyl	0.3	334
1,1-Dimethylcyclohexane	methyl	0.7	322
Isopropylcyclohexane	isopropyl	1.5	318
tert-Butylcyclohexane	tert-butyl	2.0–2.5	297

two compounds a rearrangement of the carbon skeleton of the fragment would be necessary [102, 162].

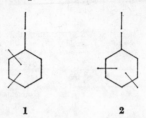

This observation can be generalized over the majority of hydrocarbons reported in the literature. In liquid-phase radiolysis, ring-scission leads to the formation of detectable amounts of only such fragment products with carbon skeletons that can be formed directly by the rupture of two C—C bonds of the original cycloalkane ring. This is not necessarily so for gas-phase reactions: under the influence of irradiation [205] or silent electric discharge [175] on cyclohexane small amounts of isobutane and isobutene were detected.

Figure 3.4 shows the fragments formed in the radiolysis of methyl-cyclopentane labelled with ^{14}C-isotope in the methyl group, as well as their radioactivities compared with that of the initial hydrocarbon [204]. It can be seen that only 80% of the methane formed contains ^{14}C, i.e. the

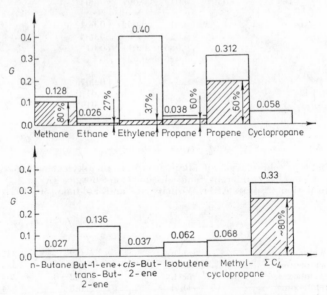

Fig. 3.4. Yields and relative radioactivities of the main products of fragmentation of ^{14}C-methylcyclopentane. The heights of the columns are proportional to the G-values and those of the shaded parts to the fraction of radioactive molecules in the individual products (radiation: ^{60}Co-γ). After Keszei et al. [204]

methane is not produced exclusively by side-chain elimination, but ring fragmentation also contributes. The CH_2 unit formed from the excited ring will be stabilized eventually as methane. If the elimination happens to occur at the tertiary carbon atom, a $^{14}CH_3—CH$ species can be formed with a subsequent rearrangement into labelled ethylene, as it is also known from other studies [206]. This species amounts only to 3.7% of the total ethylene fraction which is, however, the most important ring scission product. The majority of the C_4 products corresponds to the formula C_4H_8 and its molar activity is about 80% of that of methylcyclopentane, For this reason the most important fragmentation of methylcyclopentane must be the following:

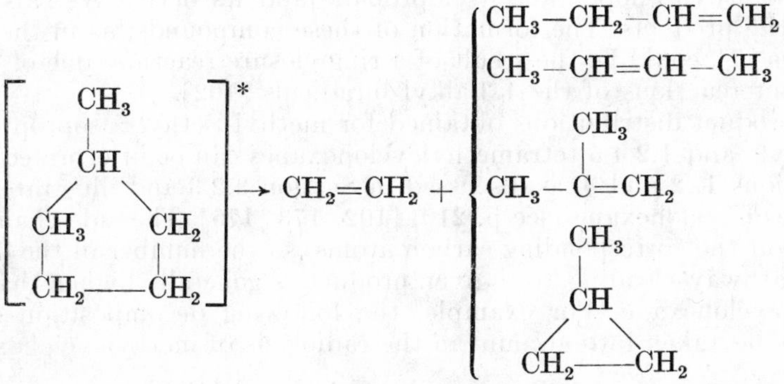

$$
\begin{bmatrix}
\quad CH_3 \\
\quad | \\
\quad CH \\
\diagup \quad \diagdown \\
CH_3 \qquad CH_2 \\
| \qquad\quad\ \\
CH_2\!\!-\!\!-\!\!CH_2
\end{bmatrix}^{*}
\rightarrow CH_2{=}CH_2 +
\begin{cases}
CH_3—CH_2—CH{=}CH_2 \\[2pt]
CH_3—CH{=}CH—CH_3 \\[4pt]
\qquad CH_3 \\
\qquad | \\
CH_3—C{=}CH_2 \\[6pt]
\qquad CH_3 \\
\qquad | \\
\qquad CH \\
\qquad \diagup \ \diagdown \\
\ \ CH_2\!\!-\!\!-\!\!CH_2
\end{cases}
$$

The relatively large yield of C_3 products and their total radioactivity of about 50% indicate that another important decomposition reaction can be postulated as follows:

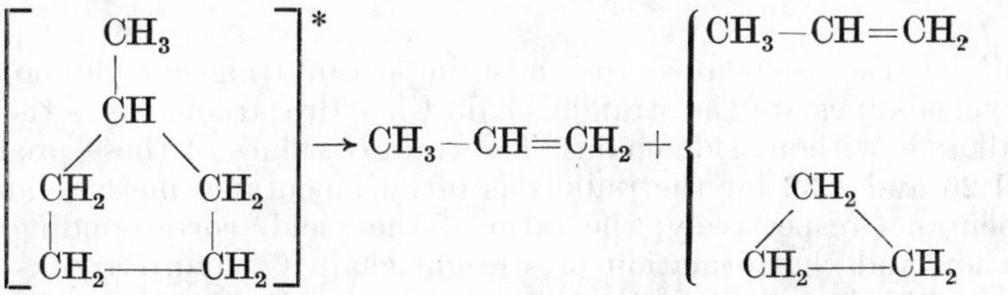

$$
\begin{bmatrix}
\quad CH_3 \\
\quad | \\
\quad CH \\
\diagup \quad \diagdown \\
CH_2 \qquad CH_2 \\
| \qquad\quad | \\
CH_2\!\!-\!\!-\!\!CH_2
\end{bmatrix}^{*}
\rightarrow CH_3—CH{=}CH_2 +
\begin{cases}
CH_3—CH{=}CH_2 \\[10pt]
\qquad CH_2 \\
\qquad \diagup \ \diagdown \\
\ \ CH_2\!\!-\!\!-\!\!CH_2
\end{cases}
$$

It is worth noting that the cyclopropane is completely inactive, i.e. it is derived entirely from the ring carbon atoms.

The fragmentation of methyl- and ethylcyclopentane proceeds mainly, via rupture of the bonds marked in the figures shown below:

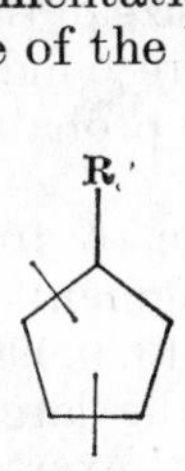

      1 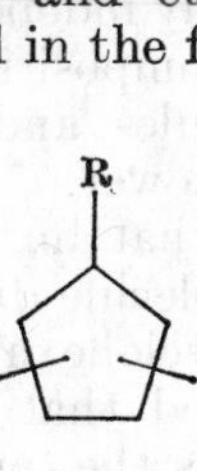   2 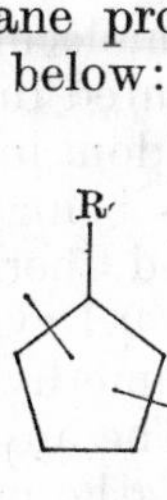   3

where R stands for methyl or ethyl groups. The ratios of decomposition pathways **1**, **2** and **3** are given in ref. [102], for methyl- and ethylcyclopentane as 50 : 17 : 33 and 48 : 17 : 35, respectively. This means that the proportion of decompositions taking place via C—C bond rupture not involving tertiary carbon atoms (Type **2**) is less than 20%. Consequently, the rupture of C—C bonds and the dissociation of the rings into C_2 and C_3 units takes place with a certain selectivity, which marks a similarity with gas-phase thermal decomposition reactions. These latter reactions involve the preferred scission of C—C bonds attached to tertiary carbon atoms, which are weaker than the other bonds in the ring [6, 7, 207].

It is remarkable, that in the radiolysis of cyclopentane, methylcyclopentane and ethylcyclopentane cyclopropane and its derivatives also appear among the products. The formation of these compounds, as in the thermal reactions [7], could be the result of a ring-closure reaction, one of the most important reactions of the 1,3-alkyl biradicals [102].

The product distributions obtained for methyl-, ethyl-, isopropyl-, 1,3,5-trimethyl- and 1,2,4,5-tetramethylcyclohexanes can be interpreted in terms of reactions **1**, **2** and **3**, as discussed in Section 3.2.3 and illustrated by the example of cyclohexane (see p. 210) [102, 173, 175]. The side-chains act as 'labels' on the corresponding carbon atoms, so the number of the decomposition pathways leading to a given product is generally higher than in the case of cyclohexane. For example, the following decomposition reactions have to be taken into account in the radiolysis of methylcyclohexane:

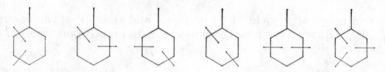

As will be discussed later, the most important ring-scission products of alkylcycloalkanes are the straight-chain C_n hydrocarbons, as is the case for cycloalkanes without side-chains. The total G-values of these products are 0.88, 1.26 and 1.32 for the radiolysis of cyclopentane, methyl- and ethylcyclopentane, respectively; the ratio of the yields corresponding to fragmentation and the formation of straight-chain C_n hydrocarbons, i.e. the ratio of multiple and single C—C bond ruptures, is 0.60, 0.38 and 0.17. The decrease of this ratio observed when side-chains are present was explained by Wojnárovits and Földiák [102, 162] as follows: the radiation energy absorbed by the molecule, being largely independent of the size of the molecule, will be distributed during the decomposition among more bonds, i.e. more degrees of freedom in larger molecules, and therefore the probability of two or more bonds being ruptured is lower.

Makarov and Chernyak found that on the absorption of 100 eV of radiation energy 0.1 cyclohexane molecule dissociate to fragments, the values for methyl-, trimethyl- and ethylcyclohexane are 0.09–0.13, 0.096 and 0.080, respectively. The authors concluded that the more or the larger the side-chains present, the more efficient is the 'protective effect' exerted by them on the scission of the ring. This effect is opposite to that observed for the alkylbenzenes, where the benzene ring 'protects' the side-chain against decomposition.

The yields of straight-chain C_n hydrocarbon products formed from methyl- and ethylcyclopentane and methyl- and ethylcyclohexane are shown in Table 3.17. The most important products are monoalkenes, as in the radiol-

Table 3.17

Yields of straight-chain C_n products from alkylcycloalkanes (radiation: ^{60}Co-γ). After Wojnárovits and Földiák [102]

Hydrocarbons	Methyl-cyclo-pentane	Methyl-cyclo-hexane	Ethyl-cyclo-pentane	Ethyl-cyclo-hexane
Products	G, molecule/100 eV			
1-C_nH_{2n}	0.75	0.34	0.40	0.22
cis-2-C_nH_{2n}	0.10	0.08	} 0.23	} 0.03
$trans$-2-C_nH_{2n}	0.23	0.20		
cis-3-C_nH_{2n}		0.02	0.13	0.05
$trans$-3-C_nH_{2n}			0.37	0.21
2-CH_3-$C_{n-1}H_{2n-3}$	0.06	0.03		
3-CH_3-$C_{n-1}H_{2n-3}$	0.04	0.07		
2-C_2H_5-$C_{n-2}H_{2n-3}$ 3-C_2H_5-$C_{n-2}H_{2n-3}$			0.09	0.08
Σn-C_nH_{2n}	1.18	0.74	1.22	0.59
n-C_nH_{2n+2}	0.08	0.08	0.10	0.04
Σn-C_n	1.26	0.82	1.32	0.63

ysis of the cycloalkanes without side-chains [50, 102]. In the radiolysis of all four alkylcycloalkanes the 1-alkanes give the highest yields. Apart from the 1-alkanes, mainly 2-alkanes are formed from the methylcycloalkanes and 3-alkenes from the ethylcycloalkanes. Accordingly, the isomerization of hydrocarbons listed in the Table producing straight-chain alkenes takes place at least in 70% as shown in reactions (3.23a) and (3.23b) by the example of propylcyclopentane:

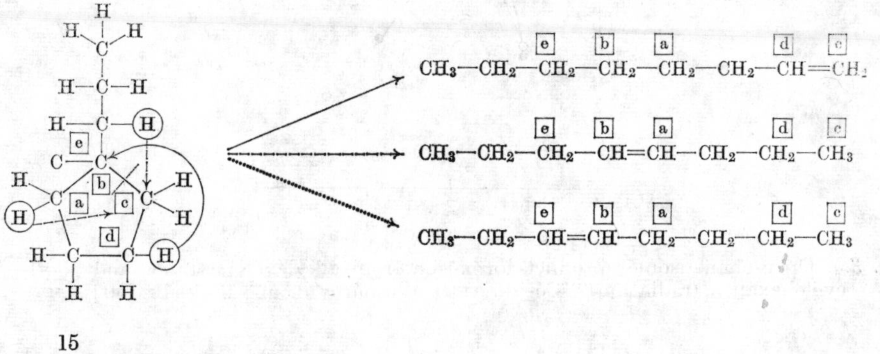

$$CH_3—CH_2—CH_2—CH_2—CH_2—CH_2—CH=CH_2 \quad (3.23a)$$

$$CH_3—CH_2—CH_2—CH=CH—CH_2—CH_2—CH_3 \quad (3.23b)$$

$$CH_3—CH_2—CH=CH—CH_2—CH_2—CH_2—CH_3 \quad (3.23c)$$

15

The reaction type (3.23c) occurs less frequently; it can play a more important role in the case of alkylcyclopentanes, but it is of negligible importance compared with reactions (3.23a) and (3.23b) in the radiolysis of alkylcyclohexanes.

Figure 3.5 shows that the total G-value for straight-chain isomeric products increases sharply along the homologous series of cyclopentane and cyclohexane derivatives, i.e. by 60 and 90%, respectively, proceeding from the cycloalkanes to the methylcycloalkanes, but thereafter continuously decreases [50]. When, for the calculation of the radiation chemical yields of isomerization, only that amount of energy is taken into account that has been primarily absorbed by the ring [$g(\Sigma$ isomer)], the yields of isomeric products of alkylcyclopentanes fall in the range 1.7–1.3, whereas those for the alkylcyclohexanes are in the range 0.9–0.7. More exactly, beginning from the ethylcycloalkanes, the yields decrease somewhat in both cases.

The observation that the g-value of ring isomerization products of alkyl-cycloalkanes is about double that for cyclopentane and cyclohexane, is in line with the behaviour of aliphatic alkanes (see e.g. Sections 2.2.9 and 2.4), in so far as in both cases the frequency of C—C bond ruptures increases as a result of chain branching. The decrease in the values of $g(\Sigma$ isomer) with increasing carbon atom number reveals a very small 'protective' effect of the alkyl chain. This behaviour is more apparent when the values of $G(\Sigma$ isomer) for propyl- and isopropylcyclohexane as well as for butyl- and tert-butylcyclohexane are compared: the value for the isopropyl and the tert-butyl compounds is not more than 40 and 18% of those for the cor-

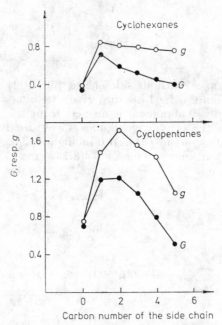

Fig. 3.5. Open-chain isomer product formation from alkylcyclopentanes and alkyl-cyclohexanes (radiation: ^{60}Co-γ). After Wojnárovits and Földiák [50]

responding n-alkylcyclohexanes, respectively. In the latter two cases, however, elimination of the alkyl group takes place with a high probability (Table 3.16). Because of the enhanced yields of C—C bond ruptures between tertiary–tertiary and quaternary–tertiary carbon atoms, ring-scission is suppressed [50].

Apart from monoalkenes, straight-chain alkanes have also been detected. As for the production of straight-chain C_n compounds, it should be noted that the C—C bonds attached to tertiary carbon atoms are more likely to rupture than the other ring C—C bonds, as mentioned above in connection with the formation of fragment products. As an example, the G-values of C—C bond rupture are shown here, as calculated for the decomposition of methylcyclopentane taking into account the yields of methyl group elimination and the straight-chain C_n products only [102]:

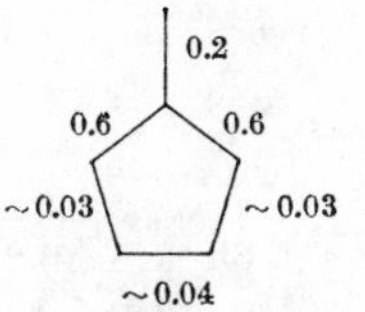

In the presence of radical acceptors, the G-values of alkene products are reduced only by about 20–40% [102]. This seems to support the suggestion that they are formed mainly via biradical intermediates, e.g.:

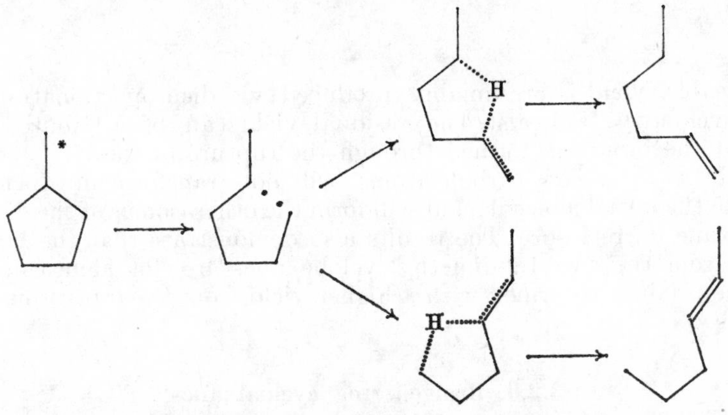

This reaction mechanism also explains the experimental results summarized in reactions (3.23a)–(3.23c). For the reactions (3.23a) and (3.23b) to proceed it is necessary for five- and six-membered ring transition states, respectively, to be formed; seven- and eight-membered rings are necessary for the reaction (3.23c) to take place. The unfavourable structure of the latter two transition states could explain the minor importance of reaction (3.23c) [50, 102, 184].

In a number of papers it is assumed that apart from rearranging into straight-chain alkenes, the biradicals can also undergo a reclosure to form cycloalkane [170, 172, 193]. The experiments of Eberhardt [193] and Woj-

15*

nárovits and Földiák [50, 208] have supplied information on this subject. According to the studies of the latter authors, the thermodynamically less stable cis-1,2-dimethylcyclohexane produces the trans form in the absence and in the presence of oxygen with a yield of 1.42 and 0.85, respectively. The opposite reaction, that is the production of the cis form from the trans, takes place with a G-value of 0.62 and 0.25, respectively, (Table 3.18). The

Table 3.18

Yields of isomers from 1,2-dimethylcyclohexanes in the absence and in the presence of oxygen radical scavenger (radiation: ^{60}Co-γ). After Wojnárovits and Földiák [208]

Hydrocarbons	cis-1,2-Dimethylcyclo-hexane		trans-1,2-Dimethylcyclo-hexane	
	Pure	10 mM O$_2$	Pure	10 mM O$_2$
Product	G, molecule/100 eV			
trans-1,2-Dimethylcyclo-hexane	1.42	0.85		
cis-1,2-Dimethylcyclo-hexane			0.62	0.25
Oct-1-ene	0.13	0.12	0.40	0.32
trans-Oct-2-ene	0.80	0.60	0.63	0.50
cis-Oct-2-ene	0.25	0.22	0.13	0.10
5-Methylhept-1-ene	0.10		0.07	

scavengeable yield is presumably produced via disproportionation of di-methylcyclohexyl radicals. The residual yield can be attributed to the fact that the biradical, formed through the rupture of the C—C bond be-tween the two tertiary carbon atoms, will not transform into octenes or reclose to the initial molecule, but will form the other isomer of the dimethyl-cyclohexane on reclosure. The results also demonstrate that the biradicals formed from the two 1,2-dimethylcyclohexanes are not identical, as the oct-1-ene product is formed with a higher yield from the trans isomer than from the cis.

3.2.6. Bridged ring cycloalkanes

Decalin (bicyclo[4,4,0]decane) exists in two conformations, cis- and trans-decalin. Their mutual transformation involves the rupture of chemical bonds. In the cis form, one of the tertiary hydrogen atoms is in an axial and the other in an equatorial position, whereas both tertiary hydrogen atoms of the trans isomer are axial. The trans isomer is the more stable, its heat of formation being 13.4 kJ mol^{-1} less than that of cis-decalin.

The yields of the liquid-phase room-temperature radiolysis of decalins are summarized in Table 3.19. According to the results of Eberhardt [193], the yield of hydrogen is 0.3–0.4 G-unit higher for trans-decalin than for the

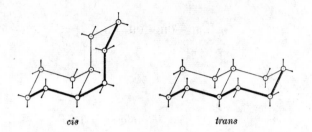

cis   *trans*

Table 3.19

Yields from liquid *cis*- and *trans*-decalin (radiation: ^{60}Co-γ)

Hydrocarbons	Mixture (*cis : trans* ratio = 32 : 68)	Mixture (*cis : trans* ratio not known)	*cis*-Decalin	*trans*-Decalin
Dose, 10^{19} eV g^{-1}	2.0	2.5–5.6	5.6–19	5.6–19
Reference	209	210, 211	193, 209	193, 209
Product	*G*, molecule/100 eV			
Hydrogen	3.51	3.57; 3.9	3.12; 4.3	3.58; 4.65
C_1-C_5 Fragments	0.016			
Butenylcyclohexane + butyl-cyclohexane			0.13	0.17
cis-Decalin				0.15
trans-Decalin			2.7	
cis-Cyclodecene			0.08	0.05
trans-Cyclodecene			0.26	0.25
9,10-Octalin				0.10
1,9-Octalin			1.2	1.0
High boiling products	1.05			

cis isomer. The author also established that in the presence of 20 mM I_2, the yields of hydrogen decrease by 1.72 and 0.72 for the *trans* and the *cis* isomer, respectively. According to his reasoning, in the case of the *cis* isomer the simultaneous elimination of the two tertiary hydrogen atoms through a four-centre transition state results in a considerable unimolecular hydrogen yield, as mentioned above in connection with 1,2-dimethylcyclohexane, whereas the same process cannot take place with the *trans* isomer. Thus, in the latter case more hydrogen is formed from hydrogen atoms.

The hydrocarbon product distribution of decalin radiolysis is not yet completely known; the *G*-values of some compounds belonging to the octalin product fraction are missing. However, the experimental data available show that skeletal isomerization, resulting in the formation of products that have the same molecular formula as the initial compound, is a reaction of major importance for both isomers [208]. Three types of this reaction have been distinguished:

(3.24a)

cis-, *trans*-cyclodecene   (3.24b)

decalin   (3.24 c)

1. By opening of one of the rings at the tertiary carbon atom butenyl-cyclohexane or butylcyclohexene is produced (3.24a);

2. By rupture of the C—C bond between the two tertiary carbon atoms, the molecule rearranges into *cis*- or *trans*-cyclodecene (3.24b);

3. *Cis*-decalin transforms into the *trans* isomer, and vice versa (3.24c).

The G-values for the transformation of *cis*-decalin into the *trans* isomer in the presence and in the absence of oxygen are 0.7 and 2.7, respectively; the values for the reverse reaction are 0.10 and 0.15, respectively. The formation of the scavengeable yield of isomer can presumably be traced back to the disproportionation reactions of tertiary decalyl radicals brought about by C—H bond ruptures, e.g.

$$cis\text{-}C_{10}H_{18}^* \longrightarrow C_{10}H_{17} + H$$

$$H + cis\text{-}C_{10}H_{18} \longrightarrow C_{10}H_{17} + H_2$$

$$2C_{10}H_{17} \Big\langle \begin{array}{l} C_{10}H_{16} + trans\text{-, or } cis\text{-}C_{10}H_{18} \\ (C_{10}H_{17})_2 \end{array}$$

The non-scavengeable yield can arise, at least in part, in unimolecular reactions, for example when the C—C bond between the two tertiary carbon atoms breaks and the subsequent reclosure of the biradical forms not the initial, but the opposite isomer. The production of *cis*- and *trans*-cyclodecene could result from stabilization of the biradical without reclosure. Radical acceptors added to the samples do not affect the yields of the latter products [50, 208].

Studying the radiolysis of *trans*-decalin, Stachowitcz [209] et al. reported that the mean molecular mass of products with high boiling points is 260, differing only very slightly from the molecular mass of the didecalyl, i.e. 274. By means of infrared spectroscopy, the authors found that the most important product among the high boiling ones is *trans, trans*-didecalyl.

The mean molecular mass of the high-boiling products of the radiolysis of *cis*-decalin is 440; thus, this fraction contains large amounts of trimers and tetramers, etc. The importance of unsaturates in this fraction was evidenced by infrared techniques.

The data summarized above clearly indicate that *cis*- and *trans*-decalin differs considerably in their radiolytic behaviour.

Adamantane is an almost unstrained hydrocarbon with a rigid structure; it contains four condensed cyclohexane rings, all of them in the chair conformation. Its structure is reminiscent of a cell of a diamond crystal. As a result of this structure the melting point of adamantane is relatively high (268°C). It is highly resistant to thermal decomposition, up to 650–670°C [212, 213].

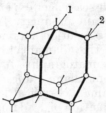

The radiolysis of adamantane has been studied by relatively few scientists. It was found that the major product of irradiation is hydrogen [214–216]. According to the investigations of Antonova et al. [216], the G-value of hydrogen for solid-phase adamantane at −196°C is 2.1; if the temperature is increased the yield increases slowly up to about −30°C and more rapidly afterwards to about 140–160°C, when it reaches a maximum of 5.5. Above this temperature it decreases again.

The authors assumed that the hydrogen formation takes place on the atomic pathway, although molecular elimination cannot be ruled out. However, this latter process, taking place via a four-centre activated state, is less likely, because the incorporation of a double bond would introduce considerable strain into the adamantane structure [216]. Therefore, in the opinion of the authors, molecular elimination would require considerable rearrangement of the molecular structure. Correspondingly, Podkhalyuzin and Vereshchinskii [215] established the formation of dicyclopentadiene, naphthalene and biphenyl.

The fact that the $G(H_2)$ of adamantane at a temperature less than 100°C is considerably lower than that of cyclohexane, was attributed by Antonova et al. [216] to unimolecular hydrogen elimination taking place more easily in the case of the latter. The decrease of the hydrogen yield above 160 °C is striking: in the view of the latter authors it is the result of the early onset of radiation-chemical–thermal cracking. The total yield of C_1-C_4 fragment products at 300°C already exceeds the value of 3 molecules/100 eV, although it is negligibly small at room temperature.

Studies dealing with the interpretation of ESR spectra of adamantane are much more numerous than those based on the determination of the yields of end-products; however, the spectra obtained during irradiation show a

structure with very few characteristic features and can be simulated by
computer only with difficulty. According to some references [217–219] the
signals of the spectra remain unchanged for several hours at room temperature.

Gee et al. [220] as well as Ferrell et al. [217] attributed the ESR spectrum
obtained after irradiation of adamantane to the 2-adamantyl radical (radical site in position 2), whereas Filby and Günther [221–223] ascribed it to
the 1-adamantyl radical (radical site in position 1). According to the detailed studies of Hyfantis and Ling [218, 219] the ESR spectrum obtained
with relatively low doses ($< 10^{19}$ eV g^{-1}) reveals really the existence of
one type of radical, namely the 1-adamantyl radical, whereas irradiation
with a higher dose leads to a more complex spectrum.

The latter authors carried out scavenger experiments in order to elucidate
the nature of the more important products. To the adamantane samples
irradiated in the solid phase, a solution was admixed that contained Br$_2$
dissolved in cyclohexane, and the adamantyl-bromides thus formed were
measured by means of gas chromatography. A value of 3 : 1 was reported
for the ratio of the 1- and 2-bromoadamantanes, indicating that the thermodynamically more stable 1-adamantyl radical had been produced in larger
amounts. The predominant formation of 1-adamantyl radical over the 2-iso-
mer also gains support from the radiation chemical alkylation experiments
of adamantane in the presence of ethylene or propene carried out in cyclohexane solution or in the gas phase [223–223C]. The G-value of 1-alkyl-
adamantane formation was found to be 4 to 10 times higher than that of
2-alkyladamantanes.

Bonazzola and Marx [224] stated that the ESR spectrum of adamantane
irradiated at 77 K did not correspond either to 1-adamantyl or to 2-ada-
mantyl radicals. They suggested that the signals are due to the substituted
1-methylcyclohexyl radical, which is formed in secondary reactions of the
1-adamantyl positive ion:

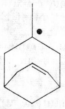

It is also possible that the spectrum is produced, at least in part, by the
presence of some impurities in the irradiated samples, e.g. some cyclohexane derivatives, as the preparation of adamantane in high purity is a
tiresome and extremely difficult task [225].

When discussing the radiolysis of adamantane, an important field of its
application should be mentioned [226, 227]. As its melting point is high and
its ESR signals are relatively weak, it can be used to investigate the signals
of several organic and inorganic radicals. The method involves the disso-
lution of the sample in solid adamantane and subsequent irradiation. The

adamantane has a dual role; it isolates the molecules of the dispersed matter, and, according to Filby and Günther [222], the radicals of the 'guest' organic molecules (S) are generated by hydrogen abstraction reactions of the 1-adamantyl radicals, e.g.

$$C_{10}H_{15} + SH \longrightarrow C_{10}H_{16} + S$$

3.3. MIXTURES OF CYCLOALKANES

Some basic problems of the radiolysis of mixtures have been discussed in Section 2.3.

Aliphatic alkane–cycloalkane mixtures

Horváth and Földiák studied the product distributions resulting from the liquid-phase irradiation of cyclopropane–propane [228] and cyclobutane–butane [229] systems. The yields of some products formed in n-butane–cyclobutane systems are shown in Fig. 3.6. It can be seen that the yield of ethylene, the most important product of cyclobutane radiolysis, exceeds over the whole composition range the value corresponding to the additivity rule, whereas the G-values of ethane are always lower than that. Ethane is the most important product of n-butane, formed by C–C bond rupture. The experimental results indicate that complex processes take place between the components of the mixtures (see Section 2.3).

Kudo and Shida have reported studies [83] connected with the determination of the yields of hydrogen and C_1-C_4 hydrocarbon products formed in 2,2,4-trimethylpentane–cyclopentane and n-hexane–cyclopentane mixtures. On the radiolysis of the former system they observed that the yield of hydrogen is lower, whereas those of the characteristic products of iso-octane are higher than the values expected on the basis of the additivity rule. As the ionization potential of cyclopentane (10.51 eV) is higher than

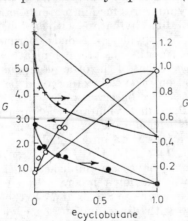

Fig. 3.6. G-values of some products from n-butane–cyclobutane mixtures.
○ C_2H_4; + C_2H_6; ● CH_4 (radiation: ^{60}Co-γ). After Horváth and Földiák [229]

that of iso-octane (9.84 eV), the results reveal a charge transfer. The yields of the n-hexane–cyclopentane system follow the additivity rule, although the ionization potentials of the components differ considerably: 10.17 and 10.51 eV, respectively. The authors interpreted these results in terms of excitation energy transfer taking place simultaneously with charge transfer but in the opposite direction, so that the two processes cancel each other out.

Claes et al. have devoted a few papers to investigations of the radiolysis of 2,2-dimethylbutane–cyclopentane and 2,2-dimethylbutane–cyclohexane mixtures at 77 K [230–232]. By comparing the ESR spectra obtained by applying UV-light as well as high-energy radiation, they concluded that in the photolysis of the mixture the yield of cyclopentyl radicals varied linearly with the composition, whereas in the radiolysis the yields of cyclopentyl and cyclohexyl radicals exceeded the values expected on the basis of a linear relationship. The observations were interpreted in terms of charge-transfer interactions directed from the neohexane to the cyclic compounds. As the gas-phase photoionization potential of cyclopentane is higher than that of neohexane (10.53 and 10.04 eV, respectively), the observations contradict the generally accepted hypothesis according to which the transfer takes place from the component having the higher ionization potential to the one with the lower ionization potential [75].

The yields from 2,3-dimethylbutane–cyclopentane and –cyclohexane mixtures irradiated at 77 K generally do not correspond to those expected on the basis of the additivity rule. According to the reasoning of Tilquin and Claes, this can be attributed to energy transfer, directed from the dimethylbutane to the cyclopentane and cyclohexane ($E_i = 10.00$, 10.53 and 9.88 eV); however, they think that a transfer in the opposite direction is also possible, at least at low concentrations [233].

György and Földiák [234, 235] have reported investigations connected with the formation of hydrogen and some fragment products in the liquid-phase radiolysis of binary mixtures made up of n-hexane, 3-methylpentane, 2,2-dimethylbutane or 2,3-dimethylbutane with cyclohexane. The deviations from the additivity rule only slightly exceeded experimental error for practically all the products. As the gas-phase ionization potential of cyclohexane (9.88 eV) is less than that of the other component in every mixture, the sensitized decomposition of cyclohexane could be expected. However, transfer to the cyclohexane was observed only with n-hexane or 3-methylpentane. With 2,2- or, 2,3-dimethylbutane the branched-chain hydrocarbons apparently 'protect' cyclohexane by their own enhanced rates of decomposition. The observations were interpreted by the authors along the lines discussed in Section 2.3.

Recently numerous papers have been published on the selective hydrogen atom abstraction by H atoms at low temperatures (e.g. [236, 237]). For instance, in cyclohexane–n-pentane solutions at 77 K the H atoms were found to react more effectively with n-pentane in the cyclo-C_6H_{12} matrix than with cyclohexane, while in the n-C_5H_{12} matrix the case was reversed [237]. It has been concluded that the solute alkane may form some active sites in the solid phase that can be attacked selectively by H-atoms.

Cycloalkane–cycloalkane mixtures

A number of papers have been devoted to the radiolysis of cyclopentane–cyclohexane mixtures [80, 238–242]. There is a great difference between the gas-phase ionization potentials of the two compounds (10.53 and 9.88 eV), and the same difference can probably be assumed to prevail in the liquid phase [100]. The yields of cyclopentene, cyclohexene and dimer products deviate from those estimated from the additivity rule. Muccini and Schuler established that the yields of cyclopentyl and cyclohexyl radicals correspond to an ideal behaviour. Accordingly, the non-ideal behaviour observed in the yields of dimer products was attributed to secondary reactions, e.g.

$$\text{cyclo-C}_6\text{H}_{11} + \text{cyclo-C}_5\text{H}_{10} \xrightarrow{k_{3.25}} \text{cyclo-C}_6\text{H}_{12} + \text{cyclo-C}_5\text{H}_9 \qquad (3.25)$$

$$\text{cyclo-C}_5\text{H}_9 + \text{cyclo-C}_6\text{H}_{12} \xrightarrow{k_{3.26}} \text{cyclo-C}_5\text{H}_{10} + \text{cyclo-C}_6\text{H}_{11} \qquad (3.26)$$

In these reactions $k_{3.25} > k_{3.26}$. This conclusion agrees well with the concept that the hydrogen atoms of cyclopentane are a somewhat eclipsed [186]. The final conclusion of the authors was that in this system no energy or charge transfer need be assumed for the data to be interpreted [238, 239]. Toma and Hamill [240] however, deduced that charge transfer occurs from cyclopentane to cyclohexane, having allowed for the fact that the yield of cyclohexene had been found to be considerably higher, and that of cyclopentene lower, than the G-values calculated from the additivity rule.

Yang and Marcus [241] carried out extensive investigations on the radiolysis of cyclo-C$_5$H$_{10}$–cyclo-C$_6$D$_{12}$ mixtures. They found the yields of cyclopentene and bicyclopentane to be considerably higher than those published in the earlier works devoted to the radiolysis of cyclo-C$_5$H$_{10}$––cyclo-C$_6$H$_{12}$ systems. On the other hand, however, the yield of cyclo-C$_6$D$_{10}$ turned out to be approximately equal to that corresponding to the energy absorption in the cyclohexane only. Taking into consideration these data they revised the results obtained in the cyclo-C$_5$H$_{10}$–cyclo-C$_6$H$_{12}$ system. According to the comments made on reactions (3.25) and (3.26), a small yield for cyclohexene and a large yield for cyclopentene can be expected for this system. Any behaviour inconsistent with this expectation could be attributed to differences in disproportionation reactions:

$$\text{cyclo-C}_6\text{H}_{11} + \text{cyclo-C}_5\text{H}_9 \xrightarrow{k_{3.27}} \text{cyclo-C}_6\text{H}_{10} + \text{cyclo-C}_5\text{H}_{10} \qquad (3.27)$$

$$\text{cyclo-C}_6\text{H}_{11} + \text{cyclo-C}_5\text{H}_9 \xrightarrow{k_{3.28}} \text{cyclo-C}_6\text{H}_{12} + \text{cyclo-C}_5\text{H}_8 \qquad (3.28)$$

The authors concluded that $k_{3.27} > k_{3.28}$ to at least the same extent as $k_{3.25} > k_{3.26}$.

Applying a careful treatment Wojnárovits and Földiák made a clear distinction between the atomic and unimolecular way of H$_2$ formation in the cyclo-C$_5$H$_{10}$-cyclo-C$_6$H$_{12}$ system. They also observed excess cyclopentyl radical formation in the mixture, but on adding cyclohexane to cyclopentane, the extent of unimolecular H$_2$ detachment from cyclopentane was found to decrease, while that from cyclohexane increased considerably. As the H_2 detachment is the typical reaction of low-energy excited unbranched hydrocarbon molecules (probably in the S_1 state), the experiment is indicative of a transfer reaction between the components of the mixture resulting in a decreased probability of formation of cyclopentane low-energy excited states and an increased production of these states in the cyclohexane molecules [242].

3.4. CONCLUSIONS

Having reported on the radiolysis of the individual cycloalkanes (Section 3.2), we shall now consider certain more general features of product distributions, the relationship of molecular structure and the radiation-induced decomposition, etc. As liquid-phase radiolysis has been studied the most extensively, the yields in this phase will be compared.

The characteristic data of the liquid radiolysis of a comparatively great number of cycloalkanes are listed in Table 3.20. We have made every effort to select data obtained under similar experimental conditions, but without success in some cases.

Processes related to C—H bond ruptures

Table 3.20 (as well as Fig. 2.8) indicates that $G(H_2)$ values of cycloalkanes increase with increasing carbon number, whereas those for n-alkanes between C_3 and C_{12} are nearly the same (4.9 ± 0.3) even if the experimental conditions are dissimilar. G-values for cyclopropane and cyclobutane [$G(H_2) = 0.91$ and 1.70, respectively] are closer to those of corresponding n-alkenes (0.8 for propene and 0.73 for but-1-ene) than to those for propane or n-butane. At the same time, above C_6, hydrogen yields from cycloalkanes are generally higher than from n-alkanes.

It is obvious from Table 3.20 and Fig. 2.8 that the hydrogen yields from monoalkylcycloalkanes are lower than those of the corresponding cycloalkanes, and the $G(H_2)$-values decrease with the number of side chains. The hydrogen yields of alkylcyclohexanes exceed those of the corresponding alkylcyclopentanes. As a result of dehydrogenation of rings larger than C_4 mainly cycloalkene and bicycloalkane products are formed. The overall yields of these latter two products are always somewhat lower than the $G(H_2)$ values. The deficit of material balance is usually reduced by taking into account the hydrogen equivalent of other products formed, although the balance is never complete [10, 85, 184]. In the course of cyclopropane radiolysis no cyclopropene and bicyclopropane formation has been observed [56]. On irradiation of cyclobutane, the formation of cyclobutene has, however, been detected, but it is not yet known whether or not any bicyclobutane is formed [56A].

The mechanism of hydrogen formation is qualitatively different on irradiation of smaller (C_3 and C_4) and larger rings. During the radiolysis of cyclopropane and cyclobutane, hydrogen is formed to a large extent as a result of the secondary decomposition of radicals and radical ions [35, 56, 60, 66].

In contrast, hydrogen is produced from C_5 and larger rings mainly by H-atom or H_2-molecule elimination from excited molecules produced directly or in the course of charge recombination.

Ausloos et al. [32] studied the gas-phase photolysis (147 nm, 8.4 eV) of cyclopropane and cyclobutane and established the quantum yield of H_2-molecule elimination as < 0.02. As increasing photon energy favours the atomic processes over the molecular ones, H_2 elimination in the radiolysis of cyclopropane and cyclobutane can presumably be neglected. The quantum yield of H_2 elimination of cyclopentane and cyclohexane in the gas phase, taken from the work cited above, is 0.7 ± 0.1, whereas during the liquid-phase irradiation of cyclopentane, cyclohexane, cycloheptane, cyclo-octane and cyclodecane with 163 nm (7.6 eV) photons, values of 0.8–0.9, 0.9, 0.85, 0.9 and 0.9, respectively, were established [36, 37, 39]. In the liquid-phase radiolysis of these cycloalkanes the G-values of unimolecular formation of cycloalkene and H_2 fall in the range of 0.8–1.6 [81, 132, 133, 184], i.e. about 20–30% of the total hydrogen yield, $G(H_2) = 5$–6.

The hydrogen atom elimination leads to alkyl radical formation. The ESR spectra obtained during the radiolysis of solid and liquid cyclopropane point to the presence of several radical species [96, 155]. Fessenden and Schuler detected mainly allyl and cyclopropyl radicals [96]. In the radiolysis of cyclopropane containing scavengers, Holroyd established the ratios for the yields of methyl, allyl and cyclopropyl radicals to be 0.25 : 1 : 1 [243]. In the ESR spectrum of irradiated cyclobutane there appeared signals for cyclobutyl and but-3-enyl radicals [96]. It has been suggested that, in both cases, the formation of aliphatic radicals is not the result of the isomerization of cyclic ones, but the two isomeric radicals (e.g. allyl and cyclopropyl) are formed independently.

In the ESR spectra of cycloalkanes containing five or more carbon atoms only cycloalkyl radicals can be observed [96, 155]. With the application of high sensitivity scavenger methods fragment alkyl radicals have also been detected in the radiolysis of cyclopentane and cyclohexane, albeit with yields smaller by an order of magnitude than those of cycloalkyl radicals [11, 94, 95, 156]. The C_5-C_{10} cycloalkyl radicals, when reacting with each other at room temperature produce approximately equal amounts of cyclo-alkenes and bicycloalkanes ($k_d/k_c \approx 1$) [11, 90A, 143, 159, 184].

On irradiation of monoalkylcycloalkanes all the possible C_nH_{2n-2} de-hydrogenation products can be observed in the product spectra, although the proportion of those that contain the π-bond at the branch point is much higher than the combined yield of other cyclic alkenes [10, 102, 190, 200]. The dimer yield of alkylcycloalkanes is generally lower than that of cyclo-alkanes. The yields of cycloalkene and bicycloalkane products could be interpreted only by assuming that the radical site of the interacting radicals is mainly in the tertiary position. The rate of disproportionation of these types of radical is several times higher than that of their recombination, and therefore high yields for cycloalkenes and low yields for bicycloalkanes aan be expected.

Table

Yields from liquid cycloalkanes

Products	Dose, 10^{19} eV g^{-1}	$G(H_2)$	G(cyclo-alkene)	G(bicyclo-alkane)	G(side chain rupture)
Cyclopropane	58.4	0.91			
Cyclobutane	58.4	1.7	0.72		
Cyclopentane	a	5.35	2.97	1.29	
Cyclohexane	a	5.6	3.2	1.76	
Cycloheptane	a	5.75	3.1	2.1	
Cyclo-octane	a	5.9	3.0	1.95	
Cyclodecane	a	5.95	3.6	1.9	
Methylcyclopentane	18.7	4.2	2.8	1.6	0.2
Ethylcyclopentane	18.7	4.2	2.5	0.9	0.25
Methylcyclohexane	18.7	4.6	2.55	0.9	0.2
Ethylcyclohexane	18.7	4.7	2.7	0.7	0.25
Isopropylcyclohexane	18.7	3.3			1.5
1,1-Dimethylcyclohexane	3.1; 18.7	3.3	1.1	1.5	0.70

a Yields extrapolated to zero dose

Processes related to C—C bond ruptures

Table 3.20 contains also the G-values of C—C bond rupture products, i.e. in the case of alkylcycloalkanes the elimination of the side chain, and the degradation of the ring, as well as decyclization into aliphatic compounds having the same carbon atom number as that of the initial hydrocarbon.

In separate column the yields of other C—C bond rupture products are given, e.g. the G-values for the transformation of cyclohexane into n-hexylcyclohexane and cyclohexylhex-1-ene. The total yields of molecules transformed into C—C bond rupture products, marked by G(C—C), are also collected in the table.

The C—C bond rupture products of the radiolysis of cyclopropane and cyclobutane are formed mainly via ion decomposition e.g. the ring-scission of positive ions, ion-molecule reactions, and perhaps ring-opening reactions by hot hydrogen atoms [55, 66]. In contrast, in the decomposition of larger rings, these reactions are, at least in the liquid phase, of secondary importance and the C—C bond rupture products are formed mainly via decomposition of excited states produced in neutralization and in direct excitation processes. The formation of these products is due in part to unimolecular reactions of biradicals, as numerous authors have suggested.

The most rapid reaction of biradicals formed with high probability from cycloalkanes is unimolecular stabilization to give aliphatic alkenes or, in the case of radicals with at least three carbon atoms, some cycloalkanes. Their bimolecular reactions are of low probability.

The hot hydrogen atoms generated during radiolysis can, in the opinion of several authors, cause ring-scission when they collide with cycloalkane

3.20

(radiation: ^{60}Co-γ)

G(ring scission)	G(straight chain C_n)	G(other C—C bond rupture)	G(C—C)	G(C—H) + G(C—C)	$j = \dfrac{G(\text{C—C})}{G(\text{H}_2)}$	Reference
2.0	1.3		3.3	4.21	3.6	56A
3.0	0.75		3.75	5.45	2.2	56A
0.53	0.88	0.2	1.6	6.95	0.30	85
0.10	0.50	0.2	0.8	6.4	0.14	102, 143
0.08	0.45	0.2	0.73	6.48	0.13	184
0.04	0.31	0.13	0.48	6.38	0.08	184
0.04	0.31	0.15	0.50	6.5	0.08	184
0.48	1.26	0.05	2.0	6.2	0.48	102
0.23	1.32	0.05	1.85	6.05	0.44	102
0.11	0.84	0.05	1.2	5.8	0.26	102
0.08	0.63	0.13	1.09	5.79	0.23	102
0.05	0.2	0.1	1.85	5.15	0.56	50, 173
0.17	1.5	0.1	2.47	5.77	0.75	50, 192

molecules. On the evidence of the product balances, these types of reaction are important only for smaller rings. Avdonina has reported that these hot atom reactions also contribute to the radiolysis of cyclohexane, but their total G-value does not exceed 0.1 [135, 137]. The results indicate that, for larger rings, these processes can be neglected.

The tendency of cycloalkanes to form fragments decreases with increasing carbon number [102, 184, 244, 245]. Relatively significant amounts of methane are produced from cyclopropane only; methane yields from higher alkane rings are between 0.06 and 0.001, decreasing monotonically with molecular weight. Normally, most of the fragments (65–90%) have the general formula C_rH_{2r}; they are mostly aliphatic alk-1-enes, mainly ethylene; this can be explained by a biradical decomposition mechanism and rapid stabilization of the fragments [162, 170, 172].

The unpaired electrons at the chain-ends of radicals weaken C—C bonds in the β-position, as their fission may lead, even in the transition state, to a partial double-bond formation that will decrease the energy of activation to a low value (a few kJ mol^{-1}). This is most evident for 1,4-biradicals where the elongation of the central C—C bond may facilitate the formation of two partial π-bonds [6].

It should be noted, however, that no unambiguous evidence has been put forward as to whether decomposition into two or more parts happens in a stepwise process via an intermediate biradical or by a simultaneous rupture of two or more C—C bonds of the rings.

Cycloalkanes with rings containing four or more carbon atoms produce some cycloalkanes and alkylcycloalkanes with smaller rings (e.g. cyclopropane), indicating ring-fission followed by ring-closure. However, no cyclobutane or alkylcyclobutane was found among the products owing to the rapid decomposition of the 1,4-C_4-biradical to give ethylene. Products of the formula C_rH_{2r} are formed from biradicals in unimolecular processes, whereas saturated aliphatic alkanes are formed from monoradicals in bimolecular processes [162, 170, 172].

C_3-C_5 cycloalkanes usually give no more than two hydrocarbon fragments with overall G-values between 1 and 6, whereas larger rings can split into three or more parts although the overall yields of fragments are much lower, not exceeding 0.25 [11, 56, 102, 162, 173].

On the basis of Fig. 3.7 it can be stated that while radiolysis of cyclopropane and cyclobutane gives more fragments than C_n products, the ratio $(C_1$-$C_{n-1})$: (C_n) is about 1 for cyclopentane and 0.5 for higher hydrocarbons. Thus, the relative yield of fragments decrease with increasing ring size. This can be explained by analogy to aliphatic alkanes, in that energy uptake is divided among more vibrational degrees of freedom in larger molecules. Thus, less energy can be concentrated in any single bond. The probability of multiple decomposition becomes lower, and consequently the yields of hydrocarbon fragments decrease in comparison with C_n products [102, 245].

The hydrocarbon products of alkylcycloalkanes are analogous to those observed for unsubstituted rings, i.e. ring fission products, but the fission

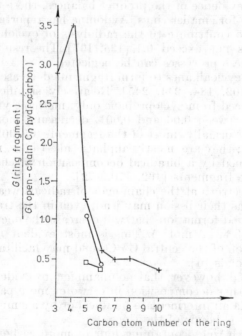

Fig. 3.7. G-value ratios of ring fragment and open-chain C_n hydrocarbon formation, as a function of the carbon atom number of the ring.
+ Cycloalkanes; ○ methylcycloalkanes; □ ethylcycloalkanes (radiation: ^{60}Co-γ).
Data taken from refs [56A, 85, 102, 184]

is directed in this case: the most favoured position is in the vicinity of the tertiary carbon atom. Aliphatic and cyclic alkanes and alkenes with branched chains (e.g. isobutane, isobutene and methylcyclopropane) are also formed in detectable amounts, but no fragments resulting from skeletal isomerization of the primary branched biradicals could be observed [102, 162].

During decomposition of cycloalkanes with substituents larger than the methyl group, any of the C—C bonds in the alkyl chain may dissociate, in addition to total dealkylation. The rate of splitting of the side-groups increases from methyl to ethyl, and even more (about by one order of magnitude) in the case of isopropyl and tert-butyl [50].

In general there is competition between the decomposition of the ring and that of the side-chain, especially if the latter contains a quaternary carbon atom and thus possesses relatively weak C—C bonds [192]. This is in agreement with observations of isoalkane radiolysis. The extent of 'protection' increases with increasing branching. The direction of 'protection' processes is opposite to that observed in alkylbenzenes, where an aromatic ring 'protects' the alkyl side-chain [173] (see Section 6.3.).

The straight-chain C_n hydrocarbon products are at least in 80% alkenes with the formula C_nH_{2n}, and consequently they can be regarded as the straight-chain isomers of the starting C_nH_{2n} cycloalkanes. For cycloalkanes without a side-chain they are alk-1-enes, whereas for alkylcycloalkanes internal straight-chain alkenes are also formed, e.g. hex-2-ene is formed from methylcyclopentane and hept-3-ene from ethylcyclopentane. As discussed in Section 3.2.5 there is a very strict law in the formation of these compounds: the alkenes containing the π-bond at a carbon atom which participated in the C—C bond rupture have the highest G-values. This rule has been interpreted to indicate a biradical decomposition mechanism [50, 102]. The biradical formed during the ring-opening can rearrange to alkenes by a single-step shift of one hydrogen atom within the carbon chain. It is supposed that hydrogen atom migration by a single step involves five- or six-membered cyclic intermediates; seven or eight-membered rings may also play a part.

The occurrence of some cis–trans isomerization indicates that ring-closure following ring-opening is not an exceptional process: e.g. the isomerization of cis-1,2-dimethylcyclohexane into its thermodynamically more stable trans-form can partly be explained in terms of such a sequence [50, 193, 208].

Alkylcycloalkanes with smaller rings can be detected among the products (e.g. methylcyclopentane from cyclohexane and methylcyclohexane from cycloheptane), but no reverse isomerization (ring enlargement) process has yet been observed, possibly owing to analytical difficulties [102, 143, 169, 184]. Skeletal isomerization leading to cyclobutane or alkylcyclobutane has not been detected either.

As opposed to aliphatic alkanes, very low amounts of hydrocarbons between C_n and C_{2n} are formed from cycloalkanes. This can be interpreted mainly in terms of the primary fragments being principally biradicals, which are more rapidly stabilized in unimolecular processes; consequently, their bimolecular combination with C_n radicals is less probable.

16

In addition to the generally accepted mechanisms of build-up reactions, the insertion of small biradicals (normally methylene) between two atoms can also be observed. These reactions are of importance only with cyclopropane and cyclobutane (e.g. the formation of methylcyclopropane observed during cyclopropane irradiation is partly the result of such a reaction), although this reaction was also detected in the radiolysis of methylcyclopentane [204].

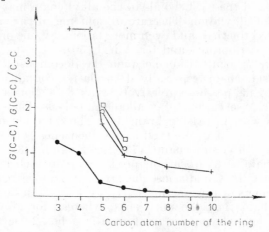

Fig. 3.8. G-values of molecules that undergo C—C bond scission, as a function of the carbon atom number in the ring.
G(C—C): □ methylcycloalkanes; ○ ethylcycloalkanes; + cycloalkanes;
● G(C—C)/C—C number: cycloalkanes (radiation: ^{60}Co-γ). Data taken from refs [56A, 85, 102, 184]

Figure 3.8 shows the *G*-values of molecules undergoing C—C bond scission and the G(C—C) values per one C—C bond. Both values decrease with increasing number of carbon atoms in the rings.

The relation between C—H and C—C bond ruptures

As reported in Section 3.2, the C—H and C—C bond rupture processes take place more or less independently from each other. For most alkanes the G(H$_2$)-values can be considered as the quantitative characteristics of molecules that undergo C—H bond ruptures, as for the formation of one H$_2$ molecule approximately one starting activated molecule has to be consumed. The C—C bond-rupture products are hydrocarbons the majority of which have C : H ratios equal to that of the initial cycloalkane (1 : 2), and which are presumably formed mainly via unimolecular processes. As a rough approximation, it can be concluded that the decomposition of a molecule activated by radiation leads either to hydrogen formation without any change in the carbon chain, or to C—C bond rupture [244, 245]. Therefore, the values of G(C—H) + G(C—C), compiled in Table

3.20, constitute a quantitative measure of the characterization of activated molecules transforming into C—H and C—C bond rupture products. This value shows hardly any change for the majority of alkanes, being in the range 6–7. However, the values for the alkylcycloalkanes are slightly lower, presumably as a result of the relatively high doses applied for the G-value determinations. The low $G(C—H) + G(C—C)$ values of cyclopropane and cyclobutane may be also caused by their greater tendency to polymerize.

A similar value, namely G(activated molecules) = 6–7 was found also in the radiolysis of n-alkanes and isoalkanes [244, 245]. Theoretical calculations suggested a somewhat higher value, 7–8, for the formation of activated alkane molecules in the gas phase [246, 247]. It is rather difficult to prove experimentally the validity of this result in gas-phase experiments, but the value seems to be reasonable. These data can be interpreted as follows: despite the differences in the early events of radiolysis in the gas and liquid phases (see Section 1.1), the effectiveness of energy consumption for activated molecules is comparable in both phases, and the dissipation of excitation energy without chemical decomposition in the liquid phase is of minor importance. The latter statement is in agreement with the very low fluorescence yields of alkanes, as it has already been mentioned before.

In the last but one column of Table 3.20, the ratios of the $G(C—C)$ and $G(C—H)$ values are listed. These numbers have been called j-values. They show a decrease of 1.5 orders of magnitude for the cycloalkanes as a function of the carbon atom number of the ring (Fig. 3.9), whereas they remain constant for the n-alkanes (0.3), and increase with increasing branching for the branched-chain alkanes.

For instance $G(C—H)$ decreases and $G(C—C)$ increases along the series cyclohexane–methylcyclohexane–1,1-dimethylcyclohexane, whereas $G(\Sigma)$ shows hardly any change, i.e. it decreases by about 10–15% only [192].

Plotting $G(C—C)$ values per C—C bond of cycloalkanes as a function of the strain energy (Fig. 3.10) reveals that the assumption that ring opening is facilitated by strain [57, 102, 244, 245] can be regarded as proven for

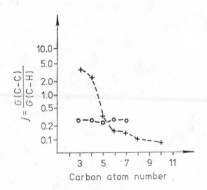

Fig. 3.9. $j = \dfrac{G(C—C)}{G(H_2)}$ of cycloalkanes, as a function of the carbon atom number.
◯ n-Alkanes; + cycloalkanes (radiation: $^{60}Co\text{-}\gamma$). See Table 3.20 for reference numbers

C_3-C_6 rings only. Between C_6 and C_{10} increasing carbon number and strain energy are accompanied by lower G-values for C—C bond rupture. The j-value characteristic of the competition between C—C and C—H bond rupture decreases between C_3 and C_{10} (Figure 3.9). For cycloalkanes higher than C_6, C—C bond rupture becomes increasingly less probable compared with n-alkanes — in spite of its becoming thermodynamically more and more favourable — but at the same time $G(H_2) = G(C—H)$ values increase with increasing strain energy. The apparent contradiction can presumably be attributed to the different origins of ring strain in the case of small and large rings [2, 4, 248]: the ring strain of cyclopropane and cyclobutane is connected with the strong distortions of C—C—C bond angles, thus favouring C—C bond ruptures and high j-values. The extremely low $G(H_2)$ values of cyclopropane and cyclobutane can also be interpreted in terms of the relatively high bond dissociation energy (439 kJ mol^{-1}) of their C—H bonds. The cleavage of a C—H bond in the parent molecule changes the hybridization state of one carbon atom from sp^3 to sp^2, thus increasing the unstrained bond angle from 109.5° to 120°. This, in turn, would increase the strain in the ring and, because of competition, makes C—H bond fission unfavourable. The decreasing j-values with increasing carbon atom number of C_6-C_{10} cycloalkanes can be related to the van der Waals repulsive interactions between the hydrogen atoms, as discussed in Section 3.1. These repulsive interactions weaken the C—H bonds, thus promoting their cleavage [184]. The continuous variation in product distribution reveals that decomposition mechanisms change with changing carbon atom number.

Alkyl substitution affects the decomposition of cycloalkanes to approximately the same extent that side-chains do in the case of aliphatic alkanes;

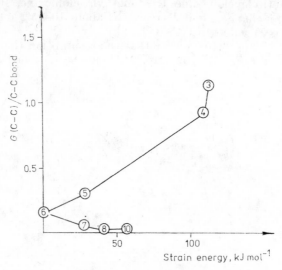

Fig. 3.10. The dependence of the $G(C—C)/C—C$ bond values on the ring-strain of cycloalkanes. The encircled characters represent the carbon atom numbers (radiation: ^{60}Co-γ). Data taken from refs [7, 56A, 85, 102, 184]

as mentioned earlier (Section 2.4), these changes in the molecular structure increase the yields of C—C bond rupture. Predominant among the products are those formed largely by the breaking of bonds attached to tertiary or quaternary carbon atoms.

Hydrogen yield and its ionic component, taken from refs [192] and [195] for the radiolysis of several alkanes, have been plotted as a function of $G(\text{C}-\text{C})$ in Fig. 3.11. It can be seen that the curves for both $G(\text{H}_2)$ and $G(\text{H}_2)_{\text{ionic}}$ fall off almost parallel to each other with increasing $G(\text{C}-\text{C})$, the distance between the two curves being about $G(\text{H}_2) = 1.4\pm0.4$. Consequently, processes based on direct excitation show little or no structure dependence [16, 192]. On the contrary, as indicated by the plot for $G(\text{H}_2)_{\text{ionic}}$ the contribution of electron-ion recombination to hydrogen formation decreases with increasing branching. As on the one hand the yield of ion pairs is about $G \approx 4$–5 both in gases and in liquids, and on the other hand this value is hardly influenced at all by molecular structure, the increase of $G(\text{C}-\text{C})$ indicates that instead of dehydrogenation, a larger and larger proportion of energy deliberated during electron-ion recombination is utilized for C—C bond fission at the branch-point.

In branched-chain alkane ions positive charge is concentrated mainly at the branch-point [249]. The correlation between positive-charge distribution and bond-fission probability has been stressed several times in mass spectrometric and radiation chemical studies, e.g. refs [249] and [250]. Investigations of the fluorescence of alkanes also indicate that the electronic excitation is highly localized in the vicinity of the branch-point [43, 44]. These facts can make the molecules extremely 'fragile' at the branching. Highly branched alkanes show extreme properties in other respects, namely they have large yields of free ions and high electron mobilities in the liquid state,

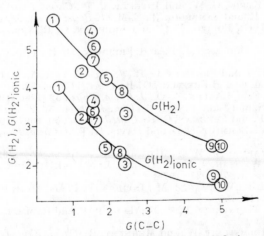

Fig. 3.11. Correlation of C—H and C—C bond fission of alkanes. *1*, Cyclohexane; *2*, methylcyclohexane; *3*, 1,1-dimethylcyclohexane; *4*, cyclopentane; *5*, methylcyclopentane; *6*, hexane; *7*, heptane; *8*, 3-methylpentane; *9*, 2,2-dimethylpropane; *10*, 2,2,4-trimethylpentane (radiation: ^{60}Co-γ). Data taken from refs [192, 195]

as a result of the increased sphericity and isotropy caused by the branches
[13, 22, 25, 249]. At present, the relationship between the high C—C bond
rupture yields and the extraordinary electric properties is not clear.

REFERENCES

1. BROWN, H. C. and ICHIKAWA, K., *Tetrahedron*, **1**, 221 (1957)
2. ELIEL, E. L., ALLINGER, N. L., ANGYAL, S. J. and MORRISON, G. A., *Conformational Analysis*, Interscience, New York, 1965
3. SCHLEYER, P. v. R., WILLIAMS, J. E. and BLANCHARD, K. R., *J. Am. Chem. Soc.*, **92**, 2377 (1970)
4. ALLINGER, N. L., TRIBBLE, M. T., MILLER, M. A. and WERTZ, D. H., *J. Am. Chem. Soc.*, **93**, 1637 (1971)
5. KÜCHLER, L., *Trans. Faraday Soc.*, **35**, 874 (1939)
6. O'NEAL, H. E. and BENSON, S. W., *J. Phys. Chem.*, **72**, 1866 (1968)
7. BENSON, S. W., *Thermochemical Kinetics*, Wiley, New York, 1968
8. STULL, D. R., WESTRUM, E. F. Jr. and SINKE, G. C., *The Chemical Thermodynamics of Organic Compounds*, Wiley, New York, 1969
9. KERR, J. A., *Chem. Rev.*, **66**, 465 (1966)
10. FREEMAN, G. R. and STOVER, E. D., *Can. J. Chem.*, **46**, 3235 (1968)
11. KESZEI, Cs., WOJNÁROVITS, L. and FÖLDIÁK, G., *Acta Chim. Acad. Sci. Hung.*, **92**, 329 (1977)
12. HUNTER, L. M. and JOHNSEN, R. H., *J. Phys. Chem.*, **71**, 3228 (1967)
13. HUMMEL, A. and SCHMIDT, W. F., *Radiat. Res. Rev.*, **5**, 199 (1974)
14. WARMAN, J. M., ASMUS, K. O. and SCHULER, R. H., *Advan. Chem. Ser.*, **82**, 25 (1968)
15. WARMAN, J. M., ASMUS, K. O. and SCHULER, R. H., *J. Phys. Chem.*, **73**, 931 (1969)
16. ASMUS, K. O., WARMAN, J. M. and SCHULER, R. H., *J. Phys. Chem.*, **74**, 246 (1970)
17. KLEIN, G. W. and SCHULER, R. H., *J. Phys. Chem.*, **77**, 978 (1973)
18. INFELTA, P. P. and SCHULER, R. H., *Int. J. Radiat. Phys. Chem.*, **5**, 41 (1973)
19. RZAD, S. J., KLEIN, G. W. and INFELTA, P. P., *Chem. Phys. Lett.*, **24**, 33 (1974)
20. FREEMAN, G. R. and SAMBROOK, T. E. M., *J. Phys. Chem.*, **78**, 102 (1974)
21. ROBINSON, M. G., FOUCHI, P. G. and FREEMAN, G. R., *Can. J. Chem.*, **49**, 3657 (1971)
22. SHINSAKA, K., DODELET, J. P. and FREEMAN, G. R., *Can. J. Chem.*, **53**, 2714 (1975)
23. ROBINSON, M. G. and FREEMAN, G. R., *J. Chem. Phys.*, **55**, 5644 (1971)
23A. ROBINSON, M. G. and FREEMAN, G. R., *Can. J. Chem.*, **52**, 440 (1974)
24. SCHMIDT, W. F. and ALLEN, A. O., *J. Chem. Phys.*, **52**, 4788 (1970)
24A. HUANG, S. S. and FREEMAN, G. R., *Can. J. Chem.*, **56**, 2388 (1978)
25. DODELET, J. P. and FREEMAN, G. R., *Can. J. Chem.*, **50**, 2667 (1972)
26. MINDAY, R. M., SCHMIDT, L. D. and DAVIS, H. T., *J. Chem. Phys.*, **50**, 1473 (1969); **54**, 3113 (1972)
27. ZÁDOR, E., WARMAN, J. M. and HUMMEL, A., *J. Chem. Phys.*, **62**, 3897 (1975)
27A. ZÁDOR, E., WARMAN, J. M., LUTHJENS, L. M. and HUMMEL, A., *J. Chem. Soc. Faraday Trans. 1.*, **70**, 227 (1974)
27B. DE HAAS, M. P., WARMAN, J. M., INFELTA, P. P. and HUMMEL, A., *Chem. Phys. Lett.*, **31**, 382 (1975)
27C. WARMAN, J. M., INFELTA, P. P., DE HAAS, M. P. and HUMMEL, A., *Can. J. Chem.*, **52**, 2249 (1977)
28. BREDE, O., HELMSTREIT, W. and MEHNERT, R., *Chem. Phys. Lett.*, **28**, 43 (1974)
28A. BREDE, O., BÖS, J., NAUMANN, W. and MEHNERT, R., *Radiochem. Radioanal. Lett.*, **35**, 85 (1978)
29. WALTER, L. and LIPSKY, S., *Int. J. Radiat. Phys. Chem.*, **7**, 175 (1975)
29A. WALTER, L., HIRAYAMA, F. and LIPSKY, S., *Int. J. Radiat. Phys. Chem.*, **8**, 237 (1976)

30. HUNT, J. W. and THOMAS, J. K., *J. Chem. Phys.*, **46**, 2954 (1967)
31. AUSLOOS, P., *Mol. Photochem.*, **4**, 39 (1972)
32. AUSLOOS, P. and LIAS, S. G., *Chemical Spectroscopy and Photochemistry in the Vacuum Ultraviolet*, D. Riedel Publishing Co. Amsterdam, 1974, p. 465
33. HOLROYD, R. A., YANG, J. A. and SERVEDIO, F. M., *J. Chem. Phys.*, **46**, 4540 (1967)
33A. HOLROYD, R. A., *J. Phys. Chem.*, **72**, 759 (1968)
34. NAFISI-MOVAGHAR, J. and HATANO, Y., *J. Phys. Chem.*, **78**, 1899 (1974)
35. SCALA, A. A. and AUSLOOS, P., *J. Chem. Phys.*, **49**, 2282 (1968)
36. WADA, T. and HATANO, Y., *J. Phys. Chem.*, **81**, 1057 (1977)
37. WOJNÁROVITS, L., unpublished results
38. COLLIN, G. J., *J. Chem. Phys.*, **74**, 302 (1977)
39. WOJNÁROVITS, L., SHINSAKA, K. and HATANO, Y., unpublished results
40. BECK, G. and THOMAS, J. K., *J. Phys. Chem.*, **76**, 3856 (1972)
41. HENRY, M. S. and HELMAN, W. P., *J. Chem. Phys.*, **56**, 5734 (1972)
41A. WARE, W. R. and LYKE, R. L., *Chem. Phys. Lett.*, **24**, 195 (1974)
41B. DELLONTE, S., GARDINI, E., BARIGELLETTI, F. and ORLANDI, G., *Chem. Phys. Lett.*, **49**, 596 (1977)
42. BAXENDALE, J. H. and MAYER, J., *Chem. Phys. Lett.*, **17**, 458 (1972)
43. HIRAYAMA, F., ROTHMAN, W. and LIPSKY, S., *Chem. Phys. Lett.*, **5**, 296 (1970)
44. ROTHMAN, W., HIRAYAMA, F. and LIPSKY, S., *J. Chem. Phys.*, **58**, 1300 (1973)
45. DOEPKER, R. D., LIAS, S. G. and AUSLOOS, P., *J. Chem. Phys.*, **46**, 4340 (1967)
46. AUSLOOS, P., *Proc. 4th Intern. Conf. Rad. Res.* Vol. 2. *Phys. and Chem.* (Eds DUPLAND, J. F. and CHAPIRO, O. A. Gordon and Breach Science Publishers, London, 1973
47. SHIDA, S. and HATANO, Y., *Int. J. Radiat. Phys. Chem.*, **8**, 171 (1976)
47A. HATANO, Y., *Bull. Chem. Soc. Jpn.*, **41**, 1126 (1968)
48. SALEM, L. and ROWLAND, C., *Angew. Chem. Internat. Edit.*, **11**, 92 (1972)
49. FREEMAN, G. R., *Can. J. Chem.*, **44**, 245 (1966)
50. WOJNÁROVITS, L. and FÖLDIÁK, G., *Acta Chim. Acad. Sci. Hung.*, **93**, 1 (1977)
51. YANG, K., *J. Phys. Chem.*, **65**, 42 (1961)
52. SMITH, C. F., CORMAN, B. G. and LAMPE, F. W., *J. Am. Chem. Soc.*, **83**, 3559 (1961)
53. UMEZAWA, H. and ROWLAND, F. S., *J. Am. Chem. Soc.*, **84**, 3077 (1962)
54. AUSLOOS, P. and LIAS, S. G., *J. Chem. Phys.*, **43**, 127 (1965)
55. VON BÜNAU, G. and KÜHNERT, P., *Ber. Busenges.*, **76**, 913 (1972)
56. HORVÁTH, ZS., FÖLDIÁK, G. and GRUBER, L., in *Proc. 3rd Tihany Symp. Rad. Chem.* (Eds DOBÓ, J. and HEDVIG, P.), Akadémiai Kiadó, Budapest, 1972, p. 239
56A. HORVÁTH, ZS. and FÖLDIÁK, G., *Acta Chim. Acad. Sci. Hung.*, **85**, 417 (1975)
56B. HORVÁTH, ZS., AUSLOOS, P. and FÖLDIÁK, G., in *Proc. 4th Tihany Symp. Rad. Chem.* (Eds SCHILLER, R. and HEDVIG, P.), Akadémiai Kiadó, Budapest, 1977, p. 57
57. DERAI, R. and DANON, J., *Chem. Phys.*, **15**, 331 (1976)
58. LESIECKI, M. L. and GUILLORY, W. A., *J. Chem. Phys.*, **66**, 4317 (1977)
59. CURRIE, C. L., OKABE, H. and MCNESBY, J. R., *J. Phys. Chem.*, **67**, 1494 (1963)
60. DOEPKER, R. D. and AUSLOOS, P., *J. Chem. Phys.*, **44**, 1641 (1966)
61. DHINGRA, A. K. and KOOB, R. D., *J. Phys. Chem.*, **74**, 4490 (1970)
62. SHIBUYA, K., OBI, K. and TANAKA, I., *Bull. Chem. Soc. Jpn.*, **48**, 1974 (1975)
62A. SHIBUYA, K., OBI, K. and TANAKA,. I., *Bull. Chem. Soc. Jpn.*, **49**, 2178 (1976)
63. SIECK, L. W. and FUTRELL, J. H., *J. Chem. Phys.*, **45**, 560 (1966)
64. ZHITNEVA, G., KOZHEMYAKINA, L. F. and PSHEZHETSKII, S. YA., *Khim. Vys. Energ.*, **8**, 181 (1974)
65. DOEPKER, R. D. and AUSLOOS, P. *J. Chem. Phys.*, **43**, 3814 (1965)
66. HECKEL, E. and HANRAHAN, R. J., *Int. J. Radiat. Phys. Chem.*, **5**, 271 (1973)
66A. HECKEL, E. and HANRAHAN, R. J., *Int. J. Radiat. Phys. Chem.*, **5**, 281 (1973)
67. OBI, K., AKIMOTO, H., OGATA, Y. and TANAKA, I., *J. Chem. Phys.*, **55**, 3822 (1971)
67A. OGATA, Y., OBI, K., AKIMOTO, H. and TANAKA, I., *Bull. Chem. Soc. Jpn.*, **44**, 2671 (1971)
68. SPITTLER, E. G. and KLEIN, G. W., *J. Phys. Chem.*, **72**, 1432 (1968)
69. SAEKI, M. and TACHIKAWA, E., *Bull. Chem. Soc. Jpn.*, **49**, 2214 (1976)
69A. NOGAR, N. S., CALLAHAN, M. B. and SPICER, L. D., *Radiochim. Acta*, **23**, 92 (1976)

70. GORDEN, G. JR. and AUSLOOS, P., *J. Res. NBS*, **75A**, 141 (1971)
71. HIROKAMI, S. and CVETANOVIC, R. J., *J. Phys. Chem.*, **78**, 1254 (1974)
72. HAUSER, W. P. and WALTERS, W. D., *J. Phys. Chem.*, **67**, 1328 (1963)
73. HUGHES, B. M. and TIERNAN, T. O., *J. Chem. Phys.*, **51**, 4373 (1969)
74. LIAS, S. G. and AUSLOOS, P., *J. Res. NBS*, **75A**, 591 (1971)
75. HARDWICK, T. J., *J. Phys. Chem.*, **66**, 2132 (1962)
76. SÖYLEMEZ, T. and SCHULER, R. H., *J. Phys. Chem.*, **78**, 1052 (1974)
77. NELSON, H. D. and DELINGNY, C. L., *Rec. Trav. Chim. Pays-Bas*, **87**, 528 (1968)
78. RABANI, J., PICK, M. and SIMIC, M., *J. Phys. Chem.*, **78**, 1049 (1974)
79. HOLROYD, R. A., *J. Phys. Chem.*, **66**, 730 (1962)
80. STONE, J. A., *Can. J. Chem.*, **42**, 2872 (1964)
81. TOMA, S. Z. and HAMILL, W. H., *J. Am. Chem. Soc.*, **86**, 1478 (1964)
82. HUGHES, B. M. and HANRAHAN, R. J., *J. Phys. Chem.*, **69**, 2707 (1965)
82A. HUGHES, B. M. and HANRAHAN, R. J., *Radiat. Res.*, **36**, 261 (1968)
83. KUDO, T. and SHIDA, S., *J. Phys. Chem.*, **71**, 1971 (1967)
84. WALKER, L. G., Ph. D. Thesis, University of Alberta, Edmonton, Canada, 1967
85. FREEMAN, G. R., *Radiat. Res. Rev.*, **1**, 1 (1968)
86. BUSI, F. and FREEMAN, G. R., *J. Phys. Chem.*, **75**, 2560 (1971)
87. KIMURA, T., FUEKI, K. and KURI, Z., *Bull. Chem. Soc. Jpn.*, **43**, 1657 (1970)
88. TILQUIN, B. and CLAES, P., *Bull. Soc. Chim. Belges*, **80**, 335 (1971)
89. KENNEDY, M. G. and STONE, J. A., *Chem. Commun.*, **1970**, 1478
90. TILQUIN, B., ALLAERT, J. and CLAES, P., *J. Phys. Chem.*, **78**, 462 (1974)
90A. TILQUIN, B., ALLAERT, J. and CLAES, P., *J. Phys. Chem.* **82**, 277 (1978)
91. HENTZ, R. R. and BRAZIER, D. W., *J. Chem. Phys.*, **54**, 554 (1971)
92. DIXON, P. S., STEFANI, A. P. and SZWARC, M., *J. Am. Chem. Soc.*, **85**, 2551 (1963)
93. HOLROYD, R. A., *J. Phys. Chem.*, **70**, 1341 (1966)
94. DAUPHIN, J., *J. Chem. Phys.*, **59**, 1207 (1962)
95. HOLROYD, R. A. and KLEIN, G. W., *J. Am. Chem. Soc.*, **87**, 4983 (1965)
96. FESSENDEN, R. W. and SCHULER, R. H., *J. Chem. Phys.*, **39**, 2147 (1963)
97. BANSAL, K. M. and SCHULER, R. H., *J. Phys. Chem.*, **74**, 3924 (1970)
98. STOCK, R. L. and GUNNIG, H. E., *Can. J. Chem.*, **28**, 2295 (1960)
99. YANG, J. Y. and MARCUS, I., *J. Chem. Phys.*, **42**, 3315 (1965)
100. WOJNÁROVITS, L., *Radiochem. Radioanal. Lett.*, **32**, 267 (1978)
101. HUANG, T. and HAMILL, W. H., *J. Phys. Chem.*, **78**, 2081 (1974)
102. WOJNÁROVITS, L. and FÖLDIÁK, G., *Acta Chim. Acad. Sci. Hung.*, **82**, 285 (1974)
103. AUSLOOS, P., SCALA, A. A. and LIAS, S. G., *J. Am. Chem. Soc.*, **89**, 3677 (1967)
104. DOEPKER, R. D. and AUSLOOS, P., *J. Chem. Phys.*, **44**, 1951 (1966)
105. BÉRCES, T., in *Comprehensive Chemical Kinetics* (Eds BAMFORD, C. H. and TIPPER, C. F. H.) Vol. 5., Elsevier, Amsterdam, 1972. p. 234
106. DOEPKER, R. D. and AUSLOOS, P., *J. Chem. Phys.*, **42**, 3746 (1965)
107. CRAMER, W. A., in *Aspects of Hydrocarbon Radiolysis* (Eds GÄUMANN, T. and HOIGNÉ, J.), Academic Press, London and New York, 1968, p. 153
108. BLACHFORD, J. and DYNE, P. J., *Can. J. Chem.*, **42**, 1165 (1964)
109. KLOTS, C. E. and JOHNSEN, R. H., *Can. J. Chem.*, **41**, 2702 (1963)
110. JONES, K. H., *J. Phys. Chem.*, **71**, 709 (1967)
111. VAN DUSEN, W. and TRUBY, F. K., *J. Am. Chem. Soc.*, **87**, 188 (1965)
112. KIMURA, T., FUEKI, K. and KURI, Z., *Bull. Chem. Soc. Jpn.*, **42**, 3088 (1969)
113. STONE, J. A., *Can. J. Chem.*, **46**, 1267 (1968)
113A. IWASAKI, M., TORIYAMA, K., MUTO, H. and NUNOME, K., *Chem. Phys. Lett.*, **56**, 494 (1978)
114. SAGERT, N. H., *Can. J. Chem.*, **46**, 89 (1968)
115. ROBINSON, M. G. and FREEMAN, G. R., *J. Chem. Phys.*, **48**, 983 (1968)
116. HOLROYD, R. A., in *Fundamental Processes in Radiation Chemistry* (Ed. AUSLOOS, P.), Interscience, New York, 1968, p. 465
117. CRAMER, W. A. and PIET, G. J., *Trans. Faraday Soc.*, **63**, 1402 (1967)
118. McCRUMB, J. L. and SCHULER, R. H., *J. Phys. Chem.*, **71**, 1953 (1967)
119. SAUER, M. C. JR. and MANI, I., *J. Phys. Chem.*, **72**, 3856 (1968)
120. HARRIS, M., ESSER, J. and STONE, J. A., in *Proc. 3rd. Tihany Symp. on Rad Chem.* (Eds DOBÓ, J. and HEDVIG, P.), Akadémiai Kiadó, Budapest, 1972, p. 347
121. ESSER, J. and STONE, J. A., *Can. J. Chem.*, **51**, 192 (1973)

121A. KENNEDY, M. G. and STONE, J. A., *Can. J. Chem.*, **51**, 149 (1973)
122. THIBAULT, R. M., HEPBURN, D. R. JR. and KLINGEN, T. J., *J. Phys. Chem.*, **78**, 788 (1974)
123. FUJISAKI, N. and GÄUMANN, T., *Ber. Bunsenges.*, **81**, 544 (1977)
124. ABRAMSON, F. P. and FUTRELL, J. H., *J. Phys. Chem.*, **71**, 3791 (1967)
125. AUSLOOS, P., REBBERT, R. E. and LIAS, S. G., *J. Phys. Chem.*, **72**, 3904 (1968)
126. STONE, J. A., *Can. J. Chem.*, **46**, 3531 (1968)
127. STONE, J. A., *Can. J. Chem.*, **43**, 809 (1965)
128. YANG, J. Y., SERVIDO, F. M. and HOLROYD, R. A., *J. Chem. Phys.*, **48**, 1331 (1968)
129. BALLENGER, M., RUF, A. and GÄUMANN, T., *Helv. Chim. Acta*, **54**, 1373 (1971)
130. DYNE, P. J. and JENKINSON, W. M., *Can. J. Chem.*, **38**, 539 (1960)
131. DYNE, P. J. and JENKINSON, W. M., *Can. J. Chem.*, **40**, 1746 (1962)
132. SAGERT, H. H. and BLAIR, A. S., *Can. J. Chem.*, **45**, 1351 (1967)
133. BURNS, W. G. and REED, C. R. V., *Trans. Faraday Soc.*, **66**, 2159 (1970)
134. AVDONINA, E. N., in *Proc. 4th Tihany Symp. Rad. Chem.* (Eds SCHILLER, R. and HEDVIG, P.), Akadémiai Kiadó, Budapest, 1977, p. 83
135. AVDONINA, E. N. and NESMEYANOV, A. N., *Radiochimiya*, **10**, 568 (1968)
136. AVDONINA, E. N., *Khim. Vys. Energ.*, **4**, 531 (1970)
137. GRACHEVA, T. A. and MAKAROV, V. I., *Int. J. Radiat. Phys. Chem.*, **7**, 425 (1975)
138. MAKAROV, V. I. and POLAK, L. S., *Int. J. Radiat. Phys. Chem.*, **8**, 187 (1976)
139. BOUILLOT, M. S., *Int. J. Radiat. Phys. Chem.*, **2**, 117 (1970)
140. SAGERT, N. H. and BLAIR, A. S., *Can. J. Chem.*, **46**, 3284 (1968)
141. THEARD, L. M., *J. Phys. Chem.*, **69**, 3292 (1965)
142. DYNE, P. J. and STONE, J. A., *Can. J. Chem.*, **39**, 2381 (1961)
143. HO, S. K. and FREEMAN, G. R., *J. Phys. Chem.*, **68**, 2189 (1964)
144. SAGERT, N. H., REID, J. A. and ROBINSON, R. W., *Can. J. Chem.*, **47**, 2655 (1969)
145. KROH, J., KAROLCZAK, S. and PIEKARSKA, J., in *Proc. 3rd Tihany Symp. Rad. Chem.* (Eds DOBÓ, J. and HEDVIG, P.), Akadémiai Kiadó, Budapest, 1972, p. 169
146. SMITH, D. R. and TOLE, J. C., *Can. J. Chem.*, **45**, 779 (1967)
147. SEVILLA, M. D. and HOLROYD, R. A., *J. Phys. Chem.*, **74**, 2459 (1970)
148. BENNETT, J. E. and THOMAS, A., *Proc. Roy. Soc.*, *A*, **280**, 123 (1964)
149. OGAWA, S. and FESSENDEN, R. W., *J. Chem. Phys.*, **41**, 994 (1964)
150. BLACKBURN, R. and CHARLESBY, A., *Proc. Roy. Soc.*, *A*, **293**, 51 (1966)
151. FESSENDEN, R. W., *J. Phys. Chem.*, **71**, 74 (1967)
152. KUWATA, K., KOTAKE, Y., INADA, K. and ONO, M., *J. Phys. Chem.*, **76**, 2061 (1972)
153. VOEVODSKII, V. V. and MOLIN, YU. N., *Radiat. Res.*, **17**, 366 (1962)
154. WILKEY, D. D., FENRICK, H. W. and WILLARD, J. E., *J. Phys. Chem.*, **81**, 220 (1977)
155. OHMAE, T., OHNISHI, S., KUWATA, K., SAKURAI, H. and NITTA, I., *Bull. Chem. Soc. Jpn.*, **40**, 226 (1967)
156. ZAITSEV, V. M., SEROVA, V. A. and TIKHONOV, V. I., *Khim. Vys. Energ.*, **7**, 174 (1973)
157. EBERT, M., KEENE, J. P., LAND, E. J. and SWALLOW, A. J., *Proc. Roy. Soc. A*, **287**, 1 (1965)
158. MAKAROV, V. I. and KABAKCHI, S. A., *Khim. Vys. Energ.*, **5**, 272 (1971)
159. CRAMER, W. A., *J. Phys. Chem.*, **71**, 1171 (1967)
160. STONE, J. A. and ESSER, J., *Can. J. Chem.*, **52**, 1253 (1974)
161. IWAHASHI, H., ISHIKAWA, Y., SATO, S. and KOYANO, K., *Bull. Chem. Soc. Jpn.*, **50**, 1278 (1977)
162. FÖLDIÁK, G. and WOJNÁROVITS, L., *Acta Chim. Acad. Sci. Hung.*, **82**, 305 (1974)
163. KABI, A., LENHERR, A. D., ORMEROD, M. G. and CHARLESBY, A., *Int. J. Radiat. Phys. Chem.*, **1**, 45 (1969)
164. BECK, P. W., KNIEBES, D. V. and GUNNING, H. E., *J. Chem. Phys.*, **22**, 672 (1954)
165. FALCONER, J. W. and BURTON, M., *J. Phys. Chem.*, **67**, 1743 (1963)
166. MAKAROV, V. I. and FILATOV, S. E., *Khim. Vys. Energ.*, **4**, 467 (1970)
167. BENNETT, J. E., GALE, L. H., HAYWARD, E. J. and MILE, B., *J. Chem. Soc. Faraday Trans. 1*, **69**, 1655 (1973)
168. GILLIS, H. A., *Can. J. Chem.*, **49**, 2861 (1971)
169. WOJNÁROVITS, L. and FÖLDIÁK, G., *Radiochem. Radioanal. Lett.*, **21**, 261 (1975)

170. GRACHOVA, T. A., MAKAROV, V. I., POLAK, L. S. and AVDONINA, E. N., Radiat. Eff., 10, 157 (1971)
171. MAKAROV, V. I. and POLAK, L. S., Khim. Vys. Energ., 4, 3 (1970)
172. GRACHOVA, T. A., MAKAROV, V. I., POLAK, L. S. and AVDONINA, E. N., in Proc. 3rd Tihany Symp. Rad. Chem. (Eds DOBÓ, J. and HEDVIG, P.), Akadémiai Kiadó, Budapest, 1972, p. 357
173. MAKAROV, V. I. and CHERNYAK, N. YA., Khim. Vys. Energ., 2, 408 (1968)
174. RAMARADHYA, J. M. and FREEMAN, G. R., J. Chem. Phys., 34, 1726 (1961)
175. SCHÜLER, H. and ARNOLD, G., Z. Naturforsch., A, 17, 670 (1962)
176. WOJNÁROVITS, L., Radiochem. Radioanal. Lett., 38, 83 (1979)
177. DUGLE, D. L. and FREEMAN, G. R., Trans. Faraday Soc., 61, 1174 (1965)
177A. SINGH, A. and FREEMAN, G. R., J. Phys. Chem., 69, 666 (1965)
178. HENTZ, R. H. and RZAD, S. J., J. Phys. Chem., 71, 4096 (1967)
179. MILON, H. and GÄUMANN, T., Int. J. Radiat. Phys. Chem., 7, 417 (1975)
180. ARAI, S., SATO, S. and SHIDA, S., J. Chem. Phys., 33, 1277 (1960)
181. KALRA, B. L. and KNIGHT, A. R., Can. J. Chem., 50, 2010 (1972)
182. GORDON, S. A., Pure Appl. Chem., 5, 441 (1962)
183. FÖLDIÁK, G. and WOJNÁROVITS, L., Int. J. Radiat. Phys. Chem., 4, 189 (1972)
184. FÖLDIÁK, G. and WOJNÁROVITS, L., Acta Chim. Acad. Sci. Hung., 82, 269 (1974)
185. COPE, A. C., MOORE, P. T. and MOORE, W. R., J. Am. Chem. Soc., 81, 3135 (1959)
186. TURNER, R. B. and MEADOR, W. R., J. Am. Chem. Soc., 79, 4133 (1957)
187. ZUCCARELLO, F., BUEMI, G. and FAVINI, G., J. Mol. Struct., 18, 295 (1973)
188. STOVER, E. D. and FREEMAN, G. R., J. Chem. Phys., 48, 3902 (1968)
189. STOVER, E. D. and FREEMAN, G. R., Can. J. Chem., 46, 2109 (1968)
190. FREEMAN, G. R., J. Chem. Phys., 36, 1534 (1962)
191. MERKLIN, J. F. and LIPSKY, S., J. Phys. Chem., 68, 3297 (1964)
192. WOJNÁROVITS, L. and FÖLDIÁK, G., Radiochem. Radioanal. Lett., 23, 257 (1975)
193. EBERHARDT, M. K., J. Phys. Chem., 72, 4509 (1968)
194. KIMURA, T., MIYAZAKI, T., FUEKI, K. and KURI, Z., Bull. Chem. Soc. Jpn., 41, 2861 (1968)
195. FÖLDIÁK, G., GYÖRGY, I. and WOJNÁROVITS, L., Int. J. Radiat. Phys. Chem., 8, 575 (1976)
196. HOLTSLANDER, W. J. and FREEMAN, G. R., Can. J. Chem., 45, 1649 (1967)
197. KIMURA, T., FUEKI, K. and KURI, Z., Bull. Chem. Soc. Jpn., 43, 3090 (1970)
198. KIMURA, T., FUKAYA, M., HADA, M., WAKAYAMA, T., FUEKI, K. and KURI, Z., Bull. Chem. Soc. Jpn., 43, 3400 (1970)
199. MIYAZAKI, T., Int. J. Radiat. Phys. Chem., 8, 57 (1976)
200. FREEMAN, G. R., J. Chem. Phys., 36, 1542 (1962)
201. GIBIAN, M. J. and CORLEY, R. C., Chem. Rev., 73, 441 (1973)
202. TILQUIN, B., LOVEAUX, M., BOMBAERT, C. and CLAES, P., Radiat. Eff., 32, 37 (1977)
203. THOMMARSON, R. L., J. Phys. Chem., 74, 938 (1970)
204. KESZEI, Cs., WOJNÁROVITS, L. and FÖLDIÁK, G., Radiochem. Radioanal. Lett., 22, 41 (1975)
205. LIAS, S. G. and AUSLOOS, P., J. Am. Chem. Soc., 92, 1840 (1970)
206. OKABE, H. and McNESBY, J. R., J. Chem. Phys., 34, 668 (1961)
207. RICE, F. O. and MURPHY, M. T., J. Am. Chem. Soc., 64, 896 (1942)
208. WOJNÁROVITS, L. and FÖLDIÁK, G., Radiochem. Radioanal. Lett., 23, 3 43 (1975)
209. STACHOWICZ, W., KECKI, Z. and MINC, S., Nukleonika, 13, 187 (1968)
210. FÖLDIÁK, G. and WOJNÁROVITS, L., unpublished results
211. DYNE, P. J. and DENHARTOG, J., Can. J. Chem., 40, 1616 (1962)
212. KAZANSKII, B. A., SHOKOVA, E. A. and KOROSTELEVA, T. V., Inv. Akad. Nauk SSSR, Ser. Khim., 11, 2640 (1968)
213. KAZANSKII, B. A., SHOKOVA, E. A. and KOROSTELEVA, T. V., Inv. Akad. Nauk SSSR, Ser. Khim., 11, 2642 (1968)
214. HATTENBACH, K. and VON BAUER, E., Atomkernenergie, 11, 262 (1966)
215. PODHALYUZIN, A .T. and VERESHCHINSKII, I. V., in Proc. 3rd Tihany Symp. Rad. Chem. (Eds DOBÓ, J. and HEDVIG, P.), Akadémiai Kiadó, Budapest, 1972, p. 115
216. ANTONOVA, E. A., DEMINA, N. A., PICHUZKIN, V. I. and SARAEVA, V. V., Khim. Vys. Energ., 8, 49 (1974)

217. FERRELL, J. R., HOLDREN, G. R., LLOYD, R. V. and WOOD, D. E., *Chem. Phys. Lett.*, **9**, 343 (1971)
218. HYFANTIS, G. J. JR. and LING, A. C., *Chem. Phys. Lett.*, **24**, 335 (1974)
219. HYFANTIS, G. J. JR., and LING, A. C., *Can. J. Chem.*, **52**, 1206 (1974)
220. GEE, D. R., FABES, L. and WANN, J. K. S., *Chem. Phys. Lett.*, **7**, 311 (1970)
221. FILBY, W. G. and GÜNTHER, K., *Chem. Phys. Lett.*, **14**, 440 (1972)
221A. FILBY, W. G. and GÜNTHER, K., *Chem. Phys. Lett.*, **17**, 150 (1972)
222. FILBY, W. G. and GÜNTHER, K., *Z. Naturforsch., B*, **27**, 1289 (1972)
223. PODKHALYUZIN, A. T., VIKULIN, V. V. and VERESHCHINSKII, I. V., *Symp. Mech. Hydr. React.* (Eds MÁRTA, F. and KALLÓ, D.). Akadémiai Kiadó, Budapest, 1975, p. 687
223A. PODKHALYUZIN, A. T., VIKULIN, V. V. and VERESHCHINSKII, I. V., *Dokl. Akad. Nauk SSSR*, **221**, 381 (1975)
223B. PODKHALYUZIN, A. T., VIKULIN, V. V. and VERESHCHINSKII, I. V., *Khim. Vys. Energ.*, **11**, 281 (1977)
223C. PODKHALYUZIN, A. T., VIKULIN, V. V., MOROZOV, V. A., NAZAROVA, M. P. and VERESHCHINSKII, I. V., *Radiat. Eff.*, **32**, 9 (1977)
224. BONAZZOLA, L. and MARX, R., *Chem. Phys. Lett.*, **8**, 413 (1971)
225. MARX, R., *Chem. Phys. Lett.*, **17**, 152 (1972)
226. LLOYD, R. V. and WOOD, D. E., *J. Am. Chem. Soc.*, **96**, 659 (1974)
227. LLOYD, R. V. and WOOD, E. D., *J. Chem. Phys.*, **60**, 2684 (1974)
228. FÖLDIÁK, G. and HORVÁTH, ZS., *Acta Chim. Acad. Sci. Hung.*, **86**, 385 (1975)
229. HORVÁTH, ZS. and FÖLDIÁK, G., *Acta Chim. Acad. Sci. Hung.*, **86**, 397 (1975)
230. DODELET, J. P., FAUQENOIT, C., SIQUET, M. and CLAES, P., *Ann. Soc. Sci. Bruxelles*, **84**, 107 (1970)
231. FAUQUENOIT C., DODELET, J. P. and CLAES, P., *Bull. Soc. Chim. Belges*, **80**, 315 (1971)
232. FAUQUENOIT, C. and CLAES, P., *Bull. Soc. Chim. Belges*, **80**, 323 (1971)
233. TILQUIN, B., MICHIELSEN, M. and CLAES, P., unpublished results
234. GYÖRGY, I., HORVÁTH, ZS., WOJNÁROVITS, L. and FÖLDIÁK, G., *Proc. 3rd Tihany Symp. Rad. Chem.* (Eds DOBÓ, J. and HEDVIG, P.), Akadémiai Kiadó, Budapest, 1972, p. 311
235. GYÖRGY, I. and FÖLDIÁK, G., *Acta Chim. Acad. Sci. Hung.*, **81**, 455 (1974)
236. IWASAKI, M., TORLYAMA, K., NUNOME, K., FUKAYA, M. and MUTO, H., *J. Phys. Chem.*, **81**, 1410 (1977)
237. MIYAZAKI, T., GUEDES, S. M.L and ., SILVA, L. G. A., *Bull. Chem. Soc. Jpn.*, **50**, 301 (1977)
238. SCHULER, R. H. and MUCCINI, G. A., *J. Am. Chem. Soc.*, **81**, 4115 (1959)
239. MUCCINI, G. A. and SCHULER, R. H., *J. Phys. Chem.*, **64**, 1436 (1960)
240. TOMA, S. Z. and HAMILL, W. J., *J. Am. Chem. Soc.*, **86**, 4761 (1964)
241. YANG, J. Y. and MARCUS, I., *J. Phys. Chem.*, **69**, 3113 (1965)
242. WOJNÁROVITS, L. and FÖLDIÁK, G., *Acta Chim. Acad. Sci. Hung.*, **105**, 27 (1980)
243. HOLROYD, R. A., private communication in ref. 96
244. FÖLDIÁK, G., CSERÉP, GY., HORVÁTH, ZS. and WOJNÁROVITS, L., in *Symp. Mech. Hydr. React.* (Eds MÁRTA, F. and KALLÓ, D.), Akadémiai Kiadó, Budapest, 1975, p. 659
245. FÖLDIÁK, G., CSERÉP, GY., GYÖRGY, I. HORVÁTH, ZS. and WOJNÁROVITS, L., *Hung. J. Ind. Chem.*, Suppl. **2**, 277 (1974)
246. SANTAR, I., in *Symp. Mech. Hydr. React.* (Eds MÁRTA, F. and KALLÓ, D.), Akadémiai Kiadó, Budapest, 1975, p. 625
247. OKAZAKI, K., YAMABE, M. and SATO, S., *Bull. Chem. Soc. Jpn.*, **50**, 1409 (1977)
248. LIEBMAN, J. F. and GREENBERG, A., *Chem. Rev.*, **76**, 311 (1976)
249. LORQUET, J. C., *Mol. Phys.*, **9**, 101 (1965)
250. LORQUET, J. C. and HALL, G. G., *Mol. Phys.*, **9**, 29 (1965)
251. WADA, T., SHINSAKA, K., NAMBA, H. and HATANO, Y., *Can. J. Chem.*, **55**, 2144 (1977)

4. ALIPHATIC ALKENES AND ALKYNES

The literature on the radiation chemistry of alkenes deals chiefly with the problems of polymerization, and considerably fewer papers have been published on their decomposition. Here we are concerned with the latter topic, because polymerization is not a subject of the present book.

We shall concentrate mainly on papers published after 1968, because those appearing before 1968 have already been reviewed in many excellent monographs. Thus the radiolysis of alkenes has been reviewed by Meisels [1] and Hardwick [2]. The book edited by Ausloos [3] contains information on the radiolytic decomposition of alkenes in the gas and solid phases, as well as the most important reactions of excited and ionized species; reviews written by Freeman [4] and Allen [5] can also be mentioned.

Polymerization induced by irradiation has been summarized by Chapiro [6] and Williams [7, 8], and the radiolysis of polymers is the subject of books by Charlesby [9] and Dole [10].

4.1. INTRODUCTORY REMARKS

The physical properties of an aliphatic unsaturated hydrocarbon are determined by the number of double or triple bonds and their location in the molecule, the length of carbon chain and the degree of branching.

In the series of alk-1-enes C_2-C_4 are gases, C_5-C_{17} are liquids and higher alkenes are solids at room temperature. The melting points and boiling points of alkenes are usually lower than those of the corresponding alkanes. The melting and boiling points are lower for terminal alkenes than for those with the double bond in an internal position. Branching usually depresses the melting and boiling points.

The differences in the heat of formation between the alkenes with the same carbon number amount to only a few kJ mol^{-1}. However, even these small differences indicate the influence of structural factors on the energy content and thermodynamic stability of isomeric molecules. For example, the ΔH_f of C_4H_8 isomers follow the order: but-1-ene < *cis*-but-2-ene < < *trans*-but-2-ene < isobutene [11]. The thermodynamic stability of *trans*-alkenes is higher than that of *cis*-isomers, so the heat of combustion and the heat of hydrogenation of the latter is smaller, although the difference is very small [11].

The conjugated dienes can be considered as the most important of the aliphatic dienes, for both theoretical and practical purposes. Their high reactivity is revealed, for example, in their addition and polymerization reactions. The thermodynamic stability of conjugated dienes is higher than that of equivalent non-conjugated dienes. The interaction between the double bonds in conjugated dienes is reflected in the fact that some physical properties, e.g. melting point, and UV absorption spectra, differ considerably from those of monoalkenes and other dienes.

The melting and boiling points of hydrocarbons containing an acetylenic bond are somewhat higher than those of the corresponding alkenes.

In aliphatic alkane molecules the carbon atoms are sp^3 hybridized, and thus the bond angles are 109.5°. The sp^2 hybridized carbon atoms next to the double bond in alkenic hydrocarbons have a planar-trigonal structure, and thus the valence angle is approximately 120°. In ethylene the scheme of the structure can be depicted as:

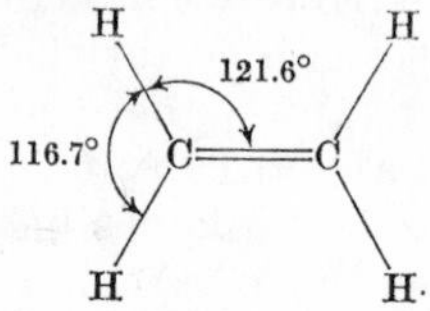

In molecules containing acetylenic bond there are two σ-bonds due to sp-hybrid orbitals of carbon atoms and the valence angle is 180°. Distortions in valence angles, in bond lengths and in the hybrid states can be observed if bulky substituents are bonded to carbon atoms, and this effect eventually causes a change in bond energies. This problem has been detailed in the monographs concerned with the stereochemistry of hydrocarbons, e.g. refs 12–14.

Characteristic absorption bands of unsaturated hydrocarbons show up in the far ultraviolet and vacuum-ultraviolet. If the complexity of the molecule, increases, i.e. if the number of substituents around the π-bond increases, the maximum of the absorption band suffers a bathochromic shift, i.e. the band maximum is shifted towards higher wavelengths. For instance, the absorption maximum of ethylene is at 173 nm and for hex-1-ene, hex-2-ene and for 2,3-dimethylbut-2-ene the maxima are at 180, 183 and 196 nm, respectively. The absorption maximum of hydrocarbons containing acetylenic bonds is also in the vacuum-UV. Photon absorption in unsaturated hydrocarbons is related to a π → π* transition.

The most characteristic reactions of alkenes can be explained by the high reactivity of the double bond in addition reactions: the addition of H-atoms and alkyl radicals is very important in photochemical, radiation-chemical and thermal reactions. These processes, especially hydrogen addition, take place relatively rapidly even at room temperature, which explains why unsaturated hydrocarbons are used as radical scavengers.

As far as the mechanism of these free radical addition reactions is concerned it is apparent that terminal addition (4.1) is preferred to non-terminal addition (4.2), because of the higher thermodynamic stability of the intermediates formed in reaction (4.1):

$$R-CH=CH_2 + Y\cdot \begin{cases} R-\overset{\cdot}{C}H-CH_2-Y \quad (4.1) \\ R-CHY-\overset{\cdot}{C}H_2 \quad (4.2) \end{cases}$$

This phenomenon can be attributed to the fact that the thermodynamic stability of the alkyl radical intermediates decreases in the order tertiary > > secondary > primary. Other factors, such as the charge distribution in the alkene molecule, seem to be of secondary importance.

The relative reactivities of alkenes of different structures in methyl radical addition, measured in iso-octane solution at 65°C, are listed in Table 4.1

Table 4.1

Relative reactivity of alkenes of different structures in methyl radical addition reactions. After Szwarc et. al [15, 16]

Hydrocarbons	Relative rate constances[a] k_{abs}/k_{add}
Ethene	26
Propene	22
Methylpropene	36
2-Methylbut-2-ene	6
cis-But-2-ene	3.4
trans-But-2-ene	7
cis-2,2,5,5-tetramethylhex-3-ene	2
trans-2,2,5,5-tetramethylhex-3-ene	0.4
Buta-1,3-diene	2015
2,3-Dimethylbuta-1,3-diene	2200
Hexa-2,4-diene	180
2,5-Dimethylhexa-2,4-diene	20

[a] abs: CH_3 + solvent → CH_4
   add: CH_3 + alkene → CH_3-alkene·

[15,16]. It can be seen that the reactivity is significantly decreased with the increasing size of the alkyl substituent. Furthermore, *trans*-isomers are generally more reactive than the corresponding *cis*-isomers.

The *cis-trans* isomerization of alkenes does not occur, at room temperature under normal conditions. However, such processes do take place at a considerable rate on the decay of electronically excited alkenes formed under the effect of irradiation. This process is generally interpreted to mean free rotation of the molecule around the single bond that remains when the second bond is temporarily broken by the radiation.

The chemical properties of isolated dienes do not differ essentially from those of monoalkenes. The reactivity of cumulated dienes, such as allene, is quite high: they readily participate in addition and isomerization reactions.

The conjugated dienes (e.g. buta-1,3-diene) are very important both from theoretical and practical point of view. The interaction of double bonds in β-positions results in an energy deficiency of 15 kJ mol^{-1} compared with the energy calculated from the classical structure. The chemical reactivity of conjugated dienes is particularly high: e.g. buta-1,3-diene can react with halogenes very easily in both 1,2- and 1,4-addition reactions.

The reactivity of hydrocarbons containing triple bonds is also very high; addition reactions proceed similarly to those of alkenes. However, the reactivities of alkenes and alkynes do differ from each other. This is particular clear in the case of alk-1-enes, which are stereochemically less screened. Although the electron density between the carbon atoms in alkynes is higher than in alkenes, alkynes react more readily with nucleophilic and less readily with electrophiles than do alkenes.

The irradiation of unsaturated hydrocarbons produces a great number of ion-electron pairs. The total ion yield (G_{tot}), calculated on the basis of the gas-phase W-value, is shown in Table 4.2. The values obtained for alkenes and alkynes of different structure and for alkanes agree within experimental error. According to literature data (see Section 1.1), the ion pair yield in the liquid phase is very close to that in the gas phase, or even a little higher ($G = 4.4-5.2$).

Table 4.2

G_{fi}^0 and the secondary electron penetration (b_{GP}) in unsaturated liquid hydrocarbons (radiation: 1.7 MeV bremmstrahlung).
After Dodelet et al. [24]

Hydrocarbons	T, K	G_{tot}	G_{fi}^0	b_{GP}, nm	$b_{GP}\varrho$, (10^{-8} g cm^{-2})
Monoalkenes					
But-1-ene	293	4.2[a]	0.093	5.4	33
trans-But-2-ene	293	4.2[a]	0.080	5.3	32
cis-But-2-ene	293	4.2[a]	0.23	7.4	46
Isobutene	293	4.1[a]	0.25	7.4	44
2-Methylbut-2-ene	292	4.3[b]	0.26	8.0	53
2,3-Dimethylbut-2-ene	293	4.3[b]	0.44	10.1	72
Hex-1-ene	293	4.2[a]	0.10	5.2	35
trans-Hex-2-ene	293	4.3[a]	0.092	5.1	35
trans-Hex-3-ene	293	4.3[b]	0.10	5.3	36
cis-Hex-3-ene	293	4.3[b]	0.13	5.6	38
Cyclohexene	293	4.3[b]	0.20	6.2	50
Dienes					
Propadiene	282	4.0[a]	0.050	4.3	26
Buta-1,3-diene	269	4.0[a]	0.038	3.9	26
Penta-1,4-diene	293	4.2[c]	0.067	4.4	29
Hexa-1,5-diene	292	4.2[c]	0.066	4.2	29
Hepta-1,6-diene	292	4.2[c]	0.066	4.2	30
Octa-1,7-diene	292	4.2[c]	0.065	4.1	31
Alkynes					
Propyne	260	4.1[b]	0.17	4.8	31
But-2-yne	293	4.2[d]	0.32	9.7	67
Hex-1-yne	253	4.2[d]	0.10	4.4	33
Hex-2-yne	293	4.3[d]	0.19	6.8	49
Hex-3-yne	293	4.3[d]	0.212	7.1	52

[a] Assumed to be equal to the average of the gas-phase values reported in refs [252−256]
[b] Assumed equal to that for hex-2-ene
[c] Assumed equal to those for but-1-ene and hex-1-ene
[d] Assumed equal to the value for the corresponding alkene

In recent years the number of papers dealing with the behaviour of free ions and electrons formed under irradiation has increased. On the basis of experiments carried out in liquid aliphatic alkanes [17–22], it seems that the more spherical the molecule, the longer the penetration range of the epithermal electrons. At the same time there is an increase in the mobility (μ_e) and yield of the thermal electrons as well as in the yield of free ions (G_{fi}). According to some results [20, 23–24], a similar trend holds for liquid alkenes (Tables 4.2 and 4.3).

Table 4.3

Thermal electron mobilities (μ_e) and activation energies (E_A) and secondary electron ranges (b_{GP}) in liquid hydrocarbons. After Dodelet et al. [23]

Hydrocarbons	T, K	μ_e, $cm^2\ V^{-1}\ s^{-1}$	b_{GP}, nm	E_A, kJ mol^{-1}
2,3-Dimethylbut-2-ene	293	5.8	10.1	8.4
cis-But-2-ene	293	2.2	7.4	15.5
Isobutene	293	1.44	7.4	
Cyclohexene	293	1.00	6.2	15.1
But-1-ene	293	0.064	5.4	
trans-2-But-2-ene	293	0.029	5.3	22.6

G_{fi} data for unsaturated hydrocarbons have been published by Schmidt and Allen [20] and Dodelet et al. [24]. G_{fi} values obtained by Dodelet are about 1.3 times higher than those for n-hexane (0.131), hex-1-ene (0.062), hex-2-ene (0.076) and cyclohexene (0.15) measured by Schmidt and Allen at 296 K. The free ion yield varies considerably with molecular structure: e.g. G_{fi} for 2,2-dimethylbut-2-ene is 0.44 whereas that for *trans*-but-2-ene is 0.08 at 293 K.

A change in temperature significantly alters these values: e.g. the G_{fi} yield is increased by an order of magnitude for hex-1-ene if the temperature increases from 152 to 383 K (0.018 and 0.22, respectively). The effect of temperature is linked to the change in density (ϱ), because the thermalization range of electrons is inversely proportional to the density of hydrocarbons.

As far as the influence of molecular structure is concerned, it has been observed that G_{fi} values can be significantly affected by certain substituents. For instance, G_{fi} is higher for hydrocarbons containing tertiary and quaternary carbon atoms, and it is lower for alkenes than for the corresponding alkanes [20]. These investigations have been extended by Dodelet et al. to hydrocarbons of a much wider variety in order to obtain a more general explanation for the structure effect [24, 24A, 24B]. They calculated the most probable range of epithermal electrons (b_{GP}) and that normalized by density ($b_{GP}\varrho$) (Table 4.2). These data indicate that for hydrocarbons of identical structure and carbon number $b_{GP}\varrho$ and G_{fi} values decrease in the sequence alkane $>$ alk-1-ene $>$ alk-1-yne. Electron range and G_{fi} are more effectively decreased by two terminal double bonds than by one triple bond. The deviation between the data for *cis-trans*-isomers, especially *cis*- and *trans*-butenes, is very marked. An even greater difference can be found with respect to the mobility of thermal electrons, where there is a difference of two orders of magnitude between the values for *cis*- and *trans*-but-2-ene (see Table 4.3). The reason for this is yet not clear.

It has been found that, for the alkenes listed in Table 4.3, the electron mobility is independent of the electric field applied, but does depend on temperature. Investigation of the temperature dependence led to the determination of the activation energies of thermal mobility for many alkenes. These values decrease with increasing electron mobility, in agreement with earlier [22, 25–28] and recent [24B] observations. New studies [24A, 24B] have shown that the Arrhenius plots of μ_e of cycloalkenes and in aliphatic alkenes for which $\mu_e > 1$ cm^2 V^{-1} s^{-1} curve downwards at about 300 K, indicating an increase in the activation energy. The larger activation energy of electron migration at lower temperatures is probably due to the formation of deeper traps for the solvated electrons.

By plotting the mobility of thermal electrons (μ_e) against the most probable range (b_{GP}) of the secondary electrons, Dodelet et al. [23] found that alkanes and alkenes can be represented by a common curve. They explained this phenomenon by assuming that the negative molecular ion states do not participate in those electron-scattering processes that control the thermal electron mobilities, nor in those that limit the epithermal electron ranges [23, 24]. However, this statement proved to be erroneous on the basis of the more recent studies of Dodelet et al. [24A, 24B]. In these studies [24A, 24B] various highly branched cyclic and polycyclic alkenes and alkadienes have been investigated. The main conclusions are as follows: increasing the molecular symmetry by adding one or more cycles to the molecule causes the density normalized range $b_{\text{GP}}\varrho$ to increase, provided that the amount of bond distortion due to strain remains small. Ranges are smaller in compounds that contain strained rings, because energy transfer from the secondary electrons to the molecules is more effective in the presence of distorted bonds. The ranges in alkenes are affected by the same factors as in alkanes, with the addition of a contribution of transient negative ion states to the energy transfer processes. In conjugated alkenes the penetration ranges are shorter than in other alkenes. This is attributed to the positive electron affinity of the conjugated alkenes and, therefore, to an increased electron energy loss interaction by way of real negative ion states.

Dodelet et al. [24] also considered the effect of molecular structure on the epithermal electron range by seeking a correlation between the most probable range and the anisotropy of molecular polarizability. They found that, for a great number of model compounds, the rate of energy loss of electrons is inversely correlated to the anisotropy of polarizability. Owing to the complexity of this phenomenon, however, some aspects have not yet been clarified.

Under the effect of ionizing radiation electronically excited molecules of different energies are also formed (see Section 1.2.1). The spinforbidden excited states of the lowest energies are triplet states the energy of which is ~ 4 eV for alkenes, and ~ 3 eV for conjugated dienes and for alkynes. Although these excited states are formed directly with very low probability, their role in the case of alkenes may be important because they are produced with high cross-section in intersystem-crossing processes. For instance, the *cis-trans* isomerization detected experimentally points to the chemical decay of triplets of low energy. The chemical and physical decay of singlet and triplet states of higher energies are similar for alkenes and for saturated hydrocarbons. The decomposition, the photoemission, the autoionization, etc. processes all may play an important role. There are no accurate data available concerning the yield of excited species or the ratio of yields produced by excitation and ionization due to irradiation.

It is to be noted, however, that Okazaki et al. [29] recently published some data concerning this problem. These data were obtained by theoretical estimation. G-values for excitation and ionization have been calculated by combining the binary-encounter-collision theory with the theory of degradation spectrum, which is based on the continuous-slowing-down approximation. The data calculated show that the yield of singlet excitation (G_s) for monoalkenes is about 2.3, except for ethylene, for which the value of about 5.0 was given. In the triplet excitation yields (G_t) there is no big difference between ethylene and other alkenes. The G_t-values for ethylene, propene, but-1-ene, *cis*- and *trans*-but-2-ene, 2-methylpropene and cyclopentene are 1.68, 1.36, 1.51, 1.55, 1.54 and 1.67, respectively. The G_t-values of the conjugated dienes are significantly higher than those of monoalkenes, e.g. the G_t-value of but-1,3-diene is 2.66. The total yield of electrons (G_e) is about 5.0–5.1. These values are somewhat higher than the experimental ones, but if dissociation from the superexcited state into the neutral fragments is taken into account the values are in agreement ($G_e = 4.1$–4.4).

The allylic interaction is of great importance in the chemistry of alkenes. For example, the high thermochemical stability of allyl-type free radicals may be an explanation for the experimental finding that in the pyrolysis and the photolysis, as well as in the radiolysis of alkenes, the most probable bond cleavage is that of the C—H or C—C bonds in the β-position with respect to the double bond. This could explain the phenomenon that the allylic radicals formed in such systems participate in radical recombination processes, whereas other free radicals promote chain reactions of the abstraction and radical decomposition type.

Hydrogen abstraction reactions occur in non-terminal alkenes to a greater extent than in terminal alkenes. This can be explained primarily by the greater number of hydrogen atoms in allylic positions and the high stability of allyl radicals formed in a high yield under irradiation. The rate of abstraction is also influenced by the order of the C—H bonds in the sequence primary allyl $<$ secondary allyl $<$ tertiary allyl.

17*

4.2. INDIVIDUAL COMPOUNDS

The radiolysis of aliphatic alkenes will be considered by dealing with individual substances in increasing carbon number. Acetylene and its homologues will be dealt with later subsequent section.

4.2.1. Ethylene

The radiation chemistry of ethylene has been investigated by many different methods under various conditions [30–67].

The yields determined in different states of aggregation are listed in Table 4.4. The complexity of the radiolysis of unsaturated hydrocarbons is reflected in the fact that even for ethylene, the alkene with the simplest structure, the number of products identified is about 30. The most important products in the gas phase are acetylene, n-butane and hydrogen and mainly polymers.

Most of the acetylene and hydrogen is formed in molecular processes when the radiation energy is either below or above the ionization potential [see reactions (4.5b), (4.6a) and (4.6c), p. 266], as has been proved by experiments carried out with ethylene–ethylene-d_4 mixtures [35, 38, 39] (see Table 4.5). The unimolecular mechanism of hydrogen formation has been investigated by Sauer and Dorfman [38], Okabe and McNesby [40] and Gorden and Ausloos [36, 37] by using gas-phase photolysis of ethylene labeled specifically with deuterium. H_2, HD and D_2 were formed from $CH_2=CD_2$, and this points to both 1,1- and 1,2-elimination processes:

$$CH_2CD_2^* \begin{array}{l} \nearrow H_2 + C_2D_2 \\ \longrightarrow HD + CHCD \\ \searrow D_2 + C_2H_2 \end{array}$$

The ratio of these three processes is 0.42 : 0.41 : 0.17, and shows no energy dependence even over a wide energy range [37, 40].

The hydrogen formed by radiolysis of *trans*-CHD=CHD contained considerable amounts of H_2 and D_2, which points to the *cis-trans* isomerization of the electronically excited ethylene molecule prior to the 1,2-elimination [40]. Recently, rotation around the C—C bond was shown to occur by Hirokami and Cvetanovic [41], on the basis of the mutual interconversion of isomers on the photolysis of *cis*- and *trans*-1,2-dideuteroethylene at 77 K using 6.7 eV photon energy.

On photolysis in the gas phase the yield of acetylene is always higher than that of molecular hydrogen: at 10 eV photon energy the ratio is $C_2H_2/H_2 = 2.8$ [37], and it is very close to the value obtained in gas-phase radiolysis (2.6) [31]. According to Gorden and Ausloos [36, 37], this can be explained by the fact that the vinyl radical formed with excess energy further decomposes [reaction (4.4)]. This seems to be supported by the fact that the CHCD fraction in the acetylene is higher than HD in the hydrogen formed in the gas-phase experiments using CH_2CD_2 (Table 4.6).

Table 4.4

Yields from ethylene (radiation: ^{60}Co-γ and electron)

Temperature, °C	25	—165	—196	
Reference	29	54	37	63
Phase	Gas	Liquid	Solid	
Products		G, molecule/100 eV		

Hydrogen	1.3	1.3		0.6
Methane	0.22			0.01
Ethane	0.85	0.68		0.6
Acetylene	3.5	1.5	1.4	0.6
Propane	0.56			
Cyclopropane			0.0098	
Propene			0.0042	
n-Butane	2.32	0.44	0.1288	0.1
Isobutane	0.11			
Cyclobutane	0.10	0.12		
Methylcyclopropane			0.0112	
But-1-ene	0.09	0.58	0.4088	0.2
trans-But-2-ene	0.03		0.0112^b	traces
cis-But-2-ene	0.03			traces
Cyclobutene			0.0763	
Butadiene	0.003			0.03
n-Pentane	0.04			
Isopentane	0.013			
Pent-1-ene	0.05			
n-Hexane	0.31			0.001
2-Methylpentane	0.02			
3-Methylpentane	0.05			0.018
Hex-1-ene	0.046	0.30		0.014
trans-Hex-2-ene	0.03	0.13		0.023
cis-Hex-2-ene				0.006
Hex-3-ene			0.042^c	
3-Methylpent-1-ene				0.017
3-Methylpent-2-ene	0.01	0.30^a		0.019
3-Ethylhexenes				0.013
3-Methylpentenes				0.03
n-Octenes				0.01

^{a} 3-Methylpent-2-ene + other hexenes
^{b} But-2-enes
^{c} C_6H_{12}

Table 4.5

Acetylene and hydrogen yields of the photolysis
and radiolysis of C_2H_4–C_2D_4 mixture (1 : 1). After Gorden and Ausloos [37]

Radiation	Energy, eV	C_2H_2	HD	C_2D_2	H_2	C_2HD	D_2
		Distribution, %					
Photolysis	8.4	59.0	3.0	34.0	61.0	7.0	36.0
Photolysis	11.6–11.8	60.5	3.5	34.5	62.0	5.0	34.5
^{60}Co-γ	$1.25 \cdot 10^6$	57.0	6.6	37.0	52.6	6.0	40.8

Table 4.6

Acetylene and hydrogen yields of the photolysis
and radiolysis of CH_2CD_2. After Gorden and Ausloos [37]

Radiation	Energy, eV	Phase	H_2	HD	D_2	C_2H_2	C_2HD	C_2D_2
					Distribution, %			
Photolysis	8.4	Gas[a]	42.0	41.0	16.9	11.0	63.9	25.1
		Solid[b]	40.1	38.0	23.9	23.1	37.3	39.6
Photolysis	10.0	Gas[a]	41.8	40.6	17.6	10.5	63.5	26.0
		Solid[b]	39.6	39.1	20.6	21.7	41.7	36.6
Photolysis	11.6–11.8	Gas[a]	41.8	41.5	16.7	11.1	62.2	26.7
		Solid[b]	42.2	41.0	16.8	14.1	53.4	32.5
[60]Co-γ	$1.25 \cdot 10^6$	Gas[a]	38.5	46.4	15.1	12.7	65.1	22.2
		Solid[b]	42.5	31.0	26.5	17.7	52.7	29.6

[a] 13 mbar;
[b] 20 K

Although ethylene itself is an excellent radical scavenger (see Section 1.4.1. [54, 62]), several radicals have been shown to exist in the radiolysis of ethylene using iodine as a radical scavenger. Among these methyl and ethyl radicals are formed in the highest yield [29], and the amount of vinyl radicals is also high. Recently, methyl, ethyl and butyl radicals were identified by Gawlowski and Herman [42] using SF_6 as electron scavenger: the SF_5 radical formed on the neutralization of SF_6^- is considered to be the radical scavenger:

$$RH^+ + SF_6^- \longrightarrow SF_5 + products$$

In the radiolysis of ethylene containing 0.4% SF_6 additive the following yields have been measured: $G(CH_3SF_5) = 0.02$, $G(C_2H_5SF_5) = 0.23$, $G(C_4H_9SF_5) = 0.27$.

The radiolysis of gas-phase ethylene has been investigated in the presence of a number of other additives. It was found that NO (which is a well known radical scavenger) has a similar effect to dimethylamine on the product yields as a charge acceptor:

$$C_4H_8^+ + NO \longrightarrow C_4H_8 + NO^+$$

This process was proposed to explain why the yield of butenes is significantly higher in presence of NO than in the radiolysis of pure ethylene (see Table 4.7) [34]. Mass spectrometry studies, however, showed the charge transfer from butene ion to NO to be negligible [43], but $C_4H_8NO^+$ ions were abundant. Butenes are probable produced from these on neutralization. According to the data reported by Kebarle et al. [43–45], NO reacts in a rather complex way in the mass spectrometer: e.g. besides the simple ionic addition, C–C bond cleavage can be caused. These experiments emphasize that if NO is used as a radical scavenger, side-reactions must be taken into account in order to avoid misleading conclusions.

Table 4.7

Effect of scavengers and the electric field on the radiolysis of gaseous ethylene. After Meisels and Sworski [31]

C₂H₄		66	66	66	67	66
NO	Pressure, mbar	—	13	13	—	
O₂			—		7	
X/p^a		—		19.5	19.9	20.4
Products		Yield, molecule/ion pair				
Methane		0.067	0.065	0.066	0.064	0.067
Ethane		0.16				0.76
Acetylene		0.87	0.90	5.8	7.4	8.7
Propane		0.18				0.29
n-Butane		0.56				3.8
But-1-ene		0.060	0.047	0.045	0.026	0.33
cis-But-2-ene			0.190	0.187	0.42	
trans-But-2-ene			0.149	0.151	0.026	
Butadiene		0.004	0.007			0.083
n-Hexane		0.036				0.302
Hex-1-ene		0.013				1.165

^{a} X/p is the ratio of the electric field (X) to the pressure (p) of C_2H_4 V cm^{-1} mbar^{-1}

Gawlowski and Niedzielski have recently published data on the gas-phase radiolysis of ethylene [46, 47]. These results support earlier results. On the other hand, it can be concluded from the dose dependence and from the data recently obtained by means of additives such as NH_3, O_2, SF_6 and iso-C_4H_8, that there is a contribution resulting from the neutralization of positive ions in the formation of atomic hydrogen and methyl radicals. Figure 4.1 shows the effect of NH_3 on the yield of some products.

It has been established by the application of O_2 and NO, and by the use of an electrostatic field that alkanes, especially methane, are formed by molecular processes [29–31, 43, 47–49] via an ionic intermediate, as was established by Meisels and Sworski:

$$C_2H_3^+ + C_2H_4 \longrightarrow (C_4H_7^+)$$

$$(C_4H_7^+) + C_2H_4 \longrightarrow (C_6H_{11}^+)$$

$$(C_6H_{11}^+) \longrightarrow CH_4 + C_5H_7^+$$

Alkanes of molecular mass higher than methane are formed in free radical reactions [24, 30, 43, 47–49]. In these processes methyl and ethyl radicals play the most important role both by direct recombination and in addition reactions with ethylene to form other radicals.

Most of the methyl radicals are formed in ion-molecule reactions, as suggested by Meisels and Sworski [31, 32]:

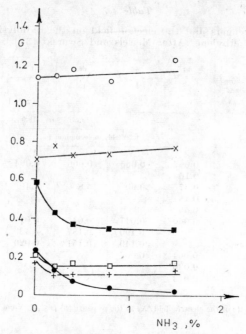

Fig. 4.1. Dependence of product yields of ethylene on the concentration of ammonia. C_2H_6 was measured in the presence of 0.3% O_2.
○ n-C_4H_{10}; × n-C_6H_{14}; ■ C_2H_6; □ C_3H_8; + C_5H_{12}; ● C_2H_6 (radiation: ^{60}Co-γ). After Gawlowski et al. [47]

$$C_2H_4^+ + C_2H_4 \longrightarrow C_3H_5^+ + CH_3$$

$$C_2H_4^+ + 2\,C_2H_4 \longrightarrow C_5H_9^+ + CH_3$$

$$C_2H_2^+ + 2\,C_2H_4 \longrightarrow C_5H_5^+ + CH_3 + H_2$$

The formation of methyl radicals by the reaction of a parent ethylene ion with an ethylene molecule has been observed mass spectrometrically [50–52]. Gawlowski and Niedzielski [46, 47] assumed that the methyl radical is produced from ions, not in an ion-molecule reaction but via the decomposition of an excited species formed by the ion-electron neutralization. Ethyl radicals are formed almost entirely by the addition of hydrogen atoms. Vinyl radicals have been observed by Fessenden and Schuler in ESR studies, formed by the primary decomposition of excited ethylene [53]. From these, as well as from the liquid-phase data of Holroyd and Fessenden [54], the presence of vinyl radicals in the gas phase also has been deduced by Meisels and Sworski [31, 32]. The yield of vinyl radicals was derived on the basis of the formation of but-1-ene, buta-1,3-diene and hex-1-ene by reactions (4.3a–c) as follows:

$$C_2H_3 + C_2H_4 \longrightarrow C_4H_7$$

$$C_4H_7 + C_2H_5 \quad\begin{cases} \nearrow & 1\text{-}C_4H_8 + C_2H_4 \\ \longrightarrow & 1,3\text{-}C_4H_6 + C_2H_6 \\ \searrow & 1\text{-}C_6H_{12} \end{cases}$$

(4.3a)

(4.3b)

(4.3c)

The reactions of vinyl radicals have been studied in vacuum UV photolysis experiments. Thus, vinyl radicals formed by irradiation with photons of energy above 8.5 eV (wavelength < 147 nm) are not stabilized, even at atmospheric pressure, but are further decomposed:

$$C_2H_4^* \longrightarrow H + C_2H_3 \qquad (4.4)$$

$$C_2H_3 \longrightarrow C_2H_2 + H$$

Stable vinyl radicals have been observed on irradiation with low-energy photons ($\lambda > 155$ nm) [55, 56]. For instance, in the flash photolysis of ethylene using wavelengths of 160–190 nm (6.5–7.8 eV) at low pressure, the formation of but-1-ene and propene was not observed by Back and Griffiths [55]. These compounds are formed in considerable amounts, however, at pressures above 12–26 mbar of ethylene. These products can be attributed to the recombination of vinyl radicals with ethyl and methyl radicals, respectively.

Of the primary ions obtained in the radiolysis of ethylene, $C_2H_4^+$ is formed in the largest amount [34], and the yields of vinyl and acetylene ions are significant. In the gas-phase radiolysis and in photolysis with energies higher than the ionization potential, the most probable reaction is between parent ions and ethylene:

$$C_2H_4^+ + C_2H_4 \longrightarrow C_4H_8^+ \qquad k = 5 \cdot 10^{11}\,\mathrm{M^{-1}\,s^{-1}}$$

There are several possible reaction pathways for the $C_4H_8^+$ dimer cation. For example, it can form products of higher molecular mass by means of ion-molecule reaction. It may also contribute to the formation of C_4H_8 products by neutralization with an electron or by charge-transfer reactions.

Meisels and Sworski [31] studied the primary decomposition of excited ethylene in the gas phase to compare the effect of secondary electrons accelerated by an electrostatic field on the data obtained from vacuum UV photolysis with that of high-energy electron irradiation. They suggested three main processes [reaction (4.6a–c)] for the primary decomposition of excited ethylene. The relative probabilities of these three processes vary significantly with energy, as shown in Table 4.8. On Hg-sensitized photolysis (excitation with energy of 4.9 eV) [57–60], ethylene mainly decomposes through long-lived triplet states. This process is not important at 66 mbar pressure because the long-lived triplet of 4.6 eV energy is largely quenched under these circumstances [57–59]. From the higher excited states, e.g. from the symmetry-forbidden 6.5 eV energy state, all three processes would occur. However, it can be seen from the Table that, with

Table 4.8

Dissociation probabilities of excited ethylene
produced by various techniques. After Meisels and Sworski [31]

Reaction	Hg-sensitized	Slow electrons[a]	Vacuum UV		Electron
			147.0 nm	123.6 nm	
		Relative probability of mechanism			
$C_2H_4 \rightarrow C_2H_2 + H_2$	0.92	0.38	0.44	0.26	0.10
$\rightarrow C_2H_2 + 2H$		0.46	0.45	0.74	0.87
$\rightarrow C_2H_3 + H$	0.08	0.16	0.11	0.0	0.03

[a]$X/p = 20.4$ V cm^{-1} mbar^{-1}, see Table 4.7

increasing energy available for excitation, reaction (4.6c) becomes predominant. To explain this phenomenon it was assumed [31] that the dissociation of ethylene at higher energy occurs through intersystem crossing
and radiationless transitions from the vibrationally excited ground state.
The energy increase favours the most endothermic process.

There are several contributions to the formation of products in the gas
phase radiolysis of ethylene. Meisels [1] has proposed the following primary
processes, energetics and radiation chemical yields:

$$C_2H_4^+ + e^- \qquad \Delta H = 10.5 \text{ eV}, \ G = 1.5$$

$$H + C_2H_3^+ + e^- \qquad \Delta H = 14.0 \text{ eV}, \ G = 1.0$$

$$C_2H_4 \qquad 2H + C_2H_2^+ + e^- \qquad \Delta H = 13.5 \text{ eV}, \ G = 0.8 \qquad (4.5a)$$

$$H_2 + C_2H_2^+ + e^- \qquad \Delta H = 13.5 \text{ eV}, \ G = 0.8 \qquad (4.5b)$$

$$H_2 + C_2H_2 \qquad \Delta H = 1.8 \text{ eV}, \ G = 0.5 \qquad (4.6a)$$

$$C_2H_4^* \qquad H + C_2H_3 \qquad \Delta H = 4.6 \text{ eV}, \ G = 0.1 \qquad (4.6b)$$

$$2H + C_2H_2 \qquad \Delta H = 6.3 \text{ eV}, \ G = 2.8 \qquad (4.6c)$$

where the heat of reaction and yield referring to reactions (4.5a) and (4.5b)
were not precisely specified.

By considering the data in several publications, Laidler and Loucks [61]
suggested the following reaction scheme for the secondary processes in the
gas-phase radiolysis of ethylene:

$$C_2H_4 + H \longrightarrow C_2H_5$$

$$C_2H_5 + C_2H_4 \longrightarrow n\text{-}C_4H_9$$

$$CH_3 + C_2H_4 \longrightarrow C_3H_7$$

$$C_2H_3 + C_2H_4 \longrightarrow C_4H_7$$

$$C_2H_4^+ + C_2H_4 \longrightarrow (C_4H_8^+)^*$$

$$(C_4H_8^+) \left\langle \begin{array}{l} C_3H_5^+ + CH_3 \\[1ex] C_4H_7^+ + H \end{array} \right.$$

$$(C_4H_8^+)^* + C_2H_4 \longrightarrow C_4H_8^+ + C_2H_4$$

$$C_3H_5^+ + C_2H_4 \longrightarrow (C_5H_9^+)^*$$

$$(C_5H_9^+)^* + C_2H_4 \longrightarrow C_5H_9^+ + C_2H_4$$

$$2C_2H_5 \left\langle \begin{array}{l} \text{n-}C_4H_{10} \\[1ex] C_2H_6 + C_2H_4 \end{array} \right.$$

$$C_2H_5 + CH_3 \left\langle \begin{array}{l} C_3H_8 \\[1ex] CH_4 + C_2H_4 \end{array} \right.$$

$$\text{n-}C_4H_9^* + C_2H_5 \left\langle \begin{array}{l} \text{n-}C_6H_{14} \\[1ex] \text{n-}C_4H_{10} + C_2H_4 \\[1ex] C_2H_6 + \text{iso-}C_4H_8 \end{array} \right.$$

$$2CH_3 \longrightarrow C_2H_6$$

$$\text{n-}C_4H_9 + C_2H_4 \longrightarrow \text{n-}C_6H_{13}$$

$$\text{n-}C_6H_{13} + C_2H_5 \left\langle \begin{array}{l} \text{n-}C_8H_{18} \\[1ex] C_2H_6 + \text{iso-}C_6H_{12} \\[1ex] \text{n-}C_6H_{14} + C_2H_4 \end{array} \right.$$

$$R + C_2H_4 \longrightarrow RCH_2CH_2 \xrightarrow{C_2H_4} \text{polymer}$$

$$C_nH_{2n}^+ + C_2H_4 \longrightarrow C_{n+2}H_{2n+4}^+ \xrightarrow{C_2H_4} \text{polymer}$$

Although most of the reactions are identical in the gas and liquid phase there are differences between the yields. Thus, the yield of n-butane is smaller than that of but-1-ene. This may be a consequence, as we have already mentioned, of the stabilization of vinyl radicals followed by chemical reactions such as combination, disproportionation, etc. By means of ESR spectroscopy, Fessenden and Schuler [53] established that ethyl, vinyl and but-3-enyl radicals are formed in the system.

Table

Yields of the photolysis and radiolysis of ethylene in the solid phase.

Radiation	Energy, eV	Additive	Hydrogen	Acetylene	Cyclopropane
Photolysis[b]	8.4	None	106	100	0.40
		Ar[a]		100	0.12
		O_2 (5%)	110	100	0.40
		CCl_4 (6.3%)		100	0.13
	10.0	None	90	100	1.1
		Ar[a]		100	1.05
		CCl_4 (6.3%)		100	0.49
	11.6–11.8	None		100	2.2
		CCl_4 (6.3%)		100	0.55
	21.2	None		100	
$^{60}Co\text{-}\gamma^{c}$	$1.25 \cdot 10^6$	None		100[d]	0.07
				(1.4)	
		CCl_4(6.3%)		100	0.65
				(1.5)	
		CCl_4(14.7%)		100	0.75
				(1.5)	

[a] $[Ar]/[C_2H_4] = 50$
[b] 20 K
[c] 77 K
[d] Values in parenthesis are G-values

Holroyd and Fessenden [54] showed that n-butane, n-hexane and some but-1-ene were formed from these radicals. From these data it seems that the vinyl and but-3-enyl radicals play a more important role in the liquid than in the gas phase.

There are only a few reports dealing with the solid-phase photolysis and radiolysis of ethylene. As shown in Table 4.4, the main products of the radiation-chemical decomposition are the same as in the liquid phase, although the yields are much smaller. In the photolysis and radiolysis of C_2H_4–C_2D_4 mixtures it was shown [37, 54] that vinyl and ethyl radicals formed in the primary decomposition may combine to form but-1-ene:

$$C_2H_3 + C_2H_5 \longrightarrow 1\text{-}C_4H_8$$

By comparing the results of the gas- and solid-phase photolysis and radiolysis of CD_2–CH_2, Gorden and Ausloos [37] found that the isotopic composition of hydrogen is generally the same in both phases (Table 4.6). It is remarkable that in photolysis experiments carried out in the solid phase at energies of 8.4 and 10 eV, the yields of acetylene and hydrogen were the same (Table 4.9). In addition, in experiments with CD_2CH_2, the complementary hydrogen and acetylene yields labelled with deuterium isotope are very

4.9

After Gorden and Ausloos [37]

Propene	n-Butane	Cyclobutane	Methylcyclopropane	But-1-ene	But-2-enes	Hexenes
Relative yields						
0.35	1.10	0.39	0.31	14.1	0.40	3.1
0.22	0.82	0.47	0.20	13.5	0.25	
0.42	0.38	0.50		11.0		
	1.08	0.39		10.1		8.4
1.4	1.90	3.05	0.63	16.0	1.15	16.9
1.5	2.07		0.60	20.7	1.20	
	2.10	3.40		20.0	0.40	47.0
1.7	3.20	7.2	0.73	23.0	1.50	37.0
	3.70	7.3		29.0	1.40	140.0
	13.0	5.1		30.0		
0.30	9.2	5.45	0.8	29.2	0.8	3.0
	9.9	6.0	1.1	32.4	1.1	6.0
	8.3	5.7	2.23	29.6	0.60	9.0

similar (C_2H_2, D_2, C_2HD, HD, C_2D_2, H_2) (see Table 4.6). This means that vinyl radicals do not decompose according to reaction (4.4) to produce excess acetylene and hydrogen atoms, but that they can be stabilized. This may be why the n-butane yield is decreased and the but-1-ene yield is increased compared with the results in the gas phase (Table 4.4).

In photolysis using high-energy (11.6–11.8 eV) photons and in radiolysis the yield of CHCD in the solid phase is again higher than that of the corresponding HD. It seems that the decomposition of vinyl radicals by radiation of higher energy may also occur in the solid phase. According to Gorden and Ausloos it is more probable that the ratio of CHCD : HD is influenced by a disproportionation between the CH_2CD or CD_2CH radicals and other radicals present in the system. However, this explanation appears to be incorrect if the results obtained in other kinetic studies on the nature and reactivity of vinyl radicals are taken into account.

On the basis of gas-phase studies, Wagner assumed [63] that but-1-ene, hexenes and products of higher molecular mass were formed mainly by ion-molecule reactions: these are initiated by $C_2H_4^+$ and are terminated by recombination of the $(C_2H_4)^+$ ion-molecule with an electron. This idea of the ionic origin of but-1-ene is only partly supported by the data reported by Ausloos and Gorden (Table 4.9), because the yield is relatively high for photolysis with 8.4 eV photons when ionization should not be important (ionization potential data are not available for the solid phase).

However, the mechanism suggested by Wagner for the formation of hexene seems to be supported by the results of Gorden and Ausloos (Table 4.9). Hexene yields increase in presence of CCl_4 as electron scavenger as well as with increasing photon energy.

Cyclopropane is produced in considerable amounts in the solid-phase photolysis and radiolysis; a carbene-addition mechanism was suggested [37]:

$$CH_2 + C_2H_4 \longrightarrow \text{cyclo-}C_3H_6$$

On this basis, the primary decomposition process in the solid-phase radiolysis of ethylene, as in the liquid phase, was proposed by Gorden and Ausloos to be:

$$C_2H_4^* \longrightarrow 2\,CH_2$$

Among the C_4H_8 products cyclobutane was formed almost exclusively through ionic ($C_4H_8^+$) intermediates [37]. This is supported to a large extent by the energy dependence of the product yield (Table 4.9), and by the isotopic distribution observed in the radiolysis and photolysis of C_2H_4–C_2D_4 mixtures.

As ethylene is one of the most important monomers in the polymer industry, its polymerization by irradiation has been extensively studied but the mechanism is still not quite clear. Hayward [64] and Laird et al. [65] have studied the process under different reaction conditions, and they obtained the same temperature dependence for the average reaction rate process (Fig. 4.2). Above the break-point shown in the Figure (123–125°C) the activation energy is 59 kJ mol^{-1}; at lower temperatures this value is about 8.4 kJ mol^{-1}. To explain this behaviour Laird et al. assumed [65]

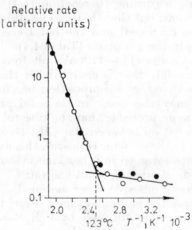

Fig. 4.2. Arrhenius plot of the relative rates in the γ-irradiated polymerization of ethylene [6].
● From ref. [64]; ○ from ref. [65]

that below 100–125°C the chain initiation was effected by a radical of high reactivity formed from ethylene, but above this temperature the radical formed possesses a smaller reactivity. However, Chapiro [6] has given a more reasonable explanation by which the rate of chain propagation and chain termination may depend on the physical properties of the polymer formed. Below 125°C the crystalline polymer is hardly swollen in ethylene and so there is only a low probability that ethylene will react on the polymer surface. Above 125°C, however, the polymer become amorphous and thus swells, so that reaction can occur inside it. Under these conditions the concentration of monomer in the vicinity of the increasing chain may be much higher than in the gas phase, and consequently the rate of reaction may increase. The concentration of monomer inside the polymer is controlled by the rate of diffusion, accordingly the high energy of activation may be rationalized.

Dauphin et al. [66] have studied in detail the γ-ray-initiated polymerization of ethylene by varying the experimental conditions over a wide range (pressure, dose rate, temperature, additives, etc.). At low temperatures ($-25°C$) the polymer formed had a high specific surface area in a fine powder form. The specific surface area decreases with increasing temperature, and at 90°C it becomes immeasurable because of the agglomeration of the particles. As the change in specific surface area was accompanied by a decrease in the rate of polymerization two parallel reactions were assumed: one of these is operative in the homogeneous phase, whereas the other occurs at the polymer surface. Of the additives applied (O_2, CO_2, H_2O, C_3H_6, NH_3, etc.) the inhibiting effect of oxygen was the most important (Fig. 4.3), this is obviously due to scavenging of the chain-initiating radicals. NH_3 increased the rate of polymerization as expected. It is known [66A] that ammonia can easily accept a proton, and thus it can increase the radical yield in the system. This fact strongly supports the radical mechanism as a reasonable explanation for the rate increase.

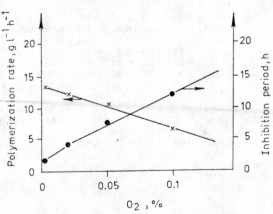

Fig. 4.3. Dependence of the rate of ethylene polymerization and of the inhibition period on the concentration of oxygen (dose rate: $7.8 \cdot 10^{15}\,eV\,g^{-1}s^{-1}$; radiation: $^{60}Co\text{-}\gamma$). After Dauphin et al. [66]

On the basis of data obtained for ethylene radiolysis, and taking the values $2 \cdot 10^3$ and $1 \cdot 10^{11}$ M^{-1} s^{-1} for the radical propagation and termination, respectively, Meisels [67] has derived an equation for ethylene polymerization.

4.2.2. Propene

The literature data for the higher homologues of the alkene series are less abundant than for ethylene. This is partly due to the fact that information concerning the primary processes is much scarcer. It is very difficult to draw exact conclusions concerning the primary processes owing to the secondary decomposition of ionized and/or vibrationally excited species formed from alkenes containing more carbon atoms.

Data for the radiolysis of propene have been reported by Wagner [68], Holroyd and Klein [69], Munari and Castello [70] and Horváth and Földiák [71]. The data referring to the decomposition products are incomplete, and the mechanism of formation has not been unambiguously identified (Table 4.10).

The gas-phase photolysis of propene has been studied using photons with energies both higher and lower than the ionization potential (Table 4.11).

Table 4.10

Yields from monoalkenes (radiation: ^{60}Co-γ)

Hydrocarbons	Propene		But-1-ene			cis-But-2-ene			trans-But-2-ene	
Reference	68	171	86	86	71	86	87	86	86	
Phase	Liquid		Gas	Liquid		Gas	Liquid		Gas	Liquid
Products	G, molecule/100 eV									
Hydrogen	0.6	0.80	1.03	0.64	0.73	1.15	1.02	0.99	1.37	0.99
Methane	0.04		0.26	0.05	0.10	0.09	0.21	0.13	0.14	0.17
Ethane			0.43	0.13	0.066	0.09	0.03	0.09	0.07	0.09
Ethylene		0.21	4.30	0.21	0.22	0.86	0.08	0.26	0.82	0.17
Acetylene		0.51	2.92	0.30	0.24	0.86	0.11	0.26	1.17	0.17
Propane	0.2		0.22	0.06	0.055	0.09		0.22	0.55	0.22
Propene			0.13	0.13	0.013	0.52	0.13	0.26	0.48	0.13
Allene (propadiene)		0.33			0.067					
n-Butane	0.02		1.94	0.30		1.29	0.83	1.03	1.51	0.95
But-1-ene	0.02	0.13				0.43	0.40	0.43	0.62	0.31
cis-But-2-ene			0.56	0.18	0.21				4.1	4.1
trans-But-2-ene	0.02	0.09	0.52	0.16	0.065	3.4	2.29	4.0		
Buta-1,3-diene							0.10		0.13	
Cyclobutene+buta-1,2-diene					0.15					
C_5 products	0.04				0.11		0.01			
C_6 products	1.0						trace			
3-Methylhex-1-ene										
4-Methylhex-1-ene										
Other C_7 products				0.02			0.01	0.01		
C_8 products			0.17	1.72			2.30	0.96		1.03

Table 4.11

Relative yields of propene photolysis

Energy, eV	6.7	8.4			10.0	
Pressure, mbar	26	53	13		13	
Reference	74	89	75	76	75	76
Products	Relative yield					
Hydrogen	0.76	0.5	0.2		0.42	
Methane	0.76	0.19	0.09	0.32	0.1	0.31
Ethane	1.00	1.00	1.00	1.00	1.00	1.00
Ethylene	0.93	0.27	0.32	0.39	0.52	0.48
Acetylene	0.58	0.18	0.19	0.3	0.12	0.33
Propane	1.00		0.19	0.39	0.19	0.56
Propyne	0.40	0.47	0.19		0.19	
Allene (propadiene)	0.20	0.63	0.19		0.19	
Isobutane	0.74	0.59	0.3		0.25	
Butenes	1.3[a]	0.1	0.09		0.13	
n-Butane		0.06				
trans-But-2-ene		0.006	0.19		0.12	
Methylcyclopropane		0.005				
cis-But-2-ene			0.05			
Isopentane		0.006				
2-Methylpentane	0.73					
4-Methylpent-1-ene	1.8					

[a] But-1-ene only

From these investigations a general conclusion can be drawn that five or six primary processes are operative:

$$C_3H_6 + h\nu \longrightarrow H + C_3H_5 \text{ (allyl)} \qquad (4.7a)$$

$$\longrightarrow H_2 + CH_2{=}C{=}CH_2 \qquad (4.7b)$$

$$\longrightarrow H_2 + CH_3{-}C{\equiv}CH \qquad (4.7c)$$

$$\longrightarrow CH_4 + C_2H_2 \qquad (4.7d)$$

$$\longrightarrow CH_3 + C_2H_3 \qquad (4.7e)$$

$$\longrightarrow CH_2 + C_2H_4 \qquad (4.7f)$$

These reactions [except (4.7f)] have already been suggested on the basis of Hg-sensitized photolysis at 237.7 nm [72, 73]. Arai et al. [74] studied the photolysis of propene and established that the formation of products can be explained also by reactions (4.7b), (4.7e) and (4.7f). Some of the C_2-C_3 species formed in reactions (4.7a–4.7f) can further decompose, owing to their excess energy. This is supported by the fact that the amount of acetylene which is the product with the highest yield among the decomposition products, exceeds that of 'molecular' methane in experiments carried out with

Table 4.12

Yields from propene (radiation: ^{60}Co-γ; temperature: 250°C). After Munari and Castello [70]

Products	G, molecule/100 eV
Hydrogen	0.24
Methane	4.20
Ethane	3.50
Ethylene	6.41
n-Butane	0.11
Isobutane	4.50
But-1-ene + but-2-ene	0.21
Neopentane	0.99
3-Methylbut-1-ene	1.10
n-Pentane	4.04
Pent-1-ene	0.85
2-Methylbut-1-ene	0.65
Pent-2-ene	1.23
2-Methylbut-2-ene	0.88
Other butenes	0.85
2,3-Dimethylbutane	3.80
2,3-Dimethylbut-1-ene	4.46
4-Methylpent-2-ene	4.16
n-Hexane	1.18
Hex-1-ene + 2-methylpent-1-ene	49.34
2,4-Dimethylpentane	2.43
Hexa-2,5-diene	4.05
Hexadienes not determined	4.32
C_7 products	4.45
C_8 products	7.87
Non-1-ene	5.80
C_9 products	35.92
$C_{10} + C_{11}$ products	11.55
C_{12} products	31.40
$C_{13} + C_{14}$ products	5.78
C_{15} products	19.36
$C_{16} + C_{17}$ products	2.96
C_{18} products	9.26
Higher products	3.22

photons of different energies. Consequently, the 'excess' acetylene should be the result of secondary decomposition of species with excess energy, such as $C_2H_3^*$ and $C_3H_5^*$ or $C_2H_4^*$. For instance:

$$C_2H_3^* \longrightarrow HC\equiv CH + H \qquad (4.8)$$

Munari and Castello [70] studied the γ-radiolysis of propene at 250°C in a stainless steel vessel (Table 4.12). The rather complex product spectrum shows that the yield of products of high molecular mass consisting of C_3 units is an order of magnitude higher than that of other products. Dimers are produced in the highest yield: hex-1-ene is particularly abundant. The product spectrum points to a radical mechanism, and for the products of higher yield a radical chain mechanism was suggested. Hex-1-ene seems to be formed by reaction between a propene molecule and an allyl radical:

$$CH_3-CH=CH_2 \rightsquigarrow CH_2-CH=CH_2 + H$$

$$CH_2=CH-CH_2 + CH_2=CH-CH_3 \longrightarrow CH_2=CH-CH_2-CH-CH_2-CH_3$$

$$CH_2=CH-CH_2-CH_2CH_3 + C_3H_6 \longrightarrow$$

$$\longrightarrow CH_2=CH-CH_2-CH_2-CH-CH_3 + C_3H_5 \text{ (allyl)}$$

A $G(-C_3H_6) = 666$ value is given for propene consumption. H and CH_3 radicals make a large contribution to product formation by initiating a series of secondary reactions, leading predominantly to the formation of ethane and propane. The larger part of these two products (in the energy range below the ionization potential) is formed by a radical pathway, as shown by the effect of NO and O_2 as radical scavengers and also by the data of Becker et al. [75], obtained using isotope mixtures.

Primary processes such as H_2 and CH_4 elimination [reactions (4.7b–4.7d)] have been elucidated by these experiments, carried out with a mixture of propene and propene-d_6 [75]. Hydrogen is formed predominantly via molecular elimination (Table 4.13): 90% of hydrogen formed from C_3H_6–C_3D_6 mixture irradiated by photons of 147.6 nm (8.4 eV) consists of H_2 and D_2. Hydrogen gas was predominantly formed from hydrogen atoms in

Table 4.13

Hydrogen yields of the photolysis of gaseous C_3H_6-C_3D_6 systems.
After Becker et al. [75]

Components	Ratio	Energy, eV	H_2	HD	D_2
			Distribution, %		
$CH_3CH=CD_2$	pure	8.4	64	27	8
$C_3H_6 : C_3D_6$	1 : 1	8.4	57	11	32
$C_3H_6 : C_3D_6 : NO$	1 : 1 : 0.15	8.4	62	3	34
$C_3H_6 : C_3D_6$	1 : 1	10.0	59	18	23

allyl positions, as was shown by photolysis of $CH_3CH=CD_2$. Methane formed from C_3H_6-C_3D_6-NO mixtures consists of CH_4 and CD_4, but in methane formed from $CH_3CH=CD_2$, CH_4 and CH_3D can be found. This shows that the overall reaction (4.7d) can be described as follows:

$$C_3H_6 + h\nu \longrightarrow CH\equiv CH + CH_4$$
$$\longrightarrow C=CH_2 + CH_4$$
$$\downarrow$$
$$CH\equiv CH$$

In order to explain the formation of the products of lower yield (HD, CH_4 and C_2D_3H) in the photolysis of C_3H_6–C_3D_6 mixtures, Becker et al. [75] proposed that hydrogen abstraction from propene occurs, and that this is also responsible for the formation of allyl radicals [see reactions (4.10), (4.11) and (4.12)].

Gordon et al. [76] have also studied the photolysis of propene in the gas phase with 147.0 and 123.6 nm photons. Contrary to the study carried out by Becker et al. their investigation was focused on the ion-molecule reactions. As the ionization potential of propene is 9.73 eV, 20% ionization takes place when photons of 123.6 nm (10.0 eV) are used:

$$C_3H_6 + h\nu \longrightarrow C_3H_6^+ + e^-$$

It was shown that at this energy, a small amount of propane is formed with ionic participation. This was indicated by the fact that 5% NO ($IP = 9.25$ eV) caused a considerable decrease in propane yield, which was further diminished by additional NO. However, this effect was not observed with additional amounts of O_2 as additive. This was explained by charge transfer from the ionic precursor of propane to NO:

$$C_3H_6^+ + NO \longrightarrow C_3H_6 + NO^+$$

In the system C_3H_6–O_2 this process cannot occur because the ionization potential of O_2 ($IP = 12.2$ eV) is higher than that of propene. The origin of that part of the propane formed under the effect of ionizing irradiation that is not decreased by O_2 as radical scavenger has been established in experiments carried out with C_3H_6–O_2 mixtures in the presence of an electric field. With increasing electric potential the yield of propane remained the same, whereas those of methane, acetylene and ethylene increased. Although the mechanism of propane formation via ionic species is not completely clear, Gorden et al. have proposed, from their own earlier results [76, 77], that propane is formed in a hydrogen-transfer reaction between propene and the product of the reaction between propene ions and propene molecules:

$$C_3H_6^+ + C_3H_6 \longrightarrow C_6H_{12}^+$$

$$C_3H_6 + C_6H_{12}^+ \longrightarrow C_3H_8 + C_6H_{10}^+$$

The $C_6H_{10}^+$ intermediate was assumed to have a cyclic structure, from energetic considerations. This latter hypothesis, however, was put in doubt by the results of experiments carried out in a mass spectrometer by Abramson and Futrell [78]. According to them, the most probable intermediate of the ion-molecule reaction between the propene parent ion and propene is the 3-hexene ion. However, a comparison between these two experiments seems to be difficult since the experimental conditions (e.g. ion source, pressure) were different.

For the interpretation of the gas phase photolysis products listed, in Table 4.11 the following secondary reactions have been proposed in addition to the primary processes (4.7a)–(4.7f):

$$CH_2 + C_3H_6 \longrightarrow C-(CH_2)_3-CH_3 \longrightarrow \text{butenes}$$

$$CH=CH_2^* + M \longrightarrow CH=CH_2 \qquad (4.9)$$

$$CH=CH_2 + C_3H_6 \longrightarrow C_2H_4 + CH_2-CH=CH_2 \qquad (4.10)$$

$$H + C_3H_6 \longrightarrow H_2 + CH_2-CH=CH_2 \qquad (4.11)$$

$$\longrightarrow \text{iso-}C_3H_7$$

$$CH_3 + C_3H_6 \longrightarrow CH_4 + CH_2-CH=CH_2 \qquad (4.12)$$

$$\longrightarrow C_4H_9$$

$$2\,CH_3 \longrightarrow C_2H_6$$

$$\text{iso-}C_3H_7 + C_3H_6 \longrightarrow C_3H_8 + CH_2-CH=CH_2$$

$$\text{iso-}C_3H_7 + CH_3 \longrightarrow \text{iso-}C_4H_{10}$$

$$CH_3 + CH_2-CH=CH_2 \longrightarrow 1\text{-}C_4H_8$$

$$\text{iso-}C_3H_7 + CH_2-CH=CH_2 \longrightarrow CH_2=CHCH_2CH(CH_3)CH_3$$

Wagner [68] studied in detail the dimer products of the liquid-phase ra-diolysis of propene at $-78°C$ (Table 4.14). On the basis of the relative yields of products, as well as of the isotope distribution of products formed from $1:1$ propene–propene-d_6 mixtures, it was established that alkanes (2-meth-ylpentane, 2,3-dimethylbutane and n-hexane) are formed through re-combination of n-propyl and isopropyl radicals. 4-Methylpent-1-ene is the

Table 4.14

Dimer yields from propene

Temperature	250°C	−76°C
References	70	68
Phase	Gas	Liquid
Products	G, molecule/100 eV	
2,3-Dimethylbutane	1.20	0.054
2-Methylpentane	2.60	0.080
2,3-Dimethylbut-1-ene	4.46	
n-Hexane	1.18	0.022
4-Methylpent-1-ene	2.06	0.375
4-Methylpent-2-ene	2.10	0.055
2-Methylpent-1-ene	49.33	0.033
Hex-1-ene		0.14
cis- and trans-Hex-1-ene	2.21	0.052
Hexa-1,5-diene	0.05	0.17
Other hexadienes	4.32	0.08

product of highest yield among the dimers. Most of these (about 72%) are a result of recombination of isopropyl and allyl radicals, and about one-third are produced by ion-molecule reactions. A similar interpretation is valid for the formation of hex-1-ene and 2-methylpent-1-ene. Hexa-1,5-diene may be formed totally by allyl radical recombination. On the other hand, hex-2-ene may be a product mostly of ion-molecule condensation.

The non-identified C_6-fraction mainly consists of branched-chain dienes in the formation of which allyl radicals may play an important role. By considering the relative amount of the products, and by assuming steady-state concentrations of different radicals and the same collision efficiency for the recombination reactions of the different radicals, Wagner has calculated the yield of propyl, isopropyl and allyl radicals and obtained values of 0.18, 0.29, and 0.58, respectively. It is of interest that the sum of the calculates concentrations of n-propyl and isopropyl radicals is almost equal to that of the concentration of allyl radicals. Two mechanisms of formation for the radicals were suggested:

$$C_3H_6^+ + C_3H_6 \longrightarrow C_3H_7^+ + C_3H_5 \text{ (allyl)} \qquad (4.13)$$

$$C_3H_7^+ + e^- \longrightarrow C_3H_7^* \qquad (4.14)$$

$$H + C_3H_6 \longrightarrow C_3H_7 \qquad (4.15)$$

Reaction (4.13) was first observed in the mass spectrometer by Tal'rose and Lyubimova [79], and later shown by Schissler and Stevenson [80] to have a large cross-section. It is formally an allyl-proton or allyl-hydrogen atom transfer. According to experiments carried out in the mass spectrometer, the cross-section of the proton-transfer reaction is higher by an order of magnitude. Reaction (4.13) has since been proved by other authors, e.g. Henis [81], who has studied the ion-molecule reaction of various alkenes.

The propyl ion, on neutralization, should give the propyl radical [reaction (4.14)]; it may be expected that the formation of isopropyl is preferred. However, Wagner [68] argued for the formation of both isomeric radicals because of the participation of processes passing through high-energy states. This idea would explain the yields quoted above. Concerning the reaction step (4.15), it is well known from work on addition processes that hydrogen addition takes place nearly exclusively to terminal carbon atoms both in the gas phase at 25°C [38] and in the liquid phase at −196°C [83]. The formation of n-propyl radicals cannot be interpreted by this process alone.

In radiolysis, however, the hydrogen atoms formed possess a wide range of energies (from hot atoms to thermal ones) therefore exclusive terminal hydrogen addition cannot necessarily be expected. Addition to the second carbon atom, cannot be excluded, owing to the higher energy. A similar conclusion can be drawn from the photolysis of but-1-ene, as reported by Collin [101] [see Section 4.2.3, reaction (4.18)].

Holroyd and Klein [69] have determined the primary radical yield on the basis of the yields of products labelled with ^{14}C. They used $^{14}CH_3I$ additive in the radiolysis of propene and butenes in the liquid phase (Table 4.15). No product pointing to vinyl radicals was found with propene, so it was assumed that direct C—C bond cleavage in propene did not occur. Methyl radicals were produced probably by a two-step reaction. According to Wagner [68], allyl radicals and propene ions are formed from propene in an ion-molecule reaction and then propyl ions are neutralized and transformed into excited propyl radicals [(4.13)–(4.14)] before decomposing to ethylene and methyl radicals in reaction (4.16):

$$C_3H_7^* \longrightarrow C_2H_4 + CH_3 \qquad (4.16)$$

The fact that Holroyd and Klein [69] did not observe products pointing to vinyl radicals, does not mean the exclusion of the latter, because the vinyl radical has a high reactivity and ethylene can be easily formed via allyl hydrogen abstraction [reaction (4.10)] or acetylene can be produced with hydrogen dissociation [reaction (4.8)]. These processes, however, must be very fast because radicals possessing excess energy can be easily stabilized in the liquid phase [reaction (4.9)]. This limitation is also valid for reaction (4.16).

Table 4.15

Radical yields from liquid aliphatic alkenes (radiation: 2.8 MeV electrons). After Holroyd and Klein [69]

Hydrocarbons	Propene	But-1-ene	cis-But-2-ene	trans-But-2-ene	Isobutene
Temperature, °C	—102	—90	—88	—88	—90
Products	Distribution, %				
Methyl	9.9	2.5	5.0	4.7	6.0
Ethyl		4.4			
n-Propyl	12.3				
Isopropyl	33.0				
n-Butyl		11.5			
sec-Butyl		24.4	27.8	31.3	
Allyl	44.8	4.2			
But-1-enyl		4.9			
But-3-enyl		6.0			
trans-1-Methylallyl		23.8	12.9	61.4	
cis-1-Methylallyl		18.2	54.2	2.6	
2-Methylallyl					62.0
1-Methylvinyl					2.0
Isobutyl					12.0

It is remarkable that Holroyd and Klein (Table 4.15) obtained data for the yield of propyl and allyl radicals that are in agreement with those of Wagner.

Recently, Guillory and Thomas [73] investigated the vacuum UV photolysis of solid propene at 174.5, 163.4, 158, 149.6 and 121.5 nm (7.1, 7.6, 7.9, 8.3 and 10.2 eV, respectively) in argon, nitrogen, carbon monoxide and CO-doped argon matrices at 8–10 K. Although the energy of the photons at 121.5 nm is higher than the ionization potential of propene (9.73 eV), products resulting from ionization processes were not observed. From these results it may be assumed that under this conditions, reactions (4.7b), (4.7c) and (4.7d) take place to the greatest extent. The ratio of these three processes varied with energy only to a small extent: at $\lambda = 121.5$ nm (4.7d) > (4.7c) > (4.7b), at the other wavelengths (4.7d) > (4.7c) $\approx$ $\approx$ (4.7b).

4.2.3. Butene isomers

The radiolysis of n-butene has been investigated in the gas phase by Hummel [85], in the gas and liquid phases by Kaufman [86] and in the liquid phase by Holroyd and Klein [69], Hatano et al. [87, 88] and Horváth and Földiák [71]. The radiolysis of isobutene has been studied in the gas phase mainly by Collin and Herman [90–92], and Herman et al. [93], and in the liquid phase by Holroyd and Klein [69].

Kaufman focused his attention on the products of high molecular mass. However, the results obtained by Hatano and co-workers differ from those of Kaufman in most cases. Tables 4.10 and 4.15 list the yields of primary radicals and compounds formed in the radiolysis of butenes.

The mechanism of hydrogen formation in the radiolysis of but-1-ene and *trans*-but-2-ene in the liquid phase was studied by Hatano et al. [39]. On the basis of isotope distribution in hydrogen formed from mixtures of protonated and perdeuterated butenes, it was established that the bimolecular reaction plays an important role for both isomers, as in the radiolysis of propene and in contrast to the data obtained from the radiation-induced decomposition of ethylene.

Kaufman found that the G-values of hydrocarbon fragments are much higher in the gas than the liquid phase. He therefore proposed that the collision stabilization of intermediates formed with excess energy in ion–molecule condensation and ion-electron recombination processes without decomposition is faster in the liquid than the gas phase.

The radical yield in the liquid-phase radiolysis of butenes was determined by means of the addition of $^{14}CH_3I$ by Holroyd and Klein [69] (Table 4.15): radiolysis of but-1-ene produced methyl, ethyl, sec- and n-butyl, but-1- and -3-enyl as well as *trans*-1- and *cis*-1-methylallyl radicals.

The formation of methyl and allyl radicals can be explained by the cleavage of the relatively weak C—C bond in the β-position with respect to the π-bond:

$$CH_2{=}CH-CH_2CH_3 \longrightarrow C_3H_5 \text{ (allyl)} + CH_3$$

whereas it was assumed that the ethyl radical arises from the n-butyl radical with excess energy [see reactions (4.13) and (4.14) for propene] because the yield of the complementary vinyl radical is very low (G-0.03). Nevertheless, the presence of vinyl radicals, albeit in small amounts, points to the fact that ethyl radicals can also be formed directly:

$$CH_2{=}CH-CH_2CH_3 \rightarrow CH_2{=}CH + CH_2CH_3$$

The methylallyl radicals may arise from an ion-molecule reaction (cf. the radiolysis of but-2-ene). C_8 dimers, formed in high yield in both the gas and the liquid phase, may be the products of ion-molecule condensation processes (see Section 4.4. and Fig. 4.15).

There is a large amount of data available on the quantum yields in the photochemical decomposition of but-1-ene at different photon energies and over a wide range of pressure. The Hg-sensitized decomposition of butene was first investigated by Gunning and Steacie [94], and later by Cvetanovic et al. [95, 96]. The initial steps were determined by Lossing et al. [97] by means of an apparatus incorporating a mass spectrometer. The vacuum UV photolysis of but-1-ene was studied at 184.9 nm (6.7 eV) by Harumya et al. [98] and Borell et al. [99, 100] and at 147 and 123.7 nm (8.4 and 10.0 eV, respectively) by Collin [101] (Table 4.16).

Table 4.16

Yields of the photolysis of gaseous but-1-ene

Energy, eV	6.7	6.7	8.4	10.0
Pressure, mbar	7.9	2.2–164	2.6	1.3
References	99	98	101	101
Products	Relative yields[a]			
Hydrogen		0.17		
Methane	0.25	0.04	0.185	0.28
Ethane	1.00	1.00	1.00	1.00
Ethylene	0.53		0.70	1.44
Acetylene	0.08	0.05	0.63	0.83
Propane	0.18	0.06		
Propene	0.23	0.35[c]	1.26	2.61
Propyne	0.16		0.11	0.22
Allene (propadiene)			1.31	1.11
n-Butene		0.03		
Isobutane	0.02			
cis-But-2-ene	0.08	trace		
trans-But-2-ene	0.03	0.02		
2-Methylbutene	0.09			
Buta-1,2-diene			0.07	0.17
Buta-1,3-diene		0.06		
Isopentane	0.26	0.18		
n-Pentane	0.11			
3-Methyl but-1-ene		0.18		
cis-Pent-2-ene	0.11			
trans-Pent-2-ene	0.08			
Pent-1-ene	0.10	0.11[c]		
C_6 products	1.07	0.58[b,c]		
4-Methylhex-1-ene		0.28		
3-Methylhexa-1,5-diene		0.31		
C_7 products	0.29			
C_8 products	0.08	trace		

[a] Normalized to ethane
[b] Hexa-1,5-diene
[c] 2.2 mbar

The radical intermediates formed during the Hg-sensitized (253.7 nm, 4.9 eV) photolysis of but-1-ene are methyl, allyl and, to a small extent, methylallyl radicals [94, 95]. This indicates that $C-H$ and $C-C$ bonds are both cleaved. Also, they are both β to the double bond.

Table 4.17 lists the primary decomposition processes as well as the quantum yields for the vacuum-UV photolysis of but-1-ene; the data of Borell et al. [99] and of Collin [101] are included. It is clear that the main decom-

Table 4.17

Primary yields of the vacuum-UV photolysis of gaseous but-1-ene

Energy, eV		6.7	8.4	10.0	11.6—11.8
References		99, 100	101	101	101
Reaction		Quantum yields			
(1) $1\text{-}C_4H_8^* \rightarrow C_3H_4$	$+ CH_3 + H$ (CH_4)	0.044	0.38	0.23	0.16
(2) $\rightarrow C_4H_6$	$+ 2 H$ (H_2)		0.32	0.36	0.17
(3) $\rightarrow 2\ C_2H_4$		0.093	0.04	0.05	
(4) $\rightarrow C_2H_3$	$+ C_2H_4 + H$		0.07	0.09	} 0.20
(5) $\rightarrow C_2H_2$	$+ C_2H_5 + H$		0.16	0.12	
(6) $\rightarrow C_2H_2$	$+ C_2H_6$	0.012	0.03	0.02	} 0.33
(7) $\rightarrow C_2H_2$	$+ CH_3$		0.04	0.05	
(8) $\rightarrow C_3H_6$	$+ CH_2$		0.03	0.045	0.05
(9) $\rightarrow C_4H_7$	$+ H$	0.12	≤ 0.05	≤ 0.05	≤ 0.05
(10) $\rightarrow C_3H_5$ (allyl) $+ CH_3$		0.71	0.0	0.0	0.0
Total		0.98	1.12	1.02	0.96

position pathway, as in Hg-sensitized photolysis, is the rupture of $C-C$ and $C-H$ bonds in the β-position. Reaction (10) in Table 4.17 takes place six (according to Harumya et al. [98]) or seven (Borell et al. [100]) times faster than reaction (9). The weakness of the $C-C$ bond β to the double bond is indicated by Collin's experimental result [101] showing that more than 80% of allene formed in the radiolysis of $CH_3CH_2CH=CD_2$ is CH_2CCD_2.

By comparing the data shown in Table 4.16 with those presented in Table 4.10, it can be established that there are significant differences between the yields of low molecular mass products of the radiolysis and the photolysis of but-1-ene, whether the radiolysis is carried out in the gas or the liquid phase. The main difference is in the decomposition products; thus, the main product of photolysis ($\lambda = 184.9$ nm) is ethane, whereas that of gas-phase radiolysis is ethylene. Moreover, radiolysis in both phases gives high yields of hydrogen, but in photolysis the hydrogen yield is very low. A similar trend can be observed for the acetylene yield.

From the data in Table 4.17 it can be established that by increasing the photon energy, the yields of other decomposition reactions (e.g. those producing H_2 and C_2-products) also increase. These processes compete at higher energies with those that predominate at lower energies (184.9 nm), such as reactions (9) and (10) in Table 4.17. On the basis of these observations as well as of the fact that in high-energy radiolysis the main products are ethylene and acetylene, it is reasonable to assume that the selectivity of decomposition decreases with increasing energy, and processes of higher energy of activation become predominant.

Collin [101] determined the relationship between the photon energy and the yield of butyl radical formed in reactions (4.17 and 4.18):

$$H + 1\text{-}C_4H_8 \begin{cases} \longrightarrow \text{sec-}C_4H_9 & (4.17) \\ \longrightarrow \text{n-}C_4H_9 & (4.18) \end{cases}$$

By irradiation with photons with $\lambda = 147.0$, 125.6 and 106.7 nm (energies of 8.4, 10 and 11.6 eV), he found that secondary butyl radicals were formed in amounts of 91, 82 and 63%, respectively. As 94.3% of the thermal hydrogen atoms participate in addition according to reaction (4.17), Collin's conclusion seems to be correct, viz. that by increasing the photon energy, more and more hydrogen atoms possess excess energy and hence undergo reaction (4.18) although this route is energetically less favourable.

The gas-phase γ-radiolysis of *trans*-but-2-ene was studied by Hummel [85] in presence of various additives at 30°C. The main object was to elucidate the mechanism of buta-1,3-diene formation, as this product was not observed in earlier investigations [86]. Hummel obtained a value of $G = 0.13$ for buta-1,3-diene from pure *trans*-but-2-ene, which is very close to the value gained in liquid *cis*-but-2-ene ($G = 0.10$) [87]. The G-value for pure *trans*-but-2-ene increased to 0.2 and 0.42, in methane–C_4H_8 and argon–C_4H_8 mixtures, respectively.

Other trends were observed when N_2O and SF_6 electron acceptors were added to the mixtures: the yield increased to 3.5 in the presence of N_2O but became nil in the presence of SF_6. Other additives (O_2, NO, NH_3, $N(CH_3)_3$) did not essentially affect the yield. On this basis, Hummel regarded the role of radical and ion-molecule reactions as negligible and concluded that the main pathway for butadiene formation is the neutralization of but-2-ene parent ion by an electron:

$$C_4H_8^+ + e^- \longrightarrow C_4H_6 + 2H$$

The significant difference between the effects of the two electron scavengers led Hummel to believe [85] that N_2O inhibits the recombination of electron and parent ion to a lesser extent, as SF_6 is a more effective electron scavenger. On the other hand, Warman's hypothesis [102] suggests the following reactions:

$$N_2O + e^- \longrightarrow N_2 + O^-$$

$$O^- + C_4H_8 \longrightarrow C_4H_7^- + OH$$

$$C_4H_7^- + N_2O \longrightarrow 1,3\text{-}C_4H_6 + N_2 + OH^-$$

Although this mechanism seems adequate to explain this phenomenon, the fact must not be neglected that (see Section 4.2.1) SF_5 radical formed from SF_6 is a very effective radical scavenger [46, 47].

The products formed in the liquid-phase radiolysis of *cis*-but-2-ene in presence of iodine and *p*-benzoquinone result partly from a unimolecular reaction and partly from a radical mechanism [87]. Thus, 60% of methane, 40% of but-1-ene and 95% of n-butane are formed in radical reactions. Hydrogen, acetylene and propene are formed in molecular processes because their yields are unchanged in the presence of radical scavengers. The complexity of the mechanism in the presence of additives is indicated by the fact that *cis–trans* isomerization is not essentially influenced by *p*-benzoquinone whereas it is greatly increased by iodine additive, possibly by a catalytic effect the details of which are not clear.

The radical yields in the liquid-phase radiolysis of *cis*- and *trans*-but-2-enes were determined by means of $^{14}CH_3I$ by Holroyd and Klein [69], and are listed in Table 4.15. Four types of radical only were determined, of which the yield of *cis*- and *trans*-methylallyl is the highest. This points to the preferential C—H bond rupture in the allylic position, as also observed for propene and but-1-ene. Owing to the presence of methyl radicals and the absence of the complementary methylvinyl radicals, the direct rupture of the C—C bond next to the double bond (α-position) is not considered to be a possible route, and the formation of methyl radicals is proposed as follows:

$$\text{sec-}C_4H_9^* \longrightarrow CH_3 + C_3H_6$$

This reaction is similar to (4.16), and the formation of the sec-$C_4H_9^*$ excited radical is presumably analogous to reactions (4.13)–(4.15). There is a 'memory' effect, found in connection with methylallyl radicals, i.e. in most cases the original skeleton is retained. The configuration of these radicals is 80–96% the same as that of the original molecule [69].

There have been several investigations of the Hg-sensitized [103–107] and the direct high-energy [108–111] photolysis of *cis*- and *trans*-butenes. Borrel and James [111] observed 32 hydrocarbon products in the direct photolysis of *cis*- and *trans*-but-2-enes at 184.9 nm (6.7 eV).

The products with the most important yields are shown in Table 4.18 and the primary decomposition reactions together with the calculated quantum yields are as follows [111]:

$$\text{cis- or trans-2-}C_4H_8 + h\nu \longrightarrow CH_3-CH=CH-CH_2 + H \qquad \Phi = 0.6$$

$$\longrightarrow HC=CH + 2\,CH_3 \qquad \Phi = 0.23$$

$$\longrightarrow CH_2=CH-CH=CH_2 + H_2 \qquad \Phi = 0.11$$

$$\text{cis-2-}C_4H_8 + h\nu \longrightarrow \text{trans-2-}C_4H_8 \qquad \Phi = 0.1$$

$$\text{trans-2-}C_4H_8 + h\nu \longrightarrow \text{cis-2-}C_4H_8 \qquad -$$

$$\text{cis- or trans-2-}C_4H_8 + h\nu \longrightarrow CH_3-CH=CH + CH_3 \qquad \Phi = 0.085$$

$$\longrightarrow CH_2=C=CH_2 + CH_4 \qquad \Phi = 0.07$$

$$\longrightarrow CH_3-C\equiv C-CH_3 + H_2 \qquad \Phi = 0.007$$

$$\longrightarrow 2\,C_2H_4 \qquad \Phi = 0.01$$

Table 4.18

Yields of the photolysis of gaseous *cis*-but-2-ene

Energy, eV	4.9	6.0–6.1	6.7	8.9
Pressure, mbar	158	28	20	11
Reference	110	257	111	108
Products	Relative yields[a]			
Hydrogen	1.67		0.9	
Methane	0.94		0.4	1.35
Ethane	1.00	1.04	0.65	1.75
Ethylene			0.1	0.6
Acetylene	0.32		1.7	1.5
Propane				1.0
Propene	1.00	1.00	1.00	1.00
Methylacetylene		0.19		0.4
Allene (propadiene)			0.35	0.6
n-Butane	0.39	0.46	0.85	
But-1-ene	1.43	2.2	1.5	
trans-But-2-ene	0.48	1.45	0.6	
Isobutene		0.22		
Buta-1,3-diene		0.19	0.8	
Buta-1,2-diene		0.14		
But-1-yne			0.05	
n-Pentane		0.23	0.05	
Isopentane		1.81	1.35	
trans-Pent-2-ene		0.98	0.6	
cis-Pent-2-ene		0.88	0.55	
3-Methylbut-1-ene		0.87	0.55	
C_6 products			0.15	
C_7 products			0.3	
C_8 products			0.5	

[a] Normalized to propene

The photolysis of *cis*-but-2-ene in the range of 180–200 nm (6.2–6.9 eV) was studied by Cundall et al. [109] and hydrogen, methane, ethylene, propene, but-1-ene and isopentane were found as decomposition products. At low pressures decomposition dominates isomerization: the ratio of isomerization to decomposition increases with increasing pressure.

Hatano and Shida [110] have studied the Hg-sensitized photolysis of *cis*-but-2-ene in the gas phase at 253.7 nm (4.8 eV) and the direct photolysis at 184.9 nm (6.7 eV). They established that in the Hg-sensitized photolysis, *trans*-but-2-ene was predominantly formed from *cis*-but-2-ene, whereas in the direct photolysis the yield of decomposition product is high compared with that of isomerization. Hatano and Shida [110] concluded that direct photolysis forms excited singlet states whereas Hg-sensitization produces excited triplet states. The photolysis data were compared with those obtained in the liquid-phase radiolysis of *cis*-but-2-ene [87]. It was concluded that the formation of geometrical isomers is promoted by the excited triplet state in the case of radiolysis, whereas decomposition occurs mainly through the excited singlet state.

Recently, Collin et al. [108, 112] published interesting data on the isomerization of but-2-enes in radiolysis and photolysis. But-2-enes were irradiated in the presence of different additives, and a surprisingly high extent of isomerization was found in the presence of some compounds containing sulphur, such as H_2S and methyl and isopropyl mercaptans. Oxygen-free *cis*-but-2-ene was largely isomerized in the presence of these additives: $G(trans\text{-but-2-ene}) = 9 \cdot 10^4$ [112]. In presence of other compounds containing sulphur (CS_2, SF_6, etc.) isomerization occurred only to a small extent. It is also remarkable that the extensive isomerization promoted by H_2S is blocked by conjugated dienes: this isomerization should be the result of a chain reaction initiated most probably by thiyl (SH) radicals. Thiyl radicals result from the abstraction reaction of other radicals present, as most of the energy is absorbed by but-2-ene:

$$H_2S + R \longrightarrow RH + SH$$

$$SH + cis\text{-}2\text{-}C_4H_8 \rightleftharpoons C_4H_8SH$$

$$C_4H_8SH \rightleftharpoons trans\text{-}2\text{-}C_4H_8 + SH$$

The radiolysis of isobutene in most cases was investigated usually in order to clarify the mechanism of polymerization, and only a few publications deal with the decomposition products [69, 90–92].

Details of the gas-phase radiolysis of isobutene were investigated by Collin and Herman [90–92]. The main decomposition products are listed in Tables 4.19 and 4.20. The formation of acetylene, ethylene, propyne and allene is

Table 4.19

Yields from gaseous isobutene (radiation: ^{60}Co-γ). After Herman and Collin [91]

Pressure, mbar	Additive	Hydrogen	Methane	Ethane	Ethylene	Acetylene	Propyne	Allene	Isobutane
					Relative yields[a]				
63 ± 4		1.3	0.76	0.81	0.60	0.54	0.99	1.01	2.09
	6% O$_2$		0.35	0.75	0.60	0.53	0.97		0.82
	15% O$_2$			0.79		0.48	0.80		0.80
	6% NO		0.28	0.76	0.59	0.54	0.98		0.39
	15% NO		0.28	0.79	0.60	0.43			0.30
	30% NO			0.75	0.59	0.54	0.95		0.31
	10% NH$_3$		0.78	0.69	0.57	0.49	0.97		1.43
	8% NH$_3$ + 4% O$_2$		0.32	0.68	0.58	0.52	0.96		0.01
128 ± 7		1.36	0.54	0.51	0.41	0.40	0.74	0.76	1.44
	6% O$_2$		0.28	0.47	0.40	0.37	0.74		0.50
	6% NO		0.20	0.46	0.39	0.38	0.71		0.22
	6% NH$_3$	1.32							

[a] Yields are normalized to G(allene) 1.01 ± 0.05 at 63 mbar, and 0.76 ± 0.05 at 128 mbar

Table 4.20

Pressure dependence of yields from gaseous isobutene. After Herman [92]

Pressure, mbar	8	16	32	62	125
Products	G, molecule/100 eV[a]				
Allene (propadiene) {	(1.00)	(1.00)	1.25 (1.00)	1.01 (1.00)	0.76 (1.00)
2-Methylbut-1-ene {	(0.64)	(0.70)	0.91 (0.73)	0.78 (0.77)	0.55 (0.72)
2-Methylbut-2-ene {	(0.23)	(0.30)	0.36 (0.29)	0.26 (0.26)	0.17 (0.23)
-1,1-Dimethylcyclo- propane {	(0.012)	(0.052)	0.08 (0.061)	0.07 (0.073)	0.07 (0.087)
3-Methylbut-1-ene {			0.02 (0.013)	0.01 (0.012)	0.007 (0.009)
Neopentane {	(0.53)	(0.46)	0.43 (0.34)		0.11 (0.015)
Isopentane {	(0.14)	(0.23)	0.33 (0.26)	0.31 (0.31)	0.31 (0.41)
n-Pentane {		(0.28)	0.04 (0.33)	0.04 (0.38)	0.02 (0.031)
Penta-1,4-diene {				0.02 (0.031)	0.02 (0.026)

[a] The relative yields are in parentheses

not affected by radical scavengers, but the yields of methane and butane are strongly influenced and that of propene is affected to a smaller extent. The presence of ammonia does not change the yield of most of the products, but it decreased the G-value of iso-C_4H_{10} to a lesser extent than that of propene. The formation of isobutane is almost completely suppressed in the presence of NH_3 and O_2. Methane is formed by radical processes and by unimolecular elimination to almost equal extents.

The following facts show that acetylene and ethylene were formed via common precursors. They are produced in nearly the same amount and their yields are not changed by radical scavengers or ammonia; acetylene and ethylene consist of mainly perprotonated and perdeuterated molecules if a 1 : 1 mixture of perprotonated and perdeuterated isobutene is irradiated. The proposed mechanism is:

$$(CH_3)_2CCH_2^* \longrightarrow H_2 + \left(CH_2=C \begin{matrix} CH_3 \\ \\ CH \end{matrix} \right)^*$$

Yields of the photolysis of gaseous isobutene.

Energy, eV	Scavengers	Hydrogen	Methane	Ethane	Ethylene
8.4	None		0.113	0.27	0.017
	O_2		0.017	0.00	0.019
	NO	0.16	0.016	0.00	0.016
10.0	None		0.115	0.16	0.024
	O_2		0.022	0.00	0.026
	NO (5%)	0.23	0.027	0.00	0.022
11.6–11.8	None		0.120	0.16	0.035
	O_2		0.028	0.00	0.033
	NO (5%)	0.28	0.027	0.00	0.034

[a] M/N_{ex} represents molecules formed per neutral excited molecule which is not autoionized or does not dissociate
[b] Numbers in parentheses represent the corresponding M/N_i- values (pressure: 6.6 mbar); M/N_i stands for mo-

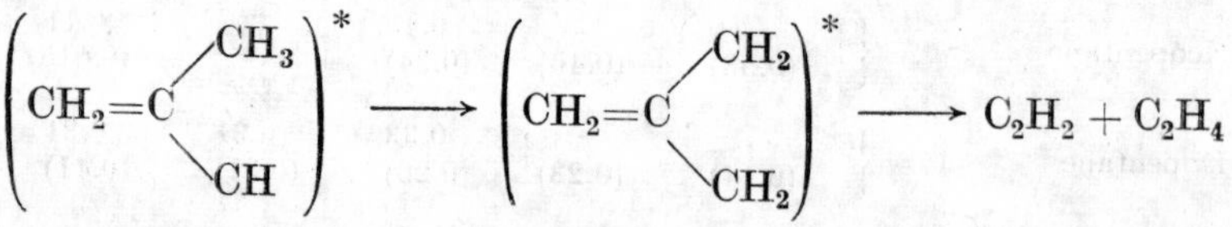

$$\left(CH_2 = C \begin{array}{c} CH_3 \\ CH \end{array}\right)^* \longrightarrow \left(CH_2 = C \begin{array}{c} CH_2 \\ CH_2 \end{array}\right)^* \longrightarrow C_2H_2 + C_2H_4$$

Allene and propyne are formed mainly from excited isobutene in a uni-molecular reaction:

$$\text{iso-}C_4H_8^* \nearrow H_2C = C = CH_2 + \begin{cases} H + CH_3 \\ \text{or} \\ CH_4 \end{cases}$$
$$\searrow HC \equiv C - CH_3 + \begin{cases} H + CH_3 \\ \text{or} \\ CH_4 \end{cases}$$

This is supported by experiments carried out in presence of an electric field [91] and those with 1 : 1 mixtures iso-C_4H_8 and iso-C_4D_8 [91, 93]. According to these latter, there is also a contribution to the formation of allene and propyne from ion-molecule reactions:

$$C_3H_3^+ + \text{iso-}C_4D_8 \longrightarrow C_3H_3D + C_4D_7^+ \tag{4.19}$$

The ratio of propene to propyne (Table 4.19) is much higher in γ-radiolysis than in photolysis (Table 4.21). This points to propene being formed via highly excited isobutene. This seems to be proved by the fact that in photolysis the propene yield drops with increasing wavelength (Table 4.21). The following reaction may explain the formation of propene from excited iso-butene:

$$\text{iso-}C_4H_8^* \longrightarrow C_3H_6 + CH_2$$

4.21

After Herman et al. [93]

Acetylene	Propene	Propadiene	Propyne	n-Butane	Isobutane	
Product yields, $M/N_{ex}{}^a$						
0.035	0.024	0.51	0.35	0.007	0.15	
0.036	0.027	0.52	0.35	0.00	0.00	
0.034	0.020	0.50	0.33	0.00	0.00	
0.040	0.042	0.52	0.34		0.141	$(0.39)^b$
0.065	0.042	0.54	0.34		0.110	(0.31)
0.055	0.036	0.53	0.31		0.076	(0.21)
0.079	0.052	0.46	0.295		0.230	(0.51)
0.073	0.055	0.44	0.32		0.130	(0.29)
0.079	0.040	0.45	0.31		0.090	(0.20)

lecules formed per ion pair to form an ion pair

However, the energy dependence of the propene yield can be rationalized in another way, namely, by the fact that e.g. ion-molecule reactions occur at higher energies:

$$C_3H_5^+ + iso\text{-}C_4D_8 \longrightarrow C_3H_5D + C_4D_7^+$$

In this process, as in reaction (4.19), hydride ion transfer may play a role.

The isobutane yield was depressed by NO to a much larger extent than by oxygen (see Tables 4.19 and 4.21). It was assumed [91–93] that besides its the radical-scavenging ability, NO is able to accept positive charge from the parent ion:

$$iso\text{-}C_4H_8^+ + NO \longrightarrow iso\text{-}C_4H_8 + NO^+$$

However, this process was proved to be very slow by mass spectrometric experiments [113], so there may be another reason, e.g. the isomerization of isobutyl ions into tert-butyl ions induced by collision with NO, as suggested by Herman et al. [93]:

$$(CH_3)_2CHCH_2^+ + NO \longrightarrow (CH_3)_3H^+ + NO$$

Isobutane is not formed in significant amounts in the reaction of tert-butyl ion with isobutene, as was established in the radiolysis of neopentane-isobutene mixtures [114, 115]. The main reaction is addition of tert-butyl ion to monomer molecule leading to chain polymerization.

In considering the role of NO a process observed in the mass spectrometer should not be neglected [261]. This is the formation of $C_4H_8NO^+$ from butylene and NO, which may lead to stable butene molecules at neutralization.

The effect of additives suggests that 60% of isobutane is formed via radicals, and 40% in ionic processes (Table 4.19). It was assumed by Herman et al. that the precursors of the radical process are tert-butyl radicals formed

19

by scavenging thermal hydrogen atoms, whereas for the ionic species is considered to be the isobutyl cation [91, 93]:

$$\text{iso-}C_4H_8^+ + \text{iso-}C_4H_8 \longrightarrow C_4H_9^+ + C_4H_7$$

$$C_4H_9^+ + \text{iso-}C_4H_8 \longrightarrow \text{iso-}C_4H_{10} + C_4H_7^+$$

Herman has determined some condensation products with low boiling points in the gas-phase radiolysis of isobutene (Table 4.21). All products except 2–methylbut–2–ene [92A] listed in the Table arise from radical processes (abstraction, addition, combination), as they were completely scavenged by 6% NO. Pentanes were supposed to be formed in the reaction of methyl radical with isobutene or with tert-butyl radicals, and pentenes are produced when methylene radicals react with isobutene:

$$(CH_3)_2C{=}CH_2 + CH_3 \longrightarrow (CH_3)_2CCH_2CH_3$$

$$(CH_3)_2CCH_2CH_3 \xrightarrow{\ (CH_3)_2CCH_2\ } \begin{array}{c} \text{iso-}C_5H_{12} \\ \\ \text{n-}C_5H_{12} \end{array}$$

$$(CH_3)_3C + CH_3 \longrightarrow \text{neo-}C_5H_{12}$$

The dimerization of gaseous isobutene induced by γ-radiation at high temperatures has been studied by Naito [115A]. The yields of liquid products were determined at temperatures between 30 and 400°C. In the lower temperature range, products of many kinds were formed, but their yields were small. The yields, in general, increased with the temperature; the largest increase was observed in the yield of 1,1,3-trimethylcyclopentane dimer. At 300°C this dimer is formed with a G-value of 50, while the other dimer, that of 2,4,4-trimethylpent-2-ene, is formed with a G-value of 5. Using various scavengers, it has been proved that the major dimer is formed by radical and the minor by ionic mechanism. The yield of the major dimer product was inversely proportional to the square-root of the absorbed dose rate and directly proportional to the density of isobutene. On this basis, Naito concluded that the major dimer was formed via a chain reaction with the participation of the 2-methylallyl radical. As the precursor of the other dimer the tert-butyl cation was suggested.

Collin has recently published valuable data [115] on the reactions of tert-$C_4H_9^+$ and iso-$C_4H_8^+$, which may be of interest with respect to the radiolysis and polymerization of isobutenes. iso-$C_4H_8^+$ ions formed in neopentane radiolysis react with isobutene in the liquid phase at least as fast as with methylcyclopentane; on the other hand, tert-butyl ion react with isobutene at least one order of magnitude faster and the main reaction is a proton transfer. In the liquid phase less than 5% of tert-$C_4H_9^+$ ions is added to isobutene, whereas in the gas phase this is the predominant reaction as proton transfer amounts to only 30% of the overall reaction: $k_{addition}/k_{transfer} = 2.4$. Deuterium transfer is about three times slower than proton transfer. The main addition reactions are the following:

$$(CH_3)_3C^+ + \text{iso-}C_4H_8 \longrightarrow C-\underset{\underset{C}{|}}{\overset{\overset{C}{|}}{C}}-C-C^+\diagup^C_{\diagdown C}$$

$$C-\underset{\underset{C}{|}}{\overset{\overset{C}{|}}{C}}-C-C^+\diagup^C_{\diagdown C} + \text{iso-}C_4H_8 \longrightarrow (CH_3)_3C^+ + C-\underset{\underset{C}{|}}{\overset{\overset{C}{|}}{C}}-C=C-\underset{\underset{C}{|}}{C}$$

$$(CH_3)_3C^+ + \text{iso-}C_4H_8 \longrightarrow C-\underset{\underset{C}{|}}{\overset{\overset{C}{|}}{C}}-\underset{\underset{C}{|}}{\overset{\overset{C}{|}}{C}}-C^+; \quad C-\underset{\underset{C}{|}}{\overset{\overset{C}{|}}{C}}-\underset{\underset{C}{|}}{\overset{\overset{C}{|}}{C}}^+-C-C$$

Radical intermediates were determined using $^{14}CH_3I$ as additive by Holroyd and Klein [69] in liquid isobutane at -90 °C. On the basis of labelled compounds formed, 6% methyl, 2% 1-methylvinyl-, 18% tert-butyl and 62% 2-methylallyl radicals were present. As the yield of isobutyl radical compared with that of tert-butyl is higher than expected, Holroyd and Klein feel it unlikely that isobutyl radicals are formed by addition, although Falconer [116] had assumed that it results from hydrogen atom addition in the spur. The probability of isomerization of tert-butyl into isobutyl radicals is very low, even if the activation energy requirement is fulfilled, because the more favourable reverse process has not occurred either to a measurable extent [114]. It was assumed that both radicals are formed via neutralization of $C_4H_9^+$. In addition, the form of $C_4H_9^+$ should be considered as the equilibrium mixture of tert-butyl and isobutyl ions. In mass spectrometric experiments it has always been considered [117–119] since the pioneering work of Tal'rose and Ljubimova [79] that tert-butyl cations are formed in the following reaction:

$$C_4H_8^+ + C_4H_8 \longrightarrow C_4H_9^+ + C_4H_7$$

Only a few works are available dealing with the decomposition and ion-molecule reactions of ions formed in the radiolysis of butenes. Sieck et al. [120] compared the reactivities of ions produced on the photoionization (10.0–10.6 and 11.6–11.8 eV) of butene isomers. It was shown that but-1-ene and *cis*-but-2-ene ions are isomerized to a large extent to *trans*-but-2-ene ions whereas isobutene ions undergo this change only to a small extent. The degree of isomerization increases with increasing photon energy. In the presence of an inert gas the rates of decomposition and isomerization decrease with increasing pressure. The following main reactions have been found:

$$(C_4H_8^+)^* \longrightarrow \begin{array}{l} C_4H_7^+ + H \qquad\qquad (4.20a) \\ C_3H_5^+ + CH_3 \qquad\quad (4.20b) \\ C_3H_4^+ + CH_4 \qquad\quad (4.20c) \end{array}$$

For but-1-ene ions reaction (4.20b) is predominant, and for but-2-ene and isobutene ions reactions (4.20a) and (4.20b) are of almost equal importance; the decomposition (4.20c) is of minor importance in all cases.

The rate of ion-molecule condensation:

$$C_4H_8^+ \text{ (thermal)} + C_4H_8 \longrightarrow (C_8H_{16}^+)^* \qquad\qquad (4.21)$$

as well as the ratio between reaction (4.21) and decomposition of $(C_8H_{16}^+)^*$ to C_4-C_7 ions, are controlled by the molecular structure. Other ion-molecule reactions of butene ions competing with reaction (4.21), have been studied by Koyano [117] and Fuchs [121] by the method of photoionization and electron-collision mass spectrometry. The cross-sections of reactions of isomeric butene ions with different carbon chains listed in Table 4.22 incorporate the data given in these two papers. The absolute and relative reaction rates of ions of different energies produced by the two methods are different. Further study of the mechanisms involved would be very useful for a better undertanding of radiolysis of butenes.

The study of the radiation-induced polymerization of isobutene was of particular interest from the theoretical point of view, because a polymer of high molecular mass can otherwise (e.g. catalytically) be produced from isobutene only by cationic initiation. Hence it is not surprising that in the history of radiation polymerization the first cationic mechanism described was the polymerization of isobutene. In these early experiments, especially above 0°C, polymers of only small molecular mass were obtained [127–129]. However, at low temperatures (−78°C), rubber of high molecular mass was formed [123, 130]. Recently, Okamoto et al. [131] have shown that if the monomer is pure and dry, polymer of high molecular mass

Table 4.22

Cross sections of ion-molecule reactions of butene ions

Hydrocarbons	But-1-ene		But-2-enes		Isobutene	
Reference	117[a]	120[b]	117[a]	120[b]	117[a]	120[b]
Pressure, mbar	$1.3-13 \cdot 10^{-6}$	0.1–4	$1.3-13 \cdot 10^{-6}$	0.1–4	$1.3-13 \cdot 10^{-6}$	0.1–4
Reaction	Cross-section, 10^{-20} m^2 molecule^{-1}					
$C_4H_8^+ + C_4H_8^+ \rightarrow C_4H_7^+ + C_4H_9$	11.1		8.3		3.3	
$\rightarrow C_4H_9^+ + C_4H_7$	7.2		3.1		21.0	
$\rightarrow C_5H_9^+ + C_3H_7$	4.7	5	2.4	3.7	3.4	6.4
$\rightarrow C_5H_{10}^+ + C_3H_6$	8.9	8				
$\rightarrow C_5H_{11}^+ + C_3H_5$	1.6					
$\rightarrow C_6H_{11}^+ + C_2H_5$	4.7	2.4				
$\rightarrow C_6H_{12}^+ + C_2H_4$	2.1	1.3				
$\rightarrow C_5H_8^+ + C_3H_8$		0.6		0.4		0.8

[a] Photoionization mass spectrometer, energy: 10.0 eV
[b] Electron impact mass spectrometer

($M = 6000$) is formed in the gas phase at 790 mbar pressure with a significant yield [$G(-C_4H_8) > 1000$]. The rate of polymerization decreased to a value of $G(-C_4H_8) < 100$ in the presence of 1 mM ammonia, and this clearly points to a cationic mechanism for the polymerization. On the other hand, a retarding effect was also exerted by NO and N_2O, and therefore the presence of radicals cannot be completely excluded. This is supported by the experiments carried out by Tabata et al. [132] and Davison et al. [123]. However, in these investigations, insufficient attention was paid to purification procedures, e.g. drying; hence there was only a small cationic contribution owing to chain termination caused by trace amounts of water.

By careful purification of isobutene, Taylor and Williams [133] succeeded in obtaining a high yield of polymerization [$G(-C_4H_8) = 6.8 \cdot 10^6$], which has not been achieved before at 0°C. The rate constant of propagation and energy of activation were $1.5 \cdot 10^8$ M^{-1} s^{-1} and 13 kJ mol^{-1}, respectively. On the basis of numerous experimental data, it was unambiguously established that the polymerization of isobutene initiated by both γ-rays and high-energy electrons under extremely high purity condition occurs by an ionic mechanism even at 0–20°C. In the course of chain initiation various radicals, particularly methylallyl, are also present [117, 134, 135], but even in this case the ionic process becomes predominant because the rate of propagation is an order of magnitude higher with cations than with radicals, and because of the low energy of activation of the cationic process. Both the tert-butyl cation and the methylallyl radical are stable intermediates owing to the allyl resonance as well as the hyperconjugation effect of methyl groups:

$$CH_3-\underset{\overset{|}{CH_3}}{C}=CH_2 \; \rightsquigarrow \; CH_3-\underset{\overset{|}{CH_3}}{\overset{+}{C}}-\dot{C}H_2 + e^-$$

$$CH_3-\overset{+}{\underset{\underset{CH_3}{|}}{C}}-\dot{C}H_2 + CH_3-\underset{\underset{CH_3}{|}}{C}=CH_2 \Bigg\langle \begin{array}{l} CH_3-\overset{+}{\underset{\underset{CH_3}{|}}{C}}-CH_3 + CH_2=\underset{\underset{CH_3}{|}}{\dot{C}}-CH_2 \quad (4.22) \\[2em] CH_3-\overset{+}{\underset{\underset{CH_3}{|}}{C}}-CH_2-CH_2-\underset{\underset{CH_3}{|}}{C}-\dot{C}H_3 \quad (4.23) \end{array}$$

Mass spectrometric experiments [118] indicate reaction (4.22) to be important in the gas phase, whereas the liquid phase the contribution of reaction (4.23) to the initiation process seems to be considerable, a proposal supported by ESR measurements [117, 135, 136].

4.2.4. Pentene isomers

Basic research on the radiolysis of pentenes has been carried out by Wagner [137, 138]. He dealt mainly with the formation of products of high yield such as those of dimerization and isomerization. Cserép's data [139] on fragment products (Table 4.23) and data obtained from the photolysis of pent-1-ene with different photon energies, are listed. Although a comparison with respect to the energy dependence is not possible owing to the entirely different experimental conditions, it seems useful to present the data in order that trends may be discussed. The higher the energy of photons the more similar are the distributions of products of small molecular weight resulting from the photolysis and radiolysis. Thus it may be assumed that most of the decomposition schemes suggested by Perrin and Collin [140] for photolysis with photons of 8.4 and 10.0 eV are also valid for radiolysis (Table 4.24).

In radiolysis, of course, ionization takes place to a larger extent and the reactions of excited ions must also be considered. Thus, e.g. for a part of the n-pentane yield, the following ion-molecule reaction should be taken into account:

$$C_5H_{10}^+ + C_5H_{10} \longrightarrow C_5H_8^+ + C_5H_{12}$$

This reaction has a large cross-section [81, 117, 141], and it therefore seems peculiar that the n-pentane yield is smaller with 10.0 eV photons than with 8.4 eV photons, whose energy is below the ionization potential (see Table 4.23).

From the data of Perrin and Collin [140] (see Table 4.23), it can be seen that in some cases there is a considerable difference between the results obtained using 8.4 and 10.0 eV photons, but generally the product distributions are similar. Nevertheless, the data from Hg-sensitized photolysis at a much lower energy (4.8 eV) show significant differences (see Table 4.23 and refs [142, 143]).

Dimer, trimer and tetramer products formed in the radiolysis of pent-1-ene in presence of various additives provide information on the reaction mech-

Table 4.23

Yields of the photolysis and radiolysis of pent-1-ene

Radiation	^{60}Co-γ	Photolysis		
Energy, eV	$1.25 \cdot 10^6$	4.8	8.4	10.0
Pressure, mbar	—	20	0.5	1.7
Reference	139	142	142	140
Phase	liquid	gas		
Products	relative yields[a]			
Hydrogen	$(0.83)^e$			
Methane	0.47		0.35	0.17
Ethane	0.4		0.18	0.31
Ethylene	1.26	0.43	1.43	1.55
Acetylene	0.8	0.19	0.18	0.35
Propane	0.3		0.25^b	0.33^b
Propene	1.00	1.00	1.00	1.00
	$(0.15)^e$		$(0.29)^c$	$(0.29)^c$
Allene (propadiene)	0.13		0.46	0.65
n-Butane		0.057	0.32	0.20
But-1-ene	0.26	0.023		
Butenes + allene			0.57^d	0.53^d
n-Pentane	2.66	0.127	0.11	0.062
Pent-2-ene		0.061		
Methylcyclobutane		0.046		
Penta-1,3 and 1,4-dienes			0.11	0.068
Hexa-1,5-diene		0.045		
Hex-1-ene		0.056		

[a] Normalized to propene
[b] Propane + propyne
[c] Quantum yields
[d] Buta-1,3-diene + but-1-yne + but-1-ene + isobutene
[e] G, molecule/100 eV

Table 4.24

Yields of the vacuum-UV photolysis of gaseous pent-1-ene. After Perrin and Collin [140]

Energy, eV	8.4	10.0
Reaction	Quantum yield	
$(C_5H_{10})^* \rightarrow C_5H_8 + H_2$ (2H)	0.06 ± 0.01	0.02
$\rightarrow C_4H_6 + CH_3 + H$ (CH$_4$)	0.15 ± 0.01	} 0.10
$\rightarrow C_4H_8 + CH_2$	0.01 ± 0.01	
$\rightarrow C_3H_4 + C_2H_5 + H$ (C$_2$H$_6$)	0.14 ± 0.01	0.21
$\rightarrow C_2H_2 + C_3H_7 + H$ (C$_3$H$_8$)	0.05 ± 0.01	0.09
$\rightarrow C_2H_4 + C_3H_5 + H$ (C$_3$H$_6$)	0.26 ± 0.01	
$\rightarrow C_3H_6 + H + C_2H_3$ (C$_2$H$_4$)	0.02 ± 0.02	} 0.43
$\rightarrow C_2H_4 + CH_3 + C_2H_3$	0.16 ± 0.02	
Total	0.85 ± 0.10	0.85

anism [137], but only the most important component of the dimer products, dec-3-ene was identified.

The yields of different dimers are influenced by changes of temperature in the range 77–300 K to different extents: the decrease in the yield of dec-3-ene with decreasing temperature is only small, whereas the other dimers are strongly affected. The absence of radical processes was deduced by Wagner from the relatively small temperature dependence, as well as from radical scavenger studies that showed no effect (tetranitromethane, benzoquinone). From a consideration of these and earlier results, he proposed that the formation of dec-3-ene is an ionic process.

The yield of other dimers, which amounts to about 70–80% of the total dimer yield, is strongly influenced by temperature and by radical scavengers.

Intermediate products of solid *cis*-pent-2-ene γ-radiolysis at 77 K have been studied by Owczarczyk and Stachowicz [143A]. Three distinct absorption maxima have been observed at 262, 310 and 610 nm. On the basis of scavenger studies, thermal annealing, photobleaching and ESR measurements, it has been established that the long wavelength maximum at 610 nm belongs to the parent radical cations formed in the primary ionization process. The maximum at 310 nm has been ascribed to pentallyl anion, the production of which was attributed to an electron capture process by the pentallyl radical; this latter intermediate is produced in relatively large amounts ($G = 1.3$) and has an absorption band maximum at 262 nm.

The radiolysis of isopentenes at 77 K with 3 MeV bremsstrahlung generated by a van de Graaff generator was also studied by Wagner [138]. In these experiments both the dose rate ($1.7 \cdot 10^{18}$ eV g^{-1} s^{-1}) and dose

Table 4.25

Yields from solid isopentenes (radiation: 3 MeV bremsstrahlung; temperature: 77 K). After Wagner [137]

Hydrocarbons	3-Methylbut-1--ene	2-Methylbut-1--ene	2-Methylbut-2--ene
Products	*G*, molecule/100 eV		
Hydrogen	0.58	0.62	1.07
Methane	0.07	0.03	0.09
C$_2$-products	0.25	0.1	0.00
C$_3$-products	0.28	0.06	0.00
C$_4$-products			
Methylbutanes	0.45[a]	0.29[b]	0.44[b]
3-Methylbutenes	0.17		
Isoprene	0.12	0.18	0.19
Methylbutenes	0.23[c]	0.66[c]	0.78[d]
Pent-2-enes	0.07	0.03	0.04
1,1-Dimethylcyclopropane	0.1	0.01	0.05
trans-1,2-Dimethylcyclopropane		0.07	0.09
Dimers	1.20	1.14	1.82

[a] 3-Methylbutane
[b] 2-Methylbutane
[c] 3-Methylbut-2-ene
[d] 2- and 3-Methylbut-1-ene

$(300–600 \cdot 10^{19}$ eV g$^{-1})$ were very high. The data are presented in Table 4.25. The products of lower molecular mass were identified only by their carbon number. It appears that the main products are hydrogen and C_5 hydrocarbons [alkanes $(G = 0.3–0.5)$, isomeric alkenes $(G = 0.2–0.8)$], alkynes and dialkenes formed by hydrogen dissociation $(G = 0.2–0.3)$ and dimers $(G = 1.1–1.8)$. Skeletal isomers (such as straight-chain cis-, and trans-pent-2-enes and alkylcyclopropanes) are formed in low yields; the suggested mechanism is as follows [138]:

$$\left(\begin{array}{c} C \\ | \\ C-C-C-C \end{array} \right)^+ \longrightarrow \left(\begin{array}{c} C \\ C-C \diagdown \diagup C-C \end{array} \right)^+ \xrightarrow{e^-} C-C-C-C-C \quad (4.24)$$

As far as the fragment products and the pathway of decomposition of isopentenes are concerned, no conclusions can be drawn from Wagner's data. Therefore we have to rely upon the vacuum photolysis experiments.

On the basis of flash photolysis of 2-methylbut-1-ene, Bayrakceken et al. [144] showed that the most sensitive part of the molecule is the C—C bond in the β-position to the double bond. The product distribution shows that the $(C-C)_\beta : (C-H)_\beta : (C-C)_\alpha$ ratio for bond rupture is $13 : 1.37 : 1$, which reflects the respective bond dissociation energies (290, 345 and 377 kJ mol^{-1}). The main reaction in the energy range 5.9–7.3 eV is the radical decomposition of excited molecules; isomerization and decomposition into molecular products are of minor importance. It was found that methyl and β-methallyl radicals disappear in second order reactions. The rate constant for methallyl combination is $(2.6 \pm 0.3) \cdot 10^{10}$ M^{-1} s^{-1} at 295 K [144].

Similar results were obtained by Gagnon et al. [145] in experiments over the energy range 8.4–11.8 eV. The main products of photolysis are allene, isoprene and ethylene, as well as methyl, ethyl and tert-pentyl radicals; the last is formed by addition of hydrogen atoms to 2-methylbut-1-ene. In the presence of radical scavengers the k_d/k_c values for methyl and for tert-pentyl radicals were determined, and these values (1.6 ± 0.2) are in excellent agreement with data from other sources. The formation of 2-methylbut-2-ene in experiments using a light of short wavelength was attributed to ionic isomerization.

There is a similar explanation for the formation of 2-methylbut-2-ene and 2-methylbut-1-ene in the 10.0 eV photolysis of 3-methylbut-1-ene [146]. The quantum yield of C_5-isomer formation becomes important at photon energies higher than the ionization potential of the irradiated alkene. However, contradictory data can also be found [142]: although the energy available is much lower than the ionization potential of the alkene, the main products of the Hg-sensitized photolysis of 2-methylbut-1-ene are still 2-methylbut-2-ene and 3-methylbut-1-ene isomers.

The main products of the vacuum-UV photolysis of 2-methylbut-2-ene are isoprene, buta-1,3-diene, propyne, allene and ethylene [148]. According to Collin and Gaucher [148] the main pathways for decomposition are as follows:

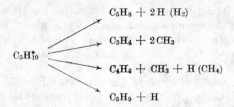

$k_d/k_c = 1.32 \pm 0.14$, calculated for methyl and tert-pentyl radicals, is in good agreement with the data given by Gagnon et al. [145].

The rates of reaction between various pentene isomeric parent ions and the respective neutral molecule were studied by Koyano et al. [141] and Henis [81] using photoionization mass spectrometry and the ion-cyclotron resonance spectrometry. It can be seen in Table 4.26 that the most reactive

Table 4.26

Total cross section (σ_{tot}) of the reaction $C_5H_{10}^+ + C_5H_{10} \rightarrow$ products.
After Koyano et al. [141]

Hydrocarbons	σ_{tot}, 10^{-20} m^2
Pent-1-ene	146
cis-Pent-2-ene	84.9
2-Methylbut-1-ene	97.6
2-Methylbut-2-ene	24.1
3-Methylbut-1-ene	24.1

ion proved to be the pent-1-ene ion. The product distribution was found to be rather complex with this isomer [141], in agreement with other data in the literature [81, 117]. In these latter papers thermochemical and kinetic parameters for ion-molecule reactions of pentene parent ions are also presented. On the basis of these data, a structure was proposed for the transition complex in the fragmentation processes occurring in the ion–molecule reaction. According to Henis [81] the reaction of pentene isomeric ions is initiated by the addition of the unpaired electron of the ion to the π-bond of the neutral molecule. Fragmentation of the adduct proceeds via the energetically most favourable pathway, i.e. rupture of the tertiary C–C bond. It was shown by Koyano et al. [141] that this mechanism is also valid for the reaction of branched chain pentene isomers; here partial rearrangement takes place in the transition complex.

By comparing the experimental data of Koyano et al. [141] and Henis [81] it can be concluded that the hydrogen transfer reactions accompanied by the appearance of parent-minus-two ions [see reaction (4.25a)] become of secondary importance with increasing kinetic energy of the ions, and that the cross section increases with increasing exothermicity of the reaction. The more branched the carbon skeleton in the reactant, the less exothermal is the reaction and, consequently, the smaller is the cross-section of the reaction. Steric hindrance may also be important. As far as the so-called parent minus-one (4.25b) and parent-plus-one (4.25c) reactions are concerned, it is not easy on the basis of these experimental results to draw exact conclusions concerning the correlation between reactivity and chemical structure, except that the proportion of products resulting from reaction (4.25b) is extremely high in the case of branched-chain pentenes:

$$C_5H_{10}^+ + C_5H_{10} \Bigg\langle \begin{array}{ll} C_5H_8^+ + C_5H_{12} & (4.25a) \\ C_5H_9^+ + C_5H_{11} & (4.25b) \\ C_5H_{11}^+ + C_5H_9 & (4.25c) \end{array}$$

4.2.5 C_6 aliphatic alkenes and alkadiene

This section summarizes the radiolysis of n-hexenes, isohexenes and hexa-1,5-diene.

The radiolysis of hex-1-ene has been studied by Chang et al. [149]. They estimated the yields of some gas products; otherwise only dimer yields have been determined in detail, and an ionic mechanism has been suggested for dimer formation. The yields of dimer and $G(H_2)$ values for all three n-hexene isomers were determined by Brasch and Golub [150], but there are significant differences between these and those given by Chang et al. [149] (Table 4.45).

The decomposition products of n-hexenes have been detailed so far only by Cserép et al. [151–156]. As Table 4.27 shows, the yields of hydrogen and

Table 4.27

Yields from liquid n-hexene isomers (radiation: $^{60}Co\text{-}\gamma$). After Cserép and Földiák [153]

Hydrocarbons	Hex-1-ene	Hex-2-ene	Hex-3-ene
Products	G, molecule/100 eV		
Hydrogen	0.090	1.37	1.64
Methane	0.030	0.132	0.233
Ethane	0.028	0.107	0.160
Ethylene	0.133	0.122	0.073
Acetylene	0.091	0.049	0.045
Propane	0.043	0.063	0.008
Cyclopropane	0.015	0.003	0.004
Propene	0.113	0.063	0.035
Allene (propadiene)	0.031	0.003	0.004
Propine	0.006	0.027	0.004
n-Butane	0.035	0.008	0.002
But-1-ene	0.044	0.022	0.074
But-2-ene	0.005	0.006	0.002
Buta-1,3-diene	0.003	0.003	0.004
Pent-1-ene	0.003	0.003	0.002
Pent-2-ene	0.004	0.001	0.040
n-Pentane	0.001	0.001	0.002

methane are greater the nearer the double bond is to the middle of molecule: it was proved that this phenomenon can be explained by the increase in the number of C—C and C—H bonds in β-positions to the double bond. Methane mainly arises from the methyl groups of the molecule. Comparison of the methane yields of different hexene isomers shows that the increase of $G(CH_4)$ for the isomer with the double bond in the inner position can be only partly due to the duplication of the methyl groups. As the number of methyl groups is identical in hex-3-ene and hex-2-ene, the greater methane yield for the former has been explained by the loosening of the terminal C—C bonds in the β-position to the π-bond. The higher than expected methane yield for hex-2-ene may be explained by assuming a transition state similar to that formed from hex-3-ene before decomposition. The formation of propene also indicates that isomerization may occur before decomposition.

The yield of ethane also increases as the double bond is shifted from the terminal position, but there is no such change in the G-values of ethylene and acetylene. Ethane production from hex-1-ene is considerably lower than from hex-2-ene. This is to be expected because in the case of hex-2-ene the formation of ethane and most of the ethylene can be explained by cleavage of the allylic C—C bond which is about 54 kJ mol^{-1} weaker than the normal one owing to the allylic interaction.

Although the π-bond in hex-1-ene promotes the formation of C$_3$ products and not the rupture of the ethyl group, the ethylene yield is still high. This is probably due to the loosening effect of the saturated chain end, which favours the fragmentation at each two carbon atoms of the molecule. It is because the C—C bond in the β-position to the free electron of alkyl radicals is weakened. This is similar to the effect of the π-bond. The yield of C$_3$ products in the case of hex-1-ene is the highest among the hexene isomers: this points to a weak C—C bond between carbon atoms 3 and 4.

The G-value of propene is exceptionally high. There are several reasons for the propene yield being three times higher than that of propane. One of these may be that along with rupture of C—C bond in the β-position, an intramolecular rearrangement also takes place in some of the decomposing ions or excited molecules:

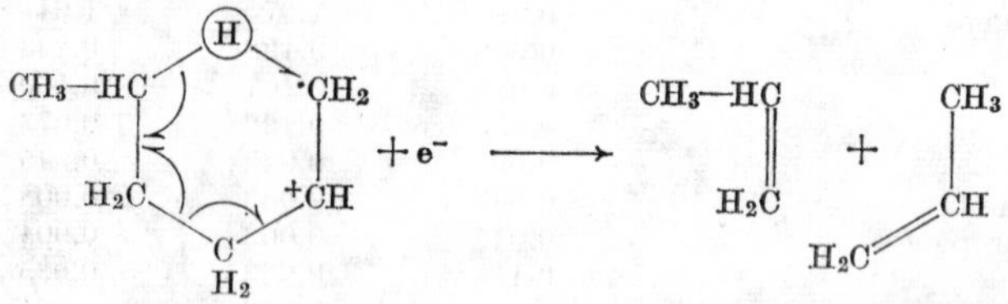

This kind of rearrangement is well known in mass spectrometry [157, 158]; Wagner [137] has suggested that this process also occurs in the liquid-phase radiolysis. A second reason could be the further reactions, such as decomposition, and disproportionation of the propyl radicals formed in the primary process. Species formed in primary decomposition possess a great deal of energy, and the decomposition of propyl radicals into ethylene and methyl radicals needs only 96 kJ mol^{-1} (1 eV) energy.

The yield of C_4 products is generally low. They are produced by cleavage of the $C—C$ bonds weakened by the π-bond in both hex-2-ene and hex-3-ene. The low yields suggest that the methylallyl radical or ion formed after cleavage of the ethyl group further decomposes or undergoes in secondary reactions. The methylallyl intermediate is stabilized by resonance, which increases the chance that it will react with other radicals or intermediates, owing to its long lifetime compared with other alkyl or alkenyl radicals. Formation of butadiene from hex-3-ene would be expected following 'simultaneous' splitting of two methyl groups in the β-position. Thus the very low butadiene yield indicates that this reaction has a very low probability. However, recent investigations using additives [156] show that this probability is somewhat higher than previously supposed [155].

It is remarkable that but-1-ene has the highest yield of C_4 products from all three hexene isomers. This may be due either to isomerization prior to decomposition or to rupture of the $C—C$ bond in the α-position to the π-bond. The latter mechanism is supported by the formation of high amounts of ethane.

The yield of C_5 is lower than that of C_4. Only pent-2-ene is produced in considerable amounts from hex-3-ene, which again points to the preferential cleavage of the $C—C$ bond in the β-position.

The distribution of fragments, to a first approximation, can be correlated with the dissociation energies (kJ mol⁻¹) of the bonds involved [155]:

$$C=C\overset{392}{\underline{\quad}}C\overset{300}{\underline{\quad}}C\overset{342}{\underline{\quad}}C\overset{354}{\underline{\quad}}C$$

It was established by Kovács et al. [156], in experiments carried out in the presence of different additives (p-benzoquinone, CCl_4 and C_2H_5OH), that the yields of most products decrease in presence of a radical scavenger (p-benzoquinone), whereas those of conjugated dienes increase. The acetylene yield was not influenced by any of these scavengers. A considerable drop was observed in the yields of methane, ethane and propane. The effect of 50 mм p-benzoquinone was to decrease the methane yield by about 50% for all three isomers, and a similar decrease was observed in the yield of ethane from hex-1-ene and hex-3-ene. The inhibition of the ethane yield is much higher in the case of hex-2-ene (about 80%). By considering these changes together with others, the authors suggested the following reaction scheme:

$$C_6H_{12}^* \longrightarrow CH_3 + C_5H_9 \qquad\qquad (4.26a)$$

$$\longrightarrow 2\,CH_3 + C_4H_6 \qquad\qquad (4.26b)$$

$$\longrightarrow C_2H_5 + C_4H_7 \qquad\qquad (4.26c)$$

$$\longrightarrow C_3H_5 + C_3H_7 \qquad\qquad (4.26d)$$

Reaction (4.26c) has also been proposed by Chesick [159] for the photolysis of cis-hex-2-ene. Most products were considered to be formed in secondary reactions of the C_2H and C_4H_7 radicals.

As the yield of acetylene was not affected by additives in the radiolysis of any of the n-hexene isomers, it was assumed to be formed in very fast processes from the original double bond.

Dimerization of hex-1-ene by ion-molecule condensation was suggested by both Chang et al. [149] and Brasch and Golub [150]. For hexenes with the double bond in an internal position, Brasch and Golub proposed that besides ionic processes the radical pathway is also important,. This is supported by the finding that the yield of dimer decreases by 66% when hex-2-ene is irradiated in the presence of iodine (see Fig. 4.4), though the scavenger effect of iodine has not been unambiguously defined; see e.g. refs [160, 161].

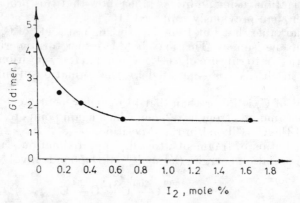

Fig. 4.4. Effect of iodine on the yield of dimer in the radiolysis of hex-2-ene (radiation: 1 MeV electrons). After Brash et al. [150]

The formation of dimers from n-hexene isomers was also investigated by Kovács et al. [156]. The results only partly support the data obtained by Chang et al. [149] and Brasch and Golub [150], for the radiolysis of hexenes in the presence of different additives (p-benzoquinone, CCl_4 and C_2H_5OH). The amount of dimer formed by the ionic mechanism is higher for hex-1-ene than for the other two isomers, for which radical processes are predominant. Nevertheless, radical processes were considered to be of greater importance than was thought by Brasch and Golub for hexenes containing an internal double bond based on the drop in yield of more than 80% in the presence of p-benzoquinone.

Ion-cyclotron resonance spectroscopy supports the radical mechanism for the internal isomers and the ionic one for hex-1-ene [81]. According to this investigation, the cross-section of the reaction

$$C_6H_{12}^+ + C_6H_{12} \longrightarrow C_{12}H_{24}^+$$

is much higher for hex-1-ene than for the other two isomers.

Roginskii and Pshezhetskii [162] studied the γ-radiolysis of hex-1-ene, hept-1-ene and hept-2-ene in the solid phase at 77 K, and determined the radicals formed by ESR spectroscopy. Allyl-type radicals were observed, as they were in a 3-methylpentane matrix. Furthermore, it was established that following irradiation with light of 3.0 eV energy, the allyl-type radicals produced an equal amount of alkyl radicals. This phenomenon is similar to that found earlier for allyl radicals [163]. The authors suggested a mechanism by which the photoreaction of allyl and allyl-type radicals occurs through hydrogen abstraction from the surrounding molecules; consequently it is not a result of an intramolecular migration of the free radical, as had been thought [164–166]. In the case of alkenes dissolved in a 3-methylpentane matrix, allyl radicals excited by light are transformed into alkyl radicals in the following way:

$$(R-CH_2 \cdots CH \cdots CH_2)^* + CH_3CH_2CH(CH_3)CH_2CH_3 \longrightarrow$$

$$\longrightarrow R-CH_2-CH=CH_2 + CH_3CH_2C(CH_3)CH_2CH_3$$

The fact that the rate of abstraction of allyl radicals excited by light [163] is practically the same as that of butyl-allyl radicals formed from heptene supports the existence of an electronically rather than vibrationally excited state of allyl and allyl-type radicals irradiated by light of appropriate wavelength. This is because the degree of vibrational freedom is much higher in larger than in smaller molecules, so the rate of abstraction should be a few orders of magnitude lower; this is not the case.

Without light the reverse process occurs probably even at 77 K, as the reaction is exothermic to the extent of 63–84 kJ mol^{-1}. This energy comes from the difference of the dissociation energies between the alkyl and allylic C$-$H bonds:

$$R \cdot + R-CH_2-CH=CH_2 \longrightarrow RH + R-CH \cdots CH \cdots CH_2$$

The behaviour of n-hexenes was also studied in the solid phase by Smith et al. [167] using ESR spectroscopy. They were able to prove the presence of trapped electrons and allyl-type radicals in the case of hex-1-ene and hex-2-ene. Ayscough et al. [168] have also investigated several alkenes under similar conditions. In a few cases considerable differences can be found between the results of these two groups. For instance, in case of hex-2-ene Ayscough and Evans [169], observed alkyl radicals whereas allyl radicals were detected by Smith et al. [167]. These data can be found in detail in Kevan's review [170]. Shida and Hamill [171] observed positive ions in glassy alkenes.

More recently Cserép et al. [171A] have observed radical cations of hex-1-ene and hex-2-ene in electron pulse-irradiated liquid alkane–alkene solutions.

The radiolysis of hexa-1,5-diene was investigated by van der Heide and Wagner [172]. As shown in Table 4.28, the yields of C$_3$, C$_6$, C$_9$ and C$_{12}$ are surprisingly higher than those of other products. The yields of C$_3$ and C$_9$ obviously result from weak C$-$C and C$-$H bonds in allylic positions. The σ-bond between carbon atoms 3 and 4 is especially weak because it is influenced by the delocalization of both π-bonds.

As far as the C$_9$ fraction is concerned, it was established by means of mass spectrometry that it consists mainly of di- and tri-alkenes, and that the

Table 4.28

Yields from liquid aliphatic dienes

Hydrocarbons	Hexa-1,5-diene	2,6-Dimethyl-octa-2,6-diene
Radiation	3.0 MeV bremsstrahlung	
Reference	172	178
Products	G, molecule/100 eV	
Hydrogen	0.45	0.80
Methane		0.052
C_2 products	0.06	0.07
C_3 products	0.09	0.023
C_4 products	0.06	0.050
C_5 products	0.02	0.068
C_6 products	0.41	
C_7 products	0.02	
C_8 products	0.02	
C_9 products	0.42	
C_{10} products	0.02	
C_{12} products	0.77	

double bonds are of the vinyl type, although vinylene (RCH=CHR) double bonds are also present to a small extent. The C_9 and C_{12} fractions are of similar structure, therefore it was assumed [172] that they are formed by ion-molecule condensation and that a part of the intermediates formed decomposes into C_3 and C_9 products. As well as this hypothesis, the possibility cannot be ruled out of the formation of C_3, C_9 and C_{12} products in secondary reactions (abstraction, combination, addition) of allyl or allyl-type radicals formed from ions and excited molecules resulting from primary processes. Thus, the data available are not adequate for elucidation of the reaction mechanism.

Among the isohexenes Wagner has studied the radiolysis of 3-methylpent-1-ene, 2-ethylbut-1-ene, *trans*-3-methylpent-2-ene, *cis*-3-methylpent-2-ene and 3,3-dimethylbut-1-ene at 77 K by using 3 MeV bremsstrahlung [138]. The results are similar to those for isopentenes. Besides hydrogen, the amount of which was determined only in the radiolysis of 3-methylpent-1-ene $[G(H_2) = 0.32]$ and 3,3-dimethylbut-1-ene $[G(H_2) = 0.34]$, the main products are:

1. Alkanes with identical carbon number formed by saturation of the double bond ($G = 0.2$–0.6);
2. Isomeric alkenes formed by migration of the double bond to the adjacent carbon atom (adjacent isomers) ($G = 0.15$–0.9);
3. Alkynes and dialkenes with the same carbon atom number as the reactant ($G = 0.14$–0.32);
4. *Cis* and *trans* isomers of 3-methylpent-2-enes ($G = 1.2$–1.3).

Dimers are formed with the highest yield in all cases ($G = 0.7$–2.8). Products from skeletal isomerization are of minor importance, the most im-

Table 4.29

Dimer yields from liquid branched chain monoalkenes (radiation: 3 MeV bremsstrahlung). After Wagner [137]

Hydrocarbons	Dimers G, molecule/100 eV
3-Methylbut-1-ene	1.20
2-Methylbut-1-ene	1.14
2-Methylbut-2-ene	1.82
3-Methylpent-1-ene	1.65
2-Ethylbut-1-ene	2.8
trans-3-Methylpent-2-ene	2.2
cis-3-Methylpent-2-ene	1.7
3,3-Dimethylbut-1-ene	0.67

portant being 1,1,2-trimethylcyclopropane ($G = 0.23$) formed from 3,3-dimethylbut-1-ene.

It can also be established from Wagner's data (Table 4.29) that the higher the ratio of allylic hydrogen in alkenes, the more dimer is formed (e.g. from 2-ethylbut-1-ene and *trans*-3-methylpent-2-ene). This is true in the reverse case as well (e.g. 3,3-dimethylbut-1-ene).

In the case of 3,3-dimethylbut-1-ene, the number and structure of dimers were different from what was expected by Wagner [138], so this alkene was further investigated (Table 4.30). Based on experiments carried out at

Table 4.30

Temperature dependence of yields from liquid 3,3-dimethylbutene (radiation: 3 MeV bremsstrahlung). After Wagner [137]

Temperature / Products	—196	—78	0
	G, molecule/100 eV		
Hydrogen	0.34		0.37
Methane	0.13		0.14
Ethane + ethylene	0.06		0.10
Isobutane	0.35		0.11
Isobutene	0.11		0.24
Isoprene	0.04		0.03
Dimethylbutane	0.24		0.24
Dimethylbutyne	0.20	0.10	0.18
1,1,2-Trimethylcyclopropane	0.23	0.08	0.23
4-Methylpent-1-ene	0.04		0.04
2-Methylpent-2-ene	0.06	0.14	0.18
2,3-Dimethylbut-1-ene	0.04		0.04
2,3-Dimethylbut-2-ene	0.03		0.03
2,2-Dimethylpentane	0.00	0.09	0.20
tert-4,4-Dimethylpent-2-ene	0.00	0.02	0.11
Unknown C₆ and C₇ products	0.17		0.13
C₁₂ products	0.67		1.29

−196 and 0°C, it is apparent that the change in temperature and phase brought about significant changes in the yields of some products, yet in most cases the change was negligible. A significant drop in the yield of 2-methylpent-2-ene and of two C_7 fractions shows that these products were formed via radicals. This was supported by the data obtained in experiments carried out in the presence of radical scavengers. Of the data obtained at −78°C, the yields of dimethylbutyne and 1,1,2-trimethylcyclopropane are very interesting: these are lower than the values obtained both at higher and at lower temperature. There is no information available about the decomposition products beyond their G-values. The significant dependence on temperature and phase of the yields of isobutene and isobutane resulting from cleavage of the tert-butyl group was also neglected (Table 4.30).

It was assumed by Wagner [138] that the formation of skeletal isomers is not due to thermal free radical reactions but to ionic processes, the existence of which is supported by a decrease of production of trimethylcyclopropane in the presence of trimethylamine. The proposed mechanism is as follows:

$$
\begin{bmatrix} \begin{array}{c} C \\ | \\ C-C-C-C \\ | \\ C \end{array} \end{bmatrix}^+ \longrightarrow \begin{bmatrix} \begin{array}{c} C \\ | \\ C-C-C-C \\ \diagup \\ C \end{array} \end{bmatrix}^+ \xrightarrow{e^-} \begin{array}{c} C-C-C-C-C \\[2mm] C-C-C-C \\ | \quad | \\ C \quad C \end{array}
$$

It has been observed in the γ-radiolysis of 2-methylpent-1-ene at 77 K by Guarino and Hamill [173] that after irradiation, the sample turned blue and a maximum at 1800 nm and a shoulder at 680 nm were found in the absorption spectrum. The maximum at 1800 nm was supposed to be caused by trapped electrons: this was supported by the effect of electron scavengers such as biphenyl (0.13 mol%), naphthalene (0.13 mol%) and CCl_4 (2.6 mol%), because the absorption band at 1800 nm disappeared, whereas the intensity of the band at 680 nm increased. This is because the latter is supposedly caused by 2-methylpent-1-ene cation: it can be eliminated by hole scavengers, such as trimethylamine, propan-1-ol, etc.

2-Methylpent-1-ene cations have also been observed by Shida and Hamill in a butyl chloride matrix at 720 nm [171]. Ayscough and Evans [169] studied the γ-radiolysis of glassy 2-methylpent-1-ene by means of ESR spectroscopy and observed allyl-type radicals with a G-value of 0.8; however, they did not detect trapped electrons. Smith and Pieroni [174] used ESR spectroscopy to support Guarino and Hamill's results obtained by optical methods. Similarly to Ayscough and Evans, they observed allyl-type radicals, for which they proposed a different structure (II) from that proposed by Ayscough and Evans (I), on the basis of nine spectral lines.

$$CH_2{=}C(CH_3)CHCH_2CH_3 \rightleftharpoons CH_2(CH_3)C{=}CHCH_2CH_3$$
(I)

$$(CH_3)_2C{=}CHCHCH_3 \rightleftharpoons (CH_3)_2CCH{=}CHCH_3$$
(II)

The formation of (II) was explained by the transformation of 2-methyl-pent-1-ene ion into 2-methylpent-2-ene ion followed by loss of a proton (H$^+$). As radical (II) is converted into (I) by UV radiation, it was assumed that this process is a simple one, and isomerization or the formation of (I) occurs via hydrogen abstraction from a 2-methylpent-1-ene molecule by a radical (II). This latter view is supported by Roginskii and Pshezhetskii [162] following experiments with n-hexene. They proposed that a hydrogen atom is abstracted from the neighbouring molecule by an allyl-type radical when irradiated with photons of 410 nm wavelength. The observations were summarized by Smith and Pieroni in the following scheme [174]:

$$2\text{-methylpent-1-ene} \xrightarrow{\;\sim\!\!\wedge\!\!\vee\;} (2\text{-methylpent-1-ene})^+ + e^-$$

$$(2\text{-methylpent-1-ene})^+ \rightarrow (2\text{-methylpent-2-ene})^+$$

$$(2\text{-methylpent-2-ene})^+ + 2\text{-methylpent-1-ene} \rightarrow$$

$$\rightarrow \text{II} + 2\text{-methylpent-1-ene/H}^+$$

$$e^- \rightarrow e^-_{trapped}$$

or

$$e^- + \text{biphenyl} \rightarrow (\text{biphenyl})^-$$

$$e^-_{trapped} \xrightarrow[\text{thermal decay}]{\text{visible light or}} e^-_{untrapped}$$

$e^-_{untrapped}$ + (2-methylpent-1-ene) H$^+$ → (2-methylpent-1-ene) + H.
H reacts with no net formation or removal of II:

$$\text{II} \xrightarrow{\;253.7 \text{ nm } (4.9 \text{ eV}) \text{ light}\;} \text{I}$$

or

$$\text{II} + 2\text{-methylpent-1-ene} \xrightarrow{\;253.7 \text{ nm } (5,9 \text{ eV}) \text{ light}\;} 2\text{-methylpent-2-ene} + \text{I}$$

II and I are thermally stable at 77 K.

4.2.6. C$_7$-C$_{16}$ aliphatic alkenes

Of alkenes with a carbon atom number higher than six, the radiolysis of heptene and octene isomers and 2,6-dimethylocta-2,6-diene and hexadecene has been studied.

The products (H$_2$ and light hydrocarbons) formed in the radiolysis of some isoheptene isomers have been reported by Cserép and Földiák [151, 153, 175] (Table 4.31). The composition of the light hydrocarbons is controlled by the position of the π-bond in the monoalkene and thus by the bond energies. This is clearly shown in the methane yield: $G(\text{CH}_4)$ values increase in the sequence 2-methylhex-1-ene > 4-methylhex-1-ene > 3-methylhex-1-ene > 5-methylhex-1-ene > 2-methylhex-3-ene. The accuracy of the dissociation energies is very low, therefore no quantitative conclusion can be drawn concerning the relationship between G-values and bond energies. However, the trend is obvious: the lower the dissociation energy, the higher the G-value.

20*

Table 4.31

Yields from liquid methylhexene isomers (radiation: $^{60}Co-\gamma$). After Cserép and Földiák [175]

Hydrocarbons	2-Methyl--hex-1-ene	3-Methyl--hex-1-ene	4-Methyl--hex-1-ene	5-Methyl--hex-1-ene	trans-2--Methyl-3--hexene
Products	G, molecule/100 eV				
Hydrogen	1.25	0.82	1.13	1.16	1.30
Methane	0.06	0.10	0.07	0.15	0.27
Ethane	0.025	0.025	0.065	0.005	0.057
Ethylene	0.079	0.140	0.158	0.087	0.032
Acetylene	0.012	0.110	0.060	0.056	0.015
Propane	0.035	0.120	0.003	0.022	0.073
Propene	0.101	0.110	0.238	0.202	0.060
n-Butane	0.029	0.010	0.075	0.002	0.001
But-1-ene	0.022	0.020	0.102	0.095	0.056
But-2-ene		0.042	0.146		
Buta-1,3-diene		0.035	0.248		
Methylcyclopropane		0.025			
Isobutane			0.001	0.112	0.001
Isobutene	0.060			0.125	

In 2-methylhex-3-ene three methyl groups are weakened by the π-bond: thus, a high methane yield can be expected. The yield of methane from 5-methylhex-1-ene is about double that from 4-methylhex-1-ene because two methyl groups are bonded to a tertiary carbon atom in the former. This is supported by the fact that the amount of ethane formed from 4-methylhex-1-ene is more than an order of magnitude higher than the amount formed from 5-methylhex-1-ene, which is the opposite of what was found for methane. The very low $G(C_2H_6)$ value for 5-methylhex-1-ene may be a consequence of the mechanism, namely, that at the dose rate applied ethane is produced mainly from the cleaved ethyl group and not via recombination of methyl radicals.

As the distance between the double bond and the methyl group increases the yield of ethylene formed from different methylhex-1-enes passes through a maximum. This phenomenon has not been fully explained. Nevertheless, it was assumed by Cserép and Földiák [175] that the maximum results from further decomposition of fragments of four carbon atoms formed by β-splitting into fragments of two carbons.

The lowest acetylene yield was measured with 2-methylhex-1-ene, the highest, however, with 3-methylhex-1-ene. The G-values of acetylene formed from 5-methylhex-1-ene and 4-methylhex-1-ene are equal, and the yield from 2-methylhex-3-ene is very low. These data strongly indicate that the acetylene is formed almost exclusively from the part of the molecule containing the double bond, and its formation is promoted by the adjacent tertiary carbon atom in 3-methylhex-1-ene. In addition, Tables 4.27, 4.31 and 4.32 clearly indicate that acetylene is preferentially formed from molecules with the double bond in the terminal position.

The yields of C_3 products, propane and propene, are relatively high from methylhex-1-ene isomers because the main products of the decomposition of these molecules are always fragments with three carbon atoms, owing to the allylic interaction, e.g.:

$$C=C-C-C-C-C \quad \underset{|}{C} \quad \longrightarrow\!\!\!\bigvee\!\!\!\!\!\longrightarrow \quad C=C-C \quad \underset{|}{C} \quad + \quad C-C-C \qquad (4.27)$$

$$C=C-C-C-C-C \quad \underset{|}{C} \quad \longrightarrow\!\!\!\bigvee\!\!\!\!\!\longrightarrow \quad C=C-C \quad \underset{|}{C} \quad + \quad C-C-C \qquad (4.28)$$

$$C=C-C-C-C-C \quad \underset{|}{C} \quad \longrightarrow\!\!\!\bigvee\!\!\!\!\!\longrightarrow \quad C=C-C \quad + \quad C \quad \underset{|}{C}-C-C \qquad (4.29)$$

$$C=C-C-C-C-C \quad \underset{|}{C} \quad \longrightarrow\!\!\!\bigvee\!\!\!\!\!\longrightarrow \quad C=C-C \quad + \quad C-C-C \quad \underset{|}{C} \qquad (4.30)$$

If the effect of a tertiary carbon atom coincides with the weakening effect of the double bond e.g. in 4-methylhex-1-ene, the rupture of the particular bond is even more probable.

In case of 2-methylhex-1-ene, rupture of the terminal molecule fragment with the double bond yields C_3 products [G(propyne) = 0.03, G(allene) = = 0.03, and a small amount of propene] but not acetylene. Propyne and allene were produced in negligible amounts from other methylhex-1-ene isomers, therefore they are not included in Table 4.31.

The propane from 5-methylhex-1-ene and 2-methylhex-3-ene isomers results not from rupture of the C—C bond in the β-position to the π-bond, but from rupture of the bond weakened by the tertiary carbon atom. This is supported by the fact that G(propane) for 2-methylhex-3-ene is nearly equal to the propene yield. Moreover, with 4-methylhex-1-ene, which cannot yield propyl radicals by either process, propane is formed only in trace amounts.

C_4 products from methylhex-1-ene isomers can be divided into several groups. Different products can be formed from secondary butyl and methylallyl radicals via for example, disproportionation or cyclization (see Table 4.31). The G-value for n-butane formation generally is low: n-butane is formed in considerable amounts only from 4-methylhex-1-ene, because the cleavage of the secondary butyl radical is promoted by both the π-bond and tertiary carbon atom. C_4 products are formed in much lower yield from 3-methylhex-1-ene, though one fragment of the molecule contains (1-methylallyl ion or radical) four carbon atoms. 1-Methylallyl radicals, however, can be stabilized by several pathways, and this is partly reflected in the product distribution. It is very probable, however, that the low yield of C_4 products is due to the fact that the major part of the methylallyl radicals become stabilized by combination [175A].

n-Butenes are produced in a high yield from 3-methylhex-1-ene, presumably via disproportionation of methylallyl radical, and from 4-methylhex-

1-ene via that of secondary butyl radicals. The amount of but-2-enes is higher than that of but-1-ene; this may be partly due to the different re- activities of primary and secondary C—H bonds and partly due to the higher thermochemical stability of the but-2-enes. Although Table 4.31 lists only the sum of the G-values of *cis-* and *trans*-but-2-enes, they were determined separately by Cserép [139]. It was found that from both 3- and 4-methylhex-1-ene, the amount of *trans*-isomer produced is 2.5 times greater than that of *cis*-isomer. Owing to the free enthalpy of formation, the *trans*-isomer is more stable than the *cis*-isomer, and the higher yield of the former can been understood on a thermochemical basis.

Branched-chain C_4 products (isobutane and isobutene) are formed prac- tically only from 2- and 5-methylhex-1-enes, i.e. only where the formation of isobutyl and 2-methylallyl radicals by allyl interaction is energetically favoured. Table 4.31 shows that 2-methylhex-1-ene yields only isobutene and that 5-methylhex-1-ene yields isobutane and isobutene in nearly equal amounts. This points to the formation of products from 5-methylhex-1-ene via disproportionation of isobutyl radicals formed by reaction (4.30).

The structure of the 2-methylhex-1-ene molecule explains why no iso- butane is produced. Isobutene may be formed by hydrogen abstraction from another molecule by the 2-methylallyl radical produced in reaction (4.27), or via intramolecular rearrangement in a unimolecular process. The latter possibility may also explain why the propene yield from 2-methyl- hex-1-ene is higher than that of propane:

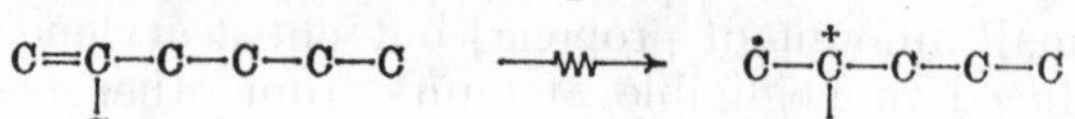

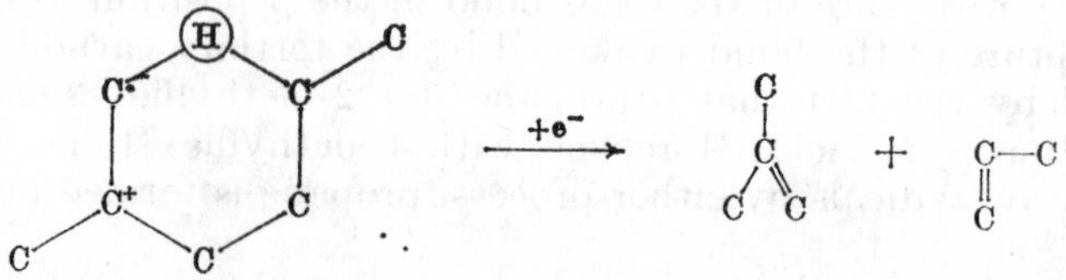

The radiolysis of octene isomers has been investigated by Cserép and Föl- diák [155] (Table 4.32); oct-1-ene has also been studied by Kourim [176]. Here (Table 4.32) the proportion of products not resulting from the direct- ing effect of the π-bond due to the longer chain increases. Thus, it is more difficult to separate the different effects.

There is an analogy concerning the methane yield between n-hexenes and n-octenes: $G(CH_4)$ values increase in the sequence oct-1-ene $<$ oct-2-ene $<$ $<$ oct-4-ene $<$ oct-3-ene. The lowest methane yield of oct-1-ene is probably due to the presence of only one methyl group. Oct-3-ene contains one methyl group in the β-position to the π-bond; the methane yield is increased by about 20% compared with that of oct-1-ene.

There is an analogy between the formation of methane and the produc- tion of C_2 hydrocarbons (mainly ethane). The ethane yield is greater when the π-bond is in an internal position. This shows that the ethane yield is pro-

Table 4.32

Yields from liquid n-octenes (radiation: ^{60}Co-γ). After Cserép and Földiák [155]

Hydrocarbons	Oct-1-ene	Oct-2-ene	Oct-3-ene	Oct-4-ene
Products		G, molecule/100 eV		
Hydrogen	0.98	1.42	1.63	1.61
Methane	0.015	0.058	0.077	0.055
Ethane	0.016	0.031	0.074	0.115
Ethylene	0.081	0.066	0.056	0.111
Acetylene	0.030	0.010	0.008	0.010
Propane	0.013	0.020	0.055	0.067
Propene	0.051	0.056	0.052	0.028
Butane	0.009	0.048	0.035	0.005
But-1-ene	0.018	0.036	0.041	0.013
Pentane	0.029	0.030	0.003	0.002
Pent-1-ene	0.029	0.022	0.013	0.045
Pent-2-ene	0.002	0.017	0.420	0.066

portional to the number of ethyl groups in the parent alkene, or in the activated transition state. G(ethane) from oct-2-ene is double that from oct-1--ene, although both isomers (in the ground state) contain the same number of ethyl groups. However, in oct-2-ene the double bond should migrate only one carbon atom inside the chain (to contain two ethyl groups), which is not the case with oct-1-ene. This may be an explanation for the different ethane yields.

The total C_2 yield is highest in oct-4-ene, which can be explained by the symmetrical position of the π-bond in the chain. In this way both ethyl groups are weakened:

$$C-C-C-C=C-C-C-C \longrightarrow C-C \; + \; C-C=C-C-C-C$$

$$2\,C-C \longrightarrow C=C + C-C$$

This phenomenon is similar to that observed for the methane yield from hex-3-ene.

The G-values for ethylene decrease in the sequence oct-4-ene > oct-1-ene > oct-2-ene > oct-3-ene. With exception of oct-4-ene, this is the sequence of decreasing chain length of the saturated part in the molecule. Therefore, the possibility of C_2 fragments being produced from the alkyl or alkenyl radicals obtained by irradiation from the original molecule is decreased.

The acetylene yield of octenes is similar to that of hexenes: i.e. acetylene is formed from the terminal double bond with higher probability than from internal ones.

The differences among the yields of C_3 products formed from n-octene isomers are small: the highest $G(C_3)$ was measured with oct-3-ene, which may be attributed to interaction of π-bond. Propane and propene were obtained in unexpectedly low amount from oct-1-ene.

The highest yield of C_4 compounds was formed from oct-2-ene, thought in lower yields than one would expect from double-bond interaction. The yields from oct-3-ene and oct-4-ene show that the probability for the simultaneous rupture of two allylic C—C bonds is very low (buta-1,3-diene is formed only in trace amounts).

C_5 hydrocarbons are produced in the highest yield from oct-3-ene, in agreement with the directing effect of the π-bond.

Reaction mechanisms can generally be elucidated by studying the influence of dose, dose rate, temperature, additives and phase. Thus, Kourim [176] has found (Table 4.33) that the diene-type dimer yield was increased by increasing the dose rate by an order of magnitude. However, the amount of saturated and monoalkene-type dimers decreased. Dimers formed in the irradiation of oct-1-ene saturated by oxygen can be divided into two groups: dienes are completely absent, the yield of alkanes drops by 80% and the yield of monoalkenes shows only a very small decrease. The effect of benzoquinone on dimer formation is similar to that of oxygen, except that the yield of n-hexadecene is unchanged. To explain this we have to consider oxygen as a radical scavenger, although interaction with ions to a smaller extent cannot be excluded [177]. The total dimer yield from oct-1-ene is not affected by NH_3, which is usually considered as positive ion scavenger, to the same extent as by radical scavengers. The most characteristic change is a decrease in the amounts of n-hexadecene and 7-methylpentadecene, whereas the yields of saturated products and dienes increase.

Table 4.33

Dimer yields from liquid oct-1-ene (radiation: ^{60}Co-γ). After Kourim [176]

Dose rate, eV $g^{-1}s^{-1}$	$8.1 \cdot 10^{12}$	$7.7 \cdot 10^{13}$	$7.7 \cdot 10^{13}$	$7.7 \cdot 10^{13}$	$7.7 \cdot 10^{13}$
Additives	—	—	10 mM O_2	50 mM benzoquinone	290 mM NH_3
Products	\multicolumn{5}{c}{G, molecule/100 eV}				
n-Hexadecane	0.055	0.064	0.015	0.018	0.066
7-Methylpentadecane	0.227	0.247	0.050	0.058	0.399
7,8-Dimethyltetradecane	0.006	0.008	0.003	0.004	0.013
n-Hexadecene	0.488	0.549	0.423	0.636	0.251
7-Methylpentadecene	0.130	0.143	0.090	0.022	0.125
6-Vinyltetradecane	0.196	0.121	0.023	0.028	0.137
7-Methyl-8-vinyltridecane	0.042	0.048	0.029	0.035	0.052
n-Hexadecadiene	0.242	0.167	0.010	0.010	0.193
6,7-Divinyldodecane	0.072	0.050	0.004	0.008	0.049
6-Vinyltetradecene	0.037	0.033	0.032	0.039	0.030
Other dimers	0.114	0.109	0.044	0.072	0.125
Total dimers	1.609	1.539	0.723	1.030	1.440
of this: dienes	0.510	0.338	0.037	0.046	0.379
monoalkenes	0.697	0.773			
saturated dimers	0.288	0.319	0.068	0.080	

The relative activities of two resonance hybrids (allyl-type radicals)

$$CH_3(CH_2)_4 - \dot{C}H - CH = CH_2 \rightleftharpoons \dot{C}H_2 - CH = CH - (CH_2)_4CH_3$$
$$\text{(a)} \qquad\qquad\qquad\qquad\qquad \text{(b)}$$

can be estimated by comparing the yields of 6,7-divinyldodecane, 6-vinyl-tetradecene and hexadecadiene formed from oct-1-ene. The ratio $b/a = 2$, shows that the free radical is twice as likely to react at the end of the chain as at carbon atom 3, owing to the different thermochemical stabilities and possibly also to steric factors.

2,6-Dimethylocta-2,6-diene radiolysis was studied by Shellberg et al. [178]. From their data (Table 4.28) a free radical mechanism was suggested for the formation of hydrogen and light hydrocarbons. The role of bond strength can be observed here in the formation of C_5 hydrocarbons, but the high yields ethylene and ethane cannot be thus explained.

In addition to the products of low molecular mass, the formation of polymers was also studied by Widmer [179]: both the free radical and the molecular mechanism can be supported by these data. Recently, Brash and Golub [180] studied this reaction mainly from the structural point of view: a decrease of unsaturation was found of $G = 7.8$–9.9. Ion-molecule condensation was suggested to be responsible for the saturation and hydrogen atom addition being of minor importance.

The γ-radiolysis of hexadec-1-ene has been studied by Collinson et al. [181], especially the polymerization, the temperature dependence and the effect of the solid–liquid phase transition. The yields of hydrogen, methane and C_2-C_8 products at room temperature were 1.31, 0.03 and 0.215, respectively. $G(H_2)$ is not influenced by dose rate or by radical scavengers. The total yield of C_9-C_{31} is $G < 0.04$. The polymer formed under irradiation contains both vinyl and trans-vinylene groups. Neither the hydrogen yield nor trans-vinylene production was influenced by temperature change and melting, the rate of polymerization, however, was strongly affected by both factors.

It was assumed that hydrogen is eliminated from long-chain cations so that trans-vinylene groups are formed. Because the rate of polymer formation was influenced by radical scavengers only to a small extent the conclusion was drawn that polymerization occurs mainly via an ionic mechanism. The experimental data were interpreted by suggesting two types of ionic species for the initiation of polymerization. One of these ions carries the positive charge in the plane of vinyl group: this is reactive in the whole temperature range; the other kind of ion has the charge located on the alkyl chain: this is inactive for initiation below the λ-temperature ($-26 \ldots -24°C$), but above it the ion rearranges into the former kind and thus becomes reactive for polymerization.

The mechanism of radiolysis of hexadec-1-ene has been studied by Ayscough et al. [182] by means of ESR spectroscopy. The overall G(radical) was found to be 3.5 at $-196°C$. The ESR spectra were very complex; however, some interpretation is possible: e.g. the signal given by a paramagnetic species at $-196°C$ is probably the $CH_3(CH_2)_{13}\dot{C}HCH_2CH_2\overset{+}{C}H(CH_2)_{13}CH_3$

radical ion formed by ionic dimerization. The signal that appeared at $-77°C$ was supposed to be due to the appearance of an allyl-type $CH_3(CH_2)_{12}CHCH=$ $=CH_2$ radical.

4.2.7. Acetylenes

The main products of acetylene radiolysis are benzene and the oligomer cuprene. Vinylacetylene, cyclo-octatetraene, phenylacetylene and butadiene are of minor importance [183, 184]. Despite of the great number of experiments on acetylene radiolysis [183–194] the mechanism of product formation, e.g. for benzene, has not been elucidated.

No benzene is formed in the radiolysis of acetylene sensitized by noble gases; in these experiments of Dorfman and Wahl, acetylene was present in the ionic state because most of the energy was absorbed by the excess noble gas, and subsequent transfer of energy or charge from the noble gas ionized the acetylene [186]. This was interpreted to mean the absence of the excited state of acetylene (probably triplet) that is responsible for the benzene formation.

A study of the pressure-dependence of acetylene radiolysis without additives has revealed that benzene formation is independent of pressure above 26–33 mbar, but decreases below this pressure. There is no effect on polymer formation. It is of interest that the pressure-dependence of the benzene yield is the same as found in direct photolysis at 184.9 nm (6.7 eV) [196] and in Hg-sensitized photolysis [195]. However, in photolysis, polymer formation also decreased. Considering these findings, Dorfman and Wahl assumed that in the radiolysis of pure acetylene, benzene is formed in molecular processes via an excited-state precursor (with long half-life: 100 μs). For polymer formation, the following ionic and free-radical pathways were suggested [186]:

$$C_2H_2^* + C_2H_2 \longrightarrow (C_2H_2)_2^* + C_2H_2 \xrightarrow{C_2H_2} C_6H_6$$

$$C_2H_2^+ + C_2H_2 \xrightarrow{C_2H_2} + C_4H_{3i}^+ H$$

$$C_2H_2 + H \longrightarrow C_2H_3 \xrightarrow{C_2H_2} polymer$$

Mains et al. [183] studied the radiolysis of 1 : 1 mixtures of C_2H_2 and C_2D_2 and found high amounts of C_6H_5D, $C_6H_3D_3$ and C_6HD_5 in the benzene fraction. On this basis together with using the result obtained in presence of radical scavengers, when no benzene was formed, a free-radical mechanism was proposed:

$$C_2H_2 + H \longrightarrow C_2H_3$$

$$C_2H_3 + C_2H_2 \longrightarrow C_4H_5$$

$$C_4H_5 + C_2H_2 \longrightarrow C_6H_7$$

$$C_6H_7 \longrightarrow C_6H_6 + H$$

Acetylene photolysis and radiolysis have also been studied by Shida et al. [191–193, 195]. Based upon the results obtained in the radiolysis of 1 : 1 mixtures of C_2H_2 and C_2D_2 using radiation of different energies it was established (Table 4.34) that the mechanism of benzene formation is in-

Table 4.34

Deuterium distribution in benzene formed in the gas-phase photolysis and radiolysis of acetylene labelled with deuterium. After Shida and Tsukoda [191]

Radiation	Photolysis		$^{60}Co\text{-}\gamma$
Energy, eV	6.7	(Mercury-sensitized) 4.9	$1.25 \cdot 10^6$
Pressure, mbar[a]	8–133	12–155	26–662
Products	Distribution, %		
C_6H_6	9.4	9.4	9.5
C_6H_5D	6.7	6.5	6.7
$C_6H_4D_2$	28.5	26.6	27.3
$C_6H_3D_3$	12.5	12.2	13.0
$C_6H_2D_4$	28.1	28.4	27.5
C_6HD_5	5.9	6.8	6.6
C_6D_6	9.0	10.3	9.4

[a] Average values measured over the given pressure interval

dependent of the type of irradiation. For benzene formation a 'modified excited molecule mechanism' was proposed, in which the hot acetylene dimer (cyclobutadiene) redissociates again, to a lower excited state that is suitable for trimerization to benzene:

$$2\ C_2H_2 \longrightarrow \begin{matrix} CH{=}CH^* \\ |\qquad| \\ CH{=}CH \end{matrix}$$

$$\begin{matrix} CH{=}CH^* \\ |\qquad| \\ CH{=}CH \end{matrix} \longrightarrow C_2H_2^{**} + C_2H_2$$

$$C_2H_2^{**} \xrightarrow{C_2H_2} C_4H_4^{**} \xrightarrow{C_2H_2} C_6H_6$$

where * denotes the higher, and ** lower electronically vibrationally excited state.

Cuprene is produced with the highest yield in acetylene radiolysis, and its formation has been interpreted by ionic, free-radical and excited-molecule mechanisms. Briggs and Back [194] suggested a four step mechanism instead of the direct polymerization of acetylene, based upon the experiments carried out in an electric field. This mechanism involves the gas phase secondary polymerization of transition polyenes (C_4-C_{20}) formed in primary polymerization, and after condensation in liquid phase. The benzene itself is considered as a transition polyene, and this is partly in agreement with earlier ideas according to which there is a common intermediate for benzene and polymers. This explanation, however, contradicts that of Dorfman and Wahl [186], who proposed to consider two different precursors for the two processes, without any kind of interaction between the two.

The following scheme is proposed for the formation of the radiolysis and photolysis products of acetylene:

$$C_2H_2 \begin{cases} C_2H_2^+ + e^- \\ C_2H^+ + H + e^- \\ C_2 + H_2 \\ C_2H_2^* \end{cases}$$

$$C_2H_2^* + C_2H_2 \longrightarrow (C_2H_2)_2^* \xrightarrow{C_2H_2} C_6H_6$$

$$C_2H_2^+ + C_2H_2 \begin{cases} C_4H_3^+ + H \\ C_4H_4^+ \end{cases} \xrightarrow{C_2H_2} \text{polymer}$$

$$C_4H_3^+ \xrightarrow{C_2H_2} \text{polymer}$$

$$2\, C_2H \longrightarrow C_4H_2$$

$$H + C_2H_2 \longrightarrow C_2H_3 \xrightarrow{C_2H_2} \text{polymer}$$

$$C_2H_3 + C_2H_2 \longrightarrow C_4H_5$$

$$C_4H_5 + C_2H_2 \longrightarrow C_6H_7$$

$$C_6H_7 \longrightarrow C_6H_6 + H$$

C_2H, C_2H_3, C_4H_5, C_6H_5, $C_6H_7 \longrightarrow$ addition, combination and disproportionation products.

Information on the radiolysis of alkylacetylenes comes from the work of Rondeau et al. [197] and Whitten and Berngruber [198]. Propyne (gas phase), but-1-yne (gas and liquid), but-2-yne, pent-1-yne, pent-2-yne and hex-3-yne (all liquid) were investigated. The G-values of the products are presented in Tables 4.35 and 4.36.

In the case of gas-phase radiolysis of propyne the main product is hydrogen, indicating a C—H bond rupture, and a few hydrocarbons resulting from C—C bond cleavage [197] (see Table 4.35). Galli et al. [199] have investigated the photolysis (206.2 nm, 6.0 eV) of propyne. The main products were hexa-1,5-diene, propene, hydrogen and acetylene; C_4H_6, C_6H_8 and C_6H_{10} were produced to a lesser extent. About 80% of the propyne undergoes polymerization, producing liquid and solid polymer. The quantum yield for the disappearance of propyne decreases with increasing pressure. In the photolysis of propyne at 147 nm (8.4 eV) the formation of atomic

Table 4.35

Yields from gaseous alkynes (pressure: 260 mbar). After Rondeau et al. [197]

Hydrocarbons Products	Prop-1-yne	But-1-yne
	Relative yields[a]	
Hydrogen	0.76	1.00
Methane	0.14	1.40
Ethane		0.10
Ethylene		0.50
Acetylene	trace	trace
Propane		0.03
Propene	0.06	0.12
Propyne		trace
Propadiene	trace	0.14
Butene		0.07
But-1-ene	0.07	1.05
Buta-1,2-diene	0.02	0.06
But-2-yne		trace
Pentenes		0.15

[a] Yields are normalized to the hydrogen yield of but-1-yne

Table 4.36

Yields of the liquid-phase radiolysis of alkynes (radiation: ^{60}Co-γ). After Rondeau et al. [197]

Hydrocarbons Products	But-1-yne	But-2-yne		Pent-1-yne	Pent-2-yne	Hex-3-yne
	G, molecule/100 eV					
Hydrogen	0.42	0.79		0.49	0.62	0.57
Methane	0.12	0.14		0.17	0.16	0.13
Ethane	trace	trace		0.15	0.15	0.05
Ethylene	trace	trace			0.05	
Acetylene	0.05	0.04				
Propane				0.14		
Propene				0.06		
Allene (propadiene)	trace	trace		trace		
Propyne	0.13	0.09				
But-2-ene		1.7				
But-1-ene	1.0					
Buta-1,2-diene	0.45	0.23			trace	
But-2-yne					0.14	
Pentenes	0.06	0.01		0.9	2.0	
Pentadienes				?	0.4	
Hexenes				trace	0.7	1.9
Dimers	1.2	0.7	$(10.86)^a$	1.2	2.8	3.1
Alkylbenzene	2.3	0.5	$(0.23)^a$	0.7	0.20	0.16
Other trimer	0.45	trace		0.9	0.6	0.4
Tetramer	0.7	0.45	$(0.45)^a$			

[a] The values in brackets are from ref. [198]

hydrogen has been proposed by Stief et al. as the primary process [200]:

$$CH_3-C \equiv CH \xrightarrow{h\nu} C_3H_3 + H$$

with subsequent formation of acetylene and ethane:

$$H + CH_3 - C \equiv CH \longrightarrow CH_3 + C_2H_2$$

$$2\ CH_3 \longrightarrow C_2H_6$$

The substitution of methyl radical by a hydrogen atom has since been observed by Safrany and Yaster [201].

In contrast to propyne, $C-C$ bond rupture is an extremely important process in the radiolysis of but-1-ene as shown by the higher yield of methane than hydrogen (see Table 4.35). Generally, yields are higher in the gas than in the liquid phase, but here the ratios are also altered to a large extent, e.g. the ratio of H_2 to CH_4 in the gas phase is 0.7, whereas in the liquid phase it is 3.5. This may be explained by the cage effect, i.e. in the liquid phase the recombination of radicals occurs more easily. The same explanation is given for the low amount of buta-1,2-diene in the gas-phase irradiation of but-1-yne:

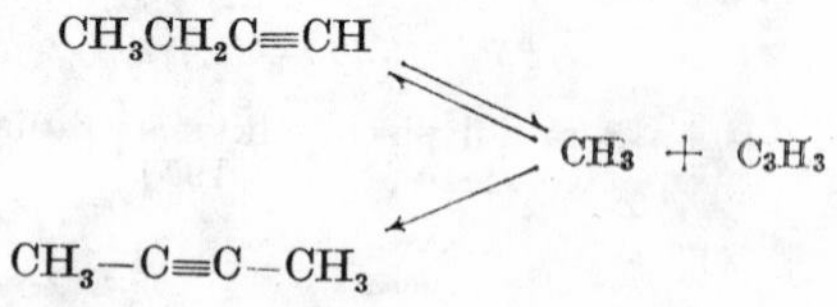

$$CH_3CH_2C\equiv CH$$
$$CH_3 + C_3H_3$$
$$CH_3-C\equiv C-CH_3$$

On the basis of quantum yields determined in the vacuum-UV photolysis of but-1-yne (Table 4.37), one of the main products is vinylacetylene [202]. As the quantum yield is not influenced either by pressure or by radical scav-

Table 4.37

Yields of the vacuum-UV photolysis of gaseous but-1-yne (pressure: 1.3 mbar). After Hill and Deopker [202]

Energy, eV	8.4		10.0	
Products	Quantum yield			
Hydrogen	0.27	0.20	0.29	0.14
Methane	0.04	0.04	0.03	0.03
Ethane	0.23	0.13	0.16	0.08
Ethylene	0.14	0.12	0.09	0.10
Acetylene	0.11	0.12	0.15	0.16
Propane	0.09	0.05	0.05	0.03
Allene (propadiene)	0.52	0.25	0.60	0.21
Vinylacetylene	0.37	0.33	0.36	0.36

engers (O_2, NO), Hill and Doepker have assumed [202] that vinylacetylene is formed by direct decomposition:

$$1\text{-}C_4H_6 + h\nu \begin{cases} \nearrow\; 2\,H + C_4H_4 \quad \Phi = 0.34 & (4.31a) \\ \searrow\; H_2 + C_4H_4^* & (4.31b) \end{cases}$$

where the C_4H_4 molecule formed in reaction (4.31b) possesses an excess energy of about 670 kJ mol^{-1}. Thus it is either decomposed to fragments or stabilized by collision. As the yield of vinylacetylene is independent of pressure, it was concluded that reaction (4.31b) is predominant.

The yields of both allene and hydrogen are decreased by radical scavengers: thus the following reactions were proposed:

$$H + 1\text{-}C_4H_6 \longrightarrow C_4H_7^* \qquad\qquad (4.32)$$

$$C_4H_7^* \begin{cases} \nearrow\; CH_3 + C_3H_4 \text{ (allene)} & (4.33a) \\ \searrow_{\!M}\; C_4H_7 & (4.33b) \end{cases}$$

It was observed by Hill and Doepker [202] that the ratio $C_2H_2 : C_2H_4$ increases with increasing photon energy. This was, however, not supported by experiments carried out using ^{60}Co γ-radiolysis (see Tables 4.35 and 4.37). This increase in ratio can be interpreted, according to Hill and Doepker, by assuming that reaction (4.34b) increases at the expense of reaction (4.34a):

$$1\text{-}C_4H_6 + h\nu \begin{cases} \nearrow\; C_2H_2 + C_2H_4 & (4.34a) \\ \longrightarrow\; C_2H_2 + C_2H_2 + H_2 \quad \Phi = 0.13 & (4.34b) \\ \searrow\; H + C_2H_4 + C_2H \quad\;\; \Phi = 0.10 & (4.34c) \end{cases}$$

Reaction (4.34b) may be also responsible for the formation of small amounts of hydrogen. Some of the ethylene may be formed by reaction (4.34c). Acetylene yield is also increased by the fragmentation of the excited $C_4H_5^*$ intermediates. In addition to reactions (4.31)–(4.34), the following reactions have been verified:

$$1\text{-}C_4H_6 + h\nu \begin{cases} CH_2 + C_3H_4 & \Phi = 0.02 \\ H_2 + C_2H + C_2H_3 & \Phi = 0.2\text{--}0.25 \\ C_2H_5 + C_2H & \Phi = 0.19 \\ CH_3 + C_3H_3 & \Phi = 0.10 \end{cases}$$

The existence of CH_3, C_2H, C_2H_3, C_2H_5 and C_3H_3 ($CH{=}C{-}CH_2$) radicals has been also proved by photolysis carried out in the presence of H_2S, D_2S and CD_3I radical scavengers.

Although there are insufficient data on the radiolysis of alkylacetylenes to establish the mechanism in detail, some general conclusions can be drawn:

1. The G-values of products formed under radiolysis from hydrocarbons containing a triple bond are similar to those found for alkenes: the G-value for the disappearance of reactant lies between 6 and 10;
2. The yields of hydrogen and fragments point to the importance of the rupture of $C-H$ and $C-C$ bonds in the β-position to the triple bond, though splitting of the α-bond is also significant.

It seems from the above facts that the presence of the acetylenic bond exerts a 'protective' effect and leads to polymerization; this is similar to the effect of double bonds. The basic difference is that aromatization reactions take place with every alkynes. Regardless of the reactant, benzenes of different structure are formed, e.g. hexamethyl benzene from but-2-yne:

$$C-C\equiv C-C^* + 2\,C-C\equiv C-C \rightarrow$$

and 1,3,5- and 1,2,4-trimethylbenzene from but-1-yne. The formation of alkylbenzenes was suggested to be a concerted reaction occurring via excited states [197]. Whitten and Berngruber [198] have investigated some of the main products (di-, tri- and tetramers) of but-2-yne radiolysis in the presence of additives. They found that dimers such as octa-2,6-diene and 3-methylhept-2-en-5-yne are formed via radical intermediates. The formation of the trimer hexamethylbenzene and the tetramer (*syn*-octamethyltricyclo[4.2.0.0$^{2.5}$]octadiene) was assumed to proceed by a step-wise process in which tetramethylbutadiene is an immediate precursor. Consideration of the effect of additives indicated that methylbutadiene participates in the process through two states of different multiplicities.

4.3. MIXTURES OF ALIPHATIC ALKENES

One of the most popular fields of study in radiolysis is the investigation of mixtures of saturated and unsaturated hydrocarbons.

As with the results reported in Section 2.3., the yields usually differ from what would expected on the basis of the linear mixing rule, owing to physicochemical and/or chemical interaction between the individual components [203, 204]. Alkenes are usually acceptors in the transfer processes, owing to their lower excitation and ionization potential, but addition of H-atoms and hydrocarbon radicals to the π-bond, and ion-molecule reactions, must also be considered.

Alkane–aliphatic alkene mixtures

$G(H_2)$ values for several alkane–alkene mixtures [203, 205] were found to be much lower than expected on the basis of the linear mixing rule. This is shown in Fig. 4.5, in which $G(H_2)$ values obtained in the radiolysis of n-hexane–hex-1-ene mixtures [205] are plotted as a function of the electron fraction of hex-1-ene.

It was found that the extent of interaction observed in the radiolysis of alkane alk-1-ene mixtures also depends on the carbon number of the alkene, which effects the ratio of σ to π-bonds. Földiák et al. [203, 214] have established that the Φ value (Fig. 4.6), like any other single character-

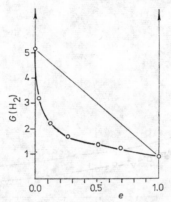

Fig. 4.5. Dependence of the $G(H_2)$-value on the electron fraction of hex-1-ene in the radiolysis of n-hexane–hex-1-ene mixtures (radiation: ^{60}Co-γ). After Földiák et al. [205]

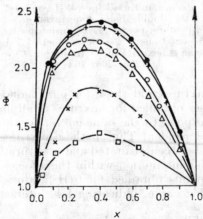

Fig. 4.6. Φ vs. x curves for mixtures of an aliphatic alkane and an aliphatic alkene
Φ-ratio of $G(H_2)$ values, calculated on the basis of the linear mixing rule and measured hydrogen yields; x mole fraction of unsaturated hydrocarbons
● C_6; + C_5; ○ C_8; △ C_7; × C_4; □ C_3 (radiation: ^{60}Co-γ). After Földiák et al. [203, 213, 214]

istic by which the extent of 'protective effect' would be qualified, is not suitable for this purpose. Because a certain transformation is needed for better evaluation, the $G(H_2)$ values were plotted not as a function of electron fraction e_E, but of the concentration of double bonds, y, i.e. of the number of π-bonds in 1 g of the mixture (a quantity introduced by Földiák et al. [205, 206]. As seen from Fig. 4.7, the points representing the $G(H_2)$ values of

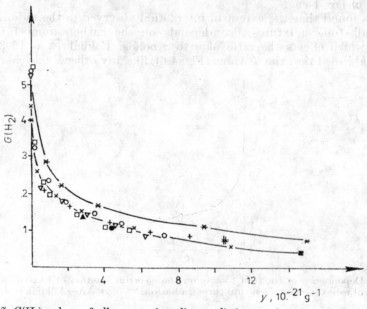

Fig. 4.7. $G(H_2)$-values of alkenes and n-alkane–alk-1-ene mixtures as a function of y, the double bond concentration.
✳ Propane-propene;　$+$ n-pentane–pent-1-ene;　○ n-hexane–hex-1-ene;　▽ n-heptane––hept-1-ene;　□ n-octane–oct-1-ene;　✕ n-octane–octa-1,7-diene;　# but-1-ene;　● dec-1-ene;　▲ hexadec-1-ene;　■ hexa-1,5-diene (radiation: ^{60}Co-γ). After Földiák et al. [205, 206, 263]

n-alkane–alk-1-ene mixtures and those for pure monoalkenes fit the same curve. The $G(H_2)$ values obtained for mixtures of aliphatic alkanes and isolated dienes also correspond to the same curve, i.e. the effects of these two π-bonds are not interdependent. Thus, in these mixtures, the extent of intermolecular interaction between saturated and unsaturated molecules, as well as of the intramolecular interaction within the alkenes, is broadly similar.

The experimental points obtained for $G(H_2)$ values of propane–propene mixtures lie above this curve, indicating less interaction. The reason for this is not clear yet.

A common curve, different from the previous one, represents the $G(H_2)$ values measured in the radiolysis of mixtures of alkanes and conjugated dienes (Fig. 4.8). This points to a more effective interaction, which may be explained by the high reactivity of conjugated dienes and their ability to dissipate energy within the molecule.

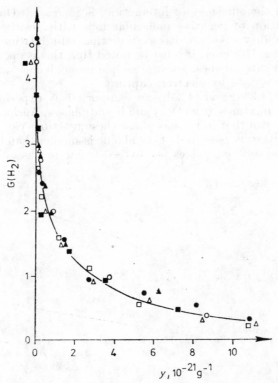

Fig. 4.8. $G(H_2)$-values of mixtures of alkanes with conjugated dienes or cyclo-octatetraene as a function of y, the double bond concentration.
○ n-Hexane–cyclohexa-1,3-diene; ■ cyclohexane–cyclohexa-1,3-diene; ▲ n-hep-tane–n-hepta-1,3-diene; ● n-octane–octa-2,4-diene; △ n-octane–octa-1,3-diene; □ n-octane–cyclo-octatetraene (radiation: ^{60}Co-γ). After Földiák and Wojnárovits [203]

The ion-molecule reactions that influence the product distribution and the shape of G-values versus composition curves in the radiolysis of alkane–alkene systems have been studied in the recent years. For example, Robinson and Freeman [207, 208] analysed the isotope content of hydrogen formed from cyclohexane–deuteroalkene mixtures under irradiation, and suggested that the electrons formed in radiolysis are added to alkene molecules with a high probability; consequently, the rate of some ion-molecule reactions are accelerated:

$$\text{cyclo-}C_6H_{12}^+ + C_3D_6 \longrightarrow \text{cyclo-}C_6H_{11}^+ + C_3D_6H \qquad (4.35a)$$

$$\longrightarrow \text{cyclo-}C_6H_{10}^+ + C_3D_6H_2 \qquad (4.35b)$$

$$\longrightarrow \text{cyclo-}C_6H_{12} + C_3D_6^+ \qquad (4.35c)$$

According to this hypothesis, the half-life of positive ions is extended by the formation of negative molecule ions, so that the probability of charge

21*

transfer, i.e. the efficiency of interaction, is increased. On the other hand, the annihilation of negative molecular ions with positive ions leads to a lower probability of bond dissociation than annihilation of positive ions with electrons. However, it must be noted that this is operative only in the case of conjugated dienes, because simple monoalkenes and isolated dienes do not form anions by electron capture.

Reactions (4.35a) and (4.35b) are supported by experiments on alkane–monoalkene mixtures in the gas and liquid phases. Ausloos et al. [209, 211, 226] have shown that in the gas phase these reactions can be observed only when the ionization potential of the alkene is smaller than that of the alkane providing the difference does not exceed a few tenths of an electron volt.

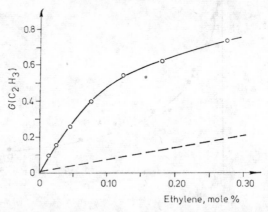

Fig. 4.9. Yields of vinyl radical from ethane–ethylene mixtures (radiation: 2.8 MeV electrons). After Fessenden and Schuler [212]

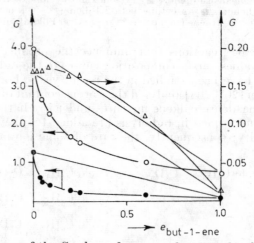

Fig. 4.10. Dependence of the G-values of some products on the electron fraction of but-1-ene in the radiolysis of n-butane–but-1-ene mixtures. △ C_3H_6; ◯ H_2; ● C_2H_6 (radiation: ^{60}Co-γ). After Földiák and Horváth [213, 214]

This statement is supported by the hypothesis of van Ingen and Cramer [204]. The probability of the H_2 transfer reaction (4.35b) increases with a decrease in the ionizing potential difference between the components, they concluded on the basis of the decrease of $G(H_2)$ values of cyclohexane irradiated in the presence of C_3-C_6 alkenes of different structures. An opposite trend was found for charge-transfer processes.

Fessenden and Schuler [212] have studied the relative yields of ethyl and vinyl radicals in the radiolysis of ethane–ethylene mixture.

The significant positive deviation of the observed vinyl yields (solid line) from the calculated values (dashed line in Fig. 4.9) can be interpreted in terms of various transfer processes (e.g. excitation and charge). The early conclusion of Fessenden and Schuler was that ethylene is more efficient in energy absorption than ethane.

Radiation chemical yields of hydrogen and several hydrocarbon products were determined by Földiák and Horváth [213, 214] in the radiolysis of

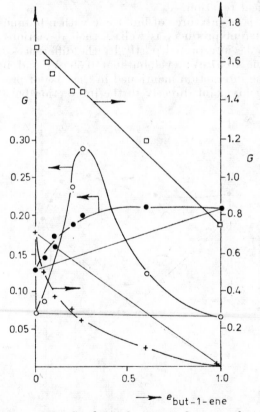

Fig. 4.11. Dependence of the G-values of some products on the electron fraction of but-1-ene in the radiolysis of cyclobutane–but-1-ene mixtures.
□ H_2; ○ *cis*-2—C_4H_8; ● *trans*-2—C_4H_8; + Cylo-C_4H_6 (radiation: ^{60}Co-γ). After Földiák and Horváth [213, 214]

binary liquid mixtures, such as C_3-alkanes–propene and C_4-alkanes–but-1-ene. It was observed not only that the plots of yields against the composition show the 'typical' curves like those in Fig. 4.5, but also that some of the functions reveal basic differences from this and from each other. The absolute or relative maxima and minima in certain cases (e.g. Figs 4.10 and 4.11), mainly when the difference between the ionization potentials of the components is relatively small, point to complex reactions.

On the basis of these values it can also be established that a part of the radiation energy absorbed by the saturated compound is transferred to propene. Thus the yields of products formed from the alkane are lower and those from propene are higher than would be expected on the basis of the linear mixing rule. This effect can be explained by the ionization potential, which is lower for propene (9.73 eV) than for both propane (11.07 eV) and cyclopropane (10.06 eV).

The curves for the yields of hydrogen and hexa-1,5-diene also point to a complex process (Fig. 4.12), consisting of a physical interaction and a bimolecular chemical reaction.

In the radiolysis of mixtures of but-1-ene with n-butane or cyclobutane, the yields of different products as well as their deviations from the linear mixing rule may be interpreted partly by the differences in ionization potentials (n-butane 10.57 eV; cyclobutane 10.56 eV; and but-1-ene 9.6 eV, and partly by the interaction mentioned in the case of propane [213, 214]). The hydrogen yields point directly to the interaction between the compo-

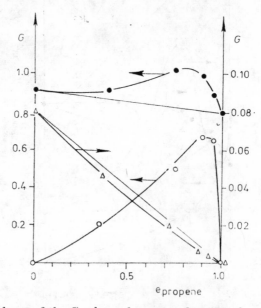

Fig. 4.12. Dependence of the *G*-values of some products on the electron fraction of propene in the radiolysis of cyclopropane–propene mixtures.
● H_2; ○ 1,5-C_6H_{10}; △ CH_3 – cylo-C_3H_5-CH_3 (radiation: ^{60}Co-γ). After Földiák and Horváth [213]

nents in the case of n-butane–but-1-ene mixtures (Fig. 4.10), but this is not so for mixtures of cyclobutane and but-1-ene: the straight line for the hydrogen yield (Fig. 4.11) in the latter case might not be the result of the the linear mixing rule, holding true, but may be caused by the compensation effect of different, opposite reactions. Therefore the authors concluded that each product requires separate investigation in elucidating the mechanism in the radiolysis of mixtures. Földiák and Horváth stated [213, 214] that the hydrogen yield is the least suitable parameter for characterization of these processes, because nearly all steps involve hydrogenation or dehydrogenation, and only the overall hydrogen yield can be measured, except if labelled compounds are applied.

A remarkable interaction is indicated by the yield of *trans*-but-2-ene in the radiolysis of mixtures of cyclobutane and but-1-ene. The positive deviation of this yield from the calculated value is much greater than that of the *cis*-but-2-ene yield. The stability of the latter lies between those of *trans*-but-2-ene and but-1-ene.

Ausloos et al. [209] studied the transfer processes in labelled hydrocarbon mixtures. These investigations were carried out in presence of radical scavengers (O_2, NO) in order to exclude the disturbing effect of neutral products formed in radical processes. In this way the analysis of neutral products gives information on the structure of the transition complex and on the stereospecificity of these reactions.

With exception of cyclopropane, H^- ion transfer is the main reaction between alkyl ions and neutral alkane molecules; in addition to this some other processes occur in the systems alkene ion–alkane, alkyl ion–alkene and alkene ion–alkene [215, 216]. This is the reason why, e.g. in alkyl ion–alkene systems hydrogen transfer occurs to a much lesser extent than in alkane systems.

The H^- transfer reaction to alkene ions has been studied by mass spectrometry and by radiolysis [211, 216–221].

$$C_nH_{2n}^+ + RH \longrightarrow C_nH_{2n+1} + R^+$$

Doepker and Ausloos [211] have studied the H^- transfer to propene ions for a number of alkanes. It was established that the rate strongly depends on the structure of the donor alkane. Generally, this process is preferred to the other competitive reactions if the donor is a branched-chain hydrocarbon, i.e. containing a hydrogen atom attached to a tertiary carbon.

H_2^- ion transfer has been also detected with several experimental techniques [209–211, 217–219]. H_2^- ion transfer to alkene ions has been studied using propene-d_6 ions produced by photolysis at a wavelength of 123.6 nm (10.0 eV) in the presence of several alkanes [219]. A transition complex of well-defined structure was detected, formed by weak bonding between specifically labelled isobutane and deuterated propene ion through which DH^- exchange takes place:

$$C_3D_6^+ + (CH_3)_3CD \longrightarrow CD_3CDHCD_3 + C_4H_8^+$$

The following H_2^- transfer reaction also takes place specifically at the positions shown:

$$C_3D_6^+ + CD_3CH_2CH_2CD_3 \longrightarrow CD_3CD_2CD_2H + CD_2CHCH_2CD_3^+$$

On the basis of the isotopic composition of the photolysis products, the relative rate of reaction (4.36a) has been calculated in presence of alkanes of different structure (Table 4.38):

$$C_3D_6^+ + RH_2 \Big\langle \begin{array}{ll} CD_3CDHCD_2H + R^+ & \text{(4.36a)} \\ C_3D_6H + RH^+ & \text{(4.36b)} \end{array}$$

The overall rate of reactions (4.36a) and (4.36b) is hardly influenced by the molecular structure but the probability of the H_2^- transfer reaction is considerably affected (Table 4.38).

Table 4.38

Relative rates of the ion-molecule reactions of propene ions
$C_3D_6^+ + $ alkane $\rightarrow C_3D_6H_2 + R^+$. After Sieck and Searles [219]

Alkane	Relative rate	
	Photoionization of C_3D_6	γ-Radiolysis of cyclo-C_5D_{10}
n-Butane	0.51	0.65
Isobutane	0.21	0.23
Cyclobutane	0.089	0.074
n-Pentane	0.96	0.112
Isopentane	0.46	0.45
Neopentane	0.00	0.000
Cyclopentane	1.0	1.0
Cyclohexane	1.24	

H_2 transfer essentially differs from H_2^- transfer only in that the positive charge is not transferred:

$$C_nH_{2n} + RH_2^+ \longrightarrow C_nH_{2n+2} + R^+$$

where C_nH_{2n} represents an unsaturated hydrocarbon or cyclopropane with n less than the carbon atom number in the alkane RH_2^+ ion. Presumably, H-atom transfer competes with this process:

$$C_nH_{2n} + RH_2^+ \rightarrow C_nH_{2n+1} + RH^+$$

Ausloos et al. have observed H_2 and H atom transfer in both gas and liquid phase radiolysis [209, 222, 223] as well as in photolysis (123.6 and 14.5 nm). The ratio of these two processes is different in the two phases. H atom and H_2 transfers have been also observed by Abramson and Futrell [224] in the mass spectrometer, and by Fujisaki et al. [225] in the radiolysis of n-butane–ethylene and n-butane–propene mixtures in the presence of oxygen as radical scavenger. The ratio of $k(H^-): k(H_2^-)$ is an order of magnitude higher for ethylene than for propene.

There is a competition between the H^+ transfer reaction of alkenes and carbonium ions:

$$C_nH_{2n} + RH_2^+ \longrightarrow C_nH_{2n+1}^+ + RH$$

and other transfer reactions. Thus, in the radiolysis of mixtures of perprotonated propane and perdeuterated propene [226] hydride ion transfer and condensation are the competing processes:

$$C_2D_5^+ + C_3H_6 \xrightarrow{D^+} C_2D_4 + C_3H_6D^+ \qquad (4.37a)$$

$$C_2D_5^+ + C_3D_8 \xrightarrow{D^-} C_2D_6 + C_3D_7^+ \qquad (4.37b)$$

$$C_2D_5^+ + C_3H_6 \xrightarrow{H^-} C_2D_5H + C_3H_5^+ \qquad (4.37c)$$

$$C_2D_5^+ + C_3H_6 \longrightarrow C_5D_5H_6^+ \qquad (4.37d)$$

The relative rates of reactions (4.37a), (4.37b), (4.37c) and (4.37d) at 40 mbar pressure are 1.07, 1.00, 0.16 and 0.93, respectively.

In a study of gas-phase neopentane radiolysis, Collin [226A] investigated the reaction of tert-$C_4H_9^+$ ions with some C_5H_{10} alkenes. He has shown that the tert-$C_4H_9^+$ ion (an intermediate product of neopentane radiolysis) reacts with *trans*-pent-2-ene and 2-methylbut-1-ene in hydrid transfer reactions, e.g.

$$\text{tert-}C_4H_9^+ + \textit{trans-}2\text{-}C_5H_{10} \longrightarrow \text{iso-}C_4H_{10} + C_5H_9^+$$

whereas, if the unsaturated additives are 2-methylbut-1-ene or 2-methylbut-2-ene, proton transfer takes place:

$$\text{tert-}C_4H_9^+ + CH_2 = C(CH_3)CH_2CH_3 \longrightarrow \text{iso-}C_4H_8 + \text{tert-}C_5H_{11}^+$$

The transition complex in condensation reactions [e.g. (4.37d)] is strongly bonded and considerable rearrangement may occur: depending on the experimental conditions, may be decomposed or stabilized by collision. In the latter case further reactions may produce condensation products of higher molecular mass [81, 227–229].

Aliphatic alkene–diene mixtures

Only very few data are available in the literature for monoalkene–diene systems. The radiolysis of hex-1-ene–*cis, trans*-hexa-2,4-diene and oct-1-ene–oct-1,7-diene systems was studied by Kovács and Földiák [226B]. It was found that the conjugated diene caused significant changes in the yields of almost every product of hex-1-ene radiolysis. By contrast, the isolated diene had a negligible influence on the products from oct-1-ene. The reason

for the different behaviour of the two dienes lies in their very unlike reactivities towards radicals and other intermediates. For instance, charge
transfer and excitation transfer to the conjugated diene are, probably, responsible for the observed decrease in the hydrogen yield and for the significant isomerization of the conjugated diene.

4.4. CONCLUSIONS

In Section 4.2 the radiolysis of individual alkenes was considered. In this
Section, we present general information on decomposition and some condensation reactions.

Decomposition reactions

The hydrogen yields of some aliphatic alkenes under radiolysis are listed
in Table 4.39. It is evident that the $G(H_2)$ values of alkenes are usually
an order of magnitude lower than those of the corresponding n-alkanes
(Table 2.43).

The mechanism of dehydrogenation of alkenes under irradiation differs
considerably from that of alkanes, because the π-bond is a very effective
scavenger of hydrogen atoms. In addition, it is also a selective energy-
acceptor. It seems to be a reasonable assumption that in alkene radiolysis
neutralization processes leading to hydrogen formation are accompanied
by fast ion-molecule reactions (e.g. condensation) and thus the probability
of C—H bond rupture is diminished.

In the irradiation of pure alkenes it is always the molecular hydrogen
yield that is measured (see Section 1.2.1.). This is because nearly all
thermal hydrogen atoms are trapped by the double bonds. This hypothesis
is verified by the fact that the hydrogen yield of monoalkenes is independent of dose over a wide range. In contrast, $G(H_2)$ of alkanes decreases
with increasing dose, which may be due to interactions between the radicals,
hydrogen atoms and unsaturated hydrocarbons produced (see Section 1.3.1.)
Hatano and Shida [88] have shown that in the liquid-phase radiolysis of
alkenes, the hydrogen yields are not significantly altered by extensive
changes in temperature (Table 4.40).

The unimolecular yield, i.e. the proportion of H_2 elimination, depends on
the molecular structure: there is a big difference between the proportions
of unimolecular H_2 production of, for example, ethylene and other alkenes
(Table 4.41). The yield of unimolecular hydrogen formed from hydrocarbons
containing more than three carbon atoms is to about 20–30% of the overall
$G(H_2)$, and the remainder of the hydrogen is formed via abstraction by hot
hydrogen atoms. If the abstraction occurs from the same molecule or radical
from which the hot hydrogen originates, this process is not experimentally
separable from the unimolecular hydrogen production.

The existence of 'hot' hydrogen atoms and their role in radiation chemistry have been proposed by many authors, e.g. Manion and Burton [230]

Table 4.39

Yields from liquid aliphatic monoalkenes and the apparent number and order of C—H bonds

Hydrocarbon	Measured $G(H_2)$, molecule/100 eV	Calculated[a]	Allylic C—H primary	secondary	tertiary	Non-allylic[b] C—H	Ref.
Propene	0.60	0.59	3			3	68
	0.80						71
But-1-ene	0.74	0.78		2		6	88
	0.73						139
cis-But-2-ene	1.02	1.06	6			2	87, 86
	0.96						
Pent-1-ene	0.83	0.88		2		8	151
Hex-1-ene	0.90	0.92		2		10	150, 151
Hex-2-ene	1.37	1.37	3	2		7	151
	1.17						150
Hex-3-ene	1.64	1.60		4		8	151
	1.61						150
Hept-1-ene	0.94	0.96		2		12	151
Oct-1-ene	0.98	1.00		2		14	151
Oct-2-ene	1.42	1.45	3	2		11	151
Oct-3-ene	1.63	1.68		4		12	151
Oct-4-ene	1.61	1.68		4		12	151
Dec-1-ene	1.10	1.08		2		18	151
Dec-5-ene	1.70	1.78		4		16	151
2-Methylhex-1-ene	1.25	1.41	3	2		9	151
3-Methylhex-1-ene	0.82	0.77			1	13	151
4-Methylhex-1-ene	1.13	1.13		2		12 (1)	151
5-Methylhex-1-ene	1.16	1.13		2		12 (1)	151
trans-2-Methylhex-3-ene	1.30	1.45		2	1	11	151
Dec-5-ene	1.70	1.78		4		16	151
Hexadec-1-ene	1.39	1.32		2		30	181

[a] Calculated by using $G(H_2)/H$ increments (Table 4.43)
[b] Figures in this column are the total number of non-allylic bonds; the number of tertiary bonds is in parenthesis

Table 4.40

Temperature dependence of hydrogen yields from liquid n-butenes (radiation: ^{60}Co-γ). After Hatano and Shida [88]

Hydrocarbon Temperature, °C	But-1-ene $G(H_2)$, molecule/100 eV	trans-But-2-ene
—78	0.70	1.24
0	0.74	1.20
30	0.74	1.22
50	0.70	1.23
70	0.79	1.23
90	0.79	1.20
110	0.79	1.22

Table 4.41

Hydrogen formation from liquid alkenes (radiation: ^{60}Co-γ). After Hatano
et al. [39]

Hydrocarbons	$G(H_2)$, molecule/100 eV	Distribution of hydrogen source, %	
		unimolecular	bimolecular
Ethylene	1.24	71	29
Propene	0.82	29	71
But-1-ene	0.74	22	78
trans-But-2-ene	1.22	18	82

and Voevodskii and Molin [231]. From the excitation spectrum of alkenes,
it was suggested by Hatano et al. that with the exception of ethylene, most
of the hot hydrogen atoms are not excited but are in the ground state and
possess a great deal of kinetic energy [88]. Following the dissociation of an
excited molecule (i.e. the rupture of a C—H bond) some of the energy of the
excited molecule may be converted into kinetic energy of fragments R and
H. As the masses of the hydrogen atom and the remaining molecular fragment are very different, and because of the conservation of momentum,
most of the kinetic energy is transferred to the hydrogen atom.

The hydrogen yield was not significantly influenced by radical scavengers,
positive ion or electron scavengers [88] (Table 4.42). Therefore, Hatano
et al. [88] concluded that in the radiolysis of alkenes, the transition products
of hydrogen formation are the directly excited molecules [reaction (4.38)],
mainly the superexcited molecules [232, 233] with energy higher than the
ionization energy of the molecule. The superexcited molecules decompose
via the very fast reactions (4.39)–(4.42):

$$RH \quad \leadsto \quad RH^* \tag{4.38}$$

$$RH^* \longrightarrow H_2 + \text{alkadiene} \tag{4.39}$$

$$RH^* \longrightarrow H^* + R \tag{4.40}$$

$$RH^* \longrightarrow 2\,H^* + \text{other products} \tag{4.41}$$

$$H^* + RH \longrightarrow H_2 + R \tag{4.42}$$

where H^* means hot hydrogen atom.

On the other hand, Kovács et al. [234] have observed a significant decrease in the hydrogen yield from cyclohexene in the presence of CCl_4 as
electron scavenger (Table 4.42). A similar effect has been observed in the
radiolysis of 2,3,3-trimethylbut-1-ene where CCl_4 at 0.05, 0.5 and 1 M concentrations lowered the initial $G(H_2) = 0.62$ value to 0.51, 0.32 and 0.25,
respectively [234A]. Though excitation energy transfer to CCl_4 cannot be
excluded, these observations point to geminate ion-electron recombination
processes as being important in the production of hydrogen.

According to Table 4.39 and Fig. 4.13, the $G(H_2)$ values of n-alk-1-enes increase monotonically with increasing number of carbon atoms. The differences between two successive n-alk-1-enes in the homologous series are small, but the deviations between structure isomers (e.g. n-hexenes) are large. Based on the explanation of this phenomenon, it has been proved by Cserép and Földiák [151] that the directing effect of the π-bond in alkenes is valid not only in thermal and photochemical decomposition, but also in radiation chemical

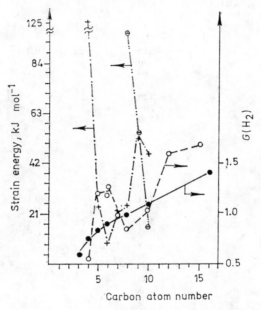

Fig. 4.13. $G(H_2)$-values of monoalkenes plotted against carbon atom numbers.
● 1-Alkenes; ○ cycloalkenes; + ring strain energies of the *cis* form; ⊖ ring straing energies of the *trans* form (radiation: ⁶⁰Co-γ). After Cserép and Földiák [151]

processes. Generally, the probability of rupture of a bond in the β-position to the double bond is an order of magnitude higher than that of one in the α-position. The σ bond in the allylic position has a determining role in the radiation-induced reactions of alkenes (e.g. decomposition, dimerization). This is supported by Wagner's data [265], according to which the formation of C_7 products in the radiolysis of mixture of pent-1-ene and n-butane can be explained by the reaction of allylic radicals. Similar results have been obtained by Ohmae et al. [235], Ayscough and Evans [169], Roginskii and Pshezhetskii [162] and Smith and Pieroni [174]. These authors have detected allyl-type radicals by ESR spectroscopy in irradiated aliphatic and cyclic alkenes. Holroyd and Klein [69], also detected allyl and allyl-type radicals in the radiolysis of C_3 and C_4 alkenes by means of $^{14}CH_3I$. These observations are opposed to one of the earlier papers of Hardwick [236], in which he stated that H atoms were mainly dissociated from a C—H bond

Table 4.42

Dependence of hydrogen yields of alkenes on scavengers (radiation: $^{60}Co-\gamma$)

Hydrocarbons		Ethylene	Propene	Cyclohexene
References		39	39	234
Scavengers	Concentration, mM	$G(H_2)$, molecule/100 eV		
Electron scavengers				
N_2O				
CCl_4	10	1.25	0.78	
	5			1.21
	40	1.24	0.78	
	50			1.16
	80	1.35	0.86	
	100			1.06
	120	1.24	0.76	
	200			0.82
	500			0.63
	1000			0.42
SF_6	4	1.30	0.82	1.19
	10			1.16
	20	1.30	0.82	
	50			0.85
Ion scavengers				
CH_3OH	100	1.52	0.82	
	200	1.42	0.82	
	500	1.52	0.86	
	1000	1.47	0.94	
C_2H_5OH	5			1.19
	50			1.21
	100			1.21
	200			1.24
NH_3	5			1.24
	10			1.25
	50			1.25
	100		0.82	
	200		0.82	1.28
	500		0.86	
	1000		0.94	
Radical scavenger				
Benzoquinone	5			1.19
	20			1.16

in vinylic position. However, this is not supported even by his own data concerning the stabilities of C—H bond [237].

The large differences between the $G(H_2)$ values of structure isomer alkenes, therefore, are the result of the different order and number of C—H bonds in the β-position to the π-bond [151—154]. However, the stabilities of both allylic and non-allylic C—H bonds follow the sequence primary > > secondary > tertiary position. This trend is also observed in the radiolysis of alkanes (see Sections 1.3.4 and 2.4). Thus, the $G(H_2)$ values of similar aliphatic alkenes can be calculated from the hydrogen increments $G(H_2)$/H

Table 4.43

Contribution of different C—H bonds in aliphatic alkenes to the hydrogen yield measured in liquid-phase radiolysis. After Cserép and Földiák [154, 258]

Type of C—H	Allylic	Non-allylic
	$G(H_2)/H$	
Primary	0.172 ± 0.009	0.010 ± 0.005
Secondary	0.359 ± 0.016	0.026 ± 0.007
Tertiary	0.520 ± 0.047	0.210 ± 0.050

(Table 4.43) obtained by Cserép and Földiák for primary, secondary and tertiary allylic and non allylic C—H bonds [154, 258].

This view is also supported by the observations made in the radiolysis of the homologous series of alk-1-enes. Here the $G(H_2)$ values can be estimated from the fact that each additional carbon atom introduces two secondary non-allylic hydrogen atoms which increase the hydrogen yield by 0.02–0.03 per hydrogen atom.

The greater probability of σ-bond decomposition being in the β-position to a π-bond is also supported by energy considerations: i.e. if a bond of this type is ruptured and an allyl-type radical is formed, then the allyl-resonance energy arising by the interaction of the unpaired electron of the unsaturated radical with the double bond may partially supply the energy required for the decomposition. This is not the case for cleavage of a C—C bond in the α-position, which explains the greater stability of the latter bond. According to Benson's data [238] the dissociation of a primary allyl C—H bond requires 69 kJ mol^{-1} less than that of a vinyl C—H bond.

Figure 4.14 is a plot of the specific hydrogen yields (increments) of different allylic and non-allylic C—H bonds as a function of the dissociation energy.

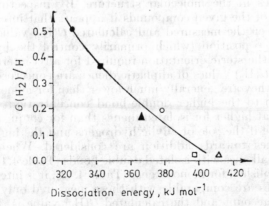

Fig. 4.14. Specific hydrogen yield increments of aliphatic alkenes as a function of the bond dissociation energy of the respective C—H bonds.
● Tertiary allyl; ■ secondary allyl; ▲ primary allyl; ○ tertiary non-allyl; △ primary non-allyl; □ secondary non-allyl (radiation: ^{60}Co-γ). After Földiák et al. [264]

The role of steric factors in $G(H_2)$ values in the radiolysis of n-alkenes has been considered by Cserép and Földiák. The results obtained for n-hexene isomers (Table 4.39) clearly show that the $G(H_2)$ value increases as the π-bond is shifted further from a terminal position. This could be explained by increased steric hindrance to H atom addition in internal positions. This assumption, however, seems to be contradicted by the results obtained from the radiolysis of octene isomers and of dec-5-ene (Table 4.39), namely that the $G(H_2)$ values for oct-3-ene and oct-4-ene (both possess four secondary allyl hydrogen atoms) are practically the same (1.63 and 1.61, respectively) and differ only slightly from that obtained for dec-5-ene (1.70). All these hydrogen yields are very close to the one calculated on the basis of the corresponding $G(H_2)/H$ values (Table 4.43). These results show that the steric influence on the hydrogen yield in the radiolysis of n-alkenes may be of minor importance.

Further confirmation of the connection between hydrogen yield and bond energies is provided by data on methylhex-1-ene isomers (Table 4.31): if the deviations in hydrogen yield were caused by unequal reactivities of π-bonds in different positions, the $G(H_2)$ values of different isomers would not differ much from each other or from that of hex-1-ene. However, the deviation is fairly large and the $G(H_2)$ values are generally in agreement with those calculated on the basis of specific $G(H_2)/H$ values of C—H bonds of different orders (Table 4.43).

Hydrogen formation in the radiolysis of alkenes has been studied by Cserép and Földiák using multi-methyl substituted compounds [258, 258A]. It has been found that there are significant differences between the measured and calculated $G(H_2)$ values of many alkenes. For example, the measured yields of 2,3-dimethylbut-2-ene, 2,4-dimethylpent-1-ene and 2,4,4-trimethylpent-1-ene are much smaller than the calculated ones, whereas in other cases, e.g. for 2,2-dimethylhex-3-ene, and 2,3,3-trimethylbut-1-ene, the two values agree very well. The reason for the disagreement found in certain cases lies in the molecular structure. By inspecting the Stuart–Briegleb model of the given compounds it appears that in cases where the difference between the measured and calculated $G(H_2)$ values is large, the C—H bonds in β-position (which primarily control the hydrogen yield) are hindered in the steric orientation required for allyl interaction.

As far as the $G(H_2)$ values of aliphatic dienes are concerned (Table 4.44), it appears that they are generally much lower than for monoalkenes, which is probably due to the higher double-bond concentration. The hydrogen yield is somewhat higher for isolated dienes than for conjugated ones. This seems reasonable if the role of allylic hydrogens and the high reactivity of conjugated dienes towards addition are considered. When calculating $G(H_2)$ values for alkenes with isolated double bonds (Table 4.44) by using the increments established for monoalkenes (Table 4.43), it is interesting to note that these values are comparable with those measured only if all π-bonds are taken into account and the calculated $G(H_2)$ value is divided by the square of the π-bond number. The reason for this is not clear.

Although all the methylhexenes listed in Table 4.31 contain one branched carbon atom, the $G(CH_4)$ values increase in the sequence 2-methylhex-1-

-ene $<$ 4-methylhex-1-ene $<$ 3-methylhex-1-ene $<$ 5-methylhex-1-ene $<$ $<$ 2-methylhex-3-ene. On this basis the role of the relative position of π-bonds with respect to methyl groups, as well as the connection between $G(CH_4)$ values and dissociation energies, may be recognized. The directing effect resulting from π-bond delocalization on the σ-bond in the β-position

Table 4.44

Hydrogen yields from liquid aliphatic dienes and hexaisoprene (radiation: ^{60}Co-γ)

Hydrocarbons	Measured	Calculated	References
	$G(H_2)$, molecule/100 eV		
Isolated dienes and polyenes			
Hexa-1,5-diene	0.45		172
	0.38	0.38	67
Octa-1,7-diene	0.54	0.41	259
	0.52		67
2,6-Dimethylocta-2,6-diene	0.77		178
	0.81	0.88	180
Hexaisoprene (Squalene)	0.58	0.31	260
Cumulated diene			
Hepta-1,2-diene	0.47		259
Conjugated dienes			
Hepta-1,3-diene	0.46		259
Octa-2,4-diene	0.38		259

is at a maximum in 3-methylhex-1-ene and 2-methylhex-3-ene: in the former compound one, and in the latter three methyl groups are loosened, so the ratio of the $G(CH_4)$ values is 1 : 3 (Table 4.31).

The results also show that in the decomposition, besides the above-mentioned effect, an important role is played also by the position of C—C bond with respect to the end of the molecule or to the position of the unpaired electrons of free radicals. The high acetylene yield from alk-1-enes can be explained in this way. In some cases these effects are revealed as 'background' effects depending on the structure of the molecule. [152, 155]. The $G(CH_4)$ values also indicate that the radiation-excited molecules isomerize to a small extent.

In the radiolysis of hydrocarbons, 'simultaneous' decomposition may also take place. Thus, for example prior to thermalization monoradicals of four carbon atoms excited or ionized can decompose into radicals of two carbon atoms. One of these is a biradical that can be stabilized as ethylene, the other is a monoradical and this can participate in a bimolecular reaction. However, the process also may be explained by ionic reactions e.g. in case of hex-1-ene [155]:

$$C=C-C-C-C-C \quad \longrightarrow \quad \overset{+}{C}-\overset{\cdot}{C}-C-C-C-C \ + \ e$$

$$\overset{+}{C}-\overset{\cdot}{C} \quad \longrightarrow \quad \overset{+}{C}-C-C-\overset{\cdot}{C} \ + \ C=C$$

$$\overset{+}{C}-C-C-\overset{\cdot}{C} \quad \overset{+e^-}{\longrightarrow} \quad 2\,C=C$$

Besides the secondary decomposition of primary radicals, there is a small probability of simultaneous formation of one biradical and two monoradicals as a consequence of the rupture of σ-bond weakened by the π-bond: the small amount of C_4 produced from hex-3-ene is an indication process [152, 155].

To sum up, it can be said that the decomposition reactions of alkenes under irradiation are very complex, and chemical processes of the highest energy may take place as a result of the high energy of radiation. This is the reason for the formation of compounds produced by rupture of the σ-bond in the α-position to the π-bond, which is not characteristic of thermal decomposition. Nevertheless, the decomposition under irradiation is a selective process despite the very high primary energy absorbed: the position of bond rupture is determined to a large extent by the molecular structure, i.e. the difference in the dissociation energies of the individual bonds, although these differences amount only to 42–85 kJ mol^{-1}.

Build-up reactions

The chemical effect induced by irradiation in unsaturated hydrocarbons becomes obvious in the formation of products of high molecular mass (dimers, oligomers, polymers). Under the usual radiation conditions, however, macromolecular polymers are not formed from most unsaturated hydrocarbons, the reaction terminating in dimers and/or oligomers. However, information on these addition reactions of non-chain character is still very important because, on the one hand, they are basic radiation chemical processes, and on the other hand, the elucidation of the mechanism of radiation-induced polymerization is very important from the industrial point of view. The mechanism of dimer formation has been clarified by considering the effects of structure, product composition, the additives, temperature, etc. It is characteristic of the complexity of the process that information about the full product spectrum is not available even for dimer products. Detailed data are available only for ethylene [29, 37, 63], propene [68] (Table 4.14), and oct-1-ene [176] (Table 4.33) in other cases only overall dimer [150]; trimer [181] and/or oligomer yields have been published (Tables 4.28, 4.29, 4.45).

Table 4.45 present data about the dimer products of some monoalkenes. It has been established that about 50–80% of the dimers are monoalkenes and that straight-chain monoalkene dimers are usually formed with higher probability from terminal alkenes than from alkenes with an internal double bond.

In the case of butenes, the yield of dimers containing dienes and saturated component decreases with increasing temperature and that of monoalkene dimers slightly increases (Table 4.45).

The preferred formation of straight-chain monoalkene dimers from terminal alkenes has been interpreted on the basis of molecular orbital calculations by Chang et al. [149]. The higher probability of their formation was explained by the unique bond formation between ions and the carbon atom originally connected to the double bond.

Table 4.45

Dimer yields from n-alkenes (radiation: $^{60}Co\text{-}\gamma$)

Hydrocarbon	References	Temperature, °C	Structure of chain		Degree of unsaturation			C=C bond per dimer molecule	G(dimer)	G(telomer)
			straight	branched	$C_{2n}H_{4n+2}$	$C_{2n}H_{4n}$	$C_{2n}H_{4n-2}$			
			%		relative measure, %					
Ethylene	63	—196	100[a]		24	67	9	1.0	0.2–0.3	2.5
Propene	68	—78	36	56	15	62	26	1.1		
But-1-ene	86	—78	41	46	9	81	10	1.0	1.34	
		+23	50	44	10	80	10		1.72	
trans-But-2-ene	86	—78	24	74	7	75	18	1.2	0.90	
		+23	22	71	17	59	24		1.03	
cis-But-2-ene	86	—78	18	73	4	77	19		0.96	
		+23	25	64	15	61	26		0.90	
Pent-1-ene	137	+30							3.0	3.0
	138								1.2	4.6
Hex-1-ene	149	—78	45	45[b]	8	85	7	1.0	1.7	3.4
Oct-1-ene	176	+23	50	42[c]	21	50	20	1.0	1.6	4.9
	149								2.0	3.5
Hexadec-1-ene	181	+20								3.9
Hex-2-ene	150	+20						1.2	4.6	
Hex-3-ene	150	+20						1.2	4.1	

[a] Telomers of increasing molecular weight become more branched with increasing molecular size
[b] About 8% cyclic compound is included
[c] About 4% cyclic compound is included

Formation of dimers and telomers from n-alkenes can be explained by the elementary reaction steps summarized in Fig. 4.15 [68, 86, 137, 149, 176]. Besides these processes the following cannot be excluded:

$$R-CH_2-CH=CH_2 \xrightarrow{\sim\hspace{-0.3em}\sim\hspace{-0.3em}\sim} R-CH-CH=CH_2 + H$$

$$H + CH_2=CH-CH_2-R \longrightarrow R-CH_2-CH-CH_3$$

Rearrangement via hydride ion (H⁻) transfer in the transition state
(pathway I) of radical ion-molecule condensation is shown in the following
scheme:

$$
\begin{array}{ccc}
\underset{RCH_2}{\underset{|}{\underset{HC}{\underset{\shortparallel}{\underset{H}{\underset{RHC}{}}}}}}\overset{H}{\underset{}{\overset{|}{\overset{\cdot}{C}}}}\!\!-\!\!CH_2 & \longrightarrow & \overset{H}{\overset{|}{\overset{\cdot}{C}}} & \longrightarrow & \overset{H}{\overset{|}{\overset{+}{C}}}
\end{array}
$$

Monoalkene dimers are formed after neutralization. Hence we should
expect to obtain hex-1-ene from propene, oct-2-ene from but-1-ene, dec-3-
ene from pent-1-ene, dodec-4-ene from hex-1-ene and hexadec-6-ene from
oct-1-ene. These products have in fact been detected from hydrocarbons of
low molecular mass (C_3-C_5) [137]. For the larger molecules (hex-1-ene,
oct-1-ene), however, the location of the π-bond in the dimer could not be
precisely determined thus on the basis of work published by Chang et al.
[149] and Kourim [176], one can conclude only that their data are not in
disagreement with the ideas given above.

Alkenes with a terminal double bond react mostly by an ionic mechanism
to form dimers and products with high molecular mass (see Fig. 4.15,
pathway A, I). However, inner alkenes (but-2-ene, hex-2-ene, hex-3-ene)
follow mostly the other pathways (A, II, B and C): the proportion of dimers
formed via recombination of radicals is much higher than in the case of
terminal alkenes [86, 152]. This is because, in inner alkenes, there are more
allylic C—H bonds and therefore the probability of radical formation is
higher. Propene behaves similarly to this group, probably because a symmetrical allyl radical is formed from propene. Consequently, the position
of the π-bond causes a change in the proportion of ionic and radical processes.

Recently, an attempt has been made by Hunter et al. [239] to elucidate
the mechanism of dimer formation by a method different from conventional
techniques (e.g. effect of additives). Hex-1-ene was frozen at 77 K in thin
layers (about 70 nm) and bombarded by electrons of various energies. The
result was surprising: a plot of the yield of dimer products against electron
energy revealed that the dimer which was thought to result from an ionic
mechanism showed a maximum at 4 eV and decreased nearly to zero at
around the ionization potential (8–9 eV). The yield began to increase again
with increasing electron energy to reach a value similar to that obtained
under the effect of γ-irradiation at the same temperature at about 50 eV.
It was thus concluded that the so-called 'ionic' component of the dimer is
not formed by ion-molecule condensation but through triplet excited states.
If this statement is supported by other experimental data, the mechanism
of dimer formation from alkenes must be reconsidered.

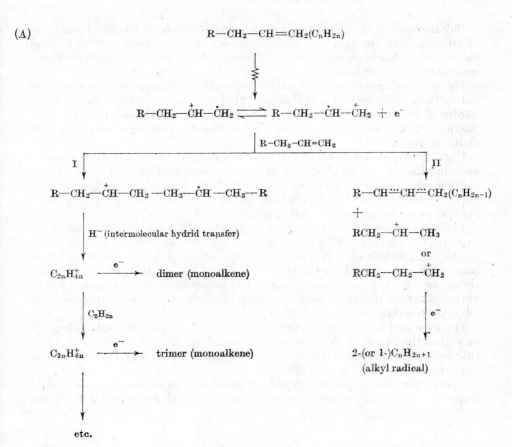

(A)

$R\!-\!CH_2\!-\!CH\!=\!CH_2(C_nH_{2n})$

$R\!-\!CH_2\!-\!\overset{+}{CH}\!-\!\dot{C}H_2 \rightleftharpoons R\!-\!CH_2\!-\!\dot{C}H\!-\!\overset{+}{C}H_2 + e^-$

$R\!-\!CH_2\!-\!CH\!=\!CH_2$

I

$R\!-\!CH_2\!-\!\overset{+}{CH}\!-\!CH_2\!-\!CH_2\!-\!\dot{C}H\!-\!CH_2\!-\!R$

H^- (intermolecular hydrid transfer)

$C_{2n}H_{4n}^+ \xrightarrow{\ e^-\ } $ dimer (monoalkene)

C_2H_{2n}

$C_{3n}H_{6n}^+ \xrightarrow{\ e^-\ } $ trimer (monoalkene)

etc.

II

$R\!-\!CH\overset{\cdots}{\cdots}CH\overset{\cdots}{\cdots}CH_2(C_nH_{2n-1})$
+
$RCH_2\!-\!\overset{+}{CH}\!-\!CH_3$

or

$RCH_2\!-\!CH_2\!-\!\overset{+}{C}H_2$

e^-

2-(or 1-)C_nH_{2n+1}
(alkyl radical)

(B)

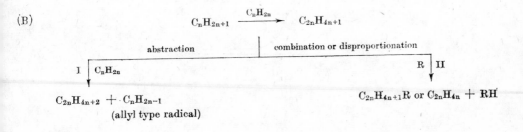

$C_nH_{2n+1} \xrightarrow{\ C_nH_{2n}\ } C_{2n}H_{4n+1}$

abstraction   combination or disproportionation

I C_nH_{2n} R   II

$C_{2n}H_{4n+2} + C_nH_{2n-1}$ $C_{2n}H_{4n+1}R$ or $C_{2n}H_{4n} + RH$
(allyl type radical)

(C) $2\,C_nH_{2n-1} \longrightarrow C_{2n}H_{4n-2}$

Fig. 4.15. Scheme of dimer and oligomer production in the radiolysis of n-alkenes

Polymerization is one of the most important radiation chemical reactions of alkenes from the practical point of view. Detailed discussion is beyond the scope of this book (see refs [6–10]), but we can mention some problems concerning its mechanism.

The first experiments on radiation induced polymerization were carried out in 1938 by Hopwood and Phillips [240], who subjected styrene to bombardment by fast neutrons. Two years later, Joliot [241] patented the production of a transparent, stress-free poly(methyl-methacrylate) by fast neutron irradiation.

Until 1957 radiation polymerization was thought to be a radical process, ionic mechanisms being observed only at low temperatures [241–246]. However, in the past 10–15 years, under carefully controlled conditions (e.g. extra drying) the mechanism was established to have an ionic component even at room temperature [247–249].

The proportion of the two reaction mechanisms can be determined by experiments carried out in the presence of additives, and by investigation of the composition of copolymers, and the effect of temperature and dose rate on the rate of polymerization (e.g. Fig. 4.16). In this Figure the data shown on the left-hand side point to the predominance of a radical mechanism, whereas the other side (experiments below 0°C) is characteristic of ionic polymerization.

The rate and extent of reaction of radicals, ions and excited molecules, as well as the proportions of these reactions, depend on the molecular structure and on experimental conditions (temperature, dose, dose rate, additives, impurities, moisture content etc.).

The polymerization in the liquid phase of isobutene, buta-1,3-diene and isoprene occurs mainly by a cationic chain reaction (Section 4.2.3). It has

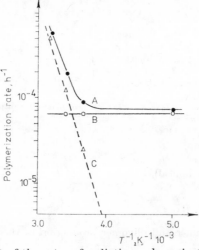

Fig. 4.16. Arrhenius plot of the rates of radiation polymerization of isoprene in bulk. Curve *A*, no additive; curve *B*, with added DPPH; curve *C*, free radical contribution derived by subtracting curve *B* from curve *A* (radiation: ^{60}Co-γ). After Tabata et al. [132] and Chapiro [251]

been shown that the reaction rate and the molecular mass of products both increase with decreasing temperature in the range from 0 to $-195°C$ [246]. Similarly, an apparently negative energy of activation was obtained for the polymerization of isoprene in the temperature range from $+45$ to $-40°C$ by Burlant and Green [250]. If, however, the polymerization of butadiene occurs in an emulsion, the main process is a radical chain reaction [247]. According to Williams [7], however, it is extremely difficult to draw any conclusions from these experiments about the extent of ionic polymerization owing to the low consumption of monomer [$G(-M) < 100$].

REFERENCES

1. MEISELS, G. G., in *The Chemistry of Alkenes* (Ed. ZABICZKY, J.), Interscience, London, 1970, Vol. 2, p. 359
2. HARDWICK, T. J., in *Actions Chimiques et Biologiques des Radiation* (Ed. HAISSINSKY, M.), Masson and Cie, Paris, 1966, Vol. X, p. 125
3. AUSLOOS, P. (Ed.)., *Fundamental Processes in Radiation Chemistry*, Wiley, New York, 1968
4. FREEMAN, G. R., *Radiat. Res. Rev.*, **1**, 1 (1968)
5. ALLEN, A. O., in *Current Topics in Radiation Research*, (Eds EBERT, M. and HOWARD, A.), North-Holland, Amsterdam, 1968, Vol. 4, p. 1
6. CHAPIRO, A., *Radiation Chemistry of Polymeric Systems*, Wiley, London, 1962
7. WILLIAMS, F., in *Fundamental Processes in Radiation Chemistry* (Ed. AUSLOOS, P.), Interscience, New York, 1968, p. 515
8. WILLIAMS, F., *J. Macromol. Sci. A*, **6**, 919 (1972)
9. CHARLESBY, A., *Atomic Radiation and Polymers*, Pergamon Press, New York, 1960, *Vol.* 1.
10. DOLE, M., *The Radiation Chemistry of Macromolecules*, Academic Press, New York, 1972
11. COX, J. D. and PILCHER, G., *Thermochemistry of Organic and Organometallic Compounds*, Academic Press, London, 1970
12. MISLOV, K., *Introduction to Stereochemistry*, Benjamin, New York, 1965
13. PAULING, L., *Nature of the Chemical Bond*, Cornell University Press, New York, 1960
14. PULLMAN, B., *The Modern Theory of Molecular Structure*, Dover, New York, 1962
15. BADER, A. R., BUCKLEY, R. P., LEVITT, F. and SZWARC, M., *J. Am. Chem. Soc.*, **79**, 5621 (1957)
16. SZWARC, M. and BINKS, J. H., *Theoretical Organic Chemistry, Kekule Symposium, 1958*, Butterworth, London, 1958 p. 262
17. TEWARI, P. H. and FREEMAN, G. R., *J. Chem. Phys.*, **49**, 4394 (1968)
18. SCHMIDT, W. F. and ALLEN, A. O., *J. Phys. Chem.*, **72**, 3730 (1968)
19. TEWARI, P. H. and FREEMAN, G. R., *J. Chem. Phys.*, **51**, 1276 (1969)
20. SCHMIDT, W. F. and ALLEN, A. O., *J. Chem. Phys.*, **52**, 2345 (1970)
21. ROBINSON, M. G., FUOCHI, P. G. and FREEMAN, G. R., *Can. J. Chem.*, **49**, 3657 (1971)
22. DODELET, J.-P. and FREEMAN, G. R., *Can. J. Chem.*, **50**, 2667 (1972)
23. DODELET, J.-P., SHINSAKA, F. and FREEMAN, G. R., *J. Chem. Phys.*, **59**, 1293 (1973)
24. DODELET, J.-P., SHINSAKA, K., KORTSCH, U. and FREEMAN, G. R., *J. Chem. Phys.*, **59**, 2376 (1973)
24A. SHINSAKA, K., DODELET, J.-P. and FREEMAN, G. R., *Can. J. Chem.*, **53**, 2714 (1975)
24B. DODELET, J.-P., SHINSAKA, K. and FREEMAN, G. R., *Can. J. Chem.*, **54**, 744 (1976)
25. SCHMIDT, W. F. and ALLEN, A. O., *J. Chem. Phys.*, **52**, 4788 (1970)
26. MINDAY, R. M., SCHMIDT, L. D. and DAVIS, H. T., *J. Chem. Phys.*, **54**, 3112 (1971)

27. MINDAY, R. M., SCHMIDT, L. D. and DAVIS, H. T., *J. Phys. Chem.*, **76**, 442 (1972)
28. FUOCHI, P. G. and FREEMAN, G. R., *J. Chem. Phys.*, **56**, 2333 (1972)
29. OKAZAKI, K., YAMABE, M. and SATO, S., *Bull. Chem. Soc. Jpn.*, **50**, 1409 (1977)
30. MEISELS, G. G., *J. Am. Chem. Soc.*, **87**, 950 (1965)
30A. MEISELS, G. G., *J. Chem. Phys.*, **42**, 3237 (1965)
31. MEISELS, G. G. and SWORSKI, T. J., *J. Phys. Chem.*, **69**, 2867 (1965)
32. MEISELS, G. G. and SWORSKI, T. J., *J. Phys. Chem.*, **69**, 815 (1965)
33. MEISELS, G. G., *Adv. Chem. Ser.*, **58**, 243 (1966)
34. MEISELS, G. G., *J. Chem. Phys.*, **42**, 2328 (1965)
35. AUSLOOS, P. and GORDEN, R., *J. Chem. Phys.*, **36**, 5 (1962)
36. GORDEN, R. and AUSLOOS, P., *J. Chem. Phys.*, **47**, 1799 (1967)
37. GORDEN, R. and AUSLOOS, P., *J. Res. Natl. Bur. Stand.*, *A* **75A**, 141 (1971)
38. SAUER, M. C. and DORFMAN, L. M., *J- Chem. Phys.*, **35**, 497 (1961); *J. Phys. Chem.*, **66**, 322 (1962)
39. HATANO, Y., SHIDA, S. and SATO, S., *Bull. Chem. Soc. Jpn.*, **41**, 1120 (1968)
40. OKABE, H. and McNESBY, J. R., *J. Chem. Phys.*, **36**, 601 (1962)
41. HIROKAMI, S. I. and CVETANOVIC, R. J., *J. Phys. Chem.*, **78**, 1254 (1974)
42. GAWLOWSKI, J. and HERMAN, J. A. *Can. J. Chem.*, **52**, 3631 (1974)
43. KEBARLE, P., HAYNES, R. M. and SEARLES, S., *Adv. Chem. Ser.*, **58**, 210 (1966)
44. KEBARLE, P. and HAYNES, R. M., *J. Chem. Phys.*, **47**, 1676 (1967)
45. KEBARLE, P. and HOGG, A. M., *J. Chem. Phys.*, **42**, 668 (1965)
46. GAWLOWSKI, J., NIEDZIELSKI, J., SOBOVSKI, J. and ZECHOWSKA, A., *Rad. Eff.*, **11**, 85 (1971)
47. GAWLOWSKI, J. and NIEDZIELSKI, J., *Int. J. Radiat. Phys. Chem.*, **5**, 419 (1973)
48. LAMPE, F. W., *Radiat. Res.*, **10**, 691 (1959)
49. YANG, K. and MANNO, P. J., *J. Phys. Chem.*, **63**, 752 (1959)
50. FIELD, F. H., *J. Am. Chem. Soc.*, **83**, 1523 (1961)
51. MELTON, C. E. and RUDOLPH, P. S., *J. Chem. Phys.*, **32**, 1128 (1960)
52. WEXLER, S. and MARSCHALL, R., *J. Am. Chem. Soc.*, **86**, 781 (1964)
53. FESSENDEN, R. W. and SCHULER, R. H., *J. Chem. Phys.*, **39**, 2147 (1963)
54. HOLROYD, R. A. and FESSENDEN, R. W., *J. Phys. Chem.*, **67**, 2743 (1963)
55. BACK, R. A. and GRIFFITHS, D. W. L., *J. Chem. Phys.*, **46**, 4839 (1967)
56. BORREL, P., CASMORE, P., CERVENKA, A. and JAMES, F. C., *J. Chem. Phys.*, **67**, 229 (1970)
57. STEACIE, E. W. R., *Atomic and Free Radical Reactions*, Reinhold, New York,1954
58. CVETANOVIC, R. J. and CALLEAR, A. B., *J. Chem. Phys.*, **23**, 1182 (1955); **24**, 873 (1956)
59. ARAI, S. and SHIDA, S., *J. Chem. Phys.*, **38**, 694 (1963)
60. SETSER, D. W., PLACZEK, D. W., CVETANOVIC, R. J. and RABINOVITCH, B. S., *Can. J. Chem.*, **40**, 2719 (1962)
61. LAIDLER, K. J. and LOUCKS, L. F., *Comprehensive Chemical Kinetics* (Eds BAMFORD, C. H. and TIPPER, C. F. H.), Elsevier, Amsterdam, 1972, Vol. 5
62. HOLROYD, R. A. and KLEIN, G. W., *Int. J. Appl. Radiat. Isotope*, **15**, 633 (1964)
63. WAGNER, C. D., *J. Phys. Chem.*, **66**, 1158 (1962)
64. HAYWARD, J. C., Dissertation, Yale Univ., 1955, NYO-3313
65. LAIRD, R. K., MORELL, A. G. and SEED, L., *Disc. Faraday Soc.*, **22**, 126 (1956)
66. DAUPHIN, J., GRESMANGING, J. and PETTIT, J. C., *Int. J. Appl. Rad. Isotope*, **18**, 297 (1967)
66A. BUSLER, W. R., MARTIN, D. H. and WILLIAMS, F., *Disc. Faraday Soc.*, **36**, 102 (1963)
67. MEISELS, G. G., *Am. Chem. Soc. Div. Polymer Chem. Preprints*, **5**, No 2, p. 896 (1964)
68. WAGNER, C. D., *Tetrahedron*, **14**, 164 (1961)
69. HOLROYD, R. A. and KLEIN, G. W., *J. Phys. Chem.*, **69**, 194 (1965)
70. MUNARI, S. and CASTELLO, G., *Ric. Sci. (Roma)*, **34** **(II-A)**; *Rendiconti* **A7**, 637 (1964)
71. HORVÁTH, Zs. and FÖLDIÁK, G., *Acta Chim. Acad. Sci. Hung.*, **85**, 417 (1975)
72. AVRAMI, M. and KEBARLE, P., *J. Phys. Chem.*, **67**, 354 (1963)
73. GUILLORY, W. A. and THOMAS, S. G. JR., *J. Phys. Chem.*, **79**, 692 (1975)
74. ARAI, S., SHIDA, S. and NISHIKAWA, *Bull. Chem. Soc. Jpn.*, **39**, 2548 (1966)
75. BECKER, D. A., OKABE, H. and McNESBY, J. R., *J. Phys. Chem.*, **69**, 538 (1965)

76. GORDEN, R. JR., DOEPKER, R. and AUSLOOS, P., *J. Chem. Phys.*, **44**, 7333 (1966)
77. AUSLOOS, P. and LIAS, S. G., *J. Chem. Phys.*, **43**, 127 (1965)
78. ABRAMSON, F. P. and FUTRELL, J. H., *J. Phys. Chem.*, **72**, 1826 (1968)
79. TAL'ROSE, T. L. and LYUBIMOVA, A. K., *Dokl. Akad. Nauk SSSR* **86**, 909 (1952)
80. SCHISSLER, D. O. and STEVENSON, D. P., *J. Chem. Phys.*, **24**, 926 (1956)
81. HENIS, J. H. S., *J. Chem. Phys.*, **52**, 282 (1970)
82. MOORE, W. J., *J. Chem. Phys.*, **16**, 916 (1948)
83. KLEIN, R. and SCHEER, M. D., *J. Phys. Chem.*, **62**, 1011 (1958)
84. HELLER, C. and GORDON, A. S., *J. Chem. Phys.*, **42**, 1262 (1965)
85. HUMMEL, R. W., *Chem. Commun.*, **23**, 1518 (1968)
86. KAUFMAN, P. C., *J. Phys. Chem.*, **67**, 1671 (1963)
87. HATANO, X., SHIDA, S. and SATO, S., *Bull. Chem. Soc. Jpn.*, **37**, 1854 (1964)
88. HATANO, Y. and SHIDA, S., *J. Chem. Phys.*, **46**, 4784 (1967)
89. TSCHUIKOV-ROUX, W., *J. Phys. Chem.*, **71**, 2355 (1967)
90. COLLIN, G. J. and HERMAN, J. A., *Can. J. Chem.* **45**, 3097 (1967)
91. HERMAN, J. A. and COLLIN, G. J., *Can. J. Chem.*, **47**, 3837 (1969)
92. HERMAN, J. A., *Can. J. Chem.*, **48**, 3446 (1970)
92A. HERMAN, J.A., *Can. J. Chem.*, **57**, 2633 (1979)
93. HERMAN, J. A., HERMAN, K. and AUSLOOS, P., *J. Chem. Phys.*, **52**, 28 (1970)
94. GUNNING, H. E. and STEACIE, E. W. R., *J. Chem. Phys.*, **14**, 581 (1946)
95. CVETANOVIC, R. J., GUNNING, H. E. and STEACIE, E. W. R., *J. Chem. Phys.*, **31**, 573 (1959)
96. CVETANOVIC, R. J., and DOYLE, L. C., *J. Chem. Phys.*, **37**, 543 (1962)
97. LOSSING, F. P., MARSDEN, D. G. H. and FARMER, J. B., *Can. J. Chem.*, **34**, 701 (1956)
98. HARUMYA, N., SHIDA, S. and ARAI, S., *Bull. Chem. Soc. Jpn.*, **38**, 142 (1965)
99. BORREL, P. and CASHMORE, P., *Ber. Bunsenges.*, **72**, 182 (1968)
100. BORREL, P., CASHMORE, P., CERVENKA, A. and JAMES, F. C., *J. Chem. Phys.*, **67**, 229 (1970)
101. COLLIN, G. J., *Can. J. Chem.*, **51**, 2853 (1973)
102. WARMAN, J. M., *J. Phys. Chem.*, **71**, 4066 (1967)
103. CVETANOVIC, R. J., GUNNING, H. E. and STEACIE, E. W. R., *J. Chem. Phys.*, **31**, 573 (1959)
104. CUNDALL, R. B. and PALME, T. F., *Trans. Faraday Soc.*, **56**, 1211 (1960)
105. KEBARLE, P. and AVRAMI, M., *J. Chem. Phys.*, **38**, 700 (1963)
106. CUNDALL, R. B., FLETCHER, F. J. and MILNE, P. G., *J. Chem. Phys.*, **39**, 3536 (1963)
107. SIGAL, G., *J. Chem. Phys.*, **42**, 1953 (1965)
108. COLLIN, G. J., PERRIN, P. M. and GAUCHER, C. M., *Can. J. Chem. Phys.*, **50**, 2391 (1972)
109. CUNDALL, R. B., FLETCHER, F. J. and MILNE, P. G., *Trans. Faraday Soc.*, **60**, 1146 (1964)
110. HATANO, Y. and SHIDA, S., *Bull. Chem. Soc. Jpn.*, **39**, 456 (1966)
111. BORREL, P. and JAMES, F. C., *Trans. Faraday Soc.*, **62**, 2452 (1966)
112. COLLIN, G. J., PERRIN, P. M. and GARNEAU, F. X., *Can. J. Chem.*, **52**, 2337 (1974)
113. SIECK, W. L. and FUTRELL, J. H., *J. Chem. Phys.*, **48**, 1409 (1968)
114. HOLROYD, R. A. and KLEIN, G. W., *J. Am. Chem. Soc.*, **87**, 4983 (1965)
115. COLLIN, G. J., *Can. J. Chem.*, **52**, 2341 (1974)
115A. NAITO, T., *Bull. Chem. Soc. Jpn.*, **50**, 795 (1977)
116. FALCONER, J. W., *Nature*, **198**, 985 (1963)
117. KOYANO, I., *J. Chem. Phys.*, **45**, 706 (1966)
118. AQUILANTI, V., GALLI, A., GIARDINI-GUIDONI, A. and VOLPI, G. G., *Trans. Faraday Soc.*, **63**, 926 (1967)
119. ABRAMSON, F. P. and FUTRELL, J. H., *J. Phys. Chem.*, **72**, 1994 (1968)
120. SIECK, L. W., LIAS, S. G., HELLNER, L. and AUSLOOS, P., *J. Res.Natl. Bur. Stand.*, *A. Phys. Chem.*, **76A** No 2. 115 (1972)
121. FUCHS, R., *Z. Naturforsch.*, **16a**, 1026 (1961)
122. SCHILDNECKT, C. E., *Polymer Processes*, Interscience, New York, 1956, p. 203
123. DAVISON, W. H. T., PINNER, S. H. and WORRALL, R., *Chem. and Ind.*, London, 1957, p. 1274; *Proc. Roy. Soc.*, **A252**, 187 (1959)

124. HOFFMAN, A. S., *J. Polymer Sci.*, **34**, 229 (1959)
125. COLLINSON, E., DAINTON, F. S. and GILLIS, H. A., *J. Phys. Chem.*, **63**, 909 (1959)
126. DALTON, F. L., *Polymer*, **6**, 1 (1965)
127. DAVISON, W. H. T., PINNER, S. N. and WORRAL, R., *Chem. and Ind.*, London 1957, p. 1274
128. MUND, W., GUIDÉE, C. and WANDERSAUVERS, J., *Bull. Acad. Roy. Belg.*, **41**, 805 (1955)
129. MUND, W. and HUYSKENS, P., *Bull. Acad. Roy. Belg.*, **36**, 610 (1950)
130. HAYWARD, J. C. and BRETTON, R. H., *Chem. Eng. Progr.*, **50**, 73 (1954)
131. OKAMOTO, H., FUEKI, K. and KURI, Z., *J. Phys. Chem.*, **71**, 3222 (1967)
132. TABATA, Y., SHIMICOWA, R. and SOBUE, H., *Polymer Sci.*, **54**, 201 (1961)
133. TAYLOR, R. B. and WILLIAMS, F., *J. Am. Chem. Soc.*, **89**, 6359 (1967); **90**, 3728 (1968)
134. KAMIYAMA, H., SHINKAWA, A., HAYASHI, K. and OKAMURA, S., *IUPAC Macro*, Budapest, 1969
135. EDLUND, O., KINELL, P. O., LUND, A. and SHIMIZU, A., *J. Polymer Sci.*, **86**, 133 (1968)
136. TSUJI, K., YOSHIDA, H., HAYASHI, K. and OKAMURA, S., *J. Polymer Sci.*, **85**, 313 (1967)
137. WAGNER, C. D., *Trans. Faraday Soc.*, **64**, 163 (1968)
138. WAGNER, C. D., *J. Phys. Chem.*, **71**, 3445 (1967)
139. CSERÉP, GY., unpublished results (1973, 1975)
140. PERRIN, P. M. and COLLIN, G. J., *Can. J. Chem.*, **51**, 724 (1973)
141. KOYANO, I., SUZUKI, Y. and TANAKA, I., *J. Chem. Phys.*, **59**, 101 (1973)
142. MAJER, J. R., PINKARD, J. F. T. and ROBB, J. C., *Trans. Faraday Soc.*, **60**, 1247 (1964)
143. PLACZEK, D. W. and RABINOVITCH, B. S., *Can J. Chem.*, **43**, 820 (1965)
143A. OWCZARCZYK, A. and STACHOWICZ, W., *Rad. Phys. Chem.*, **10**, 319 (1977)
144. BAYRAKCEKEN, F., BROPHY, J. H., FINK, R. D. and NICHOLAS, J. E., *J. Chem. Soc. Faraday Trans.* 1., **69**, 228 (1972)
145. GAGNON, H., COLLIN, G. J. and BERTRAND, C., *J. Phys. Chem.*, **78**, 98 (1974)
146. COLLIN, G. J. and BERTRAND, C., *J. Photochem.*, **3**, 123 (1974)
147. MAJER, J. R., MILE, B. and ROBB, J. C., *Trans. Faraday Soc.*, **57**, 1336 (1961)
148. COLLIN, G. J. and GAUCHER, C. M., *Can. J. Chem.*, **52**, 34 (1974)
149. CHANG, P. C., YANG, N. V. and WAGNER, C. D., *J. Am. Chem. Soc.*, **81**, 2060 (1959)
150. BRASCH, J. L. and GOLUB, M. A., *Can. J. Chem.*, **46**, 594 (1968)
151. CSERÉP, GY. and FÖLDIÁK, G., *Acta Chim. Acad. Sci. Hung.*, **77**, 407 (1973)
152. FÖLDIÁK, G., CSERÉP, G., JAKAB, A., STENGER, V. and WOJNÁROVITS, L., Szénhydrogén rendszerek radiolizise. (Radiolysis of hydrocarbon systems.) *A kémia újabb eredményei*, Akadémiai Kiadó, Budapest 1971, Vol. 4, p. 11
153. CSERÉP, GY. and FÖLDIÁK, G., *Int. J. Radiat. Phys. Chem.*, **5**, 235 (1973)
154. CSERÉP, GY. and FÖLDIÁK, G., in *Proc. 3rd Tihany Symp. Rad. Chem.* (Eds DOBÓ, J. and HEDVIG, P.), Akadémiai Kiadó, Budapest 1972, Vol. 1, p. 287
155. CSERÉP, GY. and FÖLDIÁK, G., *Acta Chim. Acad. Sci. Hung.*, **83**, 171 (1974)
156. KOVÁCS, A., CSERÉP, GY. and FÖLDIÁK, G., *Acta Chim. Acad. Sci. Hung.*, **92**, 223 (1977)
157. MCLAFFERTY, F. W., *Interpretation of Mass Spectra.*, Benjamin, W. A., Inc., New York 1967
158. LOUDON, A. G. and MACCOOLL, A., in *Chemistry of Alkenes* (Ed. ZABICZKY, J.), Interscience, London 1970, Vol. 2, p. 2
159. CHESICK, P. J., *J. Chem. Phys.*, **45**, 3934 (1966)
160. KUDO, T., *J. Phys. Chem.*, **71**, 3681 (1967)
161. MEISELS, G. G., in *Fundamental Processes in Radiation Chemistry* (Ed. AUSLOOS, P.), Wiley, New York, 1968, p. 362
162. ROGINSKII, V. A. and PSHEZHETSKII, S. Ya., *Khim. Vys. Energ.*, **4**, 240 (1970)
163. ROGINSKII, V. A. and PSHEZHETSKII, S. Ya., *Khim. Vys. Energ.*, **3**, 140 (1969)
164. MILINCHUK, B. K. and PSHEZHETSKII, S. Ya., *Dokl. Akad. Nauk SSSR*, **152**, 665 (1963)
165. OHNISHI, S., SUGIMOTO, S. and NITTA, I., *J. Chem. Phys.*, **39**, 2647 (1963)

166. TUPIKOV, V. I. and PSHEZHETSKII, S. Ya., *Dokl. Akad. Nauk SSSR*, **156**, 114 (1964)
167. SMITH, D. R., OKENKA, F. and PIERONI, J. J., *Can. J. Chem.*, **45**, 833 (1967)
168. AYSCOUGH, P. B., COLLILS, R. G. and DAINTON, F. S., *Nature*, **205**, 965 (1965)
169. AYSCOUGH, P. B. and EVANS, H. E., *Trans. Faraday Soc.*, **60**, 801 (1964)
170. KEVAN, K. L., in *Actions Chimiques et Biologiques des Radiation* (Ed. HAISSINSKY, M.), Masson and Cie., Paris, 1971, Vol. XV, p. 111
171. SHIDA, S. T. and HAMILL, W. H., *J. Am. Chem. Soc.*, **88**, 5376 (1966)
171A. CSERÉP, GY., BREDE, O. and MEHNERT, R. *Radiochem. Radioanal. Lett.*, in the press
172. VAN DER HEYDE, H. B. and WAGNER, C. D., *J. Phys. Chem.*, **66**, 1746 (1962)
173. GUARINO, J. P. and HAMILL, W. H., *J. Am. Chem. Soc.*, **86**, 777 (1964)
174. SMITH, D. R. and PIERONI, J. J., *J. Phys. Chem.*, **70**, 2379 (1966)
175. CSERÉP, GY. and FÖLDIÁK, G., *Acta Chim. Acad. Sci. Hung.*, **83**, 185 (1974)
175A. KLEIN, R. and KELLEY, R. D., *J. Phys. Chem.*, **79**, 1780 (1975)
176. KOURIM, P., *Int. J. Radiat. Phys. Chem.*, **1**, 345 (1968)
177. WILLIAMS, F., in *Fundamental Processes in Radiation Chemistry* (Ed. AUSLOOS, P.), Wiley, New York, 1968, p. 575
178. SHELLBERG, W. E., PESTANER, F. J. and YAHIKU, R. A., *Nature*, **200**, 254 (1963)
179. WIDMER, H., Paper presented at the IUPAC Meeting, Moscow, USSR, July 1965
180. BRASH, J. L. and GOLUB, M. A., *Can. J. Chem.*, **45**, 101 (1967)
181. COLLINSON, E., DAINTON, F. S. and WALKER, D. C., *Trans. Faraday Soc.*, **57**, 1732 (1961)
182. AYSCOUGH, P. B., McCANN, A. P., THOMSON, C. and WALKER, D. C., *Trans. Faraday Soc.*, **57**, 1487 (1961)
183. MAINS, G. J., NIKI, H. and WIJNEN, M. H. H., *J. Phys. Chem.*, **67**, 11 (1963)
184. FIELD, F. H., *J. Phys. Chem.*, **68**, 1039 (1964)
185. ROSENBLUM, C., *J. Phys. Chem.*, **52**, 474 (1948)
186. DORFMAN, L. M. and WAHL, A. C., *Radiat. Res.*, **10**, 680 (1956)
187. RUDOLPH, P. S. and MELTON, C. E., *J. Phys. Chem.*, **63**, 916 (1959)
188. DORFMAN, L. M. and SHIPKO, F. J., *J. Am. Chem. Soc.*, **77**, 4723 (1955)
189. LIND, S. C., *Radiation Chemistry of Gases*, Reinhold, New York, 1961
190. FUTRELL, J. H. and SIECK, L. W., *J. Phys. Chem.*, **69**, 892 (1965)
191. SHIDA, S., TSUKADA, M., FUJISAKI, N. and OKA, T., *Bull. Chem. Soc. Jpn.*, **43**, 3314 (1970)
192. SHIDA, S. and TSUKADA, M., *Bull. Chem. Soc. Jpn.*, **43**, 2740 (1970)
193. TSUKADA, M. and SHIDA, S., *Bull. Chem. Soc. Jpn.* **43**, 3621 (1970)
194. BRIGGS, J. P. and BACK, R. A., *Can. J. Chem.*, **49**, 3789 (1971)
195. SHIDA, S., KURI, Z. and FURUOYA, T. J., *J. Chem. Phys.*, **28**, 131 (1958)
196. ZELIKOV, M. and ASCHENBRAND, L. M., *J. Chem. Phys.*, **24**, 1034 (1965)
196A WILLIS, C., BACK, R. A. and MORRIS, R. H., *Can. J. Chem.*, **55**, 3288 (1977)
197. RONDEAU, R. E., HARRAH, L. A., NOVITT, T. D., BARBER, H. H. and SCHAFFER, R., *Air Force Technical Report*, AFML-TR-65-236, 1965
198. WHITTEN, G. D. and BERNGRUBER, V., *J. Am. Chem. Soc.*, **93**, 3204 (1971)
199. GALLI, A., HARTEK, P. and REEVES, R. R., *J. Phys. Chem.*, **71**, 2719 (1967)
200. STIEF, L. Y., DeCARLO, V. J. and MATALONI, R. J., *J. Chem. Phys.*, **42**, 3113 (1965)
201. SAFRANY, D. R. and YASTER, W., *J. Phys. Chem.*, **72**, 3323 (1968)
202. HILL, K. L. and DOEPKER, R. D., *J. Phys. Chem.*, **76**, 1112 (1972)
203. FÖLDIÁK, G. and WOJNÁROVITS, L., *Acta Chim. Acad. Sci. Hung.*, **65**, 59 (1970); **67**, 209 (1971); **67**, 221 (1971); **70**, 23 (1971)
204. VAN INGEN, J. W. F. and CRAMER, E. A., *Trans. Faraday Soc.*, **66**, 857 (1970)
205. FÖLDIÁK, G., CSERÉP, GY., STENGER, V. and WOJNÁROVITS, L., *Kémiai Közlemények*, **31**, 413 (1969)
206. FÖLDIÁK, G., CSERÉP, GY., FEJES, P., STENGER, V. and WOJNÁROVITS, L., *Izotópkémiai kutatások*. MTA Izotóp Intézete, Budapest, 1969, p. 83
207. ROBINSON, M. G. and FREEMAN, G. R., *J. Chem. Phys.*, **48**, 983 (1968)
208. ROBINSON, M. G. and FREEMAN, G. R., *J. Phys. Chem.*, **72**, 1870 (1968)
209. AUSLOOS, P. and LIAS, S. G., *J. Chem. Phys.*, **43**, 127 (1965)
210. COLLIN, G. J. and AUSLOOS, P., *J. Am. Chem. Soc.*, **93**, 1336 (1971)
211. DOEPKER, R. D. and AUSLOOS, P., *J. Chem. Phys.*, **44**, 1951 (1966)

212. FESSENDEN, R. W. and SCHULER, R. H., *Disc. Faraday Soc.*, **36**, 147 (1963)
213. FÖLDIÁK, G. and HORVÁTH, Zs., *Acta Chim. Acad. Sci. Hung.*, **86**, 385 (1975)
214. HORVÁTH, Zs. and FÖLDIÁK, G., *Acta Chim. Acad. Sci. Hung.*, **86**, 397 (1975)
215. FIELD, F. H. and LAMPE, F. W., *J. Am. Chem. Soc.*, **80**, 5987 (1958)
216. SIECK, L. W., SEARLES, S. K. and AUSLOOS, P., *J. Res. Natl. Bur. Stand.*, *A. Phys. Chem.*, **75**, 147 (1971)
217. SIECK, L. W. and FUTRELL, J. H., *J. Chem. Phys.*, **45**, 560 (1966)
218. ABRAMSON, F. P. and FUTRELL, J. H., *J. Phys. Chem.*, **71**, 3791 (1967)
219. SIECK, L. W. and SEARLES, S. K., *J. Am. Chem. Soc.*, **92**, 2937 (1970)
220. SCALA, A. A., LIAS, S. G. and AUSLOOS, P., *J. Am. Chem. Soc.*, **88**, 5701 (1966)
221. AUSLOOS, P., SCALA, A. A. and LIAS, S. G., *J. Am. Chem. Soc.*, **89**, 3677 (1967)
222. DOEPKER, R. D. and AUSLOOS, P., *J. Chem. Phys.*, **42**, 3746 (1965)
223. AUSLOOS, P., SCALA, A. A. and LIAS, S. G., *J. Am. Chem. Soc.*, **88**, 1583 (1966)
224. ABRAMSON, F. P. and FUTRELL, J. H., *J. Phys. Chem.*, **71**, 1233 (1967)
225. FUJISAKI, N. I., FUJIMOTO, Y. and HATANO, Y., *J. Phys. Chem.*, **76**, 1260 (1972)
226. AUSLOOS, P. and LIAS, S. G., *Disc. Faraday Soc.*, **38**, 36 (1965)
226A COLLIN, G. J., *Can. J. Chem.*, **54**, 3050 (1976)
226B KOVÁCS, A. and FÖLDIÁK, G., in *Proc. 4th Tihany Symp. Rad. Chem.* (Eds HEDVIG, P. and SCHILLER, R.), Akadémiai Kiadó, Budapest, 1977, p. 125
227. FIELD, F. H., *J. Am. Chem. Soc.*, **83**, 1523 (1961)
228. KEBARLE, P. and HOGG, A. M., *J. Chem. Phys.*, **42**, 668 (1965)
229. SZABÓ, I., *Arkiv für Fyzik.*, **33**, 57 (1967)
230. MANION, J. P. and BURTON, M., *J. Phys. Chem.*, **56**, 560 (1950)
231. VOEVODSKII, V. V. and MOLIN, Y. N., *Radiat. Res.*, **17**, 366 (1962)
232. PLATZMAN, R., *Radiat. Res.*, **17**, 419 (1962)
233. PLATZMAN, R. L., *Vortex.*, **23**, 1 (1962)
234. KOVÁCS, A., CSERÉP, GY. and FÖLDIÁK, G., *Acta Chim. Acad. Sci. Hung.*, **92**, 117 (1977)
234A CSERÉP, GY., unpublished results
235. OHMAE, T., OHNISHI, S., KUWATA, K., SAKURAI, H. and NITTA, I., *Bull. Chem. Soc. Jpn.*, **40**, 226 (1967)
236. HARDWICK, T. J., *J. Phys. Chem.*, **66**, 291 (1962)
237. HARDWICK, T. J., *J. Phys. Chem.*, **65**, 101 (1961)
238. BENSON, S. W., *Thermochemical Kinetics: Methods for the Estimation of Thermochemical Data and Rate Parameters*, Wiley, New York, (1968)
239. HUNTER, M. L., MATSUSHIGE, R. and HAMILL, W. H., *J. Phys. Chem.*, **74**, 1883 (1970)
240. HOPWOOD, F. L. and PHILLIPS, J. R., *Proc. Phys. Soc.*, **50**, 438 (1938); *Nature*, **143**, 640 (1959); HOPWOOD, F. L., *Brit. J. Radiol.*, **13**, 221 (1940)
241. JOLIOT, F., *Fr. Pat.*, **966**, 760 (1940)
242. OKAMURA, S., OISHI, S. Y. and INAGAKI, H., *Bull. Inst. Chem. Research*, Kyoto Univ., **14**, 103 (1957)
243. DAVISON, W. H. T., PINNER, S. H. and WORRAL, R., *Chem. and Ind.* (London), 1274 (1957)
244. OKAMURA, S., HIGASHIMURA, T. and FUTAMI, S., *Isotopes and Radiation* (Japan) **1**, 216 (1958)
245. TABATA, Y., SHOMOCOWA, R. and SOBUE, H., *Polymer Sci.*, **54**, 201 (1961)
246. ANDERSON, W. S., *J. Phys. Chem.*, **63**, 765 (1959)
247. BATES, T. H., BEST, J. V. F. and WILLIAMS, F., *J. Chem. Soc.*, **1962**, 1531; *Nature*, **188**, 469 (1960)
248. UENO, K., SHINKAWA, A., HAYASHI, K. and OKAMURA, S., *Bull. Chem. Soc. Jpn.*, **40**, 421 (1967)
249. TAYLOR, R. B. and WILLIAMS, F., *J. Am. Chem. Soc.*, **89**, 6359 (1967); **91**, 3728 (1968)
250. BURLANT, W. J. and GREEN, D. H., *J. Polymer Sci.*, **31**, 227 (1958)
251. CHAPIRO, A., *Makromol. Chem.*, **175**, 1181 (1974)
252. MEISELS, G. G., *J. Chem. Phys.*, **41**, 51 (1964)
253. LEBLANC, R. M. and HERMAN, J. A., *J. Chem. Phys.*, **63**, 1055 (1966)
254. ALDER, P. and BOTHE, H. K., *Z. Naturforsch.*, **A20**, 1700 (1968)
255. COOPER, R. and MOORING, R. M., *Aust. J. Chem.*, **21**, 2417 (1968)

256. STONCHAM, T. A., ENTHRIDGE, D. R. and MEISELS, G. G., *J. Chem. Phys.*, **54**, 4054 (1971)
257. CHESICK, J. P., *J. Chem. Phys.*, **45**, 3934 (1966)
258. CSERÉP, GY. and FÖLDIÁK, G., in *Proc. 4th Tihany Symp. Rad. Chem.* (Eds HEDVIG, P. and SHILLER, R.), Akadémiai Kiadó, Budapest, 1977, p. 147
258A CSERÉP, GY. and FÖLDIÁK, G., *Radiat. Phys. Chem.*, **15**, 215 (1979)
259. FÖLDIÁK, G., CSERÉP, GY. and STENGER, V., *Proc. 2nd Tihany Symp. Rad. Chem.* (Eds DOBÓ, J. and HEDVIG, P), Akadémiai Kiadó, Budapest, 1967, p. 381
260. DANON, J. and GOLUB, M. A., *Can. J. Chem.*, **42**, 1577 (1964)
261. KEBARLE, P., HAYNES, R. M. and SEARLES, S., *Adv. Chem. Ser.*, **58**, 210 (1966)
262. BUSLER, W. R., MARTIN, D. H. and WILLIAMS, F., *Disc. Faraday Soc.*, **36**, 102 (1963)
263. FÖLDIÁK, G. and WOJNÁROVITS, L., *Int. J. Radiat. Phys. Chem.*, **4**, 189 (1972)
264. FÖLDIÁK, G., CSERÉP, GY., GYÖRGY, I., HORVÁTH, Zs. and WOJNÁROVITS, L., *Hung. J. Ind. Chem.*, **2**, 277 (1974)
265. WAGNER, C. D., *J. Phys. Chem.*, **67**, 1793 (1963)

5. CYCLOALKENES

Cycloalkenes have been used in a number of radiation chemical studies as model compounds, but their radiolysis has been studied less systematically. The main reason for this is that they have been used in the majority of studies only as additives. Consequently, data on the radiolysis of even cyclopentene and cyclohexene (which have been investigated in relatively more detail) are not entirely complete, and the interpretation of reaction mechanisms is frequently contradictory.

Most papers in this field have been published by Freeman and Wakeford [1–5], Holroyd et al. [6–7], Sakurai and co-workers [8–11] and Cserép et al. [12–20]. Freeman and Wakeford have concentrated mainly on products with molecular mass greater than that of the parent hydrocarbon, and Cserép et al. on hydrogen and fragment formation. There are a few papers on the radiation chemistry of methylcycloalkenes and methylenecyclohexane. Very few data are available on the radiation chemistry of cyclic hydrocarbons with two or more double bonds.

As far as we know, no comprehensive review or comparative study has previously been published on the radiolysis of cycloalkenes.

5.1. INTRODUCTORY REMARKS

As can be seen from handbooks of physicochemical constants of organic compounds, the boiling points of both unsubstituted and substituted cycloalkenes are, as a rule, lower than those of semicyclic alkylidenecycloalkanes, on the other hand, they are higher than those of the corresponding saturated compounds.

The molecular structure and chemical reactivity of cycloalkenes can be related to cycloalkanes, on the one hand, and to aliphatic alkenes on the other. The cyclic structure and the presence of sp^2 hybrid carbon atoms creat a strain energy in the molecule, the extent of which depends on the carbon number. This phenomenon has been discussed in detail in Section 3.1 on cycloalkanes; it was pointed out that distortion of $C-C-C$ valence angles, bond elongation, and van der Waals repulsion between atoms can contribute to the overall strain energy, with a minimum occurring for a certain carbon number. In most cases, strain energy values for cycloalkanes and cycloalkenes are similar: a significant difference can be observed between cyclopropane and cyclopropene (115 and 225 kJ mol $^{-1}$, respectively) and for hydrocarbons with more than eight carbon atoms (see Section 3.1 and Fig. 4.13). For rings of more than seven carbon atoms there exist separate *cis* and *trans* configurations with considerably different strain energies: the *cis*-isomer is more stable, especially with C_8-C_9 rings.

In spite of the significant ring strain, cycloalkenes are generally stable compounds: even cyclopropene, although it has an extremely high strain energy, can be synthesized.

Cyclobutadiene (being the lowest possible cyclic diene) is also very unstable. Of compounds containing cumulated double or triple bonds, as far as we know, only those with rings larger than C_7 have been synthesized.

5.2. INDIVIDUAL COMPOUNDS

In this section, cyclomonoalkenes, cyclodienes and cyclopolyenes will be dealt with, each group in the order of increasing molecular mass.

5.2.1. Cyclobutene and cyclopentene

The only authors who have published detailed results on the radiolysis of cyclobutene are Horváth and Földiák [21]. In view of the extremely high strain energy of cyclobutene (125 kJ mol^{-1}), a high degree of ring decomposition can be expected and was actually found, but is is remarkable that the decomposition in liquid-phase γ-radiolysis gives mainly unsaturated C_2 products, namely ethylene and acetylene (Table 5.1). Gas-phase thermal [22, 23] and photochemical decomposition [24, 25] produce mainly butadiene, via fission of the allylic C—C bond.

Table 5.1 indicates that the $G(H_2)$ value of cyclobutene is very small: this may be explained by competing decomposition of C—C bonds. Radiation energy absorbed by the molecule will cause predominantly the rupture of C—C bonds, enhanced by the high strain energy.

A correlation can be observed between the decomposition of cyclobutene induced by high-energy radiation and vacuum-UV light. Deleon and Doepker [24] observed that the quantum yield of buta-1,3-diene decreases with increasing total pressure and with increasing energy of the incident light, whereas those of ethylene and acetylene increase with increasing energy, but seem to be independent of pressure. Horváth and Földiák carried out their experiments [21] with γ-radiation in the liquid phase (i.e. with 'extreme high energy' and 'extreme high pressure' as compared with gas phase vacuum-UV experiments): the agreement between the two sets of data is reasonable. On the basis of the pressure dependence of the buta-1,3-diene quantum yield, Delon and Doepker suggested [24] a bimolecular pathway for the formation of the diene starting with the production of a 'hot' cyclo-C_4H_7 radical and a subsequent decomposition. According to the Woodward—Hoffmann [26] theory, the process leading to the diene, at least in part, could be equally well explained by the photochemically allowed disrotatory ring-opening of the electronically excited cyclobutene (Fig. 5.1).

Table 5.1

Yields from liquid cycloalkenes (radiation: ^{60}Co-γ)

Hydrocarbons	Cyclo-butene	Cyclo-pentene	Cyclo-hexene	Cyclo-heptene	Cyclo-octene	Cyclo-decene	Cyclo-dodecene
Reference	75	14	15	15	15	15	15
Products	G, molecule/100 eV						
Hydrogen	0.55	1.20	1.19	0.98	0.85	1.02	1.60
Methane	0.02	0.008	0.005	0.007	0.004	0.002	0.002
Ethane		0.001	0.001	0.001	0.001	0.002	0.001
Ethylene	1.27	0.15	0.14	0.07	0.054	0.035	0.02
Acetylene	1.62	0.06	0.021	0.01	0.006	0.009	0.005
Propane		0.01	0.001			0.004	0.001
Cyclopropane				0.007	0.002		0.001
Propene	0.004	} 0.05	0.005	0.01	0.006	0.017	0.002
Allene (propadiene)		0.10		0.002	0.001	0.001	0.001
n-Butane			0.001			0.002	0.001
Butanes			0.006	0.003	0.005	0.011	0.002
Buta-1,3-diene			0.06	0.015	0.009	0.014	0.005
But-1-ene-3-yne	0.13						
C_5 products	0.09			0.017	0.007	0.006	

A biradical mechanism was suggested for the formation of ethylene and acetylene [21]:

$$(C_4H_6)^* \longrightarrow \begin{array}{l} C_2H_2 + C_2H_4 \\ 2\,C_2H_2 + H_2 \end{array}$$

There is no direct evidence, however, for the existence of the species $(C_4H_6)^*$. Moreover, the production of ethylene and acetylene from an electronically excited cyclobutene molecule by a single-step multicentred reaction is an orbital symmetry allowed process (see Section 3.2.1. for the analogous decomposition of cyclobutane).

The spectrum of radiolysis products of liquid cyclopentene and their G-values demonstrate a marked difference as compared with the correspond-

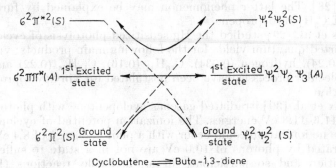

Cyclobutene $\rightleftharpoons$ Buta-1,3-diene

Fig. 5.1. Orbital correlation diagram of the cyclobutene–buta-1,3-diene system

ing data for cyclobutene (see Table 5.1). For example, the yield of hydrogen from cyclopentene is more than twofold, and that of ring fragments barely a tenth of the corresponding yields from cyclobutene. This can be attributed mainly to the almost tenfold difference in ring-strain energies, together with the fact that the contributions of the various components of the strain energy are not the same in the two cases (see Fig. 3.1 in Section 3.1). Stabilization of fragments from the rupture of the cyclopentene ring in the liquid phase leads to the formation of mainly ethylene and allene (Table 5.1) [17].

It is reasonable to assume that one of the allylic C—C bonds breaks first, and the subsequent rearrangement of electrons and one H-atom leads ultimately to the formation of the two products mentioned:

$$\text{(5.1)} \quad C{=}C + C{=}C{=}C$$

Further processes have also been proposed [17] to occur, with about half the probability of reaction (5.1):

$$C{\equiv}C + C{-}C{-}C$$
$$C{\equiv}C + C{=}C{-}C$$

Of the yields of the complementary hydrocarbons that of C_3 is lower, but the difference between actual and predicted values is not great.

On comparing the products of radiation-chemical and thermal decomposition, it has been found that any analogy between the products depends mainly on the conditions of pyrolysis. The products obtained from pyrolysis of short contact time and high temperature [27] are almost the same as those observed from radiolysis (although there is less C_3H_4 among the products of pyrolysis), whereas cracking at lower temperatures (480–520°C) carried out in a static system gives ethylene and propene as the main products [28]. The latter phenomenon may be explained by further conversion of C_3H_4 into propene.

Gibbons et al. [29] studied the Hg-sensitized photolysis of cyclopentene. They reported quantum yields for the following main products: vinylcyclopropane (0.24), hydrogen (0.034), $C_{10}H_{14}$ (0.16), $C_{10}H_{16}$ (0.22) and $C_{10}H_{18}$ (0.05). Molecular processes have been postulated for vinylcyclopropane and H_2 formation.

Lesclaux et al. [30] irradiated gaseous cyclopentene with photons of 8.4, 10.0 and 11.6–11.8 eV energies. The ionization potential of cyclopentene is 9.01 eV, so no ionization could occur with a photon energy of 8.4 eV. Parent ions generated by photons of 10.0 eV are not in a state to suffer further decomposition and so participate in ion-molecule reactions (Fig. 5.2). The quantum yield of ionization is as low as 0.16, i.e. 84% of cyclopentene

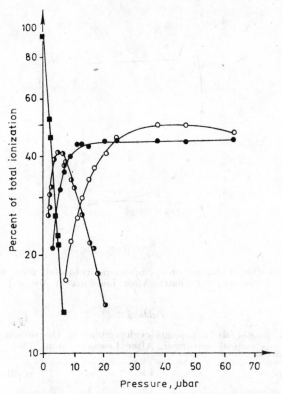

Fig. 5.2. Effect of pressure on the relative abundances of ions observed in cyclopentene irradiated with 10.0 eV photons.
○ $C_{10}H_{14}^+$; ● $C_{10}H_{16}^+$; ◑ $C_5H_6^+$; ■ $C_5H_8^+$. (Source: high-pressure photoionization mass spectrometer). After Lesclaux et al. [30]

molecules that absorb a 10.0 eV photon are 'superexcited' without ionization; this process is followed by dissociation.

Table 5.2 contains yields of products in photolysis carried out in a closed system, in the presence of 5% oxygen. The yields from excited cyclopentene molecules depend on the photon energy. The relative yields of acetylene and allene fragments show the largest increase, with a simultaneous twofold decrease of the cyclopentadiene yield, as the photon energy exceeds the ionization potential. Cyclopentane was formed only at photon energies above the ionization potential (Table 5.2), i.e. this product could be formed exclusively in ionic processes in the presence of oxygen as radical scavenger:

$$\text{cyclo-}C_5H_8^+ + \text{cyclo-}C_5H_8 \longrightarrow C_5H_6^+ + \text{cyclo-}C_5H_{10} \quad \Delta H = -0.54 \text{ eV} \quad (5.2)$$

Reaction (5.2) and the conservation of the cyclic structure in the $C_5H_8^+$ parent ion have been supported by experiments using dimethylamine as a positive charge scavenger (Fig. 5.3).

23*

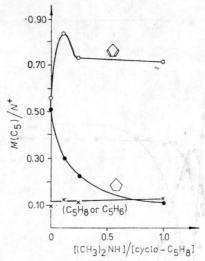

Fig. 5.3. Effect of dimethylamine on cyclopentene irradiated with 10.0 eV photons.
(Pressure: 1.3 mbar). After Lesclaux et al. [30]

Table 5.2

Yields of the photolysis of gaseous cyclopentene in the presence of 5% O_2
as radical scavenger. After Lesclaux et al. [30]

Energy, eV	8.4	10.0	11.6–11.8
Pressure, mbar	1.3	1.3	4
Products	relative yields		
Methane	n. d.	2.5	2.0
Ethylene	100	100 $(0.20)^a$	100 $(0.34)^a$
Acetylene	27.0	42.9	50.2
Propene	7.6	5.1	5.0
Allene	16.6	19.1	22.3
Propyne	8.7	6.7	7.8
Vinylacetylene	15.2	9.1	10.1
Buta-1,3-diene	3.0	4.0	3.0
Cyclopentadiene	96.9	44.6	51.5
Cyclopentane	0.0	40.5	45.6
Pent-2-ene	0.0	0.0	10.
C_5H_6 + penta-1,2-dienes	6.4	9.7	12.8

a Quantum yields in parantheses

The processes are similar when the photon energy is 11.6–11.8 eV, but
the excess energy of the parent ions leads to about 15–20% ring opening.
These experiments do not solve the problem of whether an H_2^- ion or an H_2
molecule is transferred in the transition complex of reaction (5.2).

Recently, Tu and Doepker [31] have reported on the photolysis of cyclopentene with 8.4 and 10.0 eV photons. They concentrated on the decompo-

sition of neutral excited molecules. 90% of the products found consisted of hydrogen, acetylene, ethylene, allene, methylacetylene and cyclopentadiene. The application of several radical scavengers (NO, O_2, H_2S, D_2S and CD_3I) led to the detection of H-atoms and vinyl and propyl radicals. Although not all these radicals could be determined quantitatively, the authors were able to suggest elementary processes for the primary decomposition of excited cyclopentene molecules, as shown in Table 5.3. They proposed an ionic mechanism for cyclopentane formation, similar to that suggested by Lesclaux et al. [reaction (5.2.)] [30].

Table 5.3

Yields of the primary decomposition processes of cyclopentene photolysis. After Tu and Doepker [31]

Energy, eV	8.4	10.0	10.0[a]
Reactions	Quantum yields		
cyclo-C_5H_8 + hν → C_4H_4 + CH_2 + H_2	0.04	0.03	0.04
→ C_2H_2 + C_3H_6	0.03	0.03	0.04
→ C_2H_2 + C_3H_4 + H_2	0.05	0.08	0.09
→ C_2H_4 + C_3H_3 + H	0.25	0.20	0.24
→ C_2H_4 + C_3H_4	0.02	—	—
→ C_5H_6 + 2 H	0.23	0.13	0.16
→ C_3H_3 + C_2H_3 + H_2	0.10	0.11	0.13

[a] Quantum yield based on excited molecules only

5.2.2. Cyclohexene

Yields of radiolysis of liquid cyclohexene show small differences from those of cyclopentene: e.g. hydrogen yields are almost the same and fragment yields are also similar, although their spectra are, of course, different.

Kovács et al. [18] have reported that $G(H_2)$ of cyclohexene is independent of dose up to $1.25 \cdot 10^{22}$ eV g^{-1}. On this basis it may be assumed that the effect of species (precursors or products), such as hydrocarbons with several double bonds, that decrease the hydrogen yield in a competitive way, is negligible owing to their very low concentrations. Whereas the hydrogen yield is independent of the dose, the yield of buta-1,3-diene decreases and that of ethane increases with increasing dose, and the benzene yield shows a maximum (Fig. 5.4). These phenomena can be interpreted in terms of different reactivities of various intermediates as well as of their partial transformation into C_2 hydrocarbons in direct or indirect processes. The yield of buta-1,3-diene as a function of the dose is described by the following equation:

$$G = G_0 \frac{1\text{-}\exp(-kt)}{kt}$$

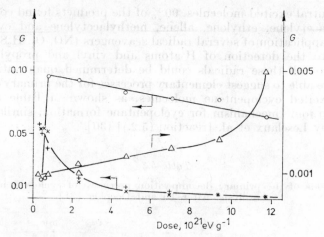

Fig. 5.4. Dose dependence of the yields of some products from cyclohexene. ○ Benzene; △ ethane; × buta-1,3-diene, measured; + buta-1,3-diene, calculated (radiation: ^{60}Co-γ). After Kovács et al. [18]

where G denotes the actual yield belonging to the actual dose, G_0 the yield extrapolated to zero conversion, k is a constant, and t is the time which is proportional to the dose if the dose rate is constant. The calculated curve together with the measured data using appropriate values of G_0 and k ($G_0 = 0.08$; $k = 4.78 \cdot 10^5$ s^{-1}) is shown in Fig. 5.4.

Freeman [1] has determined the individual yields of methane, ethylene and acetylene as well as the total yield of C_4 products, and proposed two types of ring opening:

$$2\,C_2H_4\ +\ C_2H_2 \tag{5.3a}$$

$$C_2H_4\ +\ C_4H_6 \tag{5.3b}$$

This scheme was proved by Cserép et al. [12, 13, 17]. They found that ethylene and butadiene were most abundant (Table 5.1). Thus, the most probable fragmentation can be explained in terms of a radiation-induced reverse Diels–Alder reaction, presumably via a multicentre transition complex. An essentially analogous decomposition takes place during pyrolysis [27–29] or photolysis [6,35] and in the mass spectrometer [36–37], as well as under the effect of 'recoil' tritium formed in the ^{3}He(n, p)^{3}H nuclear process [38–39]. The 'forward' Diels–Alder reaction (e.g. the formation of cyclohexene from ethylene and butadiene) is an important reaction in organic chemistry from both theoretical and practical points of view.

According to Table 5.1, $G(C_2H_4)$ is more than double that of buta-1,3-diene this may be explained by reaction (5.3a), and also by the fact that buta-1,3-diene is less stable than ethylene and may react further. The rate of addition of H-atoms is one order of magnitude higher to butadiene than to monoalkenes [40–41], and that of methyl radicals more than two orders of magnitude higher [42]. Furthermore, during irradiation, cyclohexene may also decompose via ionized states. This process leads predominantly to the formation of one ethylene molecule and a butadiene ion, the latter being a very reactive intermediate stabilized by resonance [37]. This suggestion is supported by the mass-spectrometric fragmentation pattern of cyclohexene, where the relative ion intensity of the butadiene ion is much higher than that of ethylene [36–37].

However, the results of Smith and Gordon [32] and Uchiyama et al. [33] indicate that the difference between the yields of complementary products (i.e. ethylene and butadiene) is much lower (up to 5–20%) in pyrolysis than in radiolysis, probably because higher energies are available in the latter case.

It is interesting to note that the G-values of ethylene and acetylene fragments formed on irradiation of cyclobutene are an order of magnitude higher than those of ethylene and butadiene produced from irradiated cyclohexene. As pointed out in Section 5.2.1. the former process is symmetry-allowed for electronically excited molecules according to the Woodward–Hoffman rules [26], whereas the latter reaction proceeds more readily when ground-state species are involved, as illustrated by orbital- and state-correlation diagrams constructed according to the specified molecular geometry (Fig. 5.5). The symmetry element maintained along the reaction co-ordinate is the symmetry plane of the cyclohexene molecule. It can be seen from the diagrams that the orbitals of the ground-state molecule can be transformed into those of the fragments, with conservation of orbital symmetry, without a significant increase of the energy along the reaction co-ordinate. The diagram suggest that electronically excited cyclohexene molecules formed under irradiation decompose into butadiene and ethylene fragments with a rather low efficiency.

In order to elucidate the nature of the elementary processes in cyclohexene radiolysis, Kovács et al. [19] have applied various additives (see Table 4.42).

The slight reductions of the hydrogen yield brought about by nitrogen oxide and p-benzoquinone, both of them effective radical scavengers, indicate that thermal hydrogen atoms are not of great importance in hydrogen formation. However, CCl_4 (commonly considered as an electron scavenger) caused a sharp decrease in the yield if applied in higher concentrations (0.2–1.0 M). But this effect may be attributed to excitation energy transfer to the additive present in high concentrations [43, 44] and to an effective competition between electron capture and geminate ion recombination. This assumption seems to be supported by the significant lowering of $G(H_2)$ in the presence of other electron scavengers (e.g. N_2O, CO_2) [45, 45A]. The yields of some C_6 and dimer products formed from cyclohexene are listed in Table 5.4. Freeman [1] suggested, on the basis of scavenger experiments, that only about one-third of cyclohexane is formed via disproportionation of cyclohexyl radicals, ion-molecule reactions being responsible for the remaining two-thirds. More recent results [19, 35] show,

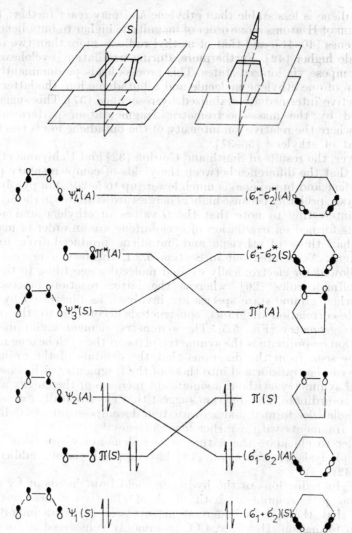

Fig. 5.5. Orbital correlation diagram of the reaction butadiene + ethylene → cyclohexene

however, that a larger fraction of cyclohexane is produced in radical processes. Most of the dimeric products are formed in radical combination rather than ionic processes [19]; the reverse is true for terminal aliphatic alkenes [46] (see e.g. Section 4.4).

Of earlier papers, that of Burns and Winter [47] can be mentioned. They discussed dose, dose rate and LET effects. The radiations used were: ^{60}Co-γ, 4 MeV electrons, 0.7–1.8 MeV protons and 1.5 MeV α-particles, with LET

Table 5.4

Yields of C_6 and C_{12} products from liquid cyclohexene (radiation: ^{60}Co-γ)

Reference	4	19
Products	\multicolumn{2}{c}{G, molecule/100 eV}	
Cyclohexane	0.95	0.96
Hexa-1,5-diene	0.02	0.07
Hexa-1,3-diene		0.05
Cyclohexa-1,4-diene	0.05	trace
Cyclohexa-1,3-diene	0.10	0.14
Benzene	0.05	
Vinylcyclobutane	0.03	
2,2-Dicyclohexenyl	1.90	
3-Cyclohexylcyclohexene	0.58	
Dicyclohexyl	0.21	0.05
Dodecahydrodiphenylene	0.32	

values of 0.2, 3.4, and 21 eV nm^{-1}, respectively. The authors reported that $G(H_2)$ increased from 1.26 for γ-rays and fast electrons through 1.75 for protons to 2.95 for α-particles, whereas $G(C_6H_{12})$ decreased from 0.85 to 0.3 and $G(2,2'$-bicyclohexenyl) from 1.4 for fast electrons to 0.4 for α-particles. The total polymer G-value decreased with increasing LET.

An attempt was made to interpret the LET effect in terms of competitive reactions of excited cyclohexene molecules within the spur:

$$C_6H_{10}^* + C_6H_{10}^* \longrightarrow H_2 + \text{unsaturated substances}$$

$$C_6H_{10}^* \longrightarrow H + C_6H_9$$

$$C_6H_{10}^* \longrightarrow C_6H_{10} \text{ (deactivation)}$$

Table 5.5

LET dependence of yield from liquid cyclohexene. After Burns and Winter [47]

Temperature, °C	Radiation	Energy, MeV	Average LET value, eV nm^{-1}	Cyclohexyl	Cyclo-hex-3-enyl	Cyclo-hex-1-enyl
				\multicolumn{3}{c}{G, radical/100 eV}		
−82	Electrons	2.0	0.02	1.60	3.41	0.16
25	Protons	2.0	3.4	0.57	1.26	0.031
25	Protons	0.5	6.8	0.53	1.23	0.040
25	^{4}He$^+$	2.3	20.0	0.255	0.655	0.027
25	^{4}He$^+$	1.0	22.9	0.25	0.65	0.026

Later Burns et al. [48] used $^{14}CH_3I$ additive to study yields of radicals generated by radiations with different LET values in liquid cyclohexene (Table 5.5). Products formed with relatively high yields (2,2'-bicyclohexenyl, 3-cyclohexylcyclohexene, bicyclohexyl and cyclohexene) were attributed to reactions of cyclohexyl and cyclohexenyl radicals with each other. Increased LET values resulted in decreased radical yields, in accordance with earlier results of Burns and Winter [47] where a similar effect was observed for C_6-C_{12} products of pure cyclohexene without additives.

G-values for radicals generated by radiation with a high LET value (Table 5.6) [48] are somewhat different from those determined from radioactivity measurements of products originating from the combination of C_6-radicals and $^{14}CH_3$ radicals. According to Burns et al. [48], this may be due to the inhomogeneity of track reactions and can be accounted for if the $^{14}CH_3$ radicals, formed possibly by electron capture, have a larger diffusion constant and a larger initial track radius than the cyclohexenyl and cyclohexyl radicals.

Table 5.6

Yields of C_6 radicals from liquid cyclohexene. After Burns et al. [48]

Radicals Radiation	$G'(C_6H_9)$	$G'(C_6H_{11})$	$G'(C_6)$	$G(C_6H_9)$	$G(C_6H_{11})$	$G(C_6)$
	radical/100 eV					
Electrons (2 MeV)	3.3	1.6	4.9	3.4	1.6	5.0
Protons (2 MeV)	2.3	1.2	3.5	1.3	0.6	1.9
$^4He^+$ ions (1 MeV)	1.6	1.1	2.7	0.65	0.25	0.9

The G'-values are calculated from the condensed product yields [47]

The ratio of the track radii of $^{14}CH_3$ and C_6 radicals has been estimated to be about 5. Another possible explanation for the deviations observed for higher LET values was put forward by Burns et al. [48], according to which the fraction of C_{12} products formed in 'molecular' processes increases with higher LET radiations. This possibility was, however, not investigated in detail.

Lesclaux et al. [35] have collected sufficient data to assume that cyclohexane and C_{12} products are formed partly in ion-molecule reactions from cyclohexene. Although these measurements were carried out in the gas phase, they suggest that the change of ratio between ionic and radical processes may have also contributed to the LET effects reported by Burns et al. [48].

To explain the formation of the major identified products of cyclohexene radiolysis, such as H_2, C_6 and C_{12} Freeman suggested the following mechanism. The alternative reactions (5.4a) and (5.4b) refer to low and high excitation energy, respectively.

Lesclaux et al. [35] irradiated cyclohexene with photons of energies 8.4, 10.0 and 11.6–11.8 eV. The ionization potential of cyclohexene is 8.9 eV,

$$\text{cyclo-}C_6H_{10} \overset{\text{\Large\leadsto}}{\underset{\text{\Large\leadsto}}{\Big\langle}} \begin{array}{l} \text{cyclo-}C_6H_{10}^{\bullet} \hfill (5.4\,\text{a}) \\[1em] \text{cyclo-}C_6H_{10}^{\bullet\bullet} \hfill (5.4\,\text{b}) \end{array}$$

$$\text{cyclo-}C_6H_{10}^{\bullet} \longrightarrow \text{cyclo-}C_6H_9 + \overset{\bullet}{H}_{\text{thermal}}$$

$$\text{cyclo-}C_6H_{10}^{\bullet\bullet} \longrightarrow \text{cyclo-}C_6H_9 + \overset{\bullet}{H}_{\text{hot}}$$

$$H_{\text{thermal}} + \text{cyclo-}C_6H_{10} \longrightarrow \text{cyclo-}C_6H_{11}$$

$$H_{\text{hot}} + \text{cyclo-}C_6H_{10} \longrightarrow H_2 + \text{cyclo-}C_6H_9$$

$$\text{cyclo-}C_6H_{10} + \text{cyclo-}\overset{\bullet}{C}_6H_{11} \longrightarrow \cdot \text{ cyclo-}C_6H_{12} + \text{cyclo-}C_6H_9$$

$$2 \text{ cyclo-}C_6H_9 \overset{}{\underset{}{\Big\langle}} \begin{array}{l} (\text{cyclo-}C_6H_9)_2 \\[1em] \text{cyclo-}C_6H_8 + \text{cyclo-}C_6H_{10} \end{array}$$

$$\text{cyclo-}C_6H_9 + \text{cyclo-}C_6H_{14} \overset{}{\Big\langle} \begin{array}{l} \text{cyclo-}C_6H_9\text{-}C_6H_{11} \\[0.8em] \text{cyclo-}C_6H_8 + \text{cyclo-}C_6H_{12} \\[0.8em] 2\text{-cyclo-}C_6H_{10} \end{array}$$

$$2 \text{ cyclo-}C_6H_{11}^{\cdot} \overset{}{\Big\langle} \begin{array}{l} (\text{cyclo-}C_6H_{11})_2 \\[0.8em] (\text{cyclo-}C_6H_{10} + \text{cyclo-}C_6H_{12}). \end{array}$$

so this energy range permitted a distinction between reactions of excited and ionized cyclohexene. This choice of energies, coupled with two different methods of product detection (ion detection in a photoionization mass spectrometer and gas chromatographic product analysis in a closed static system) resulted in data on the mechanism of formation of various products from cyclohexene.

The yields of parent ions and important product ions as a function of the pressure are shown in Fig. 5.6 for 10.0 eV photons and in Fig. 5.7 for 11.6–11.8 eV photons. It is clear from these that the ionizing energy plays a decisive role in determining the fate of the parent ions. The lower energy (10.0 eV) is close to the ionization potential, thus the parent ion is not in a state suitable for further decomposition (Fig. 5.6). Photons with energies of 11.6–11.8 eV, on the other hand, bring about parent ion decomposition as follows:

$$C_6H_{10}^{+} \longrightarrow C_6H_9^{+} + H$$
$$\longrightarrow C_4H_6^{+} + C_2H_4$$
$$\longrightarrow C_5H_7^{+} + CH_3$$

($C_4H_6^{+}$ is not shown in Fig. 5.7).

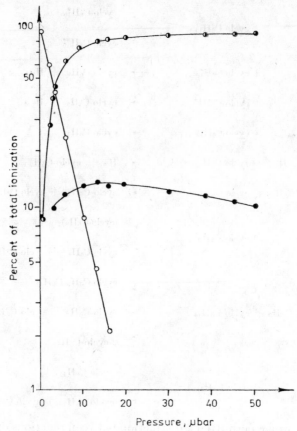

Fig. 5.6. Composite mass spectrum obtained from the photoionization of cyclohexene at 10.0 eV, as a function of chamber pressure.
○ $C_6H_{10}^+$; ● $C_6H_8^+$; ◑ $C_{12}H_{20}^+$. After Lesclaux et al. [35]

The overall rate constant for disappearance of cyclohexene parent ions is $k = 2.9 \cdot 10^{11}$ $M^{-1} s^{-1}$; the main components of this overall value are:

$$C_6H_{10}^+ + \text{cyclo-}C_6H_{10} \longrightarrow C_{12}H_{20}^+ \qquad k = 2.5 \cdot 10^{11} \ M^{-1} s^{-1}$$
$$\longrightarrow C_6H_8^+ + C_6H_{12} \quad k = 0.4 \cdot 10^{11} \ M^{-1} s^{-1}$$

The fragment ion with highest yield ($C_5H_7^+$) reacts further in a condensation reaction:

$$C_5H_7^+ + \text{cyclo-}C_6H_{10} \longrightarrow C_{11}H_{17}^+ ; \quad k = 1.5 \cdot 10^{11} \ M^{-1} s^{-1}$$

Similarly, the butadiene ion participates in a condensation reaction, as indicated by the presence of $C_{10}H_{16}$ ion at 37 mbar:

$$C_4H_6^+ + \text{cyclo-}C_6H_{10} \longrightarrow C_{10}H_{16}^+$$

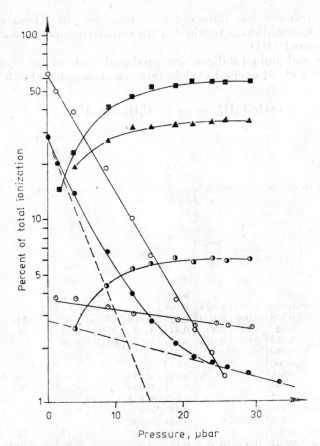

Fig. 5.7. Composite mass spectrum of the major ions obtained from the photoionization of cyclohexene at 11.7 eV, as a function of pressure. (Broken curves indicate resolved decay curves for the two $C_6H_{10}^+$ ions produced at this energy).
■ $C_{11}H_{17}^+$; ▲ $C_{12}H_{20}^+$; ◑ $C_6H_8^+$; ⊙ $C_6H_8^+$; ○ $C_5H_7^+$; ● $C_6H_{10}^+$. After Lesclaux et al. [35]

The rate constant of this latter reaction ($k = 2.4 \cdot 10^{12} \text{ M}^{-1} \text{ s}^{-1}$) is near to that corresponding to the collision rate.

Cyclohexene containing deuterium in specific positions has been photolysed using 10.0 or 11.6–11.8 eV photons in the presence of O_2 as additive; its reactions in photoionization mass spectrometer have also been studied. 75% of the cyclohexane produced had the composition cyclo-$C_6D_5H_7$. This suggested the following mechanism for the ion-molecule reaction:

H_2 and D_2 transfer was observed to a lesser extent. These experiments supplied no information as to whether the transferred species was uncharged (HD) or ionized (HD^-).

Ethylene and buta-1,3-diene are produced with highest yields in the photolysis 8.4 eV of excited cyclohexene, as shown in Table 5.7:

$$\text{cyclo-}C_6H_{10}^* \longrightarrow 1,3\text{-}C_4H_6 + C_2H_4 \qquad (5.5)$$

Table 5.7

Yields from the photolysis of gaseous cyclohexene (pressure: 1.3 mbar). After Lesclaux et al. [35]

Energy, eV	8.4			10.0			11.6–11.8		
Additive	—	10% O_2	10% H_2S	—	10% O_2	10% H_2S	—	10% O_2	10% H_2S
Products	M/N_{ex} value[a]								
Hydrogen	0.15			0.25					
Methane	0.008	0.005	0.078	0.009		0.17	0.013		0.58 (0.27)[b]
Ethane	0.016	0.005	0.009	0.026	0.006	0.008	0.035	0.010	0.98
Ethylene	0.82	0.83	0.84	0.84	0.84	0.84	0.98	0.98	0.98
Acetylene	0.015	0.01	0.015	0.078	0.078	0.077	0.19		0.21
Propene	0.010			0.017				0.025	
CH_3CCH (?)	0.006			0.012				0.019	
CH_3CCH (?)	0.008			0.032	0.046		0.063	0.061	
Vinylacetylene	0.008			0.017	0.030		0.060	0.054	
Buta-1,2-diene	0.007			0.017	0.014				
Buta-1,3-diene	0.72	0.70		0.38	0.39		0.33	0.25	0.26
C_5H_6(?)	0.033	0.035	0.040	0.034	0.026				
Cyclopentene	0.025			0.053	0.048		0.12	0.12	0.11
Cyclohexene	0.027	0.000	n. d.	0.17	0.029	0.61	0.40	0.10–0.15	0.96

[a] Molecules of product formed (M) per neutral excited molecule (N_{ex}) in the system
[b] 13 mbar

Ethylene and butadiene are also produced with photon energies above the ionization potential, but whereas the yield of ethylene increases slightly with increasing energy, that of buta-1,3-diene decreases markedly: with photon energies of 11.6–11.8 eV, the ratio of these two products was close to that observed in radiolysis [12, 13, 18]. This proves the importance of secondary reactions consuming butadiene and its precursor.

In order to study reaction (5.5) in a more detailed way, Lesclaux et al. [35] photolysed 3,3,6,6-cyclohexene-d_4. On the basis of the isotope distribution in buta-1,3-diene and ethylene, they concluded that, with 8.4 eV photons, 86–90% of reaction (5.5) takes place according to (5.6a) and the probability of (5.6b) is only 10–14%:

$$\text{[structure]} \longrightarrow C_2H_4 + CD_2=CH-CH=CD_2 \qquad (5.6a)$$

$$\text{[structure]} \longrightarrow C_2H_2D_2 + C_4H_4D_2 \qquad (5.6b)$$

It is worth mentioning that in the solid state, 8.4 eV photons produce exclusively C_2H_4 and buta-1,3-diene-d_4 [reaction (5.6a)].

Sevilla and Holroyd [6] investigated the vacuum-UV photolysis of cyclohexene by $^{14}CH_3$ radical sampling and ESR spectroscopy. They observed cyclohex-3-enyl, cyclohexyl, methyl and allyl radicals as the main products under the effect of 8.4 eV photons, together with smaller amounts of ethyl and propyl radicals. Ethylene and butadiene were found to be the most important stable products of the photolysis. A broad, unresolved band was observed in the ESR spectrum at $-190°C$, which transformed into a well-resolved spectrum on warming up to $-115°C$. This was also recorded by Ohnishi and Nitta [49], who attributed it to cyclohex-3-enyl radical. The fact that this was the only radical to be observed at this higher temperature was ascribed to the hydrogen abstraction by other, more reactive radicals to form more stable allyl-type radicals:

$$R + C_6H_{10} \longrightarrow C_6H_9 + RH.$$

Fee et al. [38] investigated decomposition and isomerization reactions of excited cyclohexene molecules in the pressure range 0.4–2 bar. The activated molecules were produced in substitution reactions of hot tritium (T)-atoms of the $^3He(n, p)T$ process, according to:

$$\text{cyclo-}C_6H_{10} + T^* \longrightarrow \text{cyclo-}C_6H_9T^* + H$$

$$CH_3\text{-cyclo-}C_6H_{10} + T^* \longrightarrow \text{cyclo-}C_6H_9T^* + CH_3$$

$$\text{cyclo-}C_6H_9T^* + M \xrightarrow{w} C_6H_9T$$

$$\text{cyclo-}C_6H_9T^* \xrightarrow{k} \begin{array}{l} C_2H_3T + C_4H_6 \qquad (5.7a) \\ C_2H_4 + C_4H_5T \qquad (5.7b) \end{array}$$

The experimental data were analysed with the aid of the Rice–Rampsperger–Kassel–Marcus (RRKM) theory of unimolecular reactions. A detailed introduction to the rather complex RRKM formulation is beyond the scope of this book [50]. The essence of the theory is that, extending the 'absolute rate theory', both the rate of energization of molecules, and the rate of their transformation into the activated complex are considered to be functions of the energy, and are calculated by quantum statistical mechanics, taking explicitly into account the various normal vibrations and external and internal rotations. It is assumed that energy randomization occurs between the various normal modes of the activated molecule prior to decomposition.

Fee et al. [38] made use of the RRKM-expression:

$$k_a = w(D/S) = z(D/S)P$$

where k_a is the rate constant, $w = zP$ is the collision frequency, P is the total pressure, z is the collision number [38], S and D are the yields of the collisionally stabilized (C_2H_3T) and the unimolecular decomposition ($C_2H_3T + C_4H_5T$) product molecules, respectively.

The rate constant is an average value, appropriate only to the decomposition of molecules with energies in the specified range involved under the conditions of the reaction studied.

The linearity of the S/D vs. P curves supports the unimolecular decomposition mechanism outlined in reactions (5.7a–b). The apparent rate constant at 135°C was $5.1 \cdot 10^6$ s^{-1}.

The average energy of the activated and tritiated cyclohexene molecules was estimated to be 5 eV molecule^{-1} for the gas-phase species undergoing this decomposition and 8–12 eV molecule^{-1} for the same decomposition in the liquid phase [38]. With the use of radical scavengers the authors determined ratios of abstraction to addition rate constants for several radicals and alkenes. The results correspond satisfactorily to those obtained by other methods.

5.2.3. C_7-C_{12} cycloalkenes

Table 5.1 shows that the value of $G(H_2)$ decreases with cycloalkenes larger than cyclohexene, but increases again from C_9 on.

Of fragments, the yield of C_2 hydrocarbons is the highest. The general tendency is that, with larger rings, fragmentation of excited and/or ionized molecules is a less and less favoured way for excess energy to be dissipated: thus, these larger molecules have a greater tendency than smaller ones for deactivation without decomposition. Molecular energetics such as weakening of σ-bonds in the β-position to the π-bond has almost no influence on the product spectra, of cyclic hydrocarbons with seven or more carbon atoms in the rings (see Section 5.4).

5.2.4. Substituted C_6 cyclic alkenes

The branched-chain cycloalkenes with six-membered rings listed in Table 5.8 can be divided into three groups according to the position of their π-bonds: the first group contains methylcyclohexenes with one π-bond in the ring (1-, 3- and 4-methylcyclohexene, respectively); 4-vinylcyclohexene represents the second group with two π-bonds, one in the ring and one in the side-chain; methylenecyclohexane, with a semicyclic π-bond, is the only member of the third group.

Comparison of the data in Tables 5.1 and 5.8 reveals that the elimination of a methyl group is practically the only source of methane as methylene-cyclohexane and 4-vinylcyclohexene, which contain no methyl groups, give very small amounts of methane. In spite of the identical chemical composition (one methyl group per molecule), the methane yields from methyl-cyclohexene isomers are different: the lowest $G(CH_4)$ was observed [16] with 1-methylcyclohexene which has a methyl group in the α-position to the π-bond; the highest $G(CH_4)$ was found with the methyl group in the β-position to the π-bond (3-methylcyclohexene). In the former compound hyperconjugation occurs, and in the latter allyl interaction.

Methylcyclohexenes and 4-vinylcyclohexene, in spite of the fact that the latter contains two double bonds, decompose similarly to cyclohexene (Section 5.2.2.), i.e. via a reverse Diels–Alder reaction. The main products of decomposition of methylcyclohexene isomers are different, depending on the position of the methyl group. 1-methylcyclohexene gives ethylene and 2-methylbuta-1,3-diene (isoprene), 3-methylcyclohexene produces ethylene and penta-1,3-diene and 4-methylcyclohexene forms propene and buta-1,3--diene [16]:

$$(5.8)$$

Of the complementary products of these main decomposition processes, the yield of that with the higher molecular mass is lower, which is similar to the radiolysis of cyclohexene. Data in Tables 5.1 and 5.8 show that decomposition of the cyclohexene molecule into three parts may not be the only reason why this difference exists; the subsequent reactions of the higher molecular mass fragments containing conjugated double bonds (buta-1,3-diene, penta-1,3-diene and 2-methylbuta-1,3-diene) have also to be considered. If the difference were exclusively due to the simultaneous fragmentation of the molecule into three parts, then the G-value of propene

24

produced from 4-methylcyclohexene in reaction (5.8) ought to be lower than that of ethylene from cyclohexene, for example, since the decomposition of the rest of the molecule would give ethylene instead of propene.

The ethylene yields of cyclohexene (Table 5.1) and 1-methyl- and 3-methylcyclohexene are, however, comparable with the yield of propene from 4-methylcyclohexene (Table 5.8). Further, the G-values of ethene and acetylene from the latter compound are as low as 0.03, and similar low yields of propene (0.02) were observed in the radiolysis of 1-methyl- and 3-methylcyclohexene. All these facts show that simultaneous fragmentation of the molecule into three parts has a low G-value (0.02–0.03) and may not be responsible for the entire deficit in the yields of the complementary hydrocarbon fragments. The latter may rather be attributed to reactions of the resonance stabilized radical-ion and/or biradical containing the original double bond of the parent molecule with each other and with other reactive species present in the system, as suggested by Cserép and Földiák for the radiolysis of cyclohexene [17]. In addition, the stabilized conjugated products react several orders of magnitude faster than the parent methylcyclohexenes.

Previously, when the reverse Diels–Alder reaction of cyclohexene was discussed (see Section 5.2.2), it was already pointed out that the same type of decomposition takes places in photochemistry and radiation chemistry and as under the effect of energy input by 'hot' tritium atoms. It is not surprising, therefore, that methylcyclohexene isomers behave similarly. Fee and Markovitz [51] investigated methylcyclohexene reactions induced by recoil tritium atoms at 135°C in the pressure range 0.4–1.7 bar. They found that tritiated methylcyclohexene molecules produced by substitution of tritium for protium have an excitation energy of about 6 eV; they undergo collision deactivation or decompose, e.g.

The unimolecular character of this decomposition was established from the pressure dependence of the product yield.

Radiolysis of 4-vinylcyclohexene yields butadiene as the only main fragmentation product [16] (Table 5.8). The most important decomposition process involves the rearrangement of unpaired electrons formed when the allylic C—C bonds are cleaved, together with the π-electrons of the double bonds, resulting in the formation of new double bonds at the expense of ruptured σ-bonds:

Table 5.8

Yields from liquid substituted cycloalkenes (radiation: ^{60}Co-γ). After Cserép and Földiák [16]

Hydrocarbons	1-Methylcyclo-hexene	3-Methylcyclo-hexene	4-Methylcyclo-hexene	4-Vinylcyclo-hexene	Methylenecyclo-hexane
Product	G, molecule/100 eV				
Hydrogen	1.30	1.30	1.18	0.79	0.75
Methane	0.016	0.059	0.038	0.002	0.003
Ethane	0.003	0.002	0.002		0.001
Ethylene	0.17	0.16	0.03	0.044	0.05
Acetylene	0.02	0.03	0.03	0.023	0.005
Propane	0.005	0.001	0.002		
Cyclopropane					0.003
Propene	0.02	0.02	0.15	0.001	0.005
Allene (propadiene)					0.007
Butanes	0.004	0.001	0.002	0.001	
But-1-ene	0.003	0.006	0.005	0.001	0.001
But-2-enes	0.005	0.008	0.002	0.001	0.001
Buta-1,3-diene	0.002	0.007	0.06	0.163	0.001
Isoprene	0.05	0.002			0.001
Penta-1,3-diene		0.077			
C₅ products	0.004	0.007	0.005	0.002	0.007

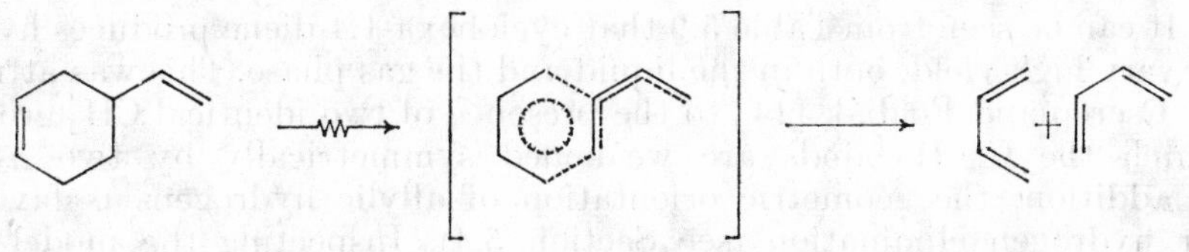

Ethylene and acetylene are also formed in significant amounts, twice as much ethylene as acetylene. They may be formed by the following processes:

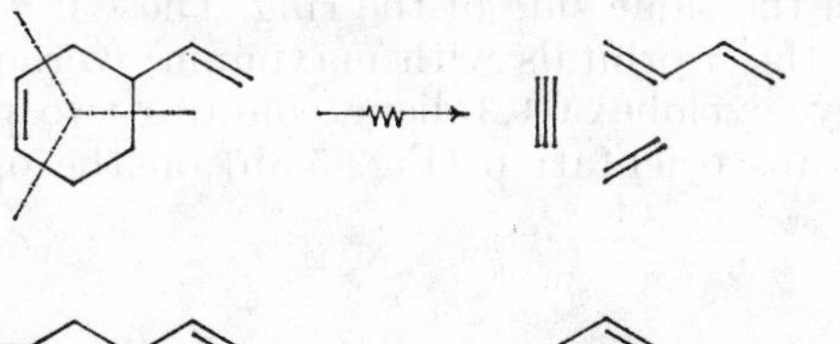

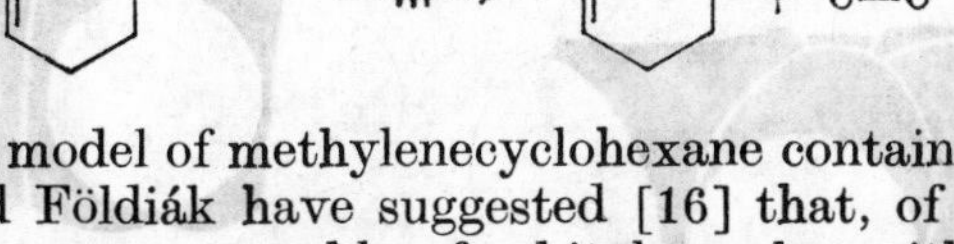

On the basis of a model of methylenecyclohexane containing a semicyclic π-bond, Cserép and Földiák have suggested [16] that, of the four allylic C—H bonds, two are more capable of orbital overlap with the π-electron system, the other two being almost in the plane of the ethylene group. Consequently, two allylic hydrogens may participate principally in hydrogen formation: the observed value of $G(H_2) = 0.75$ seems to be in agreement with this suggestion.

24*

Fragmentation of methylenecyclohexane into C_1-C_5 products is also very different from that of cyclohexene and its alkyl substituted homologues, as indicated by the very low yields.

The very small amount of fragmentation products obtained on γ-irradiation of methylenecyclohexane is rather surprising in view of the experimental fact that the lower homologues, e.g. methylenecyclopropane and methylenecyclobutane, readily decompose when exposed to vacuum-UV photons [53, 54]. On the other hand, this behaviour is analogous to that observed in methylenecyclopentane photolysis [52] where, in addition to a small amount of hydrogen, only traces of other decomposition products (e.g. ethylene and allene) are formed. This phenomenon indicates that the excited or ionic states involved must be extremely stable with respect to decomposition processes, and are deactivated with high probability to the ground state or react in some other way, e.g. forming isomers or high molecular weight products. It should be mentioned, however, that as with the cycloalkane and cycloalkene homologues, there are great differences in ring-strain energy between the C_3-C_4 and the C_5-C_6 members of methylenecycloalkane series. This phenomenon may well account for the observed effects.

5.2.5. Cyclohexadiene isomers

It can be seen from Table 5.9 that cyclohexa-1,4-diene produces hydrogen in very high yield, both in the liquid and the gas phase. This was attributed by Cserép and Földiák [14] to the presence of two identical CH_2 groups, in which the C—H bonds are weakened symmetrically by two π-bonds; in addition, the geometric orientation of allylic hydrogens is favourable for hydrogen elimination (see Section 5.4). Inspecting the model of the cyclohexa-1,4-diene molecule (Fig. 5.8a), it can be seen that two of the allylic C—H bonds are perpendicular to the plane of the —CH=CH— groups and are on the same side of the ring. These C—H bonds, therefore, can interact with the π-orbitals with maximum overlap.

On the contrary, cyclohexa-1,3-diene contains two perpendicular allylic C—H bonds in trans orientation (Fig. 5.8b) on the opposite sides of the

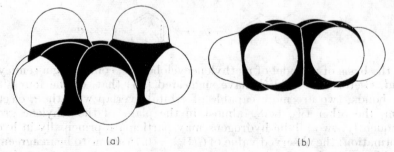

(a)   (b)

Fig. 5.8. Stuart–Briegleb models of (a) cyclohexa-1,4-diene and (b) benzene

ring, which interact only with one double bond each. This hypothesis seems to be supported by pyrolysis studies of the two isomers: Ellis and Frey [55] and Benson and Show [56] reported that cyclohexa-1,4-diene decomposes to give hydrogen and benzene in a unimolecular reaction, whereas cyclohexa-1,3-diene is stable at identical temperatures (310–350°C).

Nakagawa et al. [8] suggested a predominantly radical pathway for hydrogen formation, on the basis of a study of cyclohexa-1,4-diene radiolysis in the gas phase with NO additive.

The same authors found that the $G(H_2)$ from cyclohexa-1,4-diene decreases with increasing dose (Fig. 5.9). This behaviour differs considerably from

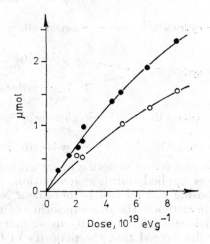

Fig. 5.9. Yields of gaseous products of the radiolysis of cyclohexa-1,4-diene vapour (55 mbar) as a function of dose.
● H_2; ○ $C_2 + C_4$ (radiation: ^{60}Co-γ). After Nakagawa et al. [8]

that of monoalkenes; e.g. cyclohexene gives a dose-independent $G(H_2)$ value up to a dose of about $1.25 \cdot 10^{22}$ eV g^{-1} [18]. The dose dependence was attributed by Nakagawa et al. to cyclohexa-1,3-diene formation during radiolysis. Their data also suggest that benzene (produced in 30 times larger yields than cyclohexa-1,3-diene) and hexa-1,3,5-triene (also produced in fairly high amounts) may have a considerable influence on the dose dependence. The yield of ring fragments, similarly to that of hydrogen, is not directly proportional to the amount of energy absorbed, owing to the effect of products of conjugated structure.

The pressure dependence of the overall radiolysis yields is similar to that observed for the photolysis [11] of cyclohexa-1,4-diene; this fact was ascribed by Nakagawa et al. [8] to a relatively long-lived transient species. As a result, the following scheme was proposed to interpret the formation of the main products in the radiolysis of cyclohexa-1,4-diene:

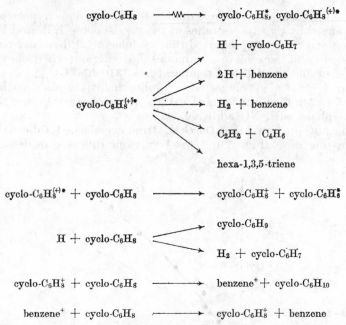

where $C_6H_8^{(+)*}$ denotes an excited molecule or an excited ion.

If an electric field is applied during γ-irradiation, information on the nature of the decomposing species can be obtained. Nakagawa et al. [8] used this method to investigate the decomposition of the neutral excited cyclohexa-1,4-diene molecule (Fig. 5.10). By using different additives (e.g. NO, N_2O, NH_3), they discovered that the majority of the C_6 products are formed in ionic chain reactions.

The state of aggregation also has an influence on product composition, e.g. twice as much benzene was formed from cyclohexadiene than cyclohexene during gas-phase radiolysis. Nakagawa et al. [8] explained this by assuming a chain termination simultaneous to neutralization involving mainly benzene ions. It may also be assumed that the ratio of the yields is closely connected to structural analogies between cyclohexa-1,4-diene and benzene. After the two axial allylic hydrogen atoms are eliminated, the carbon skeleton can readily transform into the planar aromatic structure. This reaction is promoted by the resonance energy of the benzene formed being as high as 151 kJ mol^{-1}. In earlier studies of cyclohexa-1,4-diene radiolysis by Eberhard et al. [57], the formation of cyclohexene, cyclohexa-1,3-diene and benzene was interpreted in terms of a radical mechanism involving subsequent hydrogen transfer.

Okada et al. [58] studied the γ-radiolysis of cyclohexa-1,4- and -1,3-dienes; they concluded from ESR studies and product determination that the main primary process for both isomers is the dissociation of allylic C—H bonds. Hydrogen atoms formed may add to double bonds or abstract other hydrogen atoms. The decompositions of the two isomers are similar but not identical, as the ESR spectrum of cyclohexa-1,4-diene showed the presence

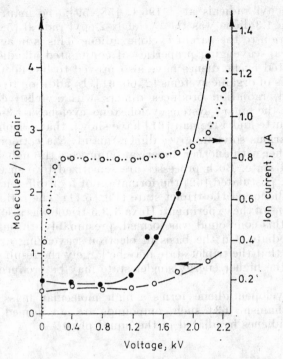

Fig. 5.10. Effect of an electric field on the relative yields of gaseous products of the radiolysis of cyclohexa-1,4-diene vapour (55 mbar).
● H_2; ○ $C_2 + C_4$ (radiation: ^{60}Co-γ). After Nakagawa et al. [10]

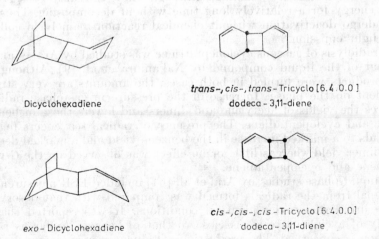

Dicyclohexadiene

*trans-,cis-,trans-*Tricyclo[6.4.0.0]
dodeca-3,11-diene

exo- Dicyclohexadiene

*cis-,cis-,cis-*Tricyclo[6.4.0.0]
dodeca-3,11-diene

Fig. 5.11. Dimer products formed in the radiolysis of cyclohexa-1,3-diene (radiation: ^{60}Co-γ). After Penner et al. [62]

of cyclohexadienyl radicals at $-196°C$ [58, 59]; the main intermediate from cyclohexa-1,3-diene was the cyclohex-2-enyl radical [58], formed via hydrogen addition to the parent cyclohexadiene. This is in agreement with the good radical-scavenging properties of conjugated alkadienes.

The conjugated cyclic dienes have also proved to be effective inhibitors in the radiolysis of organic systems [2, 60, 61]. In addition to the inhibiting effect expected, radiation produces dimers with a relatively high yield ($G = 6.3$) from the binary system cyclohexene–cyclohexa-1,3-diene. Recent studies by Schutte and Freeman [61] have shown that cyclohexa-1,3-diene dissolved in various solvents gives dimers mainly via cationic Diels–Alder addition, together with another dimerization process through a (presumably triplet) excited state. Both processes are sensitized by benzene.

Penner et al. postulated [62] the formation of four different dimers from cyclohexa-1,3-diene via the triplet state (Fig. 5.11). The ratio is influenced by the conditions of the experiment. Hexa-1,3,5-triene was also found among the products; this compound was formed, presumably, through an excited singlet intermediate. On the basis of electron-scavenging experiments, it was concluded that the triplet state may be largely the result of neutralization, whereas the higher energy singlet state may be produced in primary excitation.

Irradiated cyclopentadiene forms a high molecular mass polymer via a cationic mechanism [63], such compounds are not formed in high yield from cyclohexadienes irradiated in the liquid phase.

5.2.6. C_7-C_{12} cyclic polyenes

Hydrogen yields of these compounds are very low (Table 5.9), in some cases no higher than that of benzene [$G(H_2) = 0.04$]. The highly conjugated systems are chemically reactive, although they can also conserve the excess energy for a relatively long time without decomposing. Thus they can undergo deactivation without chemical reaction, namely in collisions or by light emission.

The radiolysis of gaseous cycloheptatriene was studied by Arai et al. [64], and that of the liquid compound by Nakamura et al. [9]. Although the same products were formed in both cases, the amounts are very strongly dependent on the phase. Increase of the pressure in gas-phase radiolysis enhances the yields of acetylene and dimers and lowers those of benzene, toluene and cyclopentadiene. The presence of radical scavengers (e.g. O_2, NO) leads to a marked increase in the benzene yield and a marked decrease of the dimer yield whereas little or no effect was observed on the G-values of toluene and cyclopentadiene.

The liquid-phase studies by Arai et al. [64] included ESR measurements. The signal from the radicals formed was compared with that observed in benzene irradiated under identical conditions. It was reported that the stability of cycloheptatriene is close to that of benzene as far as radical formation is concerned. The production of toluene was explained in terms of isomerization of excited cycloheptatriene molecules.

This assumption was also made by Nakamura et al. [9] for the gas-phase radiolysis of cycloheptatriene on the basis of the pressure dependence of the products. They interpreted the increase in the benzene yield by the following reaction:

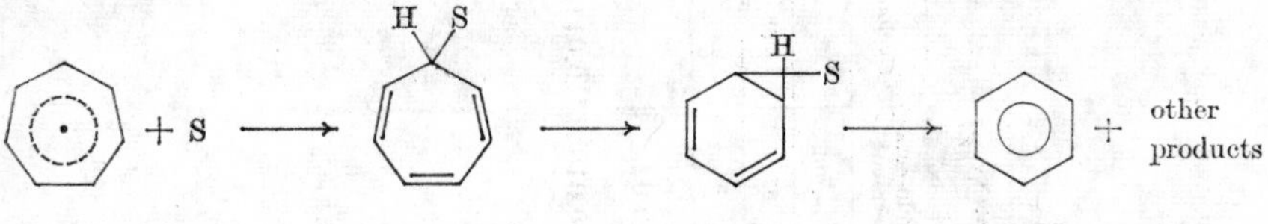

where S denotes a radical scavenger.

The formation of cyclopentadiene and acetylene was not particularly affected by various radical or charge scavengers consequently, their formation may be explained by direct decomposition of excited cycloheptatriene:

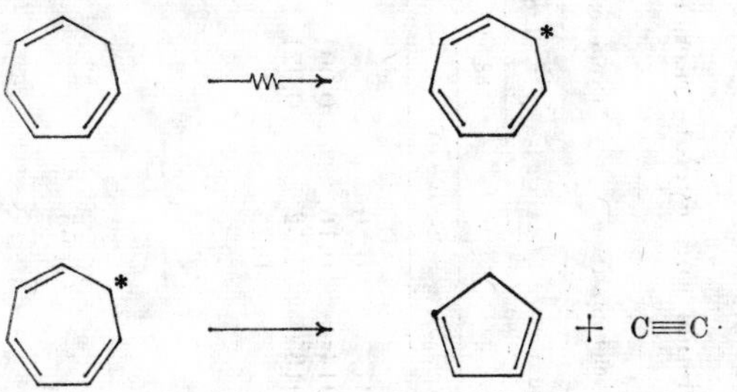

It may be assumed that the energy of the excited state of cycloheptatriene in this reaction is higher than that of the precursor of toluene, as neither the acetylene nor the cyclopentadiene yield is pressure-dependent. The existence of such an excited state seems to be supported by the absence of acetylene and cyclopentadiene in the photolysis [64, 65] and the pyrolysis [66] of cycloheptatriene.

As more acetylene than cyclopentadiene is produced, the existence of another pathway was tentatively proposed by Nakamura et al. [9]. The explanation of this phenomenon may be similar to that offered for the yields of the complementary fragments ethylene and butadiene from cyclohexene [see Section 5.2.2., reactions (5.3a) and (5.3b)], i.e. the higher molecular mass fragment (cyclopentadiene) would undergo further reaction.

Shida et al. [67] reported already more than 20 years ago that the radiation resistance of cyclo-octatetraene is even greater than that of benzene. In addition to gaseous products shown in Table 5.9, benzene (complementary to acetylene) was also formed, although not determined quantitatively. The yield of polymer was also very low ($G=0.64$).

Although cyclododeca-1,5,9-triene contains no conjugated double bonds, it is also very resistant to radiation. Its hydrogen yield is about an order of magnitude higher than that of conjugated cycloalkatrienes, e.g. benzene and cycloheptatriene (Table 5.9). This is in agreement with the suggestions about the importance of allylic C—H bonds in hydrogen elimination,

Table 5.9

Yields from liquid cyclodienes, cyclotrienes and cyclotetraenes (radiation: ^{60}Co·γ)

Hydrocarbons	Cyclohexa-1,3-diene	Cyclohexa-1,4-diene			Cycloheptatriene		Cyclo-octa-1,3-diene		Cyclo-octa-tetraene	Cyclododeca-1,5,9-triene	
Phase	Liquid	Liquid	Liquid	Gas	Liquid	Gas	Liquid	Liquid	Liquid	Liquid	Liquid
References	88	77	58	8	64	9	88	77	67	68	69
Products	*G*, molecule/100 eV										
Hydrogen	0.22	0.94	1.18	1.95	0.06	0.50	0.001	0.24	0.02	0.44	
Methane	0.0015			trace	0.0014	0.16					
$C_2 + C_4$ products	0.0079	0.08		1.25[a]	0.0016[b]	1.52[c]	0.021		0.018[c]	0.0114[d]	0.576[e]
Cyclopentadiene						0.29					
Cyclohexene				14.0							
Cyclohexa-1,3-diene				1.0							
Hexa-1,3,5-triene				1.0							
Benzene			2.3	28.5	0.5	0.65					
Toluene			2.90								

[a] Acetylene and buta-1,3-diene
[b] Mainly ethylene;
[c] Mainly acetylene
[d] Mainly acetylene and ethylene
[e] Mainly butadiene

(Section 4.4.): there are allylic hydrogen atoms in cyclododeca-1,5,9-triene (as in cyclohexa-1,4-diene), but cyclo-octatetraene contains none.

The low probability of ring fragmentation may have several causes, e.g. the presence of three π-bonds in the molecule or the increased probability of competing $C-H$ bond fission. Wiley et al. [68] found no butadiene among the products, although cyclododeca-1,5,9-triene is the trimer of butadiene. Wojnárovits [69], however, has detected considerable amount of butadiene (Table 5.9).

5.3. MIXTURES OF CYCLOALKENES

A number of papers deal with the radiolysis of mixtures of cycloalkenes with alkanes and alkenes. All these investigations indicated the existence of more than one kind of interaction between the components: these are more or less identical with those discussed in Sections 2.3, 3.3 and 4.3 [70–76, 80].

Alkane–cycloalkene mixtures

Although the character of interactions is essentially the same here as in mixtures of alkanes with aliphatic alkenes the G-values of individual mixtures, however, show much larger deviation from a common curve than in the case of alkane–aliphatic alkenes (see Figs 4.8 and 5.12). This can be attribut-

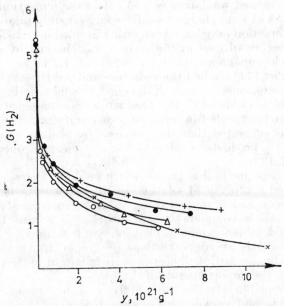

Fig. 5.12. $G(H_2)$-values of mixtures of aliphatic alkanes with cycloalkene, as a function of y, the double bond concentration.
$+$ C_5; $\bullet$ C_6; $\triangle$ C_7; $\times$ C_4; $\circ$ C_8 (radiation: ^{60}Co-γ)

ed to the cyclic structures, which involve different energeties, display different kinetic behaviour, etc.

Considerable interaction was observed, e.g. in mixtures of cyclohexane with cyclohexene. Whereas hydrogen yields were lower than those excepted on the basis of the addition rule, dicyclohexenyl yields were higher.

Cher et al. [73] interpreted their results in terms of the addition of hydrogen atoms to the π-bonds. However, several other authors have attributed the interaction not only to the capture of H-atoms but also to various transfer processes [71, 74, 77, 78]. This latter suggestion is confirmed, e.g. by the results of Toma and Hamill [79]: irradiated mixtures of cyclohexane and cyclohexene-d_{10} give more HD and D_2 than expected on the basis of additivity. This is in agreement with the concept of 'self-sacrificing' of alkenes to 'protect' alkanes [1, 74, 79].

It has been shown by Ausloos et al. [70, 80] and by van Ingen and Cramer [81] that in liquid-phase radiolysis, H_2 can be transferred from cyclo-$C_6H_{12}^+$ to some alkenes, whereas such H_2 transfer to cyclohexene is negligible. This was established by Cramer and Piet [74] on the basis of the observation that the yield of ^{14}C-cyclo-C_6H_{12} which is as high as $G = 0.6$ in a 76 mM solution of ^{14}C-cyclo-C_6H_{10}, is reduced by adding oxygen as radical scavenger to a value of $G = 0.05$.

Yang et al. [7] produced experimental evidence for such energy transfer in photolysis. They irradiated among others, liquid mixtures of cyclohexane and cyclohexene with photons of 8.4 eV energy, and found that the hydrogen yield was somewhat more reduced than during radiolysis. As the energy of photons used was lower (8.4 eV) than the ionization potential of cyclohexane (9.88 eV), no charge transfer could possibly take place; consequently, the interaction may have been either excitation transfer or H-atom capture. The absence of ions in the system was checked by other experiments using cyclopropane as additive (see Section 1.4.1).

Cramer and Piet [74] studied the radiolysis and the Hg-sensitized photolysis of solutions of inactive and ^{14}C-labelled cyclohexene in cyclohexane. The relatively high yields of $C_{12}H_{18}$ products in the presence of a number of additives were interpreted in terms of a reactivity transfer from solvent to solute. It was suggested that this proceeds by positive charge transfer. The G-values of C_{12} products as a function of cyclohexene concentration are shown in Fig. 5.13.

Comparison of specific radioactivities of C_{12} products formed by radiolysis and Hg-sensitized photolysis of solutions of ^{14}C-cyclohexene in cyclohexane has led to the conclusion that scavenging of homogeneously distributed cyclohexyl radicals is complete at [cyclo-C_6H_{10}] $\geq$ 30 mM. In radiolysis, however, a considerable amount of bicyclohexyl is formed by combination of cyclohexyl radicals at concentrations of cyclo-C_6H_{10} between 30 and 110 mM: this indicates an inhomogeneous distribution of these radicals in the solutions, i.e. the important role of the spurs.

Yang and Marcus [72] studied the radiolysis of mixtures of cyclopentane with cyclopentene and cyclohexane with cyclohexene. From the almost complete product analysis they concluded that the nature of the solvent–solute interaction must consist of both H-atom scavenging and energy or

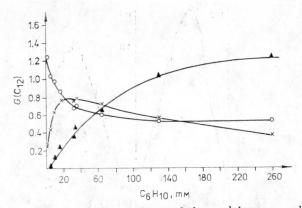

Fig. 5.13. Yields of some C_{12}-products of the cyclohexane–cyclohexene system.

▲ $C_{12}H_8$; ● $C_{12}H_{22}$; × $C_{12}H_{20}$ (radiation: ^{60}Co-γ). After Cramer et al. [74]

charge transfer. High yields of diene dimers were observed, even at an alkene concentration as low as 1%; these were attributed largely to non-radical processes following preferential energy localization in the alkenes. It was suggested that H-atom scavenging processes predominate at very low alkene concentrations, but physical interaction becomes more important as the alkene concentration is increased beyond about 1% [72]. According to Yang and Marcus, the rapid increase in dialkenic product yields as a function of the concentration of the cycloalkene component can be attributed to energy transfer followed by allylic C—H bond rupture and recombination of the resulting cycloalkenyl radicals. However, the production of the same compound ($C_{12}H_{18}$) was interpreted by Cher et al. [73] as follows:

$$C_6H_{11} + C_6H_{10} \longrightarrow C_6H_{12} + C_6H_9$$

$$2C_6H_9 \longrightarrow C_{12}H_{18}$$

It seems reasonable to assume that both the suggested [72, 73] mechanisms are operating in the formation of the products in question.

Cserép et al. [81A] have published data on the direct observation of cyclohexene cations in liquid hydrocarbon mixtures. It was possible to prove by means of pulse radiolysis that simple monoalkenes such as cyclohexene and its methyl-substituted derivatives efficiently capture positive (but not negative) charge in liquid alkanes. In solutions of cyclohexene or 1-, 3- and 4-methylcyclohexenes in cyclohexane or n-heptane, irradiated by a pulsed beam of electrons, structureless optical absorption bonds between 600 and 1000 nm have been observed (Fig. 5.14). These bands are attributed to monomeric radical cations of the cyclohexenes. From the measured growths of absorption, pseudo-first order rate constants for cation formation have been calculated (Table 5.10). From Fig. 5.14 and Table 5.10 it is seen that the substitution of hydrogen by a methyl group causes a bathochromic shift of the band maximum. This effect decreases as the methyl

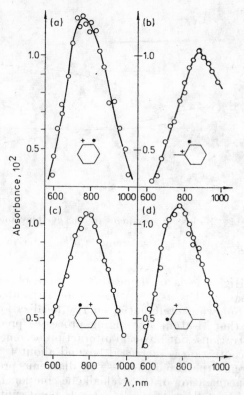

Fig. 5.14. Transient absorption spectra obtained for electron pulse irradiated solutions of cyclohexenes in n-heptane containing 1 mM CCl_4
(a) 10 mM cyclohexene; (b) 10 mM 1-methylcyclohexene; (c) 10 mM 3-methylcyclo-hexene; (d) 10 mM 4-methylcyclohexene (radiation: 1 MeV electron). After Cserép et al. [81A]

Table 5.10

Band maxima and pseudo-first order rate constants for alkene cation formation in n-heptane–cycloalkene mixtures containing 1M CCl_4 (radiation: 1 MeV electron). After Cserép et al. [81A]

Compound	Band maximum		Cation formation rate constant, $M^{-1}s^{-1}$
	Solid[a]	Liquid[b]	
	nm		
Cyclohexene	708	750	$2 \cdot 10^{10}$
1-Methylcyclo-hexene	820	870	$1.5 \cdot 10^{10}$
3-Methylcyclo-hexene	745	790	$2 \cdot 10^{10}$
4-Methylcyclo-hexene	715	770	$1 \cdot 10^{10}$

[a] Ref. [81B]
[b] Ref. [81A]

group is more remote from the double bond. These observations are in agreement with the results of solid state γ-radiolysis [81B].

Ion-molecule reactions of cyclohexenes and their radical cations formed by charge transfer processes have also been studied [81C].

In order to obtain rate constants for these reactions, e.g.

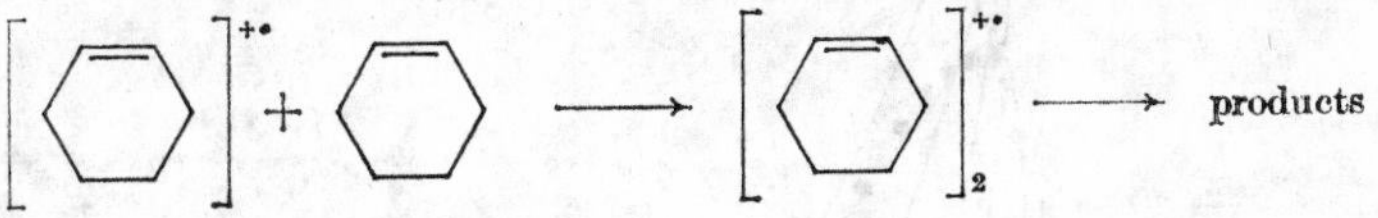

$$\left[\bighexagon \right]^{+\bullet} + \bighexagon \longrightarrow \left[\bighexagon \right]^{+\bullet}_2 \longrightarrow \text{products} \qquad (5.9)$$

the decay of transient absorption was measured at various concentrations of the monomer. Figure 5.15 shows the results for 1-methylcyclohexene. Similar decay curves were obtained also in the case of the other cyclohexenes. From the fits used to simulate the measured decay curves, the overall rate constants for the ion-molecule reactions have been calculated (Table 5.11).

Table 5.11

Pseudo-first order rate constant for reaction (5.9) (radiation: 1 MeV electrons). After Cserép et al. [81C]

Compound	k, $M^{-1}s^{-1}$
Cyclohexene	$6 \cdot 10^6$
1-Methylcyclohexene	$6.5 \cdot 10^7$
3-Methylcyclohexene	$1.8 \cdot 10^8$
4-Methylcyclohexene	$1.8 \cdot 10^7$

Table 5.12

Pseudo-first order rate constant for reaction (5.10) and proton affinites (Pa) of alcohols. (radiation: 1 MeV electrons). After Cserép et al. [81C]

Alcohol	k, $M^{-1}s^{-1}$	Pa, $kJmol^{-1}$
Methanol	$1.7 \cdot 10^9$	757
Ethanol	$2.0 \cdot 10^9$	782
Isopropanol	$3.5 \cdot 10^9$	811
tert-Butanol	$1.0 \cdot 10^{10}$	857

The decay of cyclohexene radical cations in the presence of alcohols has also been studied [81C]. Alcohols have rather high proton-affinities; therefore it is expected that the decay of the cyclohexene cation in the presence of a suitable alcohol is faster and the following reaction occurs:

$$\overset{\bullet}{\underset{}{\bighexagon}}^{+} + ROH \longrightarrow \bighexagon{}^{\bullet} + ROH_2^{+} \qquad (5.10)$$

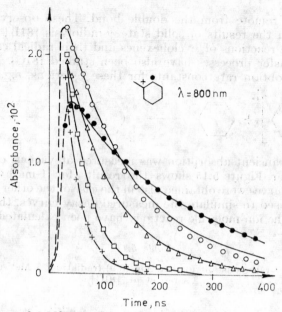

Fig. 5.15. Decay of 1-MCH cation absorption after electron pulse irradiation of cyclohexane containig 1mM CCl_4.
1-MCH concentrations: ● 10 mM; ○ 50 mM; △ 100 mM; □ 300 mM; + 500 mM (radiation: 1 MeV electron). Curves: calculated values. After Cserép et al. [81C]

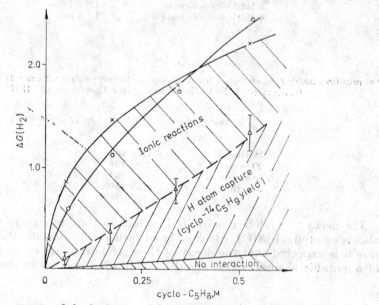

Fig. 5.16. Components of the hydrogen yield of the radiolysis of cyclopentane–cyclopentene mixtures.
× Measured overall yield; ○ calculated overall yield; △ calculated yield of H-capture + dilution effect (radiation: ^{60}Co-γ). After Keszei et al. [82]

The pseudo-first order rate constants deduced from the decay curves and the proton affinities (Pa) for the given alcohols are compiled in Table 5.12. The close correlations between the rate constants and proton affinities of alcohols strongly suggest that the increased rate of disappearance of the cyclohexene cation observed in the presence of alcohols can be attributed to proton transfer.

Keszei et al. [82] have studied the radiolysis of mixtures of cyclopentane and cyclopentene with ^{14}C tracers and NH_3 as positive ion scavenger. They have shown that, in spite of general view [13, 14], the interaction between alkanes and alkenes cannot be confined to radical processes even at low cyclopentene concentrations: ionic processes (charge transfer) must also take place (Fig. 5.16). They used a modified Schuler equation:

$$\Delta G(H_2) = \frac{G_i \sqrt{\alpha[E]}}{1 + \sqrt{\alpha[E]}} \sigma \cdot \eta$$

where $[E]$ is the molar concentration of cyclopentene, σ is the efficiency of ionic hydrogen formation, η is a correction factor for hydrogen formation from alkenes themselves, and G_i is the total ion yield. The authors estimated an α factor characteristic of the extent of the ionic interaction: $\alpha = 1.0 \pm \pm 0.2$ M^{-1}. This value is close to the α-value for other positive ions.

Horváth and Földiák [75] investigated the liquid-phase radiolysis of several alkane–cycloalkene systems. They found that the correct treatment of the reaction mechanism requires the consideration of as many individual

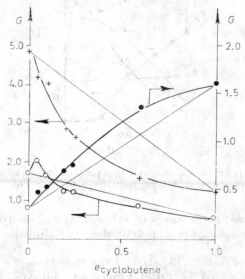

Fig. 5.17. Yields of some products of the cyclobutane–cyclobutene mixtures ● Acetylene; + ethylene; o hydrogen (radiation: ^{60}Co-γ). After Horváth et al. [75]

products as possible. The yields of individual products from the same binary mixture may be completely different when plotted as a function of the composition: one product may exhibit a positive deviation from additivity and another a negative one, and, what is more, even inflections are observed, i.e. the same curve may have two maximum deviations from the straight line, one in the positive and one in the negative direction (Figure 5.17). This latter phenomenon is generally observed when the difference between the ionization potentials of the two components is small. In such cases the observed resultant of the partial processes will depend on the composition and the sign of the deviation may be reversed in another concentration range. Consequently, the eventual linearity of the yields as a function of composition cannot be regarded as a proof of the lack of interaction between components: it may indicate that the effects of several partial processes are co-incidentally cancelling each other.

Alkene–cycloalkene mixtures

Mixtures of but-1-ene and cyclobutene have been investigated by Horváth and Földiák. They based their studies on the evaluation of the hydrocarbon fragments (Fig. 5.18): the yields of some products as a function of the composition are linear, while others are not [75]. The reasons for this are

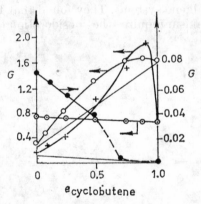

Fig. 5.18. Radiolysis of the but-1-ene–cyclobutene mixtures
$+$ 1,2-C_4H_6; $\bigcirc$ C_2H_2; $\bullet$ C_2H_6; $\odot$ H_2 (radiation: ^{60}Co-γ). After Horváth et al. [75]

analogous to those in the cases of other mixtures of hydrocarbons which have similar photoionization potentials (e.g. aliphatic hexane systems, Section 2.3.).

Wakeford and Freeman [3] studied the radiolysis of cyclohexene in the presence of cyclohexa-1,3- and -1,4-diene as additives. It was established that cyclohexa-1,3-diene decreases the yield of all major products from cyclohexene, and that cyclohexa-1,4-diene influences mainly the yields of hydrogen and cyclohexane, decreasing the former and increasing the latter slightly. Energy or charge transfer from cyclohexene to cyclohexa-1,3-diene manifested itself in dimerization of the diene with a yield of $G \approx 6$. The curves for these systems are similar to the typical curves observed for saturated–unsaturated systems exhibiting negative deviations. Systems containing alkenes and conjugated dienes behave similarly to those consisting of alkanes and monoalkenes.

5.4. CONCLUSIONS

As noted in Section 5.1, data on corresponding cycloalkanes (Chapter 3) and aliphatic alkenes (Chapter 4) can give useful pointers to the radiation chemical behaviour of individual cycloalkenes. Bearing this in mind, we will now attempt a general evaluation of the phenomena discussed in detail in Section 5.2.

Decomposition reactions

$G(H_2)$ values of cycloalkenes have been summarized in Table 5.1 and Fig. 4.13, this latter indicates the importance of stereochemical factors in cycloalkene radiolysis. Whereas $G(H_2)$ values of n-alkene homologues increase slightly but monotonously with increasing carbon number, there is no such correlation in the cycloalkene series. For example, both cyclohexene and hex-3-ene contain four secondary hydrogen atoms in the β-position to the π-bond, but $G(H_2)$ is 1.19 (Table 5.1) for cyclohexene against 1.64 for hex-3-ene, from either calculated or measured data (Table 4.27). If $G(H_2)$ values are plotted as a function of the ring-strain energy (Fig. 5.19), the correlation is negative but not linear. One reason for this may be

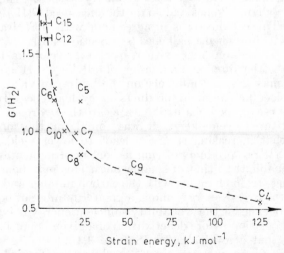

Fig. 5.19. $G(H_2)$-values of cycloalkenes as a function of the ring-strain energies of the more stable isomers (radiation: ^{60}Co-γ). After Cserép et al. [20A]

the fact that strain energy is a composite quantity, and the contributions of its components (angle distortion, bond elongation, torsion and van der Waals repulsions) to the total are different for different ring sizes [20].

Cserép and Földiák [14] suggested that the correlation between cyclo-alkene ring size and $G(H_2)$ may be at least partly attributed to the ability of the molecule to produce an allyl-type radical permitting maximum orbital overlap, this species being probably the most favourable transition state for hydrogen elimination (see Section 4.4). This requires a spatial arrangement favourable for the formation of a true allylic position, i.e. the C−H bond in the β-position should be perpendicular to the plane of the ethylene group. The electron orbitals of the π-bond and the electrons of the allylic hydrogen are then parallel, this being the most favourable configuration for orbital overlap, i.e. for delocalization. Free rotation of the allylic CH_2 group of aliphatic alkenes, except when hindered by bulky substituents, makes such an orientation feasible [20].

However, the cyclic structure of cycloalkenes allows only incomplete overlapping. Delocalization of π-electrons facilitates hydrogen elimination only to a limited extent by comparison with open-chain hydrocarbons (e.g. n-alkenes); the larger the extent of the deviation between the actual orientation of the σ-bonds of the β-hydrogens and the optimum perpendicular direction, the smaller is the effect of delocalization.

By using molecular models it can be demonstrated that the rings containing more than six carbon atoms (up to cyclononene) become more and more rigid, asymmetrically twisted and crowded. Thus the direction of allylic C−H bonds will deviate more and more from the perpendicular orientation, resulting in decreased allyl resonance, a higher energy of activation of hydrogen elimination, and lower hydrogen yields. On further increase in ring size, the molecular structures again become more favourable for allylic interactions of hydrogen. The geometry of the C_{12} ring approaches that of the open-chain alkenes and so do the measured $G(H_2)$ values (Table 5.1, Fig. 4.13). This conclusion is in agreement with that drawn from the radiolysis of multi-branched isoalkenes (see Section 4.4.).

As described in Section 5.2.6 with respect to, cyclohexa-1,3- and -1,4--dienes, not only the number and order of allylic C−H bonds but also their geometry may have an influence on the hydrogen yield.

Yields of molecular fragment products of cycloalkenes (Table 5.1) are different from those of open-chain alkenes, as are the hydrogen yields. In the case of unsubstituted cycloalkenes, it was found for the molecular mass range studied that increasing ring size leads to lower yields of the light hydrocarbons that are products of ring fission, i.e. fragmentation of larger rings is less probable than that of smaller ones. This may be a consequence partly of the ring-strain changing with the carbon number and partly of the distribution of the excess energy among a greater number of bonds in larger molecules, thus decreasing the probability sufficient energy being concentrated to cause dissociation before deactivation of a given bond. This behaviour is analogous to that of cycloalkanes (Section 3.4.).

Amongst light fragments, mostly C_2 products are formed, ethylene in particular. This may be attributed to several reasons, such as the weakening effect of π-bonds on σ-bonds in the β-position to them, and free electrons of alkyl radicals leading to C_2 biradicals, the free electrons of which rearrange very rapidly by forming an ethylenic π-bond. But it seems more likely that no C_2 biradical can even form and that the orbitals already overlap in the transition state. On the other hand, ethane formation occurs via disproportionation of ethyl radicals as well as by combination of methyl radicals. As the probability of these processes is fairly low in the case of cycloalkenes by comparison with open-chain hydrocarbons, the low yield of ethane is not surprising.

In some cases (cyclohexene and its substituted derivatives), where the reversed Diels–Alder reaction is possible, higher molecular mass decomposition products such as propene, butadiene and pentadienes are formed in significant amounts.

Of 'complementary' products, the yield of that with higher molecular mass is generally lower; the almost equal C_2 and C_3 yields from cyclopentene (see Section 5.2.1. and Table 5.1) are an exception. This can be explained, as with aliphatic alkenes, in terms of further reactions of intermediates larger than C_3, resulting mainly in C_2 formation.

Similar observations have been made in high-temperature pyrolysis with short residence times, but the difference between complementary yields is lower. For example, in the radiolysis of cyclohexene the ratio of ethylene to buta-1,3-diene is about 3 : 1, whereas in pyrolysis it is about 1 : 0.8–0.95. The difference can be ascribed to the occurrence of ionic processes in radiation chemistry, whereas the energy intake in pyrolysis is insufficient for ionization.

Build-up reactions

Much less is known about the build-up reactions (dimerization, polymerization) of cycloalkenes than about those of open-chain alkenes. Yields of individual dimers have been reported in the radiolysis of liquid cyclohexene, but the full product spectrum is unknown even here. According to the studies of Burns and Winter [47], Wakeford and Freeman [2, 3], and Kovács et al. [19], dimerization can be attributed mainly to radical processes (addition, combination). The G-value of polymer formation is about 8–10; this is relatively large compared with the total G-value of any other product, but negligible compared with the G-values of aliphatic alkenes, the radiation chemical polymerization of which may be of interest from the practical point of view.

For cyclohexa-1,4-diene, Nakagawa et al. [10] found $G(\text{polymer}) \approx 160$ in the gas phase with a cationic mechanism of polymerization. Cyclopentadiene irradiated at $-78°C$ readily gives macromolecular polymers [83, 86]. For example, Bonin et al. [83] have reported $G(-\text{cyclo-}C_5H_6) = 2.1 \cdot 10^4$ for polymerization. Various studies (e.g. with NH_3) have indicated unambiguously that polymerization takes place via a cationic mechanism [87].

In general, the tendency of cyclic molecules to dimerize or polymerize depends on their structure and therefore no general rule can be adopted for these reactions.

REFERENCES

1. FREEMAN, G. R., *Can. J. Chem.*, **38**, 1043 (1960)
2. WAKEFORD, B. R. and FREEMAN, G. R., *J. Phys. Chem.*, **68**, 2635 (1964)
3. WAKEFORD, B. R. and FREEMAN, G. R., *J. Phys. Chem.*, **68**, 2992 (1964)
4. FREEMAN, G. R., *Radiat. Res. Rev.*, **1**, 1 (1968)
5. WAKEFORD, B. R. and FREEMAN, G. R., *J. Phys. Chem.*, **68**, 3214 (1964)
6. SEVILLA, M. D. and HOLROYD, R. A., *J. Phys. Chem.*, **74**, 2459 (1970)
7. YANG, J. J., SERVEDIO, F. M. and HOLROYD, R. A., *J. Chem. Phys.*, **48**, 1331 (1968)
8. NAKAGAWA, T., TAKAMUKU, S. and SAKURAI, H., *Bull. Chem. Soc. Jpn.*, **40**, 2081 (1967)
9. NAKAMURA, K., TAKAMUKU, S. and SAKURAI, H., *Bull. Chem. Soc. Jpn.*, **44**, 2090 (1971)
10. NAKAGAWA, T., TAKAMUKU, S. and SAKURAI, H., *Bull. Chem. Soc. Jpn.*, **40**, 2773 (1967)
11. HISADA, H., TAKAMUKU, S. and SAKURAI, H., *18th Annual Meeting of the Chemical Society of Japan*, Osaka, April, 1965
12. FÖLDIÁK, G., CSERÉP, G., JAKAB, A., STENGER, V. and WAJNÁROVITS, L., *A kémia újabb eredményei*. Szénhidrogén rendszerek radiolízise. (Radiolysis of hydrocarbon systems.) Akadémiai Kiadó, Budapest, 1971, Vol. 4, p. 11
13. CSERÉP, GY. and FÖLDIÁK, G., in *Proc. 3rd Tihany Symp. Rad. Chem.* (Eds DOBÓ, J. and HEDVIG, P.), Akadémiai Kiadó, Budapest, 1972, Vol. 1, p. 287
14. CSERÉP, GY. and FÖLDIÁK, G., *Acta Chim. Acad. Sci. Hung.*, **77**, 407 (1973)
15. CSERÉP, GY. and FÖLDIÁK, G., *Int. J. Rad. Phys.*, **5**, 235 (1973)
16. CSERÉP, GY. and FÖLDIÁK, G., *Acta Chim. Acad. Sci. Hung.*, **83**, 173 (1974)
17. CSERÉP, GY. and FÖLDIÁK, G., *Acta Chim. Acad. Sci. Hung.*, **83**, 185 (1974)
18. KOVÁCS, A., CSERÉP, GY. and FÖLDIÁK, G., *Radiochem. Radioanal. Lett.*, **22**, 221 (1975)
19. KOVÁCS, A., CSERÉP, GY. and FÖLDIÁK, G., *Acta Chim. Acad. Sci. Hung.*, **92**, 117 (1977)
20. CSERÉP, GY. and FÖLDIÁK, G., in *Proc. 4th Tihany Symp. Rad. Chem.* (Eds HEDVIG, P. and SCHILLER, R.), Akadémiai Kiadó, Budapest, 1977, p. 147
21. HORVÁTH, ZS. and FÖLDIÁK, G., *Acta Chim. Acad. Sci. Hung.*, **85**, 417 (1975)
22. CARR, R. W. and WALTERS, W. D., *J. Phys. Chem.*, **69**, 1073 (1965)
23. COOPER, W. and WALTERS, W. D., *J. Am. Chem. Soc.*, **80**, 4220 (1958)
24. DELEON, A. and DOEPKER, R. D., *J. Phys. Chem.*, **75**, 3656 (1971)
25. HALLER, I. and SRINIVASAN, R., *J. Am. Chem. Soc.*, **88**, 3694 (1966)
26. WOODWARD, R. B. and HOFFMAN, R., *J. Am. Chem. Soc.*, **87**, 2045 (1965)
27. RICE, F. O. and MURPHY, M. T., *J. Am. Chem. Soc.*, **64**, 896 (1942)
28. KNOCKT, D. A., *Diss. Abst.*, **B29**, 1627 (1968)
29. GIBBONS, W. A., ALLENE, W. F. and GUNNING, H. W., *Can. J. Chem.*, **40**, 568 (1962)
30. LESCLAUX, R., SEARLES, S., SIECK, L. W. and AUSLOOS, P., *J. Chem. Phys.*, **54** 3411 (1971)
31. TU, C. K. and DOEPKER, R. D., *J. Photochem.*, **3**, 13 (1974/75)
32. SMITH, S. R. and GORDON, A. S., *J. Phys. Chem.*, **65**, 1124 (1961)
33. UCHIYAMA, M., TOMIOKA, T. and AMANO, A., *J. Phys. Chem.*, **68**, 1878 (1964)
34. TSAG, W., *J. Phys. Chem.*, **76**, 143 (1972)
35. LESCLAUX, R., SEARLES, S., SIECK, L. W. and AUSLOOS, P., *J. Chem. Phys.*, **53**, 3336 (1970)
36. BUDZIKIEWICZ, H., BRAUMAN, J. I. and DJERASSI, C., *Tetrahedron*, **21**, 1855 (1965)
37. McLAFFERTY, F. W., *Interpretation of Mass Spectra*, W. A. Benjamin Inc., New York, 1967
38. FEE, C., SAMUEL, S. and MARKOWITZ, S. S., *J. Phys. Chem.*, **78**, 347 (1974)
39. WEEKS, R. W. and GARLAND, J. K., *J. Am. Chem. Soc.*, **93**, 2380 (1971)
40. DABY, E. E., NIKI, H. and WEINSTROCK, B., *J. Phys. Chem.*, **75**, 1601 (1971)
41. CVETANOVIC, R. J. and DOYLE, L. C., *J. Chem. Phys.*, **50**, 4705 (1969)
42. COLLIN, G. J., PERRIN, M. and GAUCHER, C. M., *Can. J. Chem.*, **50**, 2391 (1972)
43. SINGH, A. and FREEMAN, G. R., *J. Phys. Chem.*, **69**, 666 (1965)
44. WADA, T. and HATANO, Y., *J. Phys. Chem.*, **81**, 1057 (1971)

45. WOJNÁROVITS, L., HIROKAMI, S. and SATO, [S., *Bull. Chem. Soc. Jpn.*, **49**, 2956 (1976)
45A HIROKAMI, S., WOJNÁROVITS, L. and SATO, S., *Bull. Chem.. Soc. Jpn.*, **52**, 299, (1976)
46. WAGNER, C. D., *Trans. Faraday Soc.*, **64**, 163 (1968)
47. BURNS, W. G. and WINTER, J. A., *Disc. Faraday Soc.*, **36**, 124 (1964)
48. BURNS, W. G., HOLROYD, R. A. and KLEIN, G. W., *J. Phys. Chem.*, **70**, 910 (1966)
49. OHNISHI, S. I. and NITTA, I., *J. Chem. Phys.*, **39**, 2848 (1963)
50. RICE, RAMSPERGER, KASSEL and MARCUS, see MARCUS, R. A., *J. Chem. Phys.*, **20**, 359 (1952); MARCUS, R. A. and RICE, O. K., *J. Phys. Colloid. Chem.*, **55**, 894 (1951)
51. FEE, D. C. and MARKOVITZ, S. S., *J. Phys. Chem.*, **78**, 354 (1974)
52. BRINTON, R. K., *J. Phys. Chem.*, **72**, 321 (1968)
53. TU, C.-K. and DOEPKER, R. D., *J. Photochem.*, **1**, 271 (1972/73)
54. HILL, L. and DOEPKER, R. D., *J. Phys. Chem.*, **76**, 3153 (1972)
55. ELLIS, R. J. and FREY, H. M., *J. Chem. Soc. A.*, 553 (1966)
56. BENSON, S. W. and SHAW, R., *Trans. Faraday Soc.*, **63**, 985 (1967); *J. Am. Chem. Soc.*, **89**, 5351 (1967)
57. EBERHARD, M. K., KLEIN, G. W. and KRIVAK, T. G., *J. Am. Chem. Soc.*, **87**, 696 (1965)
58. OKADA, T., TAKAMUKU, S. and SAKURAI, H., (Osaka, Univ.) *Nippon Kagaku Zasshi.*, **86**, 1118 (1965)
59. FESSENDEN, R. W. and SCHULER, R. H., *J. Chem. Phys.*, **38**, 773 (1963)
60. STOVER, E. D. and FREEMAN, G. R., *Can. J. Chem.*, **46**, 2109 (1968)
61. SCHUTTE, R. and FREEMAN, G. R., *J. Am. Chem. Soc.*, **91**, 3715 (1969)
62. PENNER, T., WHITTEN, D. G. and HAMMOND, G. S., *J. Am. Chem. Soc.*, **92**, 2861 (1970)
63. BUSLER, W. R., MARTIN, D. H. and WILLIAMS, F., *Disc. Faraday Soc.*, **36**, 102 (1963)
64. ARAI, S., MAOMORI, M., YAMAGUCHI, K. and SHIDA, S., *Bull. Chem. Soc. Jpn.*, **36**, 590 (1963)
65. SRINIVASAN, R., *J. Am. Chem. Soc.*, **84**, 3432 (1962)
66. WOODS, W. G., *J. Org. Chem.*, **23**, 110 (1958)
67. SHIDA, T., YAMAZAKI, H. and ARAI, S., *J. Chem. Phys.*, **29**, 245 (1958)
68. WILEY, R. H., LIPSCOMB, N. T. and RIVERA, W. H., *AT/40-1/-2055:NSA*, 1919 (1965) 36462
69. WOJNÁROVITS, L., unpublished results
70. AUSLOOS, P. and LIAS, S. G., *J. Chem. Phys.*, **43**, 127 (1965)
71. MANION, J. P. and BURTON, M., *J. Phys. Chem.*, **56**, 560 (1952)
72. YANG, J. Y. and MARCUS, I. J., *J. Chem. Phys.*, **42**, 3315 (1965)
73. CHER, M., HOLLINGSWORTH, C. S. and BROWNING, B., *J. Chem. Phys.*, **41**, 2270 (1964)
74. CRAMER, W. A. and PIET, G. J., *Trans. Faraday Soc.*, **66**, 850 (1970)
75. HORVÁTH, Zs. and FÖLDIÁK, G., *Acta Chim. Acad. Sci. Hung.*, **86**, 397 (1975)
76. FÖLDIÁK, G. and WOJNÁROVITS, L., *Int. J. Radiat. Phys. Chem.*, **1**, 189 (1972)
77. FÖLDIÁK, G., CSERÉP, GY., STENGER, V. and WOJNÁROVITS, L., *Kémiai Közlemények*, **31**, 413 (1969)
78. FÖLDIÁK, G. and WOJNÁROVITS, L., *Acta Chim. Acad. Sci. Hung.*, **65**, 59 (1970); **67**, 209 (1971); **67**, 221 (1971); **70**, 23 (1971)
79. TOMA, S. Z. and HAMILL, W. H., *J. Am. Chem. Soc.*, 86, 4761 (1964)
80. AUSLOOS, P., SCALA, A. A. and LIAS, S. G., *J. Am. Chem. Soc.*, **89**, 3677 (1967)
81. VAN INGEN, J. W. and CRAMER, W. A., *Trans. Faraday Soc.*, **66**, 857 (1970)
81A. CSERÉP, GY., BREDE, O., HELMSTREIT, W. and MEHNERT, R., *Radiochem. Radioanal. Lett.*, **32**, 15 (1978)
81B. SHIDA, T. and HAMILL, W. H., *J. Am. Chem. Soc.*, **88**, 5376 (1966)
81C. CSERÉP, GY., BREDE, O., HELMSTREIT, W. and MEHNERT, R., *Radiochem. Radioanal. Lett.*, **34**, 383 (1978)
82. KESZEI, Cs., WOJNÁROVITS, L. and FÖLDIÁK, G., *Khim. Vys. Energ.*, **11**, 316 (1977)
83. BONIN, M. A., BUSLER, W. R. and WILLIAMS, F., *J. Am. Chem. Soc.*, **84**, 2895 (1962)

84. BUSLER, W. R., MARTIN, D. H. and WILLIAMS, F., *Disc. Faraday Soc.*, **36,** 102
    (1963)
85. BONIN, M. A., BUSLER, W. R. and WILLIAMS, F., *J. Am. Chem. Soc.*, **87,** 199 (1965)
86. BATES, T. H., *Nature.*, **197,** 1101 (1963)
87. WILLIAMS, F., in: *Fundamental Processes in Radiation Chemistry* (Ed. AUSLOOS, P.),
    Wiley, New York, 1968, p. 515
88. CSERÉP, GY.. FÖLDIÁK, G. and WOJNÁROVITS, L., unpublished results

6. AROMATIC HYDROCARBONS

The radiation-induced reactions of aromatic compounds are broadly similar to those of other hydrocarbons. However, aromatics have an extraordinarily high resistance to radiation in condensed phases. Thus, for instance, the gross G-value for the decomposition of benzene is about one seventh of that of cyclohexane, the corresponding saturated ring compound. Multicyclic and condensed ring aromatics show even greater stabilities. This stability is observed to a greater or lesser degree not only with the 'pure' aromatics, but also with aromatics having alkyl side-chains.

The radiation chemical yields of aromatic hydrocarbons are rather low, so product analysis can only be carried out using high-sensitivity techniques. The formation of 'polymer' causes some difficulties: the term 'polymer' covers products formed by the attachment of variable amounts of aromatic rings to partially hydrogenated ones.

Owing to the experimental difficulties just mentioned, the investigation of the radiation chemistry of aromatic compounds appears to be much more limited than that of the aliphatics, for example. Only benzene and some very simple alkylbenzenes have been thoroughly studied.

The radiolysis of biphenyl and terphenyl compounds has been studied mostly for technical reasons: in the 1950s, aromatic hydrocarbons were recommended as suitable moderators and coolants in nuclear reactors, owing to their low volatility and high resistance [1]. Investigations carried out on test reactors working with terphenyl coolant showed, however, that the products formed spoil the cooling effect to such an extent that the procedure becomes highly uneconomical.

The lowest molecular mass aromatic hydrocarbon, benzene, has six carbon atoms and is liquid at room temperature; biphenyl, which consists of two phenyl groups, is solid.

6.1 INTRODUCTORY REMARKS

The high stability to radiation of aromatics in condensed phases can be interpreted in a number of ways, but it can surely be attributed to the particular π-electron structure. Owing to the non-localized π-electrons, the energy absorbed by the molecule during collision with a high-energy particle can readily be redistributed over the whole molecule, instead of one particular bond. Thus, although enough energy may be absorbed to break a bond, owing to the rapid redistribution, the probability of energy localization on a single bond is very low.

The high resistance to radiation can also be explained by the total conjugation of the six π-electrons, leading to higher efficiency of photon-emitting or collision-induced decay processes, compared with other types of compound.

In vapour phase radiolysis the stability is much lower. The yield of decomposition of benzene vapour, $G(-C_6H_6) \approx 4\text{--}6$ [2, 3], is approximately equal to that of cyclohexane ($G \approx 5,7$; see Section 3.2.3). The stabilization processes described above seem to be effective only in the presence of surrounding molecules. It is probable that the primary products of vapour-phase radiolysis (excited molecules and ions) have longer lifetimes than in condensed phases, and consequently decompose, rearrange and take part in ion-molecule reactions with greater efficiency; radical recombination processes are less important.

From a comparison of the radiation chemical yields obtained in vapour-phase radiolysis with some results of mass spectrometry it seems probable that, in contrast to saturated hydrocarbons, the molecule-ion attachment is specific to aromatic compounds. The energy released in exoergic ion-molecule reactions can readily be absorbed by the π-electron system of the aromatic ring, thus preventing the intermediate ion-molecule complex from undergoing dissociation [4, 5].

As with other kinds of compounds, irradiation of liquid-phase aromatic hydrocarbons gives rise to the ionization and/or excitation of molecules to form positive ions, electrons and 'directly' produced excited molecules. Collisions of secondary electrons, geminate recombination of molecular ions with electrons, and neutralization of positive ions by negative ions also result in the formation of excited molecules.

It is characteristic of aromatic compounds that they can readily transfer electronic excitation energy to other aromatics present in the irradiated system in low concentrations if the acceptor has lower excitation levels. The transferred energy reappears mostly in the form of a luminescent radiation emitted by the acceptor molecules with a higher efficiency than by the donor compounds. This phenomenon has been exploited in the liquid scintillator system applied, for instance, to the detection of β-radiation. These systems consist generally of a monocyclic aromatic solvent (e.g. toluene, ethylbenzene, p-xylene) and a multicyclic scintillator solute, called 'fluor' (e.g. naphtalene, p-terphenyl), that has the lower ionization potential [6].

Theoretical considerations lead to the conclusion [7, 8] that there is a great difference between aromatics and alkanes with respect to the relative amounts of charged and neutral intermediates. It has been suggested that during the slowing-down of electrons, relatively more excited molecules are formed in aromatic compounds than in alkanes.

On the relaxation of superexcited states, the probability of internal conversion and exciton (see Section 1.1.) dissociation, relative to that of autoionization, is greater for aromatic compounds than for saturated hydrocarbons. The same rule holds when the rate of slowing-down of electrons and the probability of geminate recombination are considered. When ion-recombination processes in aromatics and alkanes are compared, it should be noted that whereas in the case of aromatics mostly electronically excited molecules are produced, in the case of alkanes mostly vibrationally excited molecules are formed. Accordingly, in the radiolysis of aromatics neutral excited molecules predominate in the primary processes, whereas charged species are chiefly formed from alkanes and cycloalkanes.

These theoretical predictions have been justified by a number of experimental results. Thus, for example, it has been found that the yield of excited states produced in aromatics under the influence of nanosecond pulses of radiation is considerably higher than in the case of saturated compounds: for benzene and toluene, G(excited states) $\approx$ 4–5.6 [8–11] whereas the same value for cyclohexane is only 0.3 [12].

With the application of the 'clearing field' method, which is successful in the determination of yields of the homogeneously distributed 'free' charges, the yield of ions in benzene was found to be $G = 0.052$ [13, 15] and 0.081 [14], whereas the same value for saturated hydrocarbons reaches 0.14–0.8, depending on the molecular structure.

The presence of N_2O, known to be a very effective electron scavenger, in liquid cyclohexane reduces considerably (to about 40%) the radiation chemical yield of hydrogen, whereas in benzene no such effect has been observed [15]. These results show that ionic processes play a less important role in the formation of the stable products from aromatic hydrocarbons than from saturated ones.

Hummel [17] has theorised that scavengers can interact not only with the homogeneously distributed free ion-pairs but also, depending on the concentration of the scavenger, with the inhomogeneously distributed geminate molecule-ion–electron pairs. The ion yields estimated by using this hypothesis for a number of saturated hydrocarbons show good agreement with those obtained using different scavengers. Thus, for example, for cyclohexane G-values of free ions and of geminate ion-pairs were found to be $G_{fi} = 0.13$ and $G_{gi} = 3.9$, respectively [18, 20]. For aromatics, the results obtained with electron scavengers were in disagreement with this hypothesis, as discussed above, but in agreement with assumption of homogeneous electron and ion distribution. It seems that geminate ion-pairs formed in aromatics can hardly be scavenged; moreover, the conclusion can be drawn that scavengers compete ineffectively with geminate recombination, which is much faster in aromatics than in alkanes [22].

The effect of magnetic field on the fluorescence from radiolytic ion recombination of anthracene in different hydrocarbon solutions shows that in alkane solvent recombination of solute radical ion pairs is the major source of solute excited state formation, but it is far less important in benzene solution [22A].

Results obtained by Beck and Thomas using picosecond pulses of radiation suggest that singlet excited states can be formed in liquid benzene in less than 10^{-11} s, via fast neutralization or direct excitation, in good agreement with data on cyclohexane. The differing subsequent chemistry in the two liquids is due to the different nature of the two excited states. In cyclohexane, this state is short-lived ($t_{1/2} = 2 \cdot 10^{-10}$ s), it has high energy ($\sim$ 7 eV), and transfers energy rapidly ($k = 2$–$4 \cdot 10^{11}$ $M^{-1} s^{-1}$). In benzene, the first excited singlet state has $t_{1/2} = 2 \cdot 10^{-8}$ s, an energy of ~ 4.7 eV and a transfer constant of $k = 2 \cdot 10^{10}$ $M^{-1} s^{-1}$. Unlike cyclohexane, the excited singlet state of benzene is very unreactive with conventional electron scavengers [23].

A characteristic property of the liquid-phase radiolysis of aromatic hydrocarbons is that the yields of its stable products increase with the LET value of the radiation, as opposed to the other types of hydrocarbons, which show decreasing G-values. The enhancement in the yields is especially significant above 30 eVnm^{-1} LET value. Several models have been proposed to interpret this effect, most of them suggesting that inhomogeneous primary energy absorption takes place in the spurs and tracks. The excited molecule diffusion model of Burns presupposes a competition between the bimolecular process leading directly to products or radicals and the first-order deactivation process. The yields estimated on the basis of the model agree well with the experimental data to about 100 eVnm^{-1} [24, 25].

On the other hand, the solvent depletion model takes account of the significance of the concentration of solvent excited molecules [26].

In the radiolysis of aromatics with α-particles and protons, the differential yields for H_2 (product yield that refers to unit energy loss in a given energy range) have been observed to increase considerably with decreasing particle energy. This effect, termed the 'end-of-track' effect, is especially significant near zero energy and has been interpreted by Schuler [27] in his 'knock-on' and 'charge displacement' models. According to the first model, a low-energy heavy charged particle that has already slowed down can lose its energy in elastic collisions with H and C atoms, instead of bringing about ionization and excitation. The subsequent chemical bond rupture (recoil atom) taking place with high probability contributes to the number of decomposed aromatic molecules. The second model explains the increase in the yields by a charge separation brought about by electron transport as a result of the reversible charge exchange between low-energy charged particles and the medium.

However, according to the investigations of Burns, the differential yields are highest at energies corresponding to the Bragg-maximum, where the LET values are also the highest. Thus, the 'end-of-track' effect is a real LET effect, so elastic collisions and charge exchange can be only of small importance in the formation of product [28].

To interpret the influence of the LET on the radiolysis yields of aromatic compounds, Yang et al. [29] applied the 'thermal spike' model. According to this model, the rapidly lost energy of ionizing particles will be absorbed as thermal energy, thus giving rise to the instantaneous development of hot regions. These are the 'thermal spikes', where, as a result of pyrolysis or of the enhanced reaction rates of intermediates, the yields increase considerably. However, the complete thermal energy absorption assumed in the model is improbable, as the energy loss of ionizing particles takes place mainly through electronic excitation leading to the production of ions and excited molecules.

6.2. INDIVIDUAL COMPOUNDS

The aromatic hydrocarbons are discussed in the order of increasing number of carbon atoms. The section ends with compounds having a condensed-ring structure.

6.2.1. Benzene

Excitation and excitation transfer processes

In the radiolysis of benzene directly excited benzene molecules

$$C_6H_6 \rightsquigarrow C_6H_6^*$$

as well as benzene ions and electrons

$$C_6H_6 \rightsquigarrow C_6H_6^+ + e^-$$

are formed. The recombination of these latter species results in the formation of excited benzene molecules

$$C_6H_6^+ + e^- \longrightarrow C_6H_6^*$$

It has been pointed out above that in the radiolysis of aromatic compounds the excited states play a more important role than in saturated hydrocarbons. That is why the determination of yields of the different excited states with different amounts of energy is so important in the elucidation of the mechanism of benzene radiolysis.

According to Platzman and a number of other authors [8, 30, 31], the radiolysis of hydrocarbons is basically influenced by the participation of so-called superexcited states with short lifetimes (10^{-15}–10^{-14}s). Benzene molecules that are in the first excited level and observable in certain experiments are produced mainly via neutralization or from the theoretically less known and experimentally unobservable superexcited states via exciton dissociation, internal conversion (IC), intersystem crossing (ISC) and triplet–triplet annihilation. A number of theoretical and experimental studies have been devoted to the determination of relative importance of these processes.

In earlier studies it was suggested that one of the most significant methods of production of the first excited singlet state of benzene ($^1B_{2u}$) is internal conversion from short-lived upper excited states. Voltz [32] discussed the possibility of observing the $^1E_{1u}$ state, and suggested that the $^1E_{1u}$ state is an exciton with a high diffusion constant, hence inspite of its very short lifetime it probably reacts in collision with $CHCl_3$ present in the system in low concentrations (10^{-3}–10^{-2} M) [60]. According to the scintillation efficiency experiments on benzene containing binaphthyldiphenyloxazol, if the concentration of the additive is higher than 0.1 M, immediate energy transfer from the upper singlet state of the aromatic solvent to the solute can readily take place [33]. Results obtained from the radiolysis of benzene containing 4-methyl-4-phenylpentan-2-one lead to the same conclusion [10].

Platzman [34] estimated the yield of the ^1E_u state by applying the theory of optical approximation: he obtained the value $G \approx 1$. Similar results were obtained by Cooper and Thomas [35] from the nanosecond pulse radiolysis of anthracene dissolved in benzene.

These results have been challenged by Sato et al. [36], on the basis of their data for the concentration dependence of $G(CO)$ and $G(phenol)$ obtained in the radiolysis of benzene–CO_2 solution. They reject the possibility of immediate energy transfer from benzene molecules in the upper excited states in view of their rather short lifetimes.

Fuchs et al. [37] tried to study the mechanism of formation of molecules in the first excited singlet state by applying vacuum-UV induced fluorescence and radioluminescence experiments using fast electrons in liquid benzene. They found that excited molecules with energies higher than 7 eV decompose via autoionization. The geminate recombination of charged particles formed in this way results — without spin relaxation — in benzene molecules in the first excited singlet state. From their experimental data and theoretical considerations they concluded that these species are formed via recombination subsequent to autoionization from upper excited states rather than via internal conversion.

The change of intensity of fluorescence excitation spectra with the energy of the irradiating electrons, obtained in the low-energy electron irradiation of a benzene film at 77 K, indicates that the ion-recombination process is the major source of fluorescence and consequently of the first excited singlet state if the energy of the impinging electrons is higher than the ionization potential of benzene [38].

Sato et al. investigated the γ-radiolysis of benzene–N_2O and benzene–N_2O –carbonyl sulfide solutions of different concentrations [36]. From the variation of the $G(N_2)$ and $G(CO)$ values they concluded that the collision of benzene molecules with secondary electrons gives rise mainly to the formation of molecules in upper excited states, the transformation of which into the $^1B_{2u}$ state takes place with very limited probability [39]. According to their estimation, the yield of $^1B_{2u}$ state formed by direct excitation of secondary electrons cannot be more than 0.2. Horrocks [33] has published a value of $G = 0.4$ for the yield of the first excited singlet state produced in direct excitation and internal conversion.

It appears from the experimental results discussed above that the hypotheses and calculations concerning yields of upper excited states and their transformations are rather different. Further developments in experimental techniques are needed for the detection of these highly activated short-lived excited states.

A quantitative determination of the molecules in the first excited singlet and triplet states is possible through the study of chemical reactions sensitized by these molecules.

The photochemistry and photophysics of benzene reviewed by Cundall et al. [39A] provide a lot of information about the properties of the first excited states of benzene. The benzene molecule transfers its energy to the additive present in the solution by reaction (6.1) and the acceptor brings about the chemical transformation observed [reaction (6.2)]. For this kind of experiment suitable chemical reactions which take place selectively as a result of this energy transfer, have to be chosen.

$$C_6H_6^* + A \longrightarrow C_6H_6 + A^* \tag{6.1}$$

$$A^* \longrightarrow products \tag{6.2}$$

Two kinds of problem can arise in such studies: radicals produced in the radiolysis can quench the excited states, and the chemical reaction that is supposed to be the result of sensitization can in part be induced directly via collision with subexcitation electrons.

Krongauz [40] has investigated the γ-radiolysis of dilute benzene solutions containing readily decomposed diacyl peroxides, oxy compounds and disulphides. He studied the fragmentation processes, and carried out a kinetic analysis of competing reactions taking place in the solutions in the presence of scintillating compounds. Comparing these results with those obtained in the photolysis of the same compounds, he demonstrated that the decomposition of the additives was initiated by energy transfer involving benzene molecules in the first singlet state. He obtained $G = 6$ for the yield of the benzene singlet, $^1B_{2u}$, formed by direct excitation and by ion recombination. The transformation of o-nitrosobenzaldehyde into o-benzoic acid brought about by irradiation of a dilute solution of the compound in benzene is the result of triplet–triplet and singlet–singet energy transfer in the case of low (< 0.2 mM) and high concentrations of the additive, respectively. Thus, for the yields of triplet and singlet states, Kronganz reported $G = 0.4$ and 4.0, respectively.

Hammond et al. [41] have established the yield of the benzene singlet, $G(^1B_{2u})$ to be 3.4, in the γ-radiolysis of solutions of benzene and tetramethyl-oxetanone.

A number of authors have used the radiation-induced isomerization of various unsaturated compounds in dilute benzene solutions (0.1 M) to determine the yield of benzene triplets: the additive triplet formed via energy transfer from the benzene triplet isomerizes. It should be noted, however, that the isomerization of the additive can also be brought about by ions, free radicals and singlet excited molecules. In aromatic solutions the triplet state of the additive can be formed by various pathways:

1. Energy transfer from the triplet solvent molecules;
2. Intersystem crossing, subsequent to energy transfer from singlet solvent molecules;
3. The recombination of the solute cation, produced in 'hole' transfer, with electron or anion.

Golub and Stephens [43] investigated the radiolysis and photolysis of benzene solutions of oct-1-ene in the absence and the presence of scavengers. They focused their attention on the isomerization of the alkene component and carried out a kinetic analysis. They concluded that the isomerization is the result of a triplet–triplet energy transfer in which the first excited triplet state of benzene ($^3B_{1u}$) acts as the donor. As the difference between the lowest triplet levels of benzene and the alkenes is rather small (0.5 eV), the probability of isomerization via direct excitation of the alkene by subexcitation electrons must be very low in dilute solutions.

Cundall and Tippett [44, 45] have established a value of $G = 4.7$ for the triplet yield of benzene in the γ-radiolysis of liquid benzene containing but-2-ene. A part of this yield results from intersystem crossing from the $^1B_{2u}$ state. Similar results have been reported for benzene-d_6. Taking into account the variation of the yield of isomerization as a function of the concentration of the N_2O additive, the authors assumed that the yield of primary excited states formed as a result of the $C_6H_6^+ + e^-$ recombination is about $G = 2$. The presence of xenon as an additive increases the spin-orbital interactions thus promoting intersystem crossing processes and increasing the yield of isomerization. From these data the yield of singlet states not undergoing intersystem crossing was estimated, namely $G = 1.3$. Thus, the total yield of excited singlet and triplet benzenes was estimated as $G(^1B_{2u} + {}^3B_{1u}) = 6$, on the basis of isomerization studies carried out with but-2-ene solutions. When 0.5 MeV protons are used in the radiolysis of benzene, lower yields of isomerization are observed as compared with the γ-radiolysis [254]. It can be assumed that, as a consequence of the higher ionization density, a higher concentration of intermediates is formed, leading to preferred annihilation of benzene triplet states.

$$^3B_{1u} + {}^3B_{1u} \longrightarrow {}^1A_{1g} + {}^1B_{2u}$$

In the radiolysis of dilute solutions of 1,2-diphenylpropene in benzene, a value of $G = 4.1$ was found for the yield of benzene triplet states that had not been formed via intersystem crossing [46].

Hentz and Perkey [47] studied the γ-radiolysis of solutions of cyclohepta-3,5-dienone in benzene. Taking into account the yields of various products, as well as the results of photo-excitation experiments, they estimated the yield of the first excited singlet state of benzene, the effectiveness of intersystem crossing and the yield of the first excited triplet states of benzene that are not produced from singlet states, as $G = 1.45$, 0.58 and 4, respectively.

A number of authors have studied benzene triplet states using the radiolytic isomerization of stilbene [21, 47–50]. Kinetic analysis of the variation of stilbene isomerization yields in the presence of various additives has been performed for benzene solutions containing 4–600 mM stilbene. From the results some authors have assumed [42, 49, 50] that at low stilbene concentrations (< 50 mM), the isomerization of stilbene is the result of triplet–triplet energy transfer, whereas at higher concentrations, the ion-neutralization of stilbene plays an important role. For the triplet yield of benzene not produced from singlet states Hentz and Perkey calculated a G-value of 4.2 [47]. To obtain these data, the authors made use of the results of the concentration dependence of the isomerization yield of stilbene as well as of photo-excitation data and the yield of singlet states, $G(^1B_{2u}) = 1.45$.

Hentz and Sherman [21] investigated the γ-radiolysis of dilute solutions of stilbene in benzene and observed a decrease in the yield of isomerization when N_2O was added. They applied the equation proposed by Schuler [20] for inhomogeneous charge scavenging, and thus estimated the probability of formation of excited states in the benzene ion-pair neutralization, $\eta =$ $= 0.85$. Taking into consideration the value of $G = 5.4$ obtained from the isomerization of stilbene, and the total yield of ion-pairs, $G = 4$, they estimated the yields of excited benzene molecules produced in ion-recombination and direct excitation as 3.4 and 2, respectively [21].

It is apparent from the above discussion that the yields of excited states of benzene calculated on the basis of radiation induced chemical reactions show a great diversity. Overall, the following set of values seems to be reasonable: $G(^1B_{2u}) = 1.5$, $G(^3B_{1u}) = 4.1$ and the effectiveness of intersystem crossing, $\eta = 0.6$.

Apart from investigations of radiation-sensitized chemical reactions, luminescence studies are also very useful for the elucidation of the nature of the excited states formed during irradiation. The luminescence of organic scintillators is caused by energy transfer from the solvent, here benzene, to the scintillator molecule. From the change in fluorescence intensity observed in the presence of ion-scavengers in the radiolysis of solutions of p-terphenyl in benzene it was concluded that more than 30% of the excited states of benzene were produced in ion-recombination processes [51].

The range of possibilities for the investigation of excited molecules has been extended considerably with the development of pulse radiolysis techniques. It is now possible to study the various excited states directly by means of absorption and emission spectroscopy, or in solution, through the observation of transient absorption and emission of certain species formed from a suitable additive as a result of energy transfer.

Cooper and Thomas [35] observed fluorescence and absorption spectra in the nanosecond pulse radiolysis of benzene below 400 nm and in the 500–600 nm region, respectively. Similar spectra were observed in the 265 nm (4.7 eV, one-photon) [65] and 347.1 nm (3.6 eV, two-photon) [64] laser excitation of benzene. The decay kinetics of both spectra were found to be independent of pulse intensity. In the presence of electron scavengers the intensity of both spectra was reduced, though their first-order decay rates remained unchanged. According to the authors, these facts seem to justify the assumption that a transient species with a half-life of 18.2 ns is responsible for both absorption and fluorescence. Comparing the half-life and the fluorescence spectrum obtained with those determined for the $^1B_{2u}$ state formed in the photolysis of benzene, the authors established that both fluorescence and absorption are caused by the $^1B_{2u}$ first excited singlet state produced in the radiolysis. Later, on the basis of the temperature dependence of the absorption spectrum as well as of theoretical considerations, it was proved that both emission and absorption are caused by the first excited singlet excimer transient, $^1B_{1g}$ which is brought about by the configurational interaction of a first excited singlet and a ground-state benzene molecule [53, 61, 62, 122, 255, 256]:

$$^1B_{2u} + {}^1A_{1g} \longrightarrow {}^1B_{1g}$$

In the picosecond pulse radiolysis of benzene the excimer singlet state of benzene was observed to appear with a delay of 10 ps, and this was considered to be the result of prior formation of the monomer singlet, followed by the complexing reaction time which was calculated to be $\sim$ 7 ps [23]. A transition from the first excimer state to an upper excimer singlet state (E_{1u}) is responsible for the transient absorption observed [122].

In the emission spectrum [52, 53] obtained from the proton and electron irradiation of liquid benzene, the fluorescence maxima observed at 279 and 285 nm were tentatively assigned to the first excited singlet monomer, and the maximum at 320 nm to the excited singlet excimer. The electron irradiation of solid benzene at 138 K led to emission mainly from the excited singlet benzene monomer and to a small extent from that of the excimer [54]. However, during irradiation of a benzene film with low-energy electrons (10 eV) at 77 K, the emission only of the singlet benzene monomer was observed with a maximum at 280 nm. Under such conditions the probability of formation of the excimer is, presumably, reduced, because of the rigidity of the crystal lattice [38].

The emission of the excited singlet state of benzene is rather weak, so its direct study gives rise to more uncertainties. Therefore, its quantitative investigation has been carried out in dilute acceptor solution (0.1 M). Some authors, for example Thomas [55], applied scintillators that give more intensive luminescence than benzene, e.g. naphthalene and anthracene. The luminescence observed in these solutions is the result of energy transfer from the first singlet of benzene to the additive. The singlet–singlet energy transfer is diffusion-controlled in the concentration range 0.1–100 mM for the additive. These studies yielded a value of $G(^1B_{2u}) = 1.6$–1.7 [35, 56, 57].

As to the origin of excited states in benzene, ion-recombination processes have been suggested as the most probable pathway [35]. It is well known from theoretical studies [58], that the ratio of triplet to singlet excited molecules formed in ion-recombination is dependent on the number of ion-pairs in the spur: with a homogeneous distribution it is 3 : 1. The phosphorescence and absorption corresponding to the triplet have not been detected directly in the pulse radiolysis of liquid benzene. However, in the presence of various amounts of naphthalene or anthracene, the absorption of the triplets of these additives could be observed in addition to the fluorescence characteristic of their singlets. The triplet of the additive is formed via energy transfer, in part from directly produced benzene triplets, and in part from benzene triplets produced via intersystem crossing of singlet excited benzene molecules. Some problems can arise with this technique: the additive triplet can be produced from an additive singlet via intersystem crossing, subsequent to energy transfer from a benzene singlet [9]. A further source of uncertainty might be the rather short lifetime of the benzene triplet thus limiting the possibilities for complete energy transfer to the additive. In order to determine the extent of contributions of the various benzene excited states to the formation of the additive triplets, Cooper and Thomas added piperylene in different concentrations to solutions of naphtalene in benzene. From the yields the values $G(^3B_{1u}) = 1.85$ and $G(^1B_{2u}) = 1.62$

were estimated [35], i. e. the yields for the singlet and triplet states were found to be nearly equal. This suggests the existence of 2–3 ion-pairs in the spur.

Baxendale and Fiti [9] investigated the nanosecond pulse radiolysis of 0.1–50 mM solutions of naphthalene, biphenyl and anthracene in benzene, in order to establish the yields of primary triplet states not involving intersystem crossing. Taking into account the amount of triplet states formed from singlets and the effectiveness of intersystem crossing ($\eta = 0.6$), and applying the Stern–Volmer kinetic expression for the remaining yields of additive triplets, they obtained a value of $G = 4.2$ for the yield of primarily produced triplet benzene molecules with all three additives. This value is in good agreement with the triplet yields estimated from the measurement of the isomerization of alkenes, studied in static irradiation. For the half-life of the benzene triplet a value of 2.3 ns was found at 20 °C.

In the pulse radiolysis of benzene, Thomas and Mani observed an absorption at 320 nm with a half-life of 112 ns, the intensity of which was considerably reduced in the presence of radical acceptors [62]. According to their suggestion, the absorption is caused by a biradical formed in the reaction of a triplet state benzene molecule with a ground state one. Richards et al. have studied the biphotonic laser-photolysis of benzene, and observed an absorption, attributable to the benzene biradical, similar to that occurring in radiolysis [64, 65].

Cundall et al. [66] have investigated the phosphorescence on the solute in the microsecond pulse radiolysis of solutions of biacetyl in benzene. They determined the dependence of the intensity of phoshorescence on the concentration of biacetyl, on the dose, on the presence of ion scavengers, and on the concentration of cyclohexene as additive, which had no observable influence on the first singlet of benzene. Kinetic analysis of the data established that the quenching of benzene singlets and triplets by biacetyl is a diffusion controlled process. The lifetime of the benzene triplet was shown to be in the nanosecond range. For the total yield of excited states of benzene they estimated a value of $G = 2.67$, and for the yields of its constituents, $G(^1B_{2u}) = 1.43$ and $G(^3B_{1u}) = 1.24$. These values are lower than those obtained from the isomerization of alkenes and stilbene, G (first excited state) $= 5$ [42, 43]. According to Cundall [42], the direct energy transfer from upper excitation levels is possible in isomerization processes, whereas with biacetyl the upper excited states of benzene cause the dissociation of the biacetyl. Thus, the phosphorescence of biacetyl can be brought about only by energy transfer from the lowest excited states.

Land and Swallow [67] determined the yields of triplets as a function of the concentration of the additives in the pulse radiolysis of benzene containing various amounts of anthracene, napthalene and benzophenone (10^{-4}–1M). They concluded that, whereas the additive triplet is formed in cyclohexane mainly through ion neutralization of the additive, in the case of benzene, at concentrations less than 0.1 M, it is produced mainly via energy transfer from singlet and triplet excited states of benzene. However, the neutralization of additive ions can contribute to the formation of additive triplets at higher concentrations, even in the case of benzene.

The spectra of the radical cation and anion of *trans*-stilbene were resolved in the pulse radiolysis studies of solutions of *trans*-stilbene in benzene [67A].

Similar conclusions were reached at by Dainton et al. [68, 69] on the basis of data on isomerization brought about by pulse radiolysis of solutions of stilbene in benzene. From the concentration dependence of the yields of additive triplets and singlets formed in the pulse radiolysis of dilute solutions of naphthalene or pyrene in benzene, and taking into account photolysis results, they established that excited states of the additive are produced in part directly as a result of energy transfer involving benzene singlets and triplets, and in part as a result of intersystem crossing subsequent to singlet energy transfer. From the results they estimated the singlet and triplet yields of benzene as $G = 2.20$ and 2.64, respectively [70].

Nishikawa and Sauer [71] studied the effect of electron scavengers on the yield of triplet naphthalene molecules formed in the pulse radiolyis of vapour-phase benzene–naphthalene systems at 1.9 bar and 120°C. They concluded that about 30% of the naphthalene triplets are formed via energy transfer from the directly produced benzene triplet ($^3B_{1u}$), instead of ionic reactions. The following processes can contribute to the non-ionic formation of $^3B_{1u}$ states:

1. Direct excitation by low-energy electrons:

$$^1A_{1g} \xrightarrow{\text{electron}} {}^3B_{1u} \; ;$$

2. Excitation to the $^1E_{1u}$ state, which has a high oscillator strength, and a subsequent internal conversion with an efficiency of 38% into the state $^1B_{2u}$, followed by an intersystem crossing to the $^3B_{1u}$ state:

$$^1A_{1g} \xrightarrow{\text{electron}} {}^1E_{1u} \, ,$$

$$^1E_{1u} \longrightarrow {}^1B_{2u},$$

$$^1B_{2u} \longrightarrow {}^3B_{1u};$$

3. Conversion of superexcited states not undergoing autoionization.

Ueno et al. studied the triplet state relaxation of benzene in the gas phase by measuring the phosphorescence of biacetyl induced by energy transfer to biacetyl from benzene in pulse radiolysis of benzene—biacetyl and benzene-d_6—biacetyl mixtures. They found the pressure-independent radiationsless transition rate constant for the triplet benzene to be $(5 \pm 3) \cdot 10^3 \, s^{-1}$, this value for benzene-d_6 being $(1.3 \pm 0.7) \cdot 10^4 \, s^{-1}$ [71A].

Ogawa et al. observed an emission spectrum, similar to the photoexcitation spectrum, when they irradiated benzene vapour at 1.3 μbar with low-energy electrons [72]. No change was observed in the spectrum when the electron energy was varied in the range 60–300 eV. The fluorescence intensity was enhanced by an increase in the electron current and the pressure of the vapour. The 250–300 nm band was assigned to the first excited singlet state of benzene, and the 370–500 nm band to an excited fragment of benzene. The dependence of the emission of the pressure indicates a low probability of collision-induced quenching, whereas the change in emission intensity when the electron current and the pressure are varied indicates that both excitation and fragmentation are probably one-electron processes:

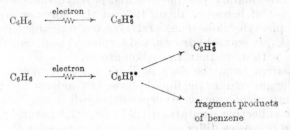

where the symbols $C_6H_6^*$ and $C_6H_6^{**}$ denote first excited and superexcited states, respectively.

Phillips and Schug [54] observed a phosphorescence, characteristic of the first triplet and excimer triplet states of benzene, when they studied the emission spectrum obtained on the irradiation of solid benzene with electrons.

Shida [73] investigated the γ-radiolysis of solid benzene at 77 K, in the presence of alkyldisulphide electron scavengers, and estimated from the change of the absorption spectrum a value of $G = 1.4$ for the ionization yield. According to several authors, the ionization yields are different for the gas and liquid phases. One possible explanation could be that the majority of ionic processes take place through the autoionization of superexcited states. On the other hand, in the solid state the probability for the decomposition of superexcited states through predissociation increases at the expense of autoionization.

The values for the yields of primary intermediates of benzene published in the literature show a great diversity. Taking into consideration the experimental errors for the first excited singlet of liquid benzene, $G(^1B_{2u}) = 1.5$–1.6 [27, 30, 35, 44, 45, 47, 55, 66] seems to be a reasonable value, from which about 1.2 and 0.4 can be accounted for ion-recombination and direct excitation, respectively [33]. For the first excited triplet, $G(^3B_{1u}) = 4.2$ [9, 44, 47, 70] looks very probable. The yield of ion-pairs obtained with nanosecond pulse radiolysis experiments is much less than this value, namely $G(\text{ion}) = 0.1$–0.3 [35, 57].

Considerations based on end-products

The stable products of radiolysis are produced through short lived intermediates in reactions of activated, i. e. excited neutral or charged benzene molecules. In the radiolysis of benzene in the liquid and gas phases about 7–16% of total products are gaseous products (hydrogen and hydrocarbons), about 1% are different cyclic compounds containing six carbon atoms, and the remainder are polymers (84–93%). The term 'polymer' means different things to different authors it denotes, generally, products having higher boiling points than the initial compound, i.e. dimers, oligomers, polymers and condensed compounds, as well as their mixtures. A complete analysis of the polymer fraction has not been carried out; it is generally accepted that there are mainly polyphenyls and polyhydrophenyls in it.

In the case of benzene, about 19% of the polymer are C_{12} compounds (biphenyl, phenylcyclohexane, and non-aromatic bicyclic hydrocarbons), about 57% C_{18} (various hydrogenated terphenyl isomers), and the remainder compounds with more than 18 carbon atoms.

It is apparent from the data in Table 6.1 that yields are several times higher in the gas than in the liquid phase. In condensed phases, presumably, both the nature of energy absorption (so-called non-localized energy absorption or collective excitation) [257] and the nature and extent of decomposition processes differ considerably from those in the gas phase.

In the vapour phase, $G(H_2)$ decreases with the density of benzene, and increases with temperature (at 260°C, 0.1; at 350°C, 1.0). Up to about $50 \cdot 10^{19}$ eV g^{-1} it is independent of the dose, but above this it decreases. The extent of the decrease is enhanced by increasing the temperature [2].

Table 6.1

Yields from benzene

Radiation	γ-Radiation from spent reactor fuel elements (average ~ 0.75 MeV)	^{60}Co-γ	^{60}Co-γ	^{90}Sr- ^{90}Y-β⁻ or ^{60}Co-γ	1.6 MeV electron
Dose rate, eV g^{-1} s^{-1}	$6.1 \cdot 10^{14}$	~$3.4 \cdot 10^{16}$		~$4 \cdot 10^{14}$	~$0.624 \cdot 10^{19}$
Dose, 10^{19} eV g^{-1}	880	1.2			624
Temperature, °C	~30	23	100	25	−113
Pressure, mbar	34.7	87	1315		
Reference	3	77	227	96	94
Phase	Vapour	Vapour	Vapour	Liquid	Solid
Product	*G*, molecule/100 eV				
Hydrogen	0.15	0.11	0.084	0.039	0.0085
Methane	0.018	0.012			0.008
Ethane		0.004			
Ethylene	0.08	0.046	0.02		
Acetylene	0.73	0.63	0.61	0.020	0.0016
Allene (propadiene)		0.041			
Butadiene		0.093			
Cyclohexene		0.013			
Cyclohexadienes		0.03		0.028^a	(−20°C) 0.061^b
Toluene	0.02				
Ethylbenzene	0.007				
Radical				0.69	0.15

^{a} From ref. [99]
^{b} From ref. [76]

The yield of hydrogen obtained in the room temperature γ-radiolysis of liquid benzene, $G(H_2) = 0.039$, is not influenced considerably by the dose and dose rate, although it changes significantly with the type of radiation and the LET value [24]. For example, the $G(H_2)$ values found for 0.69 MeV protons, 0.4 MeV He-ions and fission products are 0.11 [74], 0.32 [75] and 2.7 [14], respectively. The extent of hydrogen formation is not affected by electron- and radical-acceptors, or by the temperature up to 200°C. However, the change of physical state exerts an essential influence on the yield: in the solid state, at -4°C, it is 0.025 [76], whereas in the vapour phase, at 22°C and 87 mbar, it is 0.11 [77].

As with aliphatic hydrocarbons, in the liquid-phase radiolysis of deuterated benzene, the yield of deuterium is only one-third of the $G(H_2)$ formed from protiated benzene, i.e. $G(D_2) = 0.0135$ [78]. This fact has been explained in some papers by the difference in zero-point energy of the C$-$H and C$-$D bonds. Taking into consideration the results indicating that the hydrogen yield is independent of temperature, it is difficult to see, how such a small difference in the bond energies is significant in the presence of the high population of electronically excited states. The composition of hydrogen formed from a mixture of protiated and fully deuterated benzene in a ratio of 1 : 1, was as follows: $H_2 : HD : D_2 = 52.1 : 33.1 : 14.8$ [79].

Hydrogen can be produced via reactions of activated benzene molecules as follows:

$$C_6H_6^* \longrightarrow H_2 + C_6H_4 \tag{6.3}$$

$$C_6H_6^* \longrightarrow H + C_6H_5 \tag{6.4}$$

$$H + C_6H_6 \longrightarrow H_2 + C_6H_5$$

$$H + H \longrightarrow H_2.$$

The reaction of hydrogen atoms brought about by pulses of radiation or by gas-discharge, with liquid or gaseous benzene [80, 82A], and the yields of hydrogen formed in the radiolysis of solutions of benzene with water [83] show that the addition of hydrogen atoms to benzene molecules occurs with a high probability:

$$H + C_6H_6 \longrightarrow C_6H_7 \tag{6.5}$$

The C_6H_7 radical thus formed has been detected in the ESR spectrum and electron-absorption spectrum of both pure benzene and benzene adsorbed on silica irradiated with electrons and ^{60}Co-γ-rays, at -196°C [84–90]. The ESR spectrum obtained in benzene vapour irradiated with low-energy heavy ions also gave evidence of the presence of C_6H_7 radicals [91]. An absorption corresponding to the C_6H_7 radical was also observed in the pulse radiolysis of benzene vapour [81]. Knutti and Bühler [92] and Hoyermann et al. [82A] determined the rate constant of H-atom addition in a fast glow flow system at low pressure, and the fate of C_6H_7 radicals with mass spectrometry and gas chromatography.

The laser-induced fluorescence of the C_6H_7 and C_6D_7 radicals were observed in the irradiated benzene and deuterated benzene polycrystals at 4.2 K [91A].

Radical pair formation was observed by means of the ESR technique in irradiated polycrystalline benzene at 77 K [92A]. It was proposed that C_6H_7 and C_6H_5 radicals are to form the radical pairs [92B].

Hardwick [93] concluded from a kinetic analysis that about 95% of hydrogen atoms add to the ring and not more than 5% form H_2. Taking into consideration the high probability of reaction (6.5), from the reaction mechanism given above, only the unimolecular reaction (6.3) can contribute to hydrogen formation. However, kinetic analyses given by some authors dealing with the formation of hydrogen gas and radicals at $-196°C$, indicate the relatively low probability of molecular hydrogen formation, via reaction (6.3) [94]. If reaction (6.3) only is contributing to hydrogen formation from the benzene–benzene-d_6 (1 : 1) mixture, H_2 and D_2 would be produced, in contradiction of the experimental results discussed above [95]. Therefore, a bimolecular process has to be assumed for HD formation. One possible pathway of bimolecular H_2 formation could be a further reaction of the C_6H_7 radical formed in reaction (6.5) [96]:

$$C_6H_7 + C_6H_6 \longrightarrow H_2 + C_6H_5-C_6H_6. \tag{6.6}$$

Although the polymer product formed simultaneously has been detected by gas chromatography, hydrogen production via reaction (6.6) is rather improbable because radical acceptors such as iodine proved to have no effect on $G(H_2)$.

Burns [74, 97] interpreted the bimolecular H_2 formation as the result of an intra-spur reaction of two excited benzene molecules

$$C_6H_6^* + C_6H_6^* \longrightarrow H_2 + \text{product or radical} \tag{6.7}$$

The excited molecules can as well be deactivated:

$$C_6H_6^* + C_6H_6 \longrightarrow 2\,C_6H_6 + \text{thermal energy} \tag{6.8}$$

Reaction (6.7) can account not only for HD formation but also for the dependence of the yield of hydrogen on the LET value. The fact that the reactions responsible for hydrogen formation and deactivation are both very fast implies that they take place in spurs, and thus the ratio of the reaction proceeding between the active species themselves and their diffusion can be looked upon as a constant in the case of radiation of low ionization density. However, when radiation of high ionization density is applied the volume elements containing primary products in high concentrations overlap. Thus, the reaction between the active species becomes more important, and diffusion less so. This implies an increase in the probability of reaction (6.7), as it is a bimolecular process of the excited molecules, at the expense of the energy dissipation reaction (6.8) when the LET value is increased. Gursky and Tsoi [98] observed a decrease in the yield of hydrogen from benzene containing luminescent solute and irradiated with low doses $(5 \cdot 10^{19}$ eV $g^{-1})$. They attributed this to hydrogen formation via reaction (6.7).

The important role of bimolecular hydrogen formation is also evidenced by the fact that under α-radiation, the isotopic composition of hydrogen formed from benzene–benzene-d_6 (1 : 1) systems is apparently the same as that obtained in irradiation with electrons [27], although the total yield of hydrogen $G(H_2 + HD + H_2)$ is ten times higher in the former case. The low $G(H_2)$ in the solid phase can also be attributed to bimolecular hydrogen production; i.e. under these experimental conditions the probability of bimolecular reactions is greatly reduced.

According to Cherniak et al. [96], the reaction taking place between singlet-state excited benzene and ground-state benzene molecules might also contribute to hydrogen formation.

$$C_6H_6^* + C_6H_6 \longrightarrow H_2 + \text{product or radical}$$

Only very low yields of hydrocarbon gases are produced from benzene, the most important being acetylene.

As with hydrogen formation, the yields of gaseous hydrocarbon products are considerably higher in the vapour than in the liquid phase (Table 6.1). The yield of ethylene is independent of the dose and the dose rate, but increases in the presence of ion- and radical scavengers [$G(C_2H_4) = 0.27$]. The yield of acetylene is independent of the dose (up to $50 \cdot 10^{19}$ eV g^{-1}), the dose rate, the presence of ion and radical scavengers and the pressure 87–1315 mbar [77]; however, the yield is sharply reduced with the dose above $50 \cdot 10^{19}$ eV g^{-1}. Contrary to hydrogen formation, $G(C_2H_2)$ decreases with temperature at high temperatures [2]. The yield of propadiene decreases with increasing dose and dose rate, as well as in the presence of ion and radical scavengers [77].

In liquid-phase radiolysis, again as with hydrogen formation, the yield of acetylene is independent of the dose, dose rate and the presence of radical scavengers, but increases with the value of LET. The ratio $G(C_2H_2)/G(H_2) = 0.5$ remains unchanged with different types of radiation [75]. On the irradiation of benzene–benzene-d_6 (1 : 1) mixtures the isotopic composition of acetylene is found to be $C_2H_2 : C_2HD : C_2D_2 = 60 : 10 : 30$ [79].

The G-values of gaseous hydrocarbon products, formed exclusively by rupture of the benzene ring, indicate that this rupture takes place with a very low probability. Acetylene can be formed in uni- and bi-molecular reactions of excited benzene molecules [74]:

$$C_6H_6^* \longrightarrow 3\,C_2H_2$$

$$C_6H_6^* + C_6H_6^* \longrightarrow C_2H_2 + \text{polymer or radical}$$

The independence of the LET value found for the ratio of the yields of acetylene and hydrogen was explained by Burns [97], who assumed that hydrogen and acetylene are formed from the same precursor, namely excited molecules in the same energy state:

$$C_6H_6^* + C_6H_6^* \nearrow\searrow \begin{matrix} H_2 + \text{polymer or radical} \quad (6.9a) \\ C_2H_2 + \text{polymer or radical} \quad (6.9b) \end{matrix}$$

The ratio of the two possible reaction pathways, (6.9a) (6.9b), is independent of the concentration of excited molecules. Thus with increasing LET value the probabilities of the two reactions in comparison with the energy dissipation reaction (6.8) increase at the same rate.

In liquid-phase radiolysis, only hydrogenated compounds (cyclohexa-1,4-diene and cyclohexa-1,3-diene) could be detected in the C_6 product fraction [99]. The yields decrease with increasing dose and in the presence of radical scavengers, but increase with increasing temperature. In the vapour phase, cyclohexene was also observed.

Cyclohexadiene isomers can be formed via reactions of reactive free radicals produced by the decomposition of excited benzene molecule [reaction (6.4) and (6.5)] [99]:

$$C_6H_5 + C_6H_6 \longrightarrow C_6H_5 - C_6H_6$$

$$C_6H_7 + C_6H_5 - C_6H_6 \longrightarrow C_6H_8 + C_6H_5 - C_6H_5$$

$$2\,C_6H_7 \longrightarrow C_6H_8 + C_6H_6$$

The polymer products of benzene (Table 6.2) are mainly unstable unsaturated compounds, highly reactive towards oxygen.

In vapour-phase radiolysis, the yield of polymers decreases slightly with increasing dose and increases with increasing temperature. At high temperatures the yield changes only to a very small extent with density. In the liquid phase, the total yield depends very slightly on the dose, whereas the composition changes considerably. The mean molecular mass of the polymer fraction formed under a dose of $\sim 900 \cdot 10^{19}$ eV g^{-1} is 250–300, and this value increases with increasing dose [100, 101]. The G(polymer) increases with the LET value, but not so significantly as the yield of hydrogen [74, 75, 97]: with γ-rays, ~ 0.8–1.0; with α-rays, 1.1–1.5 and with reactor radiation, ~ 1.3. Radical scavengers reduce the yields by about 60%. In liquid-phase radiolysis, the radical yield obtained with radical scavengers was $G(R) \approx \approx 0.7$ [102], approximately equal to the total yield of polymers [96]. In the solid phase, $G(R) \approx 0.15$ [84, 94], was obtained using ESR techniques.

Most of the polymer fraction is presumably produced from reactions of H and C_6H_5 radicals formed in the decomposition of excited benzene molecules [reaction (6.4)].

Recently spin trapping technique provides concrete evidence that C_6H_5 radicals are intermediates in the radiolysis of benzene [102A, 102B]. Results obtained in photochemical and thermal systems [144, 258, 259] show that the rate constant of the recombination reaction of phenyl radicals is rather low:

$$C_6H_5 + C_6H_5 \longrightarrow (C_6H_5)_2 \qquad (6.10)$$

The low stationary concentration of phenyl radicals contributes also to the slowness of reaction (6.10), but they add to benzene molecules very readily:

$$C_6H_5 + C_6H_6 \longrightarrow C_6H_5 - \dot{C}_6H_6$$

The $C_{12}H_{11}$ radical thus formed has been detected both by ESR techniques and by optical spectroscopy [88, 103].

Table 6.2
Yields of polymers from benzene

Radiation	^{60}Co-γ	^{60}Co-γ	^{90}Sr-^{90}Y-β	^{60}Co-γ
Dose rate, eV g^{-1} s^{-1}	$\sim3.4\cdot10^{16}$	$0.87\cdot10^{16}$	$\sim4\cdot10^{14}$	$\sim1.7\cdot10^{16}$
Dose, 10^{18} eV g^{-1}	~21.6	6—62		
Temperature, °C	23	25	25	−20
Pressure, mbar	87			
Reference	77	99	96	76
Phase	Gas	Liquid		Solid
Products	G, molecule/100 eV			
Phenylcyclohexene Phenylcyclohexadienes $\Big\}$	0.006	0.058	0.022	0.052
Cyclohexadienylcyclohexene and dicyclohexadienes		0.020	0.018	0.01
Biphenyl	0.17	0.048	0.072	0.036
m-Terphenyl			0.022	
p-Terphenyl			0.021	
p-Polyphenyls			0.0054	
G (polymer)	6.0^{a}	1.1	0.8 0.94^{b}	0.755

^{a} From ref. [3]
^{b} From ref. [25]

Identified biphenyl and hydrogenated products of the radiolysis of liquid benzene (Table 6.2) can presumably result from reactions of C_6H_7 and $C_{12}H_{11}$ radicals [96, 99]:

$$C_6H_5-C_6H_6 + C_6H_6 \longrightarrow C_6H_5-C_6H_5 + C_6H_7$$

$$C_6H_7 + C_6H_7 \longrightarrow C_6H_5-C_6H_7 + H_2$$

$$C_6H_7 + C_6H_7 \longrightarrow (C_6H_7)_2$$

$$C_6H_7 + C_6H_7 \longrightarrow C_6H_5-C_6H_9$$

$$C_6H_7 + C_6H_5-C_6H_6 \longrightarrow C_6H_6 + C_6H_5-C_6H_7$$

$$2\,C_6H_5-C_6H_6 \longrightarrow (C_6H_5)_2 + C_{12}H_{12}$$

$$C_6H_7 + C_6H_5-C_6H_6 \longrightarrow C_6H_8 + (C_6H_5)_2$$

The recombinations of various radicals can account for the formation of polyphenyl and polyhydrophenyl compounds, e.g:

$$C_6H_7 + C_6H_5-\dot{C}_6H_6 \longrightarrow C_6H_7-C_6H_6-C_6H_5$$

$$2\,C_6H_5-\dot{C}_6H_6 \longrightarrow C_6H_5-C_6H_6-C_6H_6-C_6H_5$$

$$C_6H_7-\dot{C}_6H_6 + C_6H_6 \longrightarrow C_6H_7-C_6H_6-\dot{C}_6H_6$$

$$C_6H_7-(C_6H_6)_n-\dot{C}_6H_6 + C_6H_7 \longrightarrow C_6H_7-(C_6H_6)_{n+1}-C_6H_7$$

One possible route for the formation of polymer products in the presence of radical acceptors is the direct fragmentation of excited molecules [99, 104]:

$$C_6H_6^* + C_6H_6^* \longrightarrow H_2 + (C_6H_5)_2$$

$$C_6H_6^* + C_6H_6 \longrightarrow C_6H_5-C_6H_7$$

Zimmerli and Gäumann [99] studied the variation of the yields of biphenyl and phenylcyclohexadienes as a function of the concentration of additives such as naphthalene, anthracene, cyclohexene and cyclohexadiene isomers in benzene solutions irradiated with low doses. On the basis of kinetic equations derived from the radical mechanism discussed above, they arrived at the same values of differences in the yields as were determined experimentally. This unambiguously indicated the predominance of reactions between phenyl and cyclohexadienyl radicals and the additives without presumption of energy transfer.

The different dependence of yields of hydrogen and polymers on the LET value was explained by Burns by the different nature of the precursors, i.e. hydrogen and acetylene are produced from a short-lived excited molecule, whereas that giving polymers is long lived [74]. According to Burns, only some of the polymers are produced from radicals formed simultaneously with gaseous products in a bimolecular reaction of very short-lived excited molecules, the majority being formed by reactions of radicals produced in pseudo-unimolecular reactions from the excited molecules [105]. In his classical work, Burton assumed also the different reactions of excited benzene molecules [104].

In the radiolysis of liquid benzene, as mentioned earlier, the excited molecules are formed directly or by ion-recombination decay into ground-state molecules with high probability, i.e. without decomposition. However, in the vapour phase processes other than the former can also contribute to product formation, because of the longer lifetimes of intermediates, e.g. decomposition and rearrangement of ions, ion-molecule reactions. Burns and Marsh [2] interpreted the product formation in the high-temperature radiolysis (260–390°C) of benzene vapour in terms of reactions of radicals produced from excited benzene molecules. Under these experimental conditions, excited molecules cannot interact with one another within the spur or track because of the high diffusion rate. Thus, in contrast to the mechanism of hydrogen formation suggested for liquid-phase radiolysis [reaction (6.7)] the authors explained hydrogen formation in the vapour phase mainly by reactions of hydrogen atoms produced in the decomposition of excited benzene molecules [reaction (6.4)]

$$H + C_6H_6 \longrightarrow H_2 + C_6H_5$$

The collisional deactivation of excited benzene molecules [reaction (6.8)] competes with this reaction even under these experimental conditions. The dependence of $G(H_2)$ on the density can be well interpreted using this mechanism. The competition between fragmentation reactions of cyclohexadienyl radicals produced in reaction (6.5) may account for the increase of $G(H_2)$ in the temperature range 260–390°C:

$$C_6H_7 \begin{cases} H + C_6H_6 \\ H_2 + C_6H_5 \end{cases}$$

The products methane, ethane and ethylene are formed partly in secondary processes, namely in reactions with radiolytic products and partly in the decomposition reactions of the latter. This is verified by the variation of the yields with density and dose. Burns and Marsh [2] attributed polymer formation in high-temperature vapour-phase radiolysis to secondary reactions of radicals produced from excited benzene molecules, as in liquid-phase radiolysis.

Hentz and Rzad [77] compared the room-temperature vapour-phase γ-radiolysis of benzene with its UV photolysis carried out with Xe and Kr resonance lamps [106]. They studied the dependence of product yields on the dose, the dose rate, the presence of ion and radical scavengers and the pressure of the benzene vapour. They concluded that above 80 mbar, hydrogen and acetylene are not formed via free radical or ion-neutralization reactions, but by molecular elimination from either superexcited states or ion-molecule reactions of $C_6H_6^+$ with C_6H_6 or both. Cyclohexadiene can, presumably, be produced through a free radical mechanism similar to that in liquid-phase radiolysis.

In order to examine the ionic mechanism of product formation suggested for the vapour phase radiolysis of benzene, the ion-molecule reactions of benzene have been investigated in several types of mass spectrometer with electron or proton rays at various pressures [4, 5, 107–113]. In low-pressure studies (1 μbar), the most important secondary ions observed were as follows: $C_{12}H_{11}^+$, $C_{12}H_{10}^+$ $C_{12}H_9^+$, $C_{12}H_8^+$, $C_{11}H_7^+$, $C_{10}H_9^+$, $C_{10}H_8^+$, $C_{10}H_7^+$, $C_9H_7^+$ [110–112]. From these data Virin et al. [112] concluded that the most important primary ion, the benzene molecule ion, which amounts to 70% of all ions does not react with benzene. The fragment ions are: $C_6H_5^+$, $C_6H_4^+$, $C_4H_4^+$, $C_3H_3^+$. According to the rate constants determined, the most probable ion-molecule reaction of the fragment ions is [112]

$$C_6H_5^+ + C_6H_6 \longrightarrow C_{12}H_{11}^+$$

Lifshitz and Reuben came to the same conclusion, assuming that the

$$C_6H_6^+ + C_6H_6 \longrightarrow C_{12}H_{12}^+$$

reaction is slightly endothermic and thus the $C_{12}H_{12}^+$ ion can be produced only at high pressures and well-defined repeller potential [110, 111]. The main reactions of fragment ions they suggest are as follows:

$$C_6H_6 \nearrow \begin{array}{c} C_6H_5^+ \xrightarrow{C_6H_6} C_{12}H_{11}^+ \nearrow \begin{array}{c} C_{10}H_9^+ + C_2H_2 \\ \searrow \\ C_{12}H_9^+ + H_2 \end{array} \\ \searrow \\ C_6H_4^+ \xrightarrow{C_6H_6} C_{12}H_{10}^+ \longrightarrow C_{12}H_8^+ + H_2 \end{array}$$

The authors used this scheme to explain the product formation.

In investigations carried out in high-pressure (0.01–1.3 mbar) mass-spectrometers at relatively low pressures (0.02 mbar) peaks of the parent ion ($C_6H_6^+$) and primary fragment ions ($C_6H_5^+$, $C_5H_3^+$, $C_3H_3^+$) have been detected [5, 109]. With increasing pressure, the intensities of these ions decrease and intensities of ions produced in ion-molecule reactions increase at the same time. At about 0.13 mbar, polymer ions ($C_{12}H_{12}^+$, $C_{12}H_{11}^+$, $C_{12}H_{13}^+$) predominate in the spectrum:

$$C_6H_6^+ + C_6H_6 \nearrow \begin{array}{c} C_{12}H_{12}^+ \\ \searrow \\ C_6H_7^+ + C_6H_5 \end{array}$$

$$C_6H_5^+ + C_6H_6 \nearrow \begin{array}{c} C_6H_6^+ + C_6H_5 \\ \searrow \\ C_{12}H_{11}^+ \end{array}$$

$$C_{12}H_{12}^+ + C_6H_6 \longrightarrow C_{12}H_{13}^+ + C_6H_5$$

The existence of the monomer and dimer benzene cation and anion has been verified by the ESR spectra obtained in the γ-radiolysis of benzene adsorbed on silica, at 77 K, and the absorption spectra obtained on irradiation of a solution of 3-methylpentane in benzene, also at 77 K [14, 115, 117]. The ESR spectra indicate a symmetrical sandwich structure for the dimer cation, similar to that of the benzene excimer [116, 117, 253]. Similar studies on cyclohexane have not indicated the presence of dimer ions, thus the molecular ion attachment is presumably characteristic of aromatic compounds in the gas phase. It appears, that the π-electron structure of the aromatic ring absorbs the energy released in the ion–molecule reaction and thus prevents the intermediate complex from dissociating.

Results obtained with high-pressure mass spectrometry convinced Wexler et al. [4, 5] that under these experimental conditions, the reactions of primary ions with benzene are fast enough to compete with the parent-ion-electron neutralization. Hence the neutralization of secondary ions can account for the formation of products that are insensitive to the presence of radical scavengers in the gas phase.

Some characteristic properties of the polymers that are the main products of the radiolysis of benzene (their viscosity and molecular mass increase, while the number of double bonds decreases, with increasing dose; the LET effect is much smaller than in the hydrogen formation) have led to the conclusion, that one important pathway for their formation, besides radical reactions, is ion-molecule reaction, as it is shown in the scheme:

$$C_6H_6 \xrightarrow{\quad\text{\tiny /\\//\\/}\quad} C_6H_6^+ + e^- \longrightarrow C_6H_5^+ + H + e^-$$

$$C_6H_6^+ + C_6H_6 \longrightarrow C_{12}H_{12}^+$$

$$C_{12}H_{12}^+ + e^- \left\{ \begin{array}{l} \longrightarrow C_{12}H_{10} + H_2 \\[4pt] \longrightarrow C_{12}H_{12} \\[4pt] \longrightarrow [C_{12}H_{12}]^* \xrightarrow{C_6H_6} C_{12}H_{14} \end{array} \right.$$

$$C_6H_5^+ + C_6H_6 \longrightarrow C_{12}H_{11}^+$$

$$C_{12}H_{11}^+ + e^- \longrightarrow [C_{12}H_{11}]^* \xrightarrow{C_6H_6} \left\{ \begin{array}{l} \nearrow C_{12}H_{10} + C_6H_7 \\[4pt] \searrow C_{12}H_{12} + C_6H_5 \end{array} \right.$$

The increase of the mean molecular weight and the constancy of polymer yields with increasing dose can be attributed to secondary reactions of $C_6H_6^+$, $C_6H_5^+$, $C_{12}H_{12}^+$ and $C_{12}H_{11}^+$ ions with products such as biphenyl, and phenylcyclohexene. Thus, the ionic mechanism may present another possibility for polymer formation, in addition to the first- and second-order reactions of excited benzene molecules and the radical reaction mechanism, detailed above.

In contrast to the results of high-pressure mass-spectrometry mentioned above, Field et al. obtained a much smaller peak for the $C_{12}H_{12}^+$ ion among other secondary ions [107, 113]. This deviation was attributed by them to the difference in the working temperatures. They observed an increase in the intensity of the monomer molecule ion and a simultaneous decrease of the dimer ions when the temperature was increased. Based on this observation the authors suggested an equilibrium

$$C_6H_6^+ + C_6H_6 \rightleftharpoons (C_6H_6)_2^+$$

that would indicate that the π-complex formed under these conditions is unable to transform into the more stable σ-complex. Their suggestion implies that the probability of product formation through the dimer molecule ion is rather low. However, the data obtained with $C_6H_5^+$ ions did not provide evidence for the equilibrium. Thus, the addition of $C_6H_5^+$ ion to benzene renders the formation of a σ-complex probable, with a subsequent neutralization resulting in the formation of end-products.

The results of investigations of tritium substitution on benzene support the ionic mechanism for the liquid-phase radiolysis of benzene. Kroh et al. [118–120] studied the substituted benzene products formed in the γ-radiolysis and UV photolysis of solutions of tritiated water in benzene, and determined the effect of electron and radical scavengers on the yields. From the results they deduced an ionic mechanism. This conclusion was justified by the observation that the yield of intermediates necessary to the production of tritiated products is $G = 0.077$, which is comparable to the free ion yield obtained with clearing field techniques carried out on benzene [G(free ion) = 0.057] [15]. α-Radiolysis carried out with ^{210}Po isotope, where the efficiency of ion-scavenging is rather low because of effective ion-recombination, gave yields of tritiated benzene products considerably lower than those from γ-radiolysis [120].

6.2.2. Alkyl- and alkenylbenzene homologues

In the radiolysis of benzenoid compounds with an aliphatic side-chain, the behaviour of the latter is usually consistent with its aliphatic nature, although its reactivity is reduced considerably by the presence of the benzene ring. The radiation stability of these compounds is lower than that of benzene, but significantly greater than that of aliphatic compounds with the same carbon atom number as that of the side-chain; e.g. the hydrogen and radical yields of octylbenzene are, $G(H_2) = 0.48$ [134], $G(R) = 0.9$ [242] and those of octane, ~ 6 and ~ 6.4 [242], respectively.

As a result of the presence in these compounds of two types of chemical entity, hydrogen, light hydrocarbons, dealkylated and alkylated compounds, and polymers are found, among the radiolysis products.

The primary processes and the product distribution have been studied extensively only for toluene and isopropylbenzene of the alkylbenzene group. For the radiolysis of alkylbenzenes with more than one alkyl substitutent, the extent of formation of gaseous products, and sometimes the polymer formation processes, have been studied in order to examine the influence of the molecular structure on the radiolysis.

Excitation and excitation transfer processes

The primary processes taking place in the radiolysis of liquid alkylbenzene are generally similar to those of benzene. The excited states produced directly and in geminate recombination of parent ions with electrons play an important role, whereas the yield of ions is low: $G(\text{ion}) = 0.3–0.5$ [57, 128]. The free ion yields of several alkylbenzenes, determined with the clearing field technique at 20°C, fall in the range $G_{fi} = 0.073–0.094$ [14].

The yields and the mechanism of formation of first excited states are not yet fully understood. Autoionization from superexcited states, with a subsequent recombination or internal conversion reaction steps, is usually considered to occur [37, 57, 72].

The quantitative determination of molecules in the first excited states, having different multiplicities, is generally carried out using techniques discussed above with respect to benzene. An investigation of the isomerization of but-2-ene in toluene, led to a yield of $G = 6$ being estimated for toluene triplets and 10 ns for their half-life [44, 45]. Fischer et al. estimated the same numerical value for the yield of first excited states of toluene, from a study of the isomerization of stilbene dissolved in toluene [50]. A maximum (at ~ 25°C) has been observed in the isomerization yield of stilbene as a function of temperature, in the γ-radiolysis of toluene solutions containing stilbene in low concentrations [121]. According to the authors, this curve may be ascribed to the sum of the effects of at least two processes — diffusional interaction between stilbene and toluene triplets and the competing deactivation process of triplets — the rates of which change with temperature in opposite directions.

Cooper and Thomas [35] irradiated liquid toluene with nanosecond pulses and observed fluorescence below 400 nm and an absorption in the 500–600 nm region, which are similar to results obtained for benzene. Both spectra were assigned to the first singlet state of toluene with a half-life of 21.6 ns. However, further experiments proved that the transient species involved in the absorption is the singlet excimer [53, 122]. Baxendale and Rasburn [11] investigated the dependence of the intensities of absorption and emission on the concentration of admixed naphthalene in toluene solutions irradiated with nanosecond pulses. Applying the Stern–Volmer relationship and taking $G(^1B_{2u}) = 1.4$ for the benzene singlet yield, as a standard, and the efficiency of toluene intersystem crossing as 0.45 [45], they estimated the yields of the first excited singlet and triplet states of toluene as 1.35 (half-life 3.5 ns) and 2.8 (half-life 17 ns), respectively.

Beck et al. detected the higher excited states of toluene using subnanosecond pulse radiolysis and two photon laser photolyses of toluene with different additives [122A, 122B]. The results indicate that the higher excited states of toluene react efficiently with chloroform and 9,10-diphenylanthracene at low temperatures (195, 233 K), but little if any reaction occurs at room temperature. It is suggested that lower temperatures enhance the excitonic motion of a higher excited state, which may lead to reaction with solutes prior to relaxation to the first excited state. This increased excitonic motion may be due to the formation of partially ordered clusters of toluene at low temperature.

Christophorou et al. [123, 124] compared the emissions of liquid toluene, ethylbenzene, n- and isopropylbenzene, and 1,3,5-trimethylbenzene (having maxima at $\sim$ 320 nm), obtained on electron impact, with those observed on irradiation with UV light. They attributed the emission to the excited singlet excimers (see Section 6.2.1). In contrast to UV photolysis, emission of excited singlet monomers has not been observed in radiolysis, presumably because of the high probability of quenching of singlet monomers in irradiation by electrons. In Table 6.3 the lifetimes and λ_{max} of excited singlet excimer states assumed are listed [123]. The lifetimes observed are in good agreement with those found for alkylbenzene excimers obtained on irradiations with UV light [125]. 1,4-Diethylbenzene, 1,4-dimethyl-2-ethylbenzene, 1,3-dimethyl-4-ethylbenzene, 1,2,3,5-tetramethylbenzene, isobutylbenzene and n-pentylbenzene give a relatively weak emission at about 500 nm on electron impact, which can be attributed to triplet excimers.

Following the pulse radiolysis of liquid isopropylbenzene, o-, m-, p-xylene, 1,3,5-trimethylbenzene and 1,2,4-trimethylbenzene at room temperature, Bensasson et al. [126] observed absorption with maxima at about 600 and 320 nm, similarly to those found in the radiolysis and photolysis of benzene [57, 127]. The absorption at 600 nm corresponds to a singlet excimer state. The absorption at 320 nm can presumably be assigned to the long-lived cyclohexadienyl-type radical, and to a strongly unstable biradical

Table 6.3

Lifetimes and λ_{max} of emitting particles of alkylbenzenes obtained by
electron impact. After Christophorou [123]

Hydrocarbons	Lifetime, ns	λ_{max}, nm
Toluene	15.65	321
Ethylbenzene	8.56	308
Isopropylbenzene	8.73	
n-Propylbenzene	12.56	
o-Xylene	11.71	318
m-Xylene	11.13	323
p-Xylene	11.9	318
1,3,5-Trimethylbenzene	22.86	326

formed in the interaction of a triplet and a ground-state alkylbenzene molecule, analogously to benzene. The very weak absorption at ~ 400 nm has
been tentatively attributed to an excited triplet transient. This has not
been observed in the radiolysis and photolysis of benzene, but it showed up
in the pulse radiolysis of these alkylbenzenes. Spectra corresponding to
excited singlet monomers have not been observed [126].

Gangwer and Thomas [57] have determined the yields of first excited
singlet and triplet states through the concentration dependence of fluorescence and absorption produced in solutions of binaphthyl and benzanthracene in p-xylene. They applied the Stern–Volmer relationship and found
values of 2.05 and 2.33 for the yields of singlet and triplet states, respectively,
similar to the results of Horrocks [33]. The half-life of the triplet, 28 ns is
about ten times greater than the corresponding one of benzene. One possible interpretation of this is that the benzene triplet may also be an excimer
which gives a highly reactive biradical with considerably higher efficiency
than the monomer. However, the probability for excimer production is much
lower in p-xylene than in benzene. The larger yield of singlet states produced
in xylene has been explained by the considerably higher efficiency for internal
conversion as compared with benzene [33, 39]. Electron scavengers reduce
the yield of singlets to about half of the inital value. Thus, it is probable
that at least 50% of the singlets are produced through ionic precursors,
$G \approx 1$ [129], via geminate ion-recombination proceeding in the picosecond
range. The remaining yield of singlets, $G \approx 1$, could be the result of internal
conversion taking place with about 100% efficiency from upper excited states
produced directly in alkylbenzenes.

Richards and Thomas [127] investigated the pulse radiolysis of liquid
isopropylbenzene at $-78°C$, and observed a considerable enhancement of
the absorption intensity due to excimer (maximum at 525 nm) and a new
absorption band at 400 nm, as compared with the room temperature spectrum. They suggested a complex ionic transient for this absorbing species.
The absorption of the excimer was considerably reduced in the glassy state,
at about 77 K. Although the fast rotation of the molecules is hindered under
these experimental conditions, the excimer can be formed, as the random

array of molecules will have several sites where the geometry is suitable for excimer formation. The singlet energy will migrate here and be trapped in the lower energy configuration of the excimer state. The 310 nm absorption corresponding to biradicals has not been observed at 77 K.

Investigations of the pulse radiolysis of isopropylbenzene at room temperature and at 77 K in the presence of aromatic additives indicates that energy transfer to the additive is much more efficient in the glassy state. The lifetime of the triplet state of liquid isopropylbenzene was determined as < 10 ns, and that in the glassy state as 2.8 μs [127].

Gangwer and Thomas [128] studied the absorption spectrum of isopropylbenzene in the glassy state brought about by γ-radiolysis. They observed maxima at 400 and 556 nm, which were reduced upon addition of ion-scavengers and could be made to vanish at room temperature. The maxima were assigned to the isopropylbenzene anion and cation. For the yield of anions G(ion) = 0.6 was estimated. This value is similar to that found for the biphenyl ion, G(ion) = 0.8, observed in biphenyl–isopropylbenzene glassy systems [128]. The shape of the luminescence spectrum of irradiated isopropylbenzene and its decay, as well as the decrease of the yield of excited states in the presence of ion scavengers, imply that a large proportion of singlet excited states of isopropylbenzene are formed via ion neutralization.

Platzner and Thomas irradiated isopropylbenzene in the glassy form at 77 K with 7 MeV electrons, and observed a post-irradiation luminescence. Its spectrum was similar to that of the excimer. They assumed that the singlet excimer formed in the post-irradiation process (G = 0.0027) was produced from ion recombination, via singlet excited states [129].

Merkel and Hamill [53] studied the emission of a toluene thin-film irradiated at 77 K with low-energy electrons (3–100 eV). The fluorescence of the monomer predominates, with a maximum at 280 nm, which is similar to the behaviour of benzene. A weak emission characteristic of the excimer has also been observed at 400 nm. In contrast to benzene, the emission spectrum of toluene contains a maximum at 480 nm. This was assigned to emission from the benzyl radicals, on the basis of photolysis experiments [130].

Philips and Schug [54] observed emission in the bands corresponding to the maxima at ∼ 280 and 400–500 nm when they irradiated solid toluene and 1,2,4,5-tetramethylbenzene with 1 MeV electrons at 103–135 K; however, in xylene only the latter emission band appeared. The emission with a maximum at ∼280 nm has been attributed to the fluorescence of the monomer, whereas that at the longer wavelengths might be the result of benzyl and 2,4,5-trimethylbenzyl radicals, and parent cations and benzyl cations. Miyazaki and Yamamoto studying the fluorescence spectra of toluene during γ-irradiation at 77 K found that the intensity and spectrum of the emission depended remarkably upon the cooling conditions of the sample. The emission from polycrystalline toluene was much stronger than that from glassy toluene [63].

According to Batekha et al. [131], the ratio of the intensities of phosphorescence and fluorescence determined in the low-dose radiolysis of solid

27*

toluene–naphthalene-d_8 agrees well with the ratio obtained from photolysis of the same system. However, at a dose of $\sim 25 \cdot 10^{19}$ eV g^{-1} the ratio undergoes a three-fold increase, which was attributed to triplet states formed on recombination of stabilized toluene ions that had already reached a stationary concentration at the dose in question.

Ogawa et al. [72] studied the radiolysis by low-energy (60–300 eV) electrons of gaseous toluene and o-, m- and p-xylene at 1 μbar. They observed that the shape of the emission curve and the dependence of the intensity on the experimental conditions were largely similar to those described for benzene. They concluded that the emission band at 270–340 nm is generally characteristic of the aromatic hydrocarbons and is caused by the transition of the first excited singlet state to the ground state.

Considerations based on end-products

The values of $G(H_2)$ obtained in the radiolysis of alkylbenzenes in the gas, liquid and solid phases are listed in Table 6.4. The hydrogen yield is considerably larger in the gas than in the liquid phase, and it increases with temperature: e.g. for isopropylbenzene, it increases from 0.43 at 173°C to 1.88 at 367°C [132]. In the liquid phase, the yield of hydrogen is independent of dose, dose rate, radical scavengers and temperature, but increases with the LET value. For instance, for toluene $G(H_2)$ from α-radiolysis is about five times higher than that from γ-radiolysis [133].

The yields increase with the length of the side-chain. Thus, the hydrogen yield corresponding to the side-chain of branched compounds is lower than that of the normal carbon chain (n-propylbenzene, isopropylbenzene, n-butylbenzene, isobutylbenzene, tert-butylbenzene) [132, 134, 134A, 134B, 134C]. The $G(H_2)$ value is essentially independent of the number of sites where the substituents are attached to the ring: i.e. for hexylbenzene, $G(H_2) = 0.40$ [134], for hexamethylbenzene, $G(H_2) = 0.43$ [135]. In the solid state, at -196°C, the yield of hydrogen from p-xylene is reduced to 0.011 [136], and that for hexamethylbenzene to 0.24 [135].

Ingalls [137] studied the liquid-phase radiolysis of toluene, completely and partially deuterated toluene, and their mixtures of 1 : 1 molar ratio. From the isotopic composition of hydrogen he concluded that hydrogen atoms from both the ring and the side-chain contribute to hydrogen formation, but the contribution of those from the methyl group is higher. Some 38% of hydrogen formed from the ring can be attributed to some unimolecular process, and about 62% to bimolecular abstraction. About 82% of the hydrogen produced from the side-chain is the result of radical processes:

$$C_6H_5-CH_3^* \longrightarrow H + C_6H_5-\dot{C}H_2$$

$$C_6H_5-CH_3^* \longrightarrow H + \dot{C}_6H_4-CH_3$$

$$C_6H_5-CH_3^* \longrightarrow H_2 + \text{product or radical}$$

$$H + C_6H_5-CH_3 \longrightarrow H_2 + C_6H_5-\dot{C}H_2$$

$$H + C_6H_5-CH_3 \longrightarrow H_2 + \dot{C}_6H_4-CH_3$$

Table 6.4

Hydrogen yields from alkylbenzenes (radiation: ^{60}Co-γ)

Phase	Gas (25°C)			Liquid (30°C)		Solid		
Hydrocarbons	Pressure, mbar	$G(H_2)$	Ref.	$G(H_2)$	Ref.	Temperature, °C	$G(H_2)$	Ref.
Toluene	14.1	0.32	145	0.14	139	−196	0.09[a]	134
				0.13[a]	260	−159	0.03	94
o-Xylene	6.2	0.65	3	0.235	136			
				0.18	3			
m-Xylene	3.9	0.65	3	0.184	136			
p-Xylene	3.8	0.84	3	0.209	136	−196	0.0112	
						−80	0.0196	136
						0	0.0493	
Ethylbenzene	10.0	0.76	3	0.158	136	−196	0.12	134
				0.176[a]	260			
n-Propylbenzene				0.23	134	−196	0.17	134
Isopropylbenzene	513	0.49 (163°C)	132	0.179	132			
o-Ethyltoluene				0.266	141			
m-Ethyltoluene				0.207	141			
p-Ethyltoluene				0.210	141			
1,3,5-Trimethylbenzene				0.24[a]	260			
n-Butylbenzene				0.27	134	−196	0.19	134
Isobutylbenzene				0.25	134B			
				0.16	134B			
				0.205	134A			
sec-Butylbenzene				0.22	134B			
				0.179	134A			
tert-Butylbenzene				0.14	134B			
				0.099	134A			
n-Amylbenzene				0.34	134	−196	0.22	134
Hexylbenzene				0.40	134	−196	0.24	134
Hexamethylbenzene				0.43 (173°C)	135	−196	0.24	135
Heptylbenzene				0.42	134	−196	0.26	134
Octylbenzene				0.48	134			
Nonylbenzene				0.59	134	−196	0.44	134
Dodecylbenzene				0.67	134			
Tridecylbenzene				0.75	134	−196	0.56	134
Heptadecylbenzene				0.95	134			

[a] 1.6 MeV electrons

In Ingalls's view, the total hydrogen yield can also be produced in radical reactions. To explain why the hydrogen yield is independent of the presence of radical scavengers and the temperature, he assumed that the hydrogen atoms are hot, i.e. the reactions take place in the so-called 'thermal spikes' [138].

On the other hand, Weiss and Collins [139, 140] consider it much more probable that hydrogen formation occurs via direct uni- or bi-molecular reactions of excited toluene molecules, as in the radiolysis of benzene. Thus, the hydrogen atoms of the ring and of the methyl group contribute about equally:

$$C_7H_8^* \longrightarrow C_7H_6 + H_2$$

$$C_7H_8^* + C_7H_8^* \longrightarrow H_2 + (C_7H_7)_2$$

$$C_7H_8^* + C_7H_8 \longrightarrow H_2 + (C_7H_7)_2$$

$$C_7H_8^* + C_7H_8 \longrightarrow 2\,C_7H_8$$

Weiss and Collins used this mechanism to explain the dependence of the hydrogen yield on the LET-value, its independence from the temperature and the ineffectiveness of radical scavengers. In solid-state radiolysis only uni-molecular hydrogen formation can occur. Accordingly, the $G(H_2)$ decreases.

Studies of the liquid-phase radiolysis of ethylbenzene, xylene, hexa-methylbenzene and some ethyltoluene isomers [135, 136, 141] confirmed the occurrence of direct uni- or bi-molecular decomposition of excited mole-cules resulting in hydrogen formation. As a consequence of the rotations of the methyl groups, in the case of hexamethylbenzene bimolecular hydrogen production is possible also in the solid state.

The G-value of hydrogen increases in the order *meta* < *para* < *ortho*, for the isomers of xylene and ethyltoluene.

Buben and Tchkheidze [142] observed ESR spectra indicating the pres-ence of $C_6H_5\dot{R}'$ and $\dot{C}_6H_6R$ radicals in the radiolysis of solid toluene, tert-butylbenzene and 1,3,5-trimethyl-benzene. In the case of toluene, about 30% of the radicals were $C_6H_6CH_3$. This result indicates the important role of H-addition reactions in the radiolysis of alkylbenzenes. The authors sug-gested the following mechanism for hydrogen formation:

$$C_6H_5R \rightsquigarrow C_6H_5R^* \left\langle \begin{array}{l} H + \dot{C}_6H_4R \\ H + C_6H_5R'' \end{array} \right.$$

$$H + C_6H_5R \longrightarrow \left\{ \begin{array}{l} \dot{C}_6H_6R \\ C_6H_5\dot{R}' + H_2 \\ \dot{C}_6H_4R + H_2 \end{array} \right.$$

Analysis of the ESR spectra and the isotope composition of the hydrogen formed in the radiolysis of $C_6H_5\text{-}CH_3$ and $C_6D_5\text{-}CH_3$ indicates that hydro-gen atoms produced as a result of ring C—H bond ruptures add to the phe-nyl ring of another molecule with rather high probability. Other H-atoms, which appear to be more reactive (presumably because of the different bond dissociation energies), are formed from the methyl groups, and contribute to the hydrogen formation via H-atom abstraction.

In the radiolysis of solid hexamethylbenzene, no ESR signals have been observed that could point to H-atom addition. It can be assumed that the steric hindrance caused by the six methyl groups lowers the probability for addition [135].

The yields of some products obtained in the gas-and liquid-phase radiolysis of toluene, isopropylbenzene, o-, m- and p-xylene, ethylbenzene and n-, iso-, sec- and tert-butylbenzene are listed in Table 6.5. In vapour-phase radiolysis the yields of gaseous hydrocarbon products are independent of the dose in the range 125–625 · 10^{19} eV g^{-1}. Acetylene is the most important product, but even its yield is lower than that obtained in the radiolysis of benzene, and increases slightly with increasing number of alkyl substituents. In the case of isopropylbenzene, the yields of hydrocarbon gases increase with temperature, with the exception of acetylene. When radical scavengers are added, ethane and propane are not produced, but the yield of acetylene remains unchanged.

In liquid-phase radiolysis, the gaseous hydrocarbon products are methane, ethane, ethylene, acetylene, propane and propene, butane and butenes, depending on the side-chain. The most important is methane, except in butylbenzenes, where the most abundant are the gaseous hydrocarbon products, indicating rupture of the C—C bond in the β-position with respect to the ring [134A, 134B, 134C]. The yield of acetylene is an order of magnitude lower than in the case of benzene. This indicates that in liquid-phase radiolysis of substituted aromatics the probability of ring scission is much lower than in that of the corresponding pure aromatics. Each of the G(gas) values is significantly lower in the liquid than in the vapour phase, i.e., the yield of methane is 10–20 times lower, and that of ethane 100 times lower. Upon addition of radical scavengers to toluene, xylene or ethylbenzene, the yield of acetylene remains unchanged, whereas that of methane is reduced to one-third of the initial value. With increasing temperature the yield of acetylene remains constant, just as the non-scavengeable methane yield does, whereas the total yield of methane increases [139, 143].

The isotopic composition of methane produced in the liquid-phase radiolysis of toluene and selectively deuterated toluene, as well as the effect of radical scavengers, indicates that at least 80% of the methane is formed by a radical reaction pathway, i. e. presumably from the abstraction reaction of methyl radicals originating in the decomposition of excited toluene molecules. The abstraction occurs mainly from the side-chain [149]:

$$C_6H_5 - CH_3^* \longrightarrow CH_3 + C_6H_5$$
$$CH_3 + C_6H_5 - CH_3 \longrightarrow CH_4 + C_6H_5 - CH_2$$

The spin trapping technique provides evidence that C_6H_5 radicals are intermediates in the radiolysis of toluene [102B]. The absence of ethane from the products shows that the methyl radicals do not recombine with each other. The non-scanvengable methane and acetylene may be formed in the direct decomposition of the excited molecule, similarly to hydrogen.

Weiss and Rao [141] found the smallest G(CH$_4$) values in the radiolysis of m-ethyltoluene, among ethyltoluene isomers. On studying the isotopic composition of products formed from selectively deuterated ethyltoluene, they established that about 85% of methane is formed through the rupture of the C—C bond in the ethyl group, while about 15% is produced via de-

Table 6.5

Yields from alkylbenzenes (radiation: ^{60}Co-γ)

Hydrocarbons	Toluene		o-Xylene		m-Xylene		p-Xylene	
Dose rate, eV g⁻¹ s⁻¹	$1.18 \cdot 10^{16}$	$4.3 \cdot 10^{16}$	$1.9 \cdot 10^{16}$	$5.7 \cdot 10^{16}$	$1.8 \cdot 10^{16}$	$3.73 \cdot 10^{15}$	$1.7 \cdot 10^{16}$	$3.97 \cdot 10^{15}$
Dose 10^{19} eV g⁻¹	1.9	311	820	820	840	20	800	300
Temperature, °C	25	39	30	30	30		20	20
Pressure, mbar	14.1	39	6.2	3	3.9	20	3.8	20
Refrence	145	139	3	3	3	136	3	136
Phase	Gas	Liquid	Gas	Liquid	Gas	Liquid	Gas	Liquid
Products				*G*, molecule/100 eV				
Methane	0.15	0.012	0.26	0.025	0.32	0.0142	0.33	0.0144
Ethane	0.07		0.39	0.0002	0.41		0.41	0.0001
Ethylene		0.0003	0.16	0.0004	0.14		0.15	0.0002
Acethylene	0.41	0.002	0.53	0.0028	0.56		0.59	0.0033
Propane + propene								
Butane + butenes								
Benzene	0.18	0.0188	0.13	0.001	0.10	< 0.05	0.15	< 0.002
Toluene			0.36	0.031	0.21	< 0.05	0.38	0.014
Ethylbenzene	0.05	0.0012	0.028	< 0.0003	0.012		0.011	
o-Xylene	0.021 [3]	0.0009			0.1		0.08	
m-Xylene	0.06 [3]	0.0015	0.18	0.012			0.26	
p-Xylene	0.07 [3]	0.001	< 0.0003	0.070	0.11			
Methylcyclohexadiene		0.062						
Dimethylcyclohexenes } Methylcyclohexenes		0.013						
Benzocyclobutene			0.08	0.002	0.011		0.004	
o-,m-,p-Ethyltoluene			0.135	0.0028	0.104		0.115	
Trimethylbenzenes			0.0049	0.092	0.09		0.121	
Propylbenzene								
Radical	6.3 [3]	2.40 [102]	7.2	2.45 [102]	6.4	0.99	6.6	1.10
G (polymer)		1.28 [260]		1.44				

Hydrocarbons	Ethylbenzene		n-Propylbenzene	Isopropylbenzene		n-Butylbenzene	Isobutylbenzene	sec-Butylbenzene	tert-Butylbenzene
Dose rate, eV g⁻¹ s⁻¹	$2.8 \cdot 10^{18}$	$3.73 \cdot 10^{15}$	$1.2 \cdot 10^{16}$	$3 \cdot 10^{16}$	$2.6 \cdot 10^{16}$	$1.2 \cdot 10^{16}$	$1.2 \cdot 10^{16}$	$1.2 \cdot 10^{16}$	$1.2 \cdot 10^{16}$
Dose, 10^{19} eV g⁻¹	170	300	218	50	740	218	218	218	218
Temperature, °C	30	20	30	163	36	30	30	30	30
Preassre, mbar	10.0			513					
Reference	3	136	134B	132	132	134B	134B	134B	134B
Phase	Gas	Liquid	Liquid	Gas	Liquid	Liquid	Liquid	Liquid	Liquid
Products				G, molecule/100 eV					
Methane	0.051	0.026	0.014	2.09	0.090	0.015	0.054	0.041	0.076
Ethane	0.92	0.0021	0.070	0.02	0.002	0.026	0.002	0.101	0.007
Ethylene	0.40	0.0071	0.092	0.13	0.002	0.057	0.003	0.117	0.004
Acetylene	0.56	0.0057	0.007	0.21	0.004	0.005	0.007	0.008	0.004
Propane + propene			0.019	0.22	0.021	0.146	0.303	0.020	0.002
Butane + butenes						0.018	0.031	0.072	0.115
Benzene	0.38	0.016		0.3		0.018	0.013	0.037	0.110
Toluene	0.34	0.018		0.1	0.05	0.019	0.028		0.012
Ethylbenzene	0.02	<0.002				0.023			
o-Xylene	0.01							0.010	
m-Xylene									
p-Xylene									
Methylcyclohexadiene									
Dimethylcyclohexenes									
Methylcyclohexenes									
Benzocyclobutene									
o-,m-,p-Ethyltoluene	0.147								
Trimethylbenzenes	0.144								
Propylbenzene									
Radical	8.7	2.82 [102]		5					
G (polymer)		1.60 [250]			1.8				

tachment of the methyl group in the α-position. More than 90% of ethane and ethylene stems from the abstraction reactions of ethyl groups produced via detachment of the ethyl group.

Hydrogenated products, such as methylcyclohexadiene and methyl-cyclohexene are also formed in the radiolysis of liquid toluene [139]. No such products have been observed in the gas phase, which indicates the low probability of H-atom addition to the phenyl ring.

In the radiolysis of liquid toluene and isopropylbenzene, only benzene of the possible dealkylated products was detected [132, 139]. The value of G(benzene) is considerably lower than the yield of methane, in both cases. In the radiolysis of xylene and ethylbenzene, toluene and benzene have been found, respectively; the yields of these products are equal to the scavenge-able methane yields [136]. With increasing temperature the yields also increase. The G-value of dealkylated products is 10–20 times higher in the gas than in the liquid phase [3, 145].

According to the product distribution, in liquid-phase radiolysis H-atom substitution by methyl groups occurs both in the ring and the side-chain. The extent of alkylation is not influenced by the position of the H-atom; the reaction seems to proceed entirely according to statistics. In the gas phase, the yield of alkylated products decreases in the order *metha* > *ortho* > *para*. In the case of ethylbenzene, alkylation of the ring and that of the side-chain take place with equal probability. Hydrogen atoms in the α-position are more likely to be substituted than those in the β-position.

In the liquid-phase radiolysis of p- and m-xylene, no ethylbenzene or isomeric xylene products has been detected; however, in the radiolysis of o-xylene the formation of m-xylene was observed, with a rather low yield, $G = 0.012$. These results show that the radiation-induced isomerization of liquid alkylbenzenes occurs with very low efficiency, taking into account detectability. On the other hand, in the gas phase isomerization is an impor-tant process ($G = 0.3$).

The formation of dimers in the gas phase has been studied most exten-sively for toluene and ethylbenzene radiolysis (Table 6.6). 3-Methyldiphenyl-methane and 3-ethyldiphenylmethane are produced in the greatest amounts; the yield of bibenzyl is considerably lower. Radical scavengers do not affect the yields of diphenylmethanes, whereas they inhibit the formation of bibenzyl [146–148].

Only liquid toluene has been studied with respect to the distribution of dimeric products. The total yield of dimers is about 0.2 [139]. In contrast to vapour-phase radiolysis, there are partially hydrogenated ring compounds in this product fraction, and the yield of bibenzyl is highest, whereas those of various dimethylbiphenyl isomers and of benzylmethylcyclohexadienes are considerably lower. On addition of radical acceptors, the yield of biben-zyl is reduced to 5% of the initial value, although the yields of dimethyl-biphenyls are much less diminished, and the G-values of methyldiphenyl-methanes remain unchanged. An increase in the LET promotes polymer formation, as well as hydrogen production. The value of G(polymer) is much lower in the liquid than in the vapour state, as noted already with respect to the radiolysis of benzene (Table 6.5).

Table 6.6

Yields of dimers from toluene and ethylbenzene (radiation: ^{60}Co-γ)

Hydrocarbons	Toluene			Ethylbenzene	
Dose rate, eV g^{-1} s^{-1}	1.18 · 10^{16}		4.3 · 10^{16}	3.1 · 10^{15}	8.7 · 10^{15}
Dose, 10^{19} eV g^{-1}	1.9		311		
Temperature, °C	25		39	25	25
Pressure, mbar	14.1			6.2	
Reference	145	150	139	146	250
Phase	Gas	Liquid		Gas	Liquid
Products	G, molecule/100 eV				
Benzylmethylcyclo-hexadiene		0.018	0.024		
Methylbiphenyl		0.0002	0.005		
Dimethylbiphenyls		0.034	0.069		
3-Methyldiphenyl-methane	0.34	0.016	0.0097		
2-Methyldiphenyl-methane	0.05	0.0075	0.0102		
4-Methyldiphenyl-methane	0.05				
Bibenzyl	0.03	0.041	0.0715	0.05	
2-Ethyldiphenyl-methane				0.08	
3-Ethyldiphenyl-methane				0.86	
4-Ethyldiphenyl-methane				0.12	
1,3-Diphenylbutane					0.0002
1,4-Diphenylbutene					0.0001
2,3-Diphenylbutane					0.0046

In liquid-phase radiolysis, scavengeable dimers [133, 139] can be formed in abstraction, recombination and disproportionation reactions of radicals produced in the decomposition of excited toluene molecules, e.g. bibenzyl:

$$C_6H_5-CH_3^* \longrightarrow C_6H_5-\dot{C}H_2 + H$$

$$2\,C_6H_5-\dot{C}H_2 \longrightarrow (C_6H_5-CH_2)_2$$

and dimethylbiphenyl:

$$H + C_6H_5-CH_3 \longrightarrow \dot{C}_6H_6-CH_3$$

$$C_6H_5-\dot{C}H_2 + \dot{C}_6H_6-CH_3 \longrightarrow CH_3-C_6H_4-C_6H_4-CH_3 + H_2$$

It has been established that on irradiation of toluene-d_3, the G-values of dimers produced via C—D bond rupture in the side-chain (e.g. bibenzyl, methyldiphenylmethane, benzylmethylcyclohexadiene) are lower, whereas the yields of dimers formed through ring C—H bond dissociation (e. g. dimethylbiphenyl) are higher than in the radiolysis of the protiated compound. Thus, ultimately, the total dimer yield remains unchanged. Hoigné and Gäumann [149] concluded that the products showing decreased yields arise from reactions of benzyl radicals. There is a contradiction, however, in this reasoning, because the formation of methyldiphenylmethane cannot be scavenged.

Weiss and Collins [140] interpreted their results obtained with deuterated toluene by assuming that the products not scavengeable by radical acceptors result from the direct bimolecular reaction of excited toluene molecules:

$$C_6H_5-CH_3^* + C_6H_5-CH_3^* \longrightarrow H_2 + CH_3-C_6H_4-CH_2-C_6H_5$$

It can be assumed that the isotopic composition affect the bimolecular dissociation process in such a way that less methyldiphenylmethane and more dimethyldiphenyl will be produced.

On the other hand, Hoigné and Gäumann [150] assigned these products to the intraspur combination of tolyl and benzyl radicals:

$$\dot{C}_6H_4-CH_3 + C_6H_5-\dot{C}H_2 \longrightarrow CH_3-C_6H_4-CH_2-C_6H_5$$

This mechanism also explains the dependence of G-values on the LET value.

Products with molecular masses higher than dimers can be formed by the combination of two dimers or in multistep addition reactions. These polymers consist mainly of cyclic and unsaturated units, but not aromatic ones.

In the vapour-phase radiolysis of alkylbenzenes, ionic processes can contribute markedly to product formation, as in the radiolysis of benzene. In order to discover some details of the ionic mechanisms of product formation, ion-molecule reactions taking place in toluene, xylene isomers [5, 109, 151, 152] and ethylbenzene [4, 152] have been investigated in electron- and proton-beam mass spectrometers at normal ($\sim$1 μbar) and high (0.04–1.3 mbar) pressures. In the low-pressure mass spectra the intensity of the $C_7H_7^+$ primary fragment ion was found to be the greatest, apart from the corresponding parent molecular ions [4, 5, 109, 151, 152]. The mass spectrometry of toluene-d_3 [153, 154], gave some evidence that the $C_7H_7^+$ species should be regarded as a tropylium ion, in which the positive charge is distributed equally among the seven carbon atoms.

With increasing pressure, the ions produced in ion-molecule reactions become more abundant, especially the $C_{14}H_{15}^+$, $C_{14}H_{16}^+$, $C_{15}H_{17}^+$, $C_{16}H_{20}^+$, $C_{16}H_{19}^+$ and $C_{24}H_{30}^+$ species.

Time-resolved studies utilizing ion-cyclotron resonance spectroscopy demonstrated that $C_7H_7^+$ ions, generated photochemically from toluene radical cation [154A] or directly from toluene by electron impact [154B], consist of two populations distinguishable by their reactivity toward toluene. From results of isotopic labelling and photodissociation experiments it was surmised that the reactive $C_7H_7^+$ species has the benzyl structure and the nonreactive one has the tropylium structure [154C, 154D]. In studies of the effect of ionizing energy on the relative abundance of unreactive $C_7H_7^+$ ion in toluene, Jackson et al. found that the unreactive component comprises 30–60% of the $C_7H_7^+$ and this decreases with increasing energy of the ionizing electrons, indicating that the rearrangement to the seven-membered ring occurs prior to the fragmentation of the parent ion [154B]. Tandem ion-cyclotron resonance study of the reactions of allyl ions with benzene and alkylbenzenes supported the hypothesis that both benzyl and tropylium ions are formed [154E].

In the ESR spectra of toluene, p-xylene and mesitylene adsorbed on silica irradiated at 77 K, the corresponding monomeric and dimeric cations could be detected [87, 155].

As an alternative to the reactions of excited molecules, the following ionic mechanism has been suggested for the formation of the non-scavengeable yield of acetylene in the vapour phase, e.g., for toluene:

$$C_7H_7^+ \longrightarrow C_5H_5^+ + C_2H_2$$

$$C_7H_7^+ + C_7H_8 \longrightarrow C_{12}H_{13}^+ + C_2H_2$$

$$C_{14}H_{15}^+ + C_7H_8 \longrightarrow C_{19}H_{21}^+ + C_2H_2$$

According to Wexler and Pobo [4], in the vapour-phase radiolysis of ethylbenzene the light hydrocarbon products can be formed in the following ion–molecule reactions:

$$C_6H_5-C_2H_5^+ + C_6H_5-C_2H_5 \longrightarrow C_{15}H_{16}^+ + CH_4$$

$$\longrightarrow C_{15}H_{17}^+ + CH_3$$

$$CH_3 + CH_3 \longrightarrow C_2H_6$$

$$C_{15}H_{17}^+ + C_6H_5C_2H_5 \longrightarrow C_{17}H_{21}^+ + C_6H_6$$

$$C_{14}H_{15}-C_2H_5^+ + C_6H_5-C_2H_5 \longrightarrow C_{22}H_{26}^+ + C_2H_4$$

$$C_6H_5-C_2H_4^+ + C_6H_5-C_2H_5 \longrightarrow C_7H_7^+ + C_6H_5-C_3H_7$$

$$\longrightarrow C_7H_7^+ + C_6H_5-CH_3 + C_2H_4$$

Yamamoto et al. [146, 148] investigated the γ-radiolysis of toluene, ethylbenzene and p-xylene vapours at 25°C. They identified the corresponding m-benzylated alkylbenzene (3-alkyldiphenylmethane) as having the highest yield among dimeric products. Considering, that $C_7H_7^+$ is a common reactive species formed in considerable yield by the irradiation of these alkylbenzenes and that aromatic hydrocarbons readily undergo electrophilic substitution, an electrophilic substitution reaction involving $C_7H_7^+$ ions was suggested, e.g. for toluene:

$$C_6H_5-CH_3 \xrightarrow{\quad\sim\sim\sim\quad} C_6H_5-CH_3^{+*} + e^-$$

$$C_6H_5-CH_3^{+*} \longrightarrow H + C_7H_7^+$$

$$C_7H_7^+ + C_6H_5-CH_3 \longrightarrow \left[\begin{array}{c} CH_3 \\ \\ \bigoplus \\ \\ H \quad CH_2-C_6H_5 \end{array} \right]^* \equiv I^*$$

$$I^* + C_6H_5-CH_3 \nearrow \begin{array}{l} MDPM + C_7H_9^+ \text{ in small quantity} \\ \\ I + C_6H_5-CH_3 \text{ in greater quantity} \end{array}$$

$$I \rightarrow \begin{array}{c} CH_3 \\ \bigoplus \\ CH_2-C_6H_5 \\ H \quad H \end{array} \xrightarrow{C_6H_5-CH_3} 3\,MDPM + C_7H_9^+$$

where MDPM stands for methyldiphenylmethane; I* for a highly excited σ-complex ion existing as a free ion with excess vibrational energy and producing three isomers of MDPM by proton transfer; I denotes a less excited σ-complex ion. The latter species transforms into the most stable m-isomer complex ion [145]. The reactions of the intermediate complex formed during the electrophilic aromatic substitution — similar to I* — were studied in the gas-phase radiolysis of propane and toluene–toluene-d_8 (1 : 1) [155A].

The above mechanism was justified by the results obtained in noble gas-sensitized radiolysis of toluene vapour [156], according to which the precursor of MDPM cannot be an excited molecule or a molecule ion, but a fragment ion.

The relatively large yield of benzylated products, and the mechanism assumed, raise the question of the structure of the $C_7H_7^+$ ion: is it a benzyl or a tropylium ion? To answer this question, the methylene to phenyl proton ratio was determined, by NMR spectroscopy, from the products of the γ-radiolysis of toluene-d_3, and the yields of methyldiphenylmethane-d_5 and methyldiphenylmethane-d_6 were determined by mass spectrometry. From the results Yamamoto et al. concluded that partial isotopic scrambling takes place in the ionic precursor of MDPM [147]:

The benzyl ion produced in reactions (6.11) and (6.12) react with toluene to produce MDPM before the isomerization to tropylium ion occurs. The tropylium ions formed via the symmetrical intermediate (reaction 6.13) also produced MDPM's in which the deuterium atoms were statistically distributed in the benzyl group, via the isomerization of the tropylium structure to a benzyl one.

The ratios mentioned above indicated that in the gas phase radiolysis of toluene about 33% of methyldiphenylmethane was produced via a symmetric intermediate, i.e. via reaction (6.13). The results of similar experiments on ethylbenzene-d_2 showed that only 15% of ethyldiphenylmethane was formed via the tropylium intermediate [157].

Concerning the $C_7H_7^+$ structure, the observed discrepancy between gas-phase radiolysis and mass spectrometry may be explained in terms of the internal energy of the molecular ion. Molecular ions having certain internal energies play an important part in the gas-phase radiolysis and produce benzyl ions before ring expansion, and the produced benzyl ions react with toluene molecules to produce MDPM's [147].

Weiss [158] attributed the different values of yields of H_2, CH_4, N_2 and NO, obtained in the radiolysis of benzene and toluene substituted with nitro and alkyl groups in different positions, to the change in the relative electron densities. The presence of a substituent modifies the electron charge distribution in the aromatic ring. The radiolytic isotope effect that appears on addition of a second substituent is, in Weiss's opinion, determined by the electron attracting or repelling nature of this latter substituent, on the one hand, and the position of substitution, on the other. Radiation stability is greatest when an electron-attracting group is attached to a carbon atom that has a relative positive charge, or when an electron-repelling group is attached to a relatively negative carbon atom. This assumption has been verified by the author with results using various isomers of dinitrotoluene, cresol and toluidine.

Of benzene derivatives with alkene side-chains, only styrene and 1-methyl-styrene have been studied, and the effort was concentrated mainly on their radiation-induced polymerization.

The radiolytic yields for gas production in styrene, irradiated with a
^{60}Co-γ-ray dose of $2.4 \cdot 10^{21}$ eV g^{-1} at 25°C, were found to be $G(H_2) = 0.030$,
$G(C_2H_2) = 0.005$, G(1-butane) ≈ 0.002 and G(but-1,3-diene) ≈ 0.002. In the
presence of benzoquinone the $G(H_2)$ decreased to half of the original value
[158A]. For styrene, when irradiated with a dose about ten times less than
above, $G(H_2) = 0.018$ and $G(C_2H_2) = 0.005$ were found, and these values
were assumed to correspond to 'molecular' products [158B].

The experiments of radiation-induced polymerization were carried out
in the liquid and solid phases, as well as in solution. The results are discussed
in detail in the monographs by Chapiro [159] and Charlesby [160].

In liquid-phase radiolysis, the rate of polymerization of styrene was found
to be directly proportional to the square root of the dose rate, in the range
of low dose rates, i.e. up to about $1.8 \cdot 10^{14}$ eV g^{-1} s^{-1} [161]. Both the molecular weight of the polymer formed and the rate of polymerization in-
crease with increasing temperature. For the apparent activation energy of
polymerization a value of 25–30 kJ mol^{-1} was calculated from the Arrhenius equation. The polymerization is inhibited by radical acceptors. All the
results discussed above indicate that the polymerization proceeds via a radical pathway. The primary yield of radicals, $G(R) = 0.69$ [162], is very
similar to that obtained for benzene, $G(R) \approx 0.77$.

It was found that the extent of polymerization increases proportionally
to the dose and that at a certain dose, the value of which depends on dose
rate and on temperature, it shows an abrupt increase. The enhanced rate
of polymerization has been attributed partly to the gel-effect, and partly
to the radiolysis of the polystyrene formed [163], as the radiation stability
of polystyrene is known to be lower than that of styrene, owing to the disappearance of the conjugated double bonds.

However, more sophisticated experiments indicate that the above results
are obtained only in the liquid phase radiolysis of inadequately dried styrene
and methylstyrene [164]. It has been observed that the rate of polymerization and its dependence on the dose rate is markedly affected by the water
content of the monomer (Fig. 6.1).

The change in the rate of polymerization upon addition of cation scavengers, as well as the ratio of reactivities of monomers, obtained from studies of styrene–1-methylstyrene copolymerization have led to the conclusion, that radiation induced polymerization of 'dry' styrene and 1-methyl-
styrene proceeds in the liquid phase mainly by a cationic pathway. Kinetic
studies have indicated, that from chain reactions initiated by radicals, by
anions and by cations, respectively, the relative rate of propagation is the
highest in the case of cationic initiation. In the presence of water, however,
ionic chain propagation is inhibited and only radical polymerization, with
a much lower reaction rate, can be observed.

In the solid state, the rate of polymerization of styrene at a temperature
slightly below the melting point is considerably higher than that observed
in liquid phase radiolysis of 'wet' styrene, but significantly lower than that
of 'dry' styrene. The presence of water does not affect the rate of solid phase
polymerization. It has been suggested that polymerization of styrene in the
solid phase proceeds via an anionic mechanism.

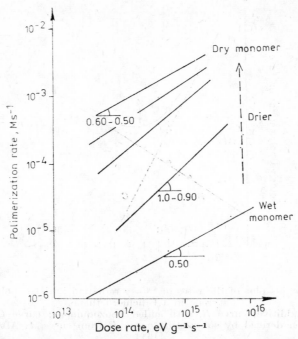

Fig. 6.1. Effect of drying on the radiation-induced polymerization of styrene. After Williams [164]

The polymerization of styrene brought about in various organic solvents has been found to take place generally via a radical mechanism. However, in the liquid-phase radiolysis of solutions of styrene in various chloroalkanes, at $-78°C$ ion-scavengers reduce the rate of polymerization, whereas radical acceptors have no effect. The rate of polymerization is directly proportional to the dose rate [165–167]. These results indicate an ionic (cationic) chain initiation. Both the temperature dependence of the rate of polymerization, and its change in the presence of radical scavengers (Fig. 6.2) indicate that in solution, at lower temperatures, cationic polymerization, which has a negative energy of activation, plays the dominant role, whereas at higher temperatures the radical mechanism becomes important. However, the degree of the ionic and radical contributions depends not only on the temperature but also on the dose rate. At low dose rates the ionic reaction pathway can be neglected, whereas at high dose rates, styrene polymerizes by an ionic mechanism, even at room temperature.

In pulse-irradiated solutions of styrene, 1-methylstyrene and 1,1'-biphenylethylene in cyclohexane at room temperature, the formation and decay of the corresponding monomer and dimer cations of the arylolefins were observed [168A].

In the mixed solvent isopentane–n-butyl chloride, the initial process of polymerization of styrene and 1-methylstyrene was studied by pulse

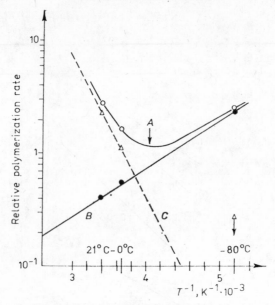

Fig. 6.2. Arrhenius plot of the rates of ^{60}Co-γ radiation initiated polymerization of styrene in 1,1-dichloroethane.
Curve *A*, no additive; curve *B*, with added benzoquinone; curve *C*, free radical contribution derived by subtracting curve *B* from curve *A*. After Chapiro
[168]

radiolysis at 108 K. The observed absorption bands were assigned to the monomeric cation radical and to, the 'associated' and 'bonded' dimeric cation radicals [168B, 168C].

6.2.3. Biphenyl

The radiolytic behaviour of biphenyl is similar to that of benzene, but its radiation stability is higher.

The fluorescence spectrum of liquid biphenyl, obtained by irradiation at 100°C with nanosecond pulses, shows the presence of at least two emitting intermediates, with half-lives of 11 and 31 ns [169]. With increasing temperature, the intensity of former at maximum of 440 nm decreases. Taking into consideration also the shape of the emission spectrum of biphenyl obtained with UV excitation Theard et al. assigned this fluorescence to the singlet excimer of biphenyl [169]. The component with longer lifetime of the fluorescence is suggested to be attributable to another excited singlet state.

The absorption spectrum shows maxima at $\sim$390 and a weaker one at $\sim$575 nm, indicating the probable existence of ionic intermediates (cation and anion) formed presumably in the following processes:

$$C_{12}H_{10} \xrightarrow{\mathlarger{\mathlarger{\sim}\!\!\!\!\sim}} C_{12}H_{10}^{+*} + e^-$$

$$C_{12}H_{10} + e^- \longrightarrow C_{12}H_{10}^-$$

In contrast with saturated compounds, the existence of molecule-anions of aromatics has been verified [262]. In biphenyl, 100 ns after nanosecond pulse, $G(\text{ion-pair}) < 0.35$. In a time interval of 20–500 ns, not more than 18% of ionic intermediates decompose. It has been assumed that 90% of ion-pairs recombine in time less than 10 ns. The neutralization of ions is expected to lead to the formation of triplet and singlet excited biphenyl molecules, in a ratio of 3 : 1 [169].

In the radiolysis of biphenyl (Table 6.7), 90% of the gaseous product consists of hydrogen and acetylene, while the remaining 10% is mainly unsaturated hydrocarbon gases. The values of $G(\text{H}_2)$ and of $G(\text{C}_2\text{H}_2)$ are independent of the dose and of the presence or absence of radical acceptors, but they are strongly dependent on the LET. With increasing temperature, the yields show moderate increases. The yield of hydrogen is drastically reduced in the solid state, to $G(\text{H}_2) = 0.0005$ [170]. All the experimental results suggest that hydrogen yield behaves very similarly to that of benzene: thus, for example, Burr et al. consider it possible that uni- and bimolecular hydrogen formation reactions occur in the radiolysis of biphenyl [171]:

$$\text{C}_{12}\text{H}_{10}^* \longrightarrow \text{H}_2 + \text{C}_{12}\text{H}_8$$

$$\text{C}_{12}\text{H}_{10}^* + \text{C}_{12}\text{H}_{10}^* \longrightarrow \text{H}_2 + \text{product or radical}$$

$$\text{C}_{12}\text{H}_{10}^* + \text{C}_{12}\text{H}_{10} \longrightarrow \text{H}_2 + \text{product}$$

If equimolar biphenyl–biphenyl-d_6 mixtures are irradiated in the liquid phase, the ratio of yields, $\text{H}_2 : \text{HD} : \text{D}_2 = 60.3 : 27.6 : 12.2$, indicates that both types of reaction contribute to hydrogen formation. Considering that the hydrogen yield is about ten times smaller in the solid phase, Scarborough and Burr suggested that 90% of the total hydrogen is formed in bimolecular reactions [170].

Table 6.7

Yields from liquid biphenyl (temperature: 82°C). After Sweeney [174]

Radiation	⁶⁰Co-γ	α(47.9 MeV)
Products	G, molecule/100 eV	
Hydrogen	0.0072	0.0188
Methane	< 0.00002	0.00001
Ethane	0.0001	
Acetylene	0.0005	0.00293
Total gas	0.008	0.022
Phenylcyclohexadienes	0.001	
Terphenyl and hydrogenated terphenyls	∼0.004	
Quaterphenyl	∼0.124	
Hydrogenated quaterphenyl	∼0.082	
Quinquephenyl and hydrogenated quinquephenyl	∼0.0005	
Hexaphenyl and hydrogenated hexaphenyl	∼0.004	
Polymers	∼0.226	0.356

The C—H bonds in biphenyl have various positions. In the radiolysis of biphenyl deuterated selectively in different positions, H_2, HD and D_2 are formed; the total amount decreases linearly with the deuterium content of the molecules, but independently of the positions carrying deuterium atoms. Burr and Scarborough [172] proposed that there are small differences between the bond dissociation energies corresponding to the *ortho-*, *meta-* and *para-*positions, but that these could have little effect on the radiolytic gas-forming processes. In addition, the results show that the probability of C—H bond rupture in partially deuterated biphenyls is about three times higher than in non-deuterated biphenyls, independently of the position of deuteration [172].

Bartonicek et al. [173] studied the variation of $G(H_2)$ with the phase and with the composition of mixtures in the radiolysis of solid and liquid biphenyl, biphenylnaphthalene and biphenyl–methylnaphtalene solutions. From the data they assumed that $G(H_2)$ comprises two kinds of hydrogen:

$$G(H_2) = G(H_2)_1 + G(H_2)_2$$

The smaller part, $G(H_2)_1$, obtained through extrapolation, is strongly dependent on the physical state, but only slightly on the temperature. This part of the hydrogen is, presumably, formed via high-energy processes like the Auger effect, bimolecular reactions of excited biphenyl molecules, and other reactions where the reactants are not in thermal equilibrium, and to a lesser extent, via H-atom abstraction of thermalized hydrogen atoms. The larger part, $G(H_2)_2$, has been suggested to originate from unimolecular decay of low-energy localized excited states of the aromatic π-electron system of biphenyl molecules. The great difference between $G(H_2)$ values in solid and liquid biphenyl has been explained, using the above mechanism, by the following argument: in the solid state, the excitation exists partly in a free and partly in a localized state (free and localized excitons); in the liquid phase, almost exclusively localized states occur (excited molecule). However, only excited molecules in localized states, can give rise to uni-molecular hydrogen formation.

Acetylene and other hydrocarbon gases are produced via ring fragmentation. In the radiolysis of biphenyl, as opposed to benzene, the ratio $G(H_2)/G(C_2H_2)$ varies with the LET value and with the phase. From this fact Sweeney et al. assumed that part of the acetylene is produced via bimolecular reactions of excited molecules in other energy states than those giving H_2 [174].

The liquid products are mainly hydrogenated compounds with rather low yields. In addition to the products listed in Table 6.7, phenylcyclohexene and various cyclohexylcyclohexene isomers have been detected [174]. These products are formed, in part, by ring-scission,

$$C_{12}H_{10}^* \longrightarrow C_6H_5 - C_2H_3 - C_4H_2$$

and, in part, by the reactions of biphenyl and phenylcyclohexadienyl radicals produced in the decomposition of excited biphenyl molecules [175]:

$$C_{12}H_{10}^* \longrightarrow H + C_{12}H_9$$

$$H + C_{12}H_{10} \longrightarrow C_{12}H_{11}$$

$$2\,C_{12}H_{11} \longrightarrow C_{12}H_{10} + C_{12}H_{12}$$

$$C_{12}H_9 + C_{12}H_{10} \longrightarrow C_{24}H_{19}$$

$$C_{12}H_{11} + C_{24}H_{19} \longrightarrow C_{24}H_{18} + C_{12}H_{12}$$

The major part of the polymer products of biphenyl radiolysis consists of polyphenyls and hydrogenated compounds, both containing an even number of phenyl rings (Table 6.7). These results indicate that the probability of rupture of C—C bonds connecting rings is rather low, and also that biphenyl and phenylcyclohexadienyl radicals play an important role in polymer formation.

G(polymer) increases with increasing temperature and LET. On irradiation of biphenyl with electrons at 150 K radicals are produced with a G-value of 0.045, whereas at 300 K the G-value observed is only 0.015, as determined by ESR spectroscopy [94, 176].

The following radical reactions were supposed to be more important:

$$2\,C_{12}H_{11} \longrightarrow C_{24}H_{22}$$

$$C_{12}H_{10} + C_{12}H_9 \longrightarrow C_{24}H_{19}$$

$$2\,C_{24}H_{19} \longrightarrow C_{24}H_{18} + C_{24}H_{20}$$

$$2\,C_{24}H_{19} \longrightarrow \text{polymer with high molecular mass}$$

$$C_{12}H_{11} + C_{12}H_{10} \longrightarrow C_{24}H_{21}$$

$$C_{24}H_{21} + C_{12}H_{11} \longrightarrow \text{polymer with high molecular mass.}$$

Direct interaction between excited molecules may also contribute to polymer formation:

$$C_{12}H_{10}^* + C_{12}H_{10}^* \longrightarrow H_2 + C_{24}H_{18}$$

$$C_{12}H_{10}^* + C_{12}H_{10}^* \longrightarrow \text{polymer}$$

The observed LET effect in the G(polymer) can be explained with these latter reactions.

Hutchison et al. [177] proposed a model to establish the relative reactivities of the three different positions on biphenyl in the generation of reactive species by ionizing radiation, and the relative reactivities of the same positions to substitution by the reactive species. The method depends on quantitative analysis of the six isomeric quaterphenyl products. Relative yields of these products from irradiation of biphenyl by electrons at 80 and 300°C have been determined and used to check the model. The results indicated that the net production of reactive species is not random; most reactive species are free radicals, but some of the species formed are more para- and less ortho-selective than free radicals. Various reactive intermediates, mainly radicals, are formed with about equal probabilities at 80°C.

At 300°C however, the formation probability of various isomeric radicals decreases in the order *meta* > *para* > *ortho*.

It thus seems probable that the majority of polymers are formed in radical reactions, accompanied by the cleavage of double bonds and hydrogenation but not by production of gaseous compounds.

6.2.4. Alkylbiphenyls and diphenylalkanes

Alkylbiphenyls and diphenylalkanes have been studied mainly because of their use in nuclear reactor techniques, and from the point of view of intramolecular energy transfer [178–180A]. Only the yields of stable products have been published.

In the liquid phase radiolysis of alkylbiphenyls [178], hydrogen is the most abundant gaseous product. The composition of the gas fraction changes with dose: the amount of H_2 and CH_4 increases, whereas that of C_2-C_4 hydrocarbons decreases with increasing dose. The latter compounds, presumably, decompose on irradiation. The total yield of gases is somewhat higher in the radiolysis of n-propylbiphenyl than in that of the corresponding branched compound, i.e. isopropylbiphenyl (Table 6.8).

Table 6.8

Yields from liquid alkylbiphenyls (mixed neutron-γ; temperature: 70°C). After Roder et al. [178]

Hydrocarbons	Isopropyl-biphenyl	n-Propyl-biphenyl	Amyl-biphenyl	Octyl-biphenyl	Dodecyl-biphenyl
Products	G, molecule/100 eV				
Hydrogen	0.094	0.107	0.21	0.7	1.97
Methane	0.0077	0.003	0.012	0.049	0.15
C_2H_n–C_4H_m	0.0087	0.0195	0.0376	0.08	0.1
Polymers	1.43	1.43	2.03	2.78	3.36

The yield of polymers decreases with increasing dose. The mean molecular mass of the polymer fraction indicates that it consists mainly of dimers. With increasing dose the average molecular mass remains unchanged.

The main gaseous product of the γ-radiolysis of diphenylmethane [180] is hydrogen; its yield is about ten times that of acetylene.

When reactor irradiation is applied, the total yield of gaseous products is about twice that obtained with γ-radiation (Table 6.9). The yields of condensed phase products of radiolysis are: G(cyclohexylbenzene) = 0.06, G(biphenyl) = 0.5, G(tetraphenylethane) = 0.04 [180] (Table 6.9).

The polymer fraction consists mainly of dimers. Benzyl, biphenyl, fluorene and terphenyl isomers could not be detected by gas-chromatography. The yield of polymers changes only slightly with LET, but it varies with dose.

Table 6.9

Yields from liquid diphenylalkanes and diphenylalkenes. After Kules et al.
[179, 180, 241]

Radiation	^{60}Co-γ					Mixed neutron-γ			
Hydrocarbons	G(gas)	G(H$_2$)	G(C$_2$H$_2$)	G(radi-cals)^a	G(poly-mer)	G(gas)	G(H$_2$)	G(C$_2$H$_2$)	G(poly-mer)
Diphenylmethane	0.047	0.030	0.004	0.06	0.504	0.097	0.035	0.036	0.896
Diphenylethane		0.056		0.1					
Diphenylethylene		0.019							
Diphenylacetylene		0.011							
Diphenylpropane	0.083	0.078		0.38	0.908	0.122	0.115		
Diphenylisopropyl-ene		0.065							
Diphenylpentane		0.118		0.60					
Diphenylheptane		0.164		0.64					
Diphenylnonane		0.241		0.63					

^{a} Temperature: 77 K

The yields of hydrogen and of radicals produced in radiolysis of diphenyl-alkanes and diphenylalkenes containing alkyl and alkenyl chains, increase with increasing carbon atom number of the non-cyclic part of the molecule: the radiation stability decreases in the same order. From a comparison of hydrogen yields of diphenylethane, diphenylethylene and diphenylacetylene it appears that with increasing number of π-electrons the rate of hydrogen production is considerably reduced. It can be assumed that the dissociability of excited molecules decreases with increasing number of non-localized electrons [179]. The mechanism of radiolysis of diphenylalkanes has been interpreted in a similar way to that of biphenyl [180].

6.2.5. Terphenyl isomers and partially hydrogenated terphenyls

As with biphenyl derivatives, the radiolysis of terphenyl isomers has been studied from the point of view of reactor techniques: thus, results obtained mainly on application of electron and of mixed reactor radiation, at high temperatures, have been published.

The radiation stability of terphenyls is even higher than that of biphenyl (Table 6.10). The products consist mainly of hydrogen and polymers formed in condensation reactions. Small amounts of hydrocarbon gases (methane, ethane, ethylene) have also been detected. The very low yields indicate the low probability of ring-scission. The *p*- and *m*-isomers of terphenyl give smaller hydrogen yields than the *o*-isomer. On reactor irradiation, the yields of polymers decrease in the order *ortho* > *meta* > *para* [97]. Insoluble tar products account for 50% of the polymers [1, 181]. The temperature dependence of yields has been studied using γ-irradiation and fast neutrons [261]. In the radiolysis of *o*-terphenyl, the yield of decomposition

Table 6.10

Yields from liquid biphenyl and terphenyl isomers (temperature: 300°C).
After Burns [97]

Hydrocarbons	1 MeV electron	Mixed neutron-γ	1 MeV electron	Mixed neutron-γ
	G(gas)		G(polymer)	
	molecule/100 eV			
Biphenyl	0.021	0.112	0.36	0.82
o-Terphenyl	0.023	0.080	0.18	0.36
m-Terphenyl	0.015	0.069	0.21	0.58
p-Terphenyl	0.015	0.062	0.18	0.47

was found to be $G = 0.2$ at 100 °C, and this value increased to 1.4 at 400 °C.
This striking increase was attributed by Boyd et al. [182] to the decomposition of thermally unstable intermediates, but, by Scarborough and Ingalls
[183] to 'thermal spikes' (see Section 6.1).

The yields of both gaseous and polymer products increase with LET:
this phenomenon was found to be of special importance above LET = 20
eV nm^{-1}. However, in the high temperature (above 380°C) radiolysis of o-terphenyl, Boyd et al. did not observe any dependence of yields on LET [182].
This can be attributed to the increased efficiency of γ-irradiation at higher
temperatures. Another explanation is offered for LET effects in terms
of decomposition and quenching of highly excited states [261].

The radiation stability of terphenyl isomers hydrogenated to different
extents has also been studied in the interests of reactor techniques [184,
185]. Just like terphenyls, they give mainly hydrogen (90% of the total
yield of gases) and polymers that are chiefly dimers.

In the radiolysis of partially hydrogenated terphenyls, the yields of gaseous
and polymer products are higher when high-ionization-density reactor
irradiation is applied, as compared with the case of ^{60}Co-γ-rays [251]. This
behaviour agrees well with that of other aromatics, however, the dependence
of the yields on the LET is considerably less pronounced than that of pure
aromatic compounds (Table 6.11).

Table 6.11

Yields from liquid hydrogenated terphenyl isomers (temperature: 240°C). After Kiss
et al. [184, 251]

Hydrocarbons	^{60}Co-γ			Mixed neutron-γ		
	G(gas)	G(H$_2$)	G(polymer)	G(gas)	G(H$_2$)	G(polymer)
	molecule/100 eV					
1,2-Dicyclohexylbenzene	0.50	0.462	0.46	0.53	0.46	0.55
1,3-Dicyclohexylbenzene	0.43	0.407	0.39	0.47	0.402	0.50
1,4-Dicyclohexylbenzene	0.56	0.533	0.22	0.68	0.605	0.28
1,4-Diphenylcyclohexane	0.155	0.134		0.190	0.138	0.14
p-Cyclohexylbiphenyl	0.140	0.11		0.179	0.133	0.13

The extent of saturation greatly affects the radiation stability. The values of G(gas) and of G(polymer) are influenced by isomerism. At equal degrees of saturation, the G-values decrease in the order *ortho* > *meta* > *para*.

6.2.6. Hydrocarbons with condensed rings

Of aromatic hydrocarbons with condensed rings, only naphthalene and its derivatives have been thoroughly studied. The radiation stability of condensed-ring aromatic hydrocarbons is similar to that of polyphenyls. Upon irradiation, small amounts of gaseous products, mainly H_2, and polymers are formed (Table 6.12).

Table 6.12

Yields from liquid aromatic hydrocarbons with condensed rings (radiation: mixed neutron-γ; temperature: 345°C). After Proksch [240]

Hydrocarbons	G(gas)	G(-M)	$\bar{M}$ (polymers)[a]
	molecule/100 eV		
Naphthalene	0.086	0.80	2.87
1-Methylnapthalene	0.153	0.43	2.38
2-Methylnaphthalene	0.111	0.45	2.34
Acenaphthene	0.111	0.37	2.21
Fluorene	0.092	0.24	2.41
Anthracene	0.025	0.50	2.58
Phenanthrene	0.034	0.42	1.80
2-Methylphenanthrene	0.068	0.33	2.22
3-Methylphenanthrene	0.060	0.41	1.77
Fluoranthrene	0.023	0.42	
Pyrene	0.027	0.38	
Crysene	0.020	0.16	2.26
2,2-Binaphthyl	0.031	0.28	2.24
Coronene	0.029	0.081	

[a] Mean molecular mass of polymers formed relative to the molecular mass of the initial compounds

By investigating the γ-radiolysis of $0.1-15 \cdot 10^{-4}$ M solutions of 4-nitro-4-methoxystilbene in naphthalene, Malan et al. [186] established that, similarly to benzene and alkylbenzenes, the isomerization of stilbene takes place in naphthalene solutions. A reaction scheme involving triplet-triplet energy transfer from naphthalene to the solute satisfies the experimental results. The efficiency of this transfer is higher in the case of naphthalene than in that of benzene.

On the basis of results obtained in the liquid-phase nanosecond pulse radiolysis of 1–20 mM solutions of benzanthracene in naphthalene, Holroyd et al. [187] concluded that the singlet and triplet excited states detected were produced as a result of singlet-singlet and triplet–triplet energy transfer from naphthalene. The triplet energy transfer was shown to be diffusion controlled, but the rate of singlet transfer turned out to be an order of magnitude higher. This, presumably, can be explained by energy transfer taking place via singlet excimers formed from naphthalene. On the other hand, the probability of formation of triplet excimers is very low [188].

The fluorescence spectrum of naphthalene obtained from liquid-phase nanosecond pulse radiolysis [189] showed, similarly to that of photoexcited naphthalene, the formation of excited singlet monomer ($\sim$365 nm) and excimer ($\sim$500 nm). The intensity of emission corresponding to the excimer decreased with increasing temperature.

The time-dependence of the absorption spectrum indicated the formation of two kinds of species. The absorption with half-lives of 36 and 24 ns, at 100 and 175°C, respectively, in agreement with those found in emission, has been assigned to naphthalene singlet. The absorption with a maximum at 420 nm has longer half-lives, namely 1.2 and 0.7 μs at 100 and 175°C, respectively; this can be attributed to the excited monomer triplet.

The total yield of triplets was determined from the 420 nm absorption as $G = 4$. On the basis of the radiolysis of naphthalene containing benzophenone, which quenches naphthalene singlets to produce triplets and hence apparently catalyses intersystem crossing (see Section 6.1.), Holroyd and Capellos [189] determined the total yield of excited molecules of naphthalene as $G(^3\text{naphthalene}) + G(^1\text{naphthalene}) = 6.2$. From the change of triplet yield in the presence of CCl_4, which quenches naphthalene singlets but does not produce naphthalene triplets, they obtained a value of 0.06 for the efficiency of intersystem crossing at 85°C. Thus $G(^3N) = 3.8$ and $G(^1N) = 2.4$ were the estimated values for the yields of first triplet and singlet states of naphthalene, respectively. Taking into consideration kinetic formulations of the depopulation possibilities of singlet states (fluorescence, intersystem crossing, internal conversion), the experimental results lead to an efficiency of 82% for the internal conversion of naphthalene singlets at 85°C. The probability of internal conversion decreases with decreasing temperature, whereas that of intersystem crossing increases simultaneously. This has been explained by the participation of naphthalene excimers. At higher temperatures, the motions of the molecule can lead to deformation in the structure of the excimer, bringing about a better overlap between vibrational wavefunctions of the two electronic states and, ultimately, an increase in the probability of internal conversion.

Ion recombination can have an important role in the formation of naphthalene excited states. The vapour-phase W-value (see Section 1.1) suggests a total ion-pair yield of 4.3, similar to that of benzene. Assuming that on neutralization of each ion-pair one excited state is formed, 70% of the excited molecules are of ionic origin. In addition to negative ions, N^-, both positive naphthalene molecular ions, N^+, and dimer cations, N_2^+, can be produced:

$$C_{10}H_8 \xrightarrow{} C_{10}H_8^+ + e^-$$

$$C_{10}H_8 + e^- \longrightarrow C_{10}H_8^-$$

$$C_{10}H_8 + C_{10}H_8^+ \longrightarrow (C_{10}H_8)_2^+$$

In the ESR and absorption spectra obtained from γ-radiolysis of solutions of naphthalene in glassy hydrocarbons at 77 K, Badger et al. [253] as well as Brocklehurst and Russel [190] observed $C_{10}H_8^+$, $C_{10}H_8^-$ and $(C_{10}H_8)_2^+$ ions. The ratio of triplet to singlet excited states formed in neutralization of molecule-ions and of dimer cations is about the same, thus suggesting a triplet yield of $G(^3N) \approx 3.5$, originating in ion neutralization. According to radioluminescence studies, excimers are formed exclusively via neutralization of dimer cations [190]. As has been already discussed, direct excitation due to low-energy electrons can also give triplet states; however, this process occurs with low probability in naphthalene.

Singlet states can also form on neutralization, through only a part of the observed yield, $G(^1N) = 2.4$, can be due to this ionic process. From the yield of naphthalene singlets, estimated via optical approximation taking into account the corresponding oscillator strengths, it can be supposed that a considerable proportion of naphthalene singlets can be formed via superexcited states not suffering autoionization. However, with regard to the radiolysis of naphthalene in condensed phases, only rough estimates can be made of the amount of excited states having ionic or non-ionic origin.

On analyzing the naphthalene singlet emission brought about by pulse radiolysis of naphthalene vapour in the presence of argon, Sauer and Mulac found that the ratio of singlet states formed in non-ionic and ionic processes varies between 1.0 and 2.6, depending on experimental conditions; the ratio decreases with increasing concentration of argon, and increases with increasing temperature [191].

The yield of triplet states (half-life 10 μs) of non-ionic origin, formed in naphthalene vapour, was found to be $G \approx 1.3$ from measurements made in the presence of SF_6 as ion scavenger. This species has been suggested to be formed mainly via intersystem crossing of singlets with considerable excess vibrational energy, or to a much lesser extent, via excitation due to subexcitation electrons. The determination of the yield of naphthalene singlets of non-ionic origin has been carried out in n-hexane solutions. The concentration dependence of singlet and triplet yields formed in energy transfer was determined, i.e. an indirect experimental method was used. The value obtained was $G \approx 0.3$ [192].

It is clear from the above discussion that further investigation is required of the qualitative and quantitative relations connected with the primary transients formed in naphthalene [263].

Investigation of the ESR and optical absorption spectra [193–196] of naphthalene single crystals, irradiated with X-rays and electrons at -196, -77 and $20°C$, revealed that at $-196°C$, approximately equal concentrations of 1-hydronaphthyl and naphthyl radicals are produced:

$$C_{10}H_8^* \longrightarrow C_{10}H_7 + H$$

$$C_{10}H_8 + H \longrightarrow C_{10}H_9$$

At $-77°C$, the ratio naphthyl:1-hydronaphthyl was found to be lower, and at room temperature only 1-hydronaphthyl radicals were detected. These results are presumably due to the stability of naphthyl radicals decreasing with increasing temperature, leading to the reaction of these radicals with naphthalene molecules:

$$C_{10}H_7 + C_{10}H_8 \longrightarrow (C_{10}H_7)_2 + H$$

At room temperature every naphthyl radical reacts thus. Optical investigations at -77 and $20°C$ revealed a new absorption band, which was assigned to the naphthalene dimer.

Radical pair formation was observed in irradiated crystalline naphthalene, anthracene and fluorene at 77 K [92A, 92B]. It was proposed that cyclohexadienyl- and phenyl-type radicals form the radical pairs.

The radical yield of naphthalene irradiated at 80 K was determined from spectroscopic measurements using the oscillator strength determined experimentally; the value 0.017 was found [196A].

From ENDOR studies of the room-temperature radiolysis of naphthalene single crystals, Böhme and Wolf established that two types of α-hydronaphthyl radical are produced, with different positions for H-atom addition. Although they are equivalent from the chemical point of view, their environments are different [197]. The 1-naphthyl radical is situated in the naphthalene crystal like a normal naphthalene molecule.

Andreev et al. [198] carried out photolysis and γ-radiolysis of various types of glassy solvent (boric acid, ethanol, diethyl ether, methylcyclohexane) containing 10^{-4}–10^{-2} M naphthalene, at 77 K, and analysed the ESR absorption and luminescence spectra. They showed that the α- and β-isomers of the hydronaphthyl radical are formed, the lowest excited states of which lie at 2.42 and 2.66 eV, respectively. The isomers can be interconverted when irradiated with light of suitable wavelength.

ESR studies carried out during γ-radiolysis of naphthalene absorbed on silica gel gave the spectra corresponding to naphthalene monomer and dimer cations, in addition to that of a cyclohexadienyl-type radical [199].

Stachowicz et al. [200] reported $G(H_2) = 0.22$ and $G(C_2H_n-C_4H_m) = 0.02$ for the γ-radiolysis of tetrahydronaphthalene, a partially hydrogenated derivative of naphthalene.

The fluorescence spectrum obtained from the room-temperature pulse radiolysis of 1-methylnaphthalene showed the appearance of singlet excimer only, with a maximum at 400 nm, whereas at 88°C the fluorescence corresponding to singlet monomer was also observed, in addition to a decrease of excimer emission intensity. The fluorescene life-time at 400 nm was reported to be 64 and 42 ns, at 25 and 87°C, respectively [189]. The absorption spectrum indicated the presence of two transients with longer lifetimes. The absorption band, with a maximum at 450 nm corresponds to methylnaphthalene triplets with a half-life of 3.5 μs, according to the photolysis data.

The yield of methylnaphthalene triplets was found to be $G = 4.55$ at 25°C, increasing to 6 in the presence of 3 mM of benzophenone and decreasing to 4.14 on addition of 30 mM CCl_4. Holroyd and Capellos [189] determined initial yields by the methods described above for naphthalene, and reported $G(^3\text{1-methylnaphthalene}) = 3.3$ and $G(^1\text{1-methylnaphthalene}) = 2.7$. The efficiency of intersystem crossing was shown to be 0.44 at 25°C, and 0.17 at 85°C. On the other hand, the efficiency of internal conversion increased with increasing temperature, and consequently, at 85°C, 73% of the singlet states were deactivated via internal conversion. The unusual temperature dependence has been attributed to a change in the structure of the excimer, similar to that of naphthalene discussed earlier. The other long-lived transient corresponding to the 390 nm absorption maximum can, presumably, be identified as the $C_{10}H_7$-CH_2-radical.

The emission spectra obtained on electron impact in 1-methylnaphthalene, 2-ethylnaphthalene, 1,2-dimethylnaphthalene and 1,6-dimethylnaphthalene show, in agreement with the photolysis data, the formation of various alkyl-naphthalene singlet excimer states, with lifetimes of 68.6, 25.4, 48.4 and 35.4 ns, respectively [123].

A qualitative description of the kinematics of excited states in antracene crystals bombarded by electrons is given by Klein [200A].

6.3. MIXTURES OF AROMATIC HYDROCARBONS

Most papers in this area are concerned with the radiolysis of mixtures containing aromatic hydrocarbons and alkanes. Mixtures of aromatics and alkenes and those composed only of aromatics are dealt with to a much lesser extent.

A general survey of the problems associated with the radiolysis of mixtures is given in Section 2.3.

Alkane–aromatic mixtures

The components of this type of system differ in their molecular structures, as well as excitation and ionization levels. Therefore, as in the radiolysis of alkane–alkene mixtures, the yields, generally, differ considerably from those estimated by the additivity rule.

Limited data are available on the radiolysis of mixtures of straight-chain alkanes with aromatic hydrocarbons. According to the results of Polak et al. on the radiolysis of n-hexane–benzene liquid mixtures, benzene reduced the yields of each product investigated (H_2, C_2-C_4 hydrocarbons, hexene, C_8-C_{12} hydrocarbons) in the same proportions. This indicates the existence of only one type of interaction [201]. At low benzene concentrations, where the reduction of yields was the greatest, the decrease of $G(H_2)$ was independent of the temperature. Thus physical interactions are evidently important [202]. Similar results were reported for $G(H_2)$ with respect to n-octane–benzene and n-butane–benzene systems [203, 204].

The most extensively studied cycloalkane–aromatic hydrocarbon system is cyclohexane–benzene. This has been investigated in all three phases. The papers devoted to this system amount to about 80–90% of those dealing with the radiolysis of alkane–aromatic mixtures (Fig. 6.3). In the liquid and vapour phases the values of $G(H_2)$, $G(C_6H_{10})$ and $G(C_{12}H_{10})$ show a neg-

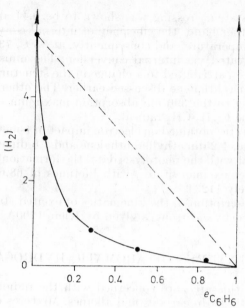

Fig. 6.3. Dependence of the hydrogen yield of cyclohexane–benzene mixtures
(radiation: 1.5 MeV electron). After Manion and Burton [205]

ative deviation from additivity over the whole concentration range [205, 206–209]. At low benzene concentrations, the yields decrease considerably, whereas at higher concentrations the decrease is relatively smaller. From theoretical considerations [205] as well as from the HD yields obtained in the radiolysis of cyclohexane–benzene-d_6 mixtures [210], Burton et al. concluded that these reductions are mainly the result of a physical interaction followed by chemical decomposition [210]:

$$C_6H_{12}^* + C_6H_6 \rightarrow C_6H_{12} + C_6H_6^* \longrightarrow \text{products of benzene}$$

where * denotes neutral or charged excited molecules. The considerable radiation stability of benzene means that the decomposition of excited benzene molecules formed by energy transfer and neutralization of benzene cations subsequent to charge transfer is relatively unimportant compared with the decomposition of cyclohexane (see Section 6.2.1).

However, one other possible reason for the reduced yields has to be considered, namely a chemical interaction, the scavenging by benzene of radicals formed in the radiolysis of cyclohexane:

$$H + C_6H_6 \longrightarrow C_6H_7 \tag{6.14}$$

$$\text{cyclo-}C_6H_{11} + C_6H_6 \longrightarrow C_6H_{11}-\dot{C}_6H_6$$

The contribution of chemical interactions is verified by the formation of certain products (e.g. phenylcyclohexane [100, 211, 212] and dicyclohexyldienyl [209]) that are not produced by irradiation of the individual pure components.

In the radiolysis of a mixture composed of benzene labelled with ^{14}C, and of cyclohexane labelled with tritium, the radioactivity of the polymer formed pointed to the scavenging by the benzene ring of T-atoms originating from cyclohexane [213]. Sauer and Mani [214] observed an absorption corresponding to cyclohexadienyl radicals formed via reaction (6.14) when they irradiated cyclohexane—benzene mixtures with electron pulses. From the dependence of the absorption intensity on the concentration of benzene, they estimated the ratio of rate constants for thermal H-atom scavenging and for H-atom abstraction, $k_{14}/k_{15} = 35$.

$$ H + C_6H_{12} \longrightarrow C_6H_{11} + H_2 \qquad (6.15) $$

The negative deviation from linear additivity in the case of biphenyl formed in the radiolysis of cyclohexane–benzene mixtures, also indicates a chemical interaction; in view of the values of the ionization potentials and excitation energies, neither charge transfer nor excitation energy transfer from benzene to cyclohexane can occur [100].

According to Burr and Goodspeed [212], the consumption of benzene and the formation of phenylcyclohexane and polymer, which indicate chemical interaction, are most pronounced in the concentration range 1–10 mole % benzene. Freeman [209], however, concluded from the concentration dependence of product formation that, at concentrations of benzene lower than 0.5 M, radical scavenging by benzene occurs, whereas at higher concentrations, physical interactions play the most important part. Stone and Dyne [211] found that, in solutions of cyclohexane containing 0.1 M benzene, the yield of benzene consumption is lower than the reduction of the yields of the decomposition products of cyclohexane. According to their calculations, only about 10% of the reduction of $G(H_2)$ can be attributed to H-atom scavenging. It follows that the chemical interaction between benzene and cyclohexane cannot be the only type of interaction, even at low concentrations. Freeman [209] proposed that the decomposition products of cyclohexane have at least two types of precursor, with different excitation levels; benzene can interact only with one of them, thus hindering its decomposition via energy transfer.

Avdonina [215, 216] determined the contribution of hot H-atoms to the formation of hydrogen and low molecular mass hydrocarbon products, on the basis of recoil tritium atom reactions carried out in cyclohexane–benzene mixtures. The hydrogen is produced, in part via abstraction by hot and thermal H-atoms, and in part via molecular elimination. Benzene has no effect on the precursor of 'hot' H-atoms, which could thus be a superexcited cyclohexane molecule, whereas it readily interact with the precursor of molecular H_2, which is thus probably an excited molecule or a molecule-ion. In solutions containing more than 30 mole % of benzene, hydrogen is formed exclusively via abstraction by 'hot' H-atoms, because, on the one hand, at 10 mole % benzene energy transfer from the precursor of molecular H_2 is already complete, and, on the other hand, the amount of H_2

formed via abstraction by thermal H-atoms is negligible at this concentration because of the high probability of H-atom addition.

These conclusions contradict those of Dyne and Jenkinson, who investigated the yield and isotopic composition of hydrogen formed in the radiolysis of cyclohexane–benzene mixture containing a few mole% of C_6D_{12}. The effect of benzene was to reduce the total yield of hydrogen, but the ratio of D_2 to HD remained unchanged [78]. Correspondingly, as D_2 can be produced exclusively in unimolecular, and HD in bimolecular reactions, it follows that benzene quenches the two processes with equal efficiency. Thus benzene presumably interacts with the common precursor of D_2 and HD, which may be of ionic origin [217]. The authors concluded, that benzene inhibits the $C_6H_{12}^+ + e^-$ neutralization reaction, which gives decomposition products, via electron capture.

By applying various light filters, Horikiri and Saigusa determined the intensity of radiation-induced luminescence emitted from each component of cyclohexane–benzene mixtures when fluorescence was excluded [218]. The luminescence intensity is proportional to the number of excited molecules. The results from this technique are not influenced by the various chemical reactions following primary processes, and it was established that the intensity of luminescence excluding fluorescence of benzene, on the one hand, and the total radical yield determined in the presence of DPPH, on the other hand, underwent the same change with composition. The variation of luminescence intensity corresponding to the individual components revealed an energy transfer from excited cyclohexane to benzene molecules. On the basis of these results, the authors considered as justified the suggestion that an energy-transfer process exists between highly excited state of two components in a mixture.

In the presence of a small amount of benzene in cyclohexane, the intensity of fluorescence from the lowest excited state of cyclohexane starts to decrease as soon as the emission from excited benzene appears [12, 23, 218A, 218B]. From nanosecond pulse radiolysis studies the rate constant of this energy transfer was estimated to be $\sim 2.10^{11} \, \text{M}^{-1} \, \text{s}^{-1}$.

The results obtained from the radiolysis of cyclohexane–benzene and methylcyclohexane–benzene mixtures, in which the molecules are differently ordered at room temperature, have been interpreted by Freeman [219] in terms of charge transfer. Toma and Hamill [220] concluded that charge transfer can also take place in the liquid phase from cyclohexane to benzene: their results also indicated the formation of benzene cations in the radiolysis of solid alkanes containing benzene.

The experimental results and theoretical considerations discussed so far indicate that the effect of benzene on the radiolysis of liquid cyclohexane is mostly physical in nature. To determine the relative importance of excitation energy transfer and charge transfer, Thomas and Mani [62] irradiated cyclohexane–benzene mixtures with nanosecond pulses in the presence of scintillators and electron acceptors. The yield of excited states of benzene thus formed was found to be considerably higher than could be obtained on direct excitation of benzene. Two possible mechanisms were proposed: first, the formation of excited cyclohexane molecules that can transfer their energy to benzene; second, the formation of benzene cations, the neutralization of which gives rise to excited benzene molecules. The yield of cyclohexane excited states produced directly by means of nanosecond pulses cannot be more than $G = 0.3$, thus the probability for the first process may be very low. On the basis of experimental results from electron scavenging, the second mechanism is considered as the more likely probable source of excited benzene molecules.

$$C_6H_{12}^+ + C_6H_6 \longrightarrow C_6H_{12} + C_6H_6^+$$

$$C_6H_6^+ + e^- \longrightarrow C_6H_6^*$$

The same conclusion was drawn by Baxendale et al. [11, 12], from the kinetic analysis of experimental results obtained using nanosecond pulses. Starting from the assumptions discussed above, and also from the formation of cyclohexadienyl radicals, albeit in low yield observed in nanosecond pulse radiolysis, Kennedy and Stone [221] explained the decrease of H_2-production as due to scavenging of cations and H-atoms. They established the cation-scavenging efficiency of benzene from the change of the yield of propane formed in the radiolysis of cyclohexane solutions containing various amounts of benzene and cyclopropane, which is a cation scavenger. They used the formula described by Warman et al. that takes into consideration the competition between two cation scavengers [19]. Given this information, they estimated the contribution of charge scavenging to the reduction of $G(H_2)$ in cyclohexane solutions containing 0–0.3 M benzene. The contribution of H-atom scavenging was calculated from the value of $G(H)$ of ionic origin and also from the probability of H-atom abstraction and addition to benzene. The estimated decrease of $G(H_2)$ was in good agreement with that observed experimentally. Thus, they verified their assumption that in cyclohexane containing low concentrations of benzene, the reduction of $G(H_2)$ could be the result of both the cation- and H-atom-scavenging properties of benzene.

Dyne and Denhartog [222] condensed vapour-phase cyclohexane–benzene mixtures and irradiated the solid mixture thus obtained at $-196°C$. They found that the 'protecting effect' of benzene did not differ from that observed in the liquid phase. Kroh and Karolczak [223] reported that in solid mixtures obtained by cooling from the liquid phase, the effect of benzene decreased at low concentrations and disappeared completely above a concentration of 40 mole% of benzene. Under these conditions the physical interaction was not effective because of the segregation of components. Dyne and Denhartog used a multilayer-technique to obtain a sandwich structure of thin layers of deuterated cyclohexane in benzene in solid phases, from which they found a value of 2–20 nm for the range of interaction [224].

In vapour-phase α-radiolysis of cyclohexane–benzene mixtures, the $G(H_2)$ values observed were lower than those estimated by simple additivity, similarly to those obtained in the liquid phase [225]. The reduction was attributed to a transfer of ionization energy from cyclohexane to benzene. Blackford and Dyne [226] studied the γ-radiolysis of vapour-phase cyclohexane–benzene mixtures containing 5 mole% C_6D_{12}. From the total yield of hy-

29

drogen and from the yields of D_2 and HD, they established that, as opposed to liquid-phase radiolysis, benzene reduces the hydrogen yield of bimolecular reactions, but does not affect unimolecular processes. To explain this, they considered two possibilities: first, that the uni- and bimolecular processes of hydrogen formation have no common precursor in the gas phase; second, that if a common precursor is formed, then its properties (e.g. its lifetime) must be entirely different from those in the liquid phase. They attributed the effect of benzene to the fact that the fragment ions (C_2-C_3) that are important in gas-phase radiolysis of cyclohexane give complex adducts with benzene, the decomposition of which does not result in hydrogen formation.

From a study of the radiolysis of cyclohexane–benzene vapor mixtures, Theard [227] established that benzene reduces the values of $G(H_2)$, $G(C_2H_6)$ and $G(C_3H_8)$ with different rates, yet does not affect at all the values of $G(C_2H_4)$, $G(C_3H_6)$ and $G(C_2H_2)$. Thus, the yields of gaseous products cannot be reduced by addition to benzene of a common precursor, the C_2-C_3 ionic fragment. The experimental results also ruled out ionization energy transfer from the cyclohexane molecular ion. The author assumed that benzene inhibits hydrogen formation via two interactions: at low benzene concentrations mainly by H-atom scavenging, and at higher concentrations by scavenging of ionic precursors of 'hot' H-atoms, in addition to the first mechanism [227].

Jones studied the effect of phase, temperature and density on the $G(H_2)$-reducing property of benzene, via the irradiation of a cyclohexane–benzene mixture containing 0.325 electron fraction of benzene [228]. The 'protective' effect increased with increasing density of the vapour-phase mixture at 285°C. However, as the density increased, the ratio of concentrations, C_6H_{12} : C_6H_6, remained unchanged, thus excluding H-atom scavenging as a possible reason for the increased interaction. In this case, energy transfer may play an impornt role. In liquid phase mixtures, $G(H_2)$ increased gradually with temperature, and at the critical temperature of cyclohexane, at the phase change, abruptly shot up (density, temperature and composition remained unchanged). This was attributed to the difference in structure of the two phases [228].

Similar interactions were observed in the radiolysis of liquid-phase methylcyclohexane–benzene, ethylcyclohexane–benzene and dicyclohexane–benzene mixtures. From the kinetic analysis of the results, these interactions were also identified as radical scavenging and energy transfer [208].

Horikiri et al. [229] investigated the radiolysis of liquid phase mixtures containing cyclohexane as one component and alkylbenzenes (such as toluene, xylene isomers and mesitylene) as the other. They determined the luminescence intensity of the individual components as a function of composition, by applying the technique described above [218] to the cyclohexane–benzene system. In the presence of alkylbenzenes the luminescence intensity of cyclohexane decreased in each case. The kinetic equation of Stern–Volmer type could not account for the variation of the emission intensities from each component. The kinetic equation derived from the assumption that energy transfer is possible only within the region of influence around excited cyclohexane gave a satisfactory explanation of the variation. From the correlation between the magnitude of this region and the ionization potentials of alkylbenzenes, it was assumed that the process is a charge transfer, wherein cyclohexane acts as electron acceptor and alkylbenzene as electron donor. The lifetime of the charge-transferring pair was estimated to be about 10^{-13} s.

Sherman [230] determined the yields of the main radiolytic products of cyclohexane and the yield of consumption of the aromatic additive in the liquid-phase radiolysis of cyclohexane solutions containing 0.1 M biphenyl, o-terphenyl, m-terphenyl, benzene or toluene. He observed that consumption of the additive had a low yield, and that the yields of all decomposition products were reduced much more by toluene, biphenyl and terphenyls than by benzene. He suggested that the protective effect of aromatics should be attributed not only to radical scavenging and energy transfer from the lowest excited state but also to energy transfer from upper excited states.

The linear correlation between the decrease of $G(H_2)$ and the electron affinity of the aromatic additive, in the irradiation of cyclohexane solutions of low concentrations (<5 mM) of phenanthrene, anthracene, stilbene and pyrene, indicates that the electron-accepting ability of the aromatics contributes to their 'protective' effect [231].

The reduced yields of cyclohexane decomposition products in the liquid-phase radiolysis of cyclohexane–hexamethylbenzene mixtures is in agreement with those observed in benzene solutions. However, at low concentrations of hexamethylbenzene, in contrast to benzene, the 'protective' effect is attributable chiefly to a positive charge transfer from cyclohexane cations to hexamethylbenzene molecules, with only a very low efficiency of excitation energy transfer [232].

In the liquid-phase radiolysis of cyclohexane–phenylcyclohexane, methylcyclohexane–1-methyl-4-phenylcyclohexane and ethylcyclohexane–2-cyclohexyl-1-phenylethane mixtures, the effect of aromatics at high concentrations is in agreement with that observed in benzene solutions, whereas the arylcycloalkanes are much less effective at low concentrations [208].

The important role of excitation energy transfer was concluded from investigation of the fluorescence of toluene in cyclopentane, cyclohexane, 2,3-dimethylbutane and 3-methylpentane solutions irradiated at 77 K [63, 232A, 232B, 232C]. The rate constants for energy migration, obtained experimentally from the kinetics, were of the same order of magnitude as those calculated by the theory of exciton migration.

Alkene–aromatic mixtures

The values of $G(H_2)$, $G(CH_4)$ and $G(C_2H_4)$ deviate from linear additivity to a much lesser extent with cyclohexene–benzene liquid mixtures than with cyclohexane–benzene systems [205, 233]. However, the very small value of $G(C_2H_2)$ is characteristic of the radiolysis of cyclohexene, and is independent of the composition of the mixture at all concentrations studied. From these results, and taking into account the ionization potentials and the energies of first excited states, Manion and Burton [205] proposed that benzene inhibits decomposition of excited cyclohexene molecules via excitation energy transfer, and also that the extent of decomposition of benzene cations is reduced by charge transfer from benzene to cyclohexene.

In contrast with liquid-phase experiments, Miyazaki and Yamamoto found that the intensity of luminescence and the $G(H_2)$-values deviated considerably from the linear additivity rule in 1-pentene–toluene and 1-hexene–toluene mixtures irradiated by γ-rays and nanosecond pulses of X-rays at 77 K [63]. Yields of excited singlet state toluene in these mixtures increase remarkably with increasing toluene concentration, while $G(H_2)$ decreases. In a study of the effect of pre-irradiation on the fluorescence of toluene and the decay curve of fluorescence intensity of toluene in mixtures, it was concluded that the luminescence was, due to the rapid excitation energy transfer from olefin, within a time period less than 10 ns. It was found that the migration distances of the energy transfer in the radiolysis of solid hydrocarbons such as aromatics, alkanes and alkenes were proportional to the lifetimes of the excited singlet states of the hydrocarbons. The authors discussed the energy transfer in terms of two models: an exciton transfer and vibrational relaxation resonance transfer. The exciton transfer offered a better explanation for the energy transfer mechanism in the radiolysis of solid hydrocarbons.

Aromatic–aromatic mixtures

Mixtures of aromatic and substituted aromatic compounds, which have similar physical and chemical properties (ionization potentials, excitation energy levels, π-electron structure, degree of saturation) show much smaller deviations from the linear additivity rule than alkane–aromatic, and alkene–aromatic systems.

In the liquid-phase radiolysis of benzene–toluene mixtures, $G(H_2)$ and $G(CH_4)$ exhibited behaviour corresponding to additivity, whereas $G(C_2H_4)$ and $G(C_2H_2)$ remained unchanged with increasing benzene concentration [205]. Photolytic studies carried out on corresponding systems indicated that no excitation energy transfer takes place between the components. From these results Manion and Burton [205] proposed that because of the different values of the ionization potentials, complete charge transfer takes place from benzene to toulene, which is why the radiolysis yields correspond to those of pure toluene. The additivity found for the $G(H_2)$ and $G(CH_4)$ values, in contrast to energy transfer, has been attributed to secondary processes (mostly radical scavenging) that are dependent on the concentrations of the components. The change of dimeric products as a function of the composition observed in the radiolysis of benzene-d_6–toluene mixtures indicated energy transfer from benzene-d_6 to toluene [234].

In the radiolysis of benzene–p-xylene mixtures, $G(H_2)$ and $G(CH_4)$ were lower than estimated by the additivity rule. This was explained by Verdin [136] as due to excitation energy transfer to the more stable benzene, because charge transfer in the same direction, or a preferred H-atom or CH_3-radical addition, can be excluded.

In the liquid-phase radiolysis of benzene–propylbenzene [136], benzene–hexamethylbenzene and toluene–hexamethylbenzene mixtures, the values of $G(H_2)$ and $G(CH_4)$ changed in proportion to the electron fractions [135].

In the radiolysis of benzene–amylbenzene systems, H_2 yields were found to be slightly lower than those predicted by the additivity rule. The shortfall is more significant in the case of benzene–nonylbenzene mixtures (Fig. 6.4). With increasing length of the side-chain, the intermolecular interaction between benzene and alkylbenzene is presumably more effective [134].

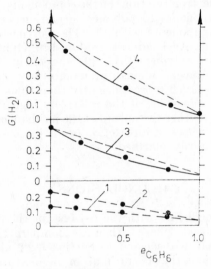

Fig. 6.4. Dependence of the hydrogen-yield of n-alkylbenzene–benzene mixtures *1*, Toluene–benzene, *2*, propylbenzene–benzene, *3*, amylbenzene–benzene, *4*, nonyl-benzene–benzene (radiation: ^{60}Co-γ). After Zeeman and Heusinger [134]

In the liquid-phase radiolysis of benzene–phenylcyclohexane, benzene–benzylcyclohexane, benzene–1-methyl-4-phenylcyclohexane and benzene–2-cyclohexyl-1-phenylethane mixtures, H_2 formation varied in proportion to the the electron fractions [208, 209]. It can be assumed that in these system arylalkanes are protected by their own phenyl rings, hence benzene molecules cannot 'protect' them more effectively.

The values of $G(H_2)$ and $G(C_2H_2)$ measured in benzene–styrene mixtures irradiated at ambient temperature were much lower than would be expect on the basis of a linear relationship [158A, 158B]. Over the whole concentration range, the acetylene yield was explained by energy transfer. The concentration dependence of hydrogen points to a more complicated interaction, i.e. besides energy transfer the radical scavenger effects cannot be ignored [158A].

In the radiolysis of benzene–styrene, toluene–styrene, ethylbenzene–styrene and xylene–styrene mixtures with a dose rate of $1.4 \cdot 10^{12}$ eV g^{-1} s^{-1}, the relative rate of conversion of styrene was shown to be proportional to the mole fraction of styrene when it was higher than 0.2. This threshold concentration increases with the dose rate. However, the relative rate of polymerization of styrene in benzene–styrene mixtures was found to increase proportionally to the mole fraction of styrene at each concentration studied [236].

According to the investigations by Proksch et al. [237] of the radiolysis of anthracene–phenanthrene and naphthalene–2-methylnaphthalene mixtures carried out with reactor radiation, the total yield of polymers changed proportionally to the electron fractions of the components. The contributions of the individual components to polymer formation were established by labelling one of the components with ^{14}C. The polymer is formed mainly via a radical mechanism. Aryl radicals produced in naphthalene–2-methylnaphthalene systems are scavenged by both components with equal probability. However, in anthracene–phenanthrene mixtures, the first component scavenges aryl radicals much more efficiently than the second one.

Thus it appears that, in spite of the relatively high number of studies performed in this field, neither the nature of the mechanism, nor the relative contribution of various types of interaction occurring in aromatic mixtures has yet been unambigously clarified.

6.4 CONCLUSIONS

In this Section, we deal first with the reasons for the radiation stability of aromatics in condensed phases, then we compare the intramolecular reactions of alkylaromatics discussed in Section 6.2.2 and 6.2.4 with the intermolecular interactions that occur in aromatic mixtures, as discussed in Section 6.3.

Radiation stability of aromatic compounds

As pointed out earlier, in the radiolysis of aromatics the π-electron structure of the rings gives rise to some special features of particular importance. As a result, a decisive role is played by excited states as the primary intermediates, the relaxation of which in many cases does not give rise to chemical changes. Thus, although the yield of primary products (ions and excited molecules) is about equal to that from saturated hydrocarbons, aromatic hydrocarbons are considerably more stable to radiation in the condensed phases. Because of considerable ring stability, the probability of C—C bond rupture is reduced, and the stable product fraction is composed mainly of H_2 and polymers.

Theories abound concerning the correlation between molecular structure and radiation stability of aromatics. According to Terenin [238], the radiation stability of aromatics is connected with the relatively low energy of their first excitation levels, and with the special behaviour of their electron orbitals, i.e. their geometry does not change much on excitation.

Voevodskii et al. [176, 239] studied radical yields in the solid-phase radiolysis (100–160 K) of different non-condensed aromatic hydrocarbons.

Starting from the experimental results, they correlated radical dissociation probabilities with the energies of the first excitation levels of the same compounds. Correspondingly the dissociation of excited molecules, to form radicals is in competition with internal conversion, which does not result in chemical decomposition. Spectroscopic data have revelaed that the conversion from upper excited states into the first level is a very fast process because of the relatively high density of energy levels. The conversion from first excitation level into the ground state is considerably slower, as the gap between these two levels is much larger. Therefore, molecular dissociation from upper excited states must be relatively unimportant, in comparison with the competing very fast internal conversion. Consequently, the yields of radicals produced from molecular dissociation are determined by the probability of dissociation occurring from the first excitation levels. As was pointed out earlier, this probability decreases with the decrease of energy of the first excitation levels consequent on the increasing aromaticity of the molecule. In solid-phase radiolysis mainly $C-H$ bonds dissociate initially, so it follows that if the energy of the first excitation level is lower than the $C-H$ bond dissociation energy, no molecular decomposition can occur from this level. This implies that in this case dissociation may occur only via upper excitation levels and with rather low probability. Table 6.13 lists the differences between energies of first excitation levels and of $C-H$ bond dissociation energy $(E_1 - D_{C-H})$ for several aromatics.

The energy difference appears mainly in the form of kinetic energy of the light H-atom (protium) produced in the dissociation. Thus with decreasing energy difference the probability of H-atom addition increases, at the expense of H-atom abstraction, the latter leading to H_2 formation.

As a consequence of the factors discussed above, in solid-phase radiolysis, where the probability of molecular H_2 formation is rather low, the ratio $G(R)/G(H_2)$ increases with increasing aromaticity.

The experimental results show that in the radiolysis of alkylaromatics, H-atoms formed via rupture of ring $C-H$ bonds add to the ring of another molecule to give $\dot{C}_6H_6-R$-type radicals. H-atoms produced from $C-H$ bond rupture in the side-chain are more reactive, and contribute with higher probability to H_2 formation via H-atom abstraction. According to the data in Table 6.13, the value of $G(R)$ generally decreases with aromaticity.

Molin et al. pointed out [239] that the radiation stability of monocyclic and isolated polycyclic aromatic compounds can be characterized by radical yields obtained in low-temperature solid phase radiolysis, where the rigid structure hinders recombination of heavy radicals. This idea assumes that in condensed phases the radiolysis of aromatic hydrocarbons takes place overwhelmingly via radical intermediates.

Proksch [240] extended the investigation of the effect of molecular structure on the radiation stability of condensed polycyclic aromatic compounds. By irradiating benzene and isolated and condensed polycyclic compounds (Table 6.14) with reactor radiation (where about 24% of the dose is due to fast neutrons), he determined the number of molecules decomposed, $G(-M)$. The values of $G(-M)$ plotted against the resonance

Table 6.13

Yields from solid aromatic hydrocarbons (radiation: 1.6 MeV electrons). Data taken from refs [94, 239]

Hydrocarbon	Temperature, K	$G(R)$	$G(H_2)$	$\dfrac{G(R)}{G(H_2)}$	$E_1\text{-}D_{O-H}$, eV
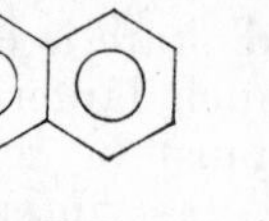 naphthalene	80	0.017^a			
 benzene	160	0.15	0.0085	18	0.3
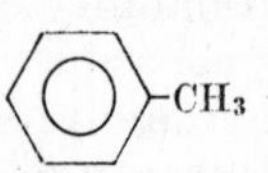 $\bigcirc$—CH₃	114	0.22	0.03	7	0.3
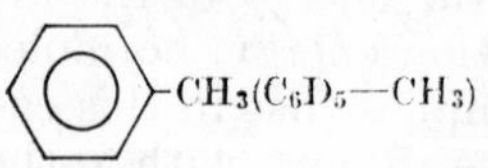 —CH₃(C₆D₅—CH₃)	137	0.11	0.014	5.5	
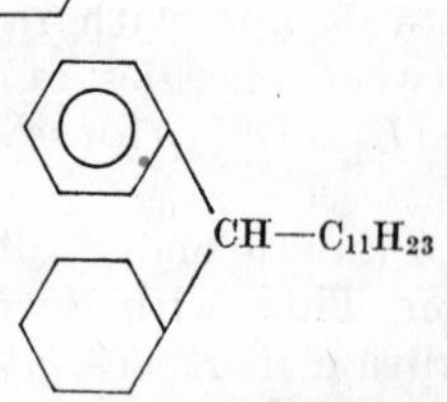 CH—C₁₁H₂₃ (phenylcyclohexyl)	157	0.48	0.5	1	
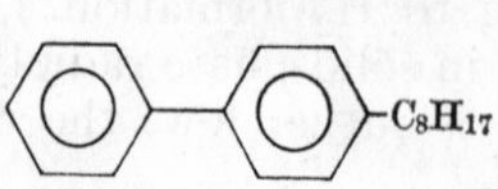 biphenyl—C₈H₁₇	156	0.37	0.23	1.6	
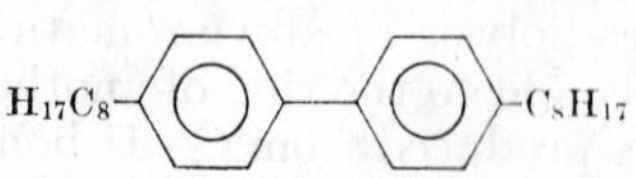 H₁₇C₈—biphenyl—C₈H₁₇	146	0.65	0.39	1.7	
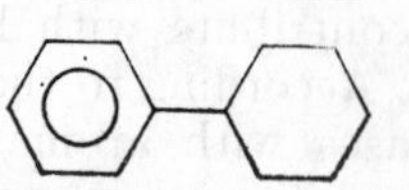 phenylcyclohexane	156	0.5			
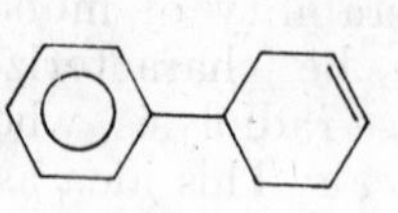 phenylcyclohexene	160	0.2			
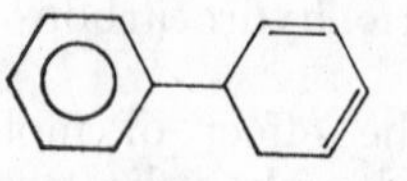 phenylcyclohexadiene	126	0.07			
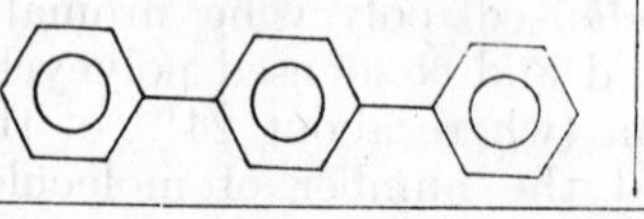 terphenyl	149	0.04—0.05			

(Cont'd on next page)

Table 6.13 (cont'd)

Hydrocarbon	Temperature, K	$G(R)$	$G(H_2)$	$\dfrac{G(R)}{G(H_2)}$	$E_1\text{-}D_{C-H}$, eV
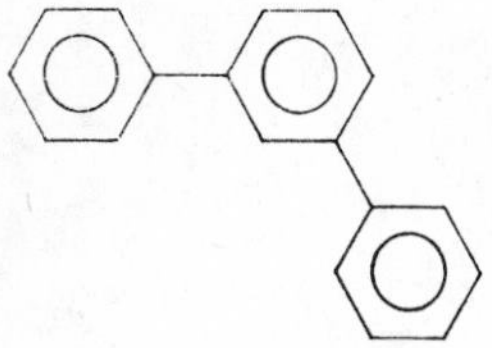 (terphenyl structure)	149	0.04–0.05			—0.2
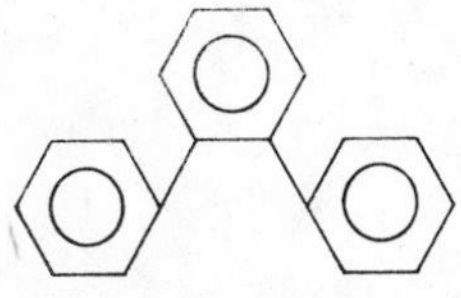 (terphenyl structure)	149	0.04—0.05			
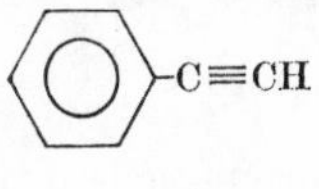 (phenyl)—C≡CH	143	0.08			0.1
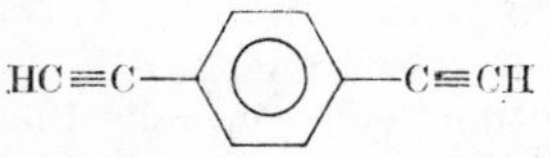 HC≡C—(phenyl)—C≡CH	143	0.05			
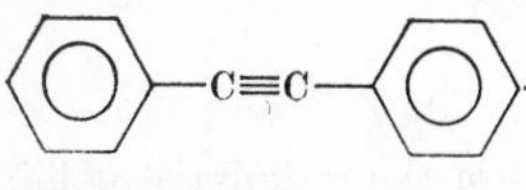 (phenyl)—C≡C—(phenyl)	143	0.04			
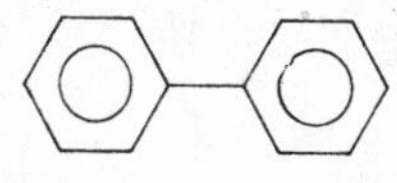 (biphenyl)	300 / 150	0.015 / 0.045	0.00052	29	—0.2

[a] Ref. [196A]

energy of the compounds (E_M) give an exponential curve (Fig. 6.5). The $G(-M)$ values corresponding to alkylaromatics are lower than those of pure aromatics. The decomposition yields, extrapolated to γ- and fast neutron radiations, corresponding to the composition of the reactor radiation, give the straight lines, 2 and 3. This implies that radiation stability depends on the resonance energy to a much larger extent in the case of γ-radiation than in the case of fast neutrons. From these results, Proksch observed a correlation between radiation stability, which is connected with the excited states of molecules and the resonance energy, which is characteristic of ground states. He explained this correlation by similar reasoning to that of Voevodskii [176] with the different rates of internal conversions involving the lowest and the upper excitation levels, respectively. These rates are inversely proportional to the energy differences (ΔE) between corresponding excitation levels. Consequently, the probability for deactivation is

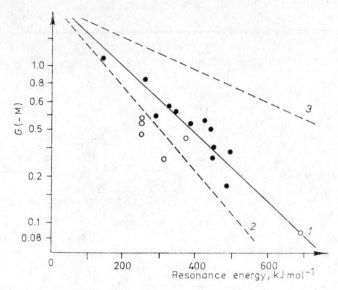

Fig. 6.5. Dependence of the radiolytic decomposition of liquid aromatic hydrocarbons on resonance energy (E_M).
1, Unalkylated aromatic hydrocarbons, fast neutron dose fraction 23.5% (●); *2*, extrapolated to pure γ-radiation (○); *3*, extrapolated to fastneutron (radiation: mixed neutron-γ). After Proksch [240]

Table 6.14

Yields of decomposition and values of resonance energy for the radiolysis of liquid-phase aromatic hydrocarbons [240] (radiation: mixed neutron-γ; temperature: 345°C)

Hydrocarbons	$G(-M)$, molecule/100 eV	Resonance energy, kJ mol⁻¹
Benzene	1.07 (25°C)	151
Naphthalene	0.8	256
1-Methylnaphthalene	0.43	256
2-Methylnachthalene	0.45	256
Acenaphthelene	0.37	256
Biphenyl	0.45	297
Fluorene	0.24	318
Anthracene	0.50	350
Phenanthrene	0.42	383
2-Methylphenanthrene	0.33	381
3-Methylphenanthrene	0.41	381
Fluoranthrene	0.42	427
Pyrene	0.38	448
m-Terphenyl	0.29	453
p-Terphenyl	0.25	453
Crysene	0.16	488
2,2'-Binaphthyl	0.28	503
Coronene	0.081	687

strongly dependent on the values of ΔE between lowest excited states. It is known that, with increasing resonance energy of the molecule, the density of triplet and singlet states increases proportionally in the proximity of the C—H bond dissociation energy, i.e. at about 4–5 eV, and, correspondingly, the ΔE value decreases. This is because with increasing resonance energy, there is a decrease in the energy not only of the ground state but also of the excited states simultaneously with a reduction of the energy difference between the individual levels. This implies that the rate of deactivational electron relaxation increases monotonously with resonance energy. It can be assumed that the rate of decomposition of excited molecules, which is in competition with the increasing deactivation, is independent of resonance energy and, ultimately, the value of $G(-M)$ is reduced.

When radiation with a high LET value is used, the deactivation without decomposition is hindered by both the high local concentrations of excited molecules and their direct reactions with each other. This implies that the dependence of total decomposition yield on the resonance energy is of less importance than when low ionization density radiation is used. It can be supposed that at on infinitely high LET value the yield of decomposition of aromatics would be equal to that of aliphatics.

The resonance energy dependence of $G(-M)$ determined for alkylaromatics was found to be different from that of pure aromatics. This can be explained in part by the formation of less reactive benzyl radicals, compared with aryl radicals and in part, by an increase in the probability of C—C bond rupture in the side-chain.

The radiolysis of alkylaromatic hydrocarbons containing both aromatic and aliphatic group is, presumably influenced by the interaction of these two types of groups.

Inter- and intra-molecular interactions

The yields of H_2, CH_4 and radicals formed from alkylaromatic hydrocarbons are considerably lower than those estimated from the rule of linear additivity, assuming that both the alkyl and the aromatic groups of the molecule decompose like independent alkyl and aromatic groups. However, the aromatic ring affects the radiolysis of the aliphatic part of the molecule [134, 178, 179, 204, 241–243, 249].

A certain fraction of the energy absorbed by the aliphatic part of the molecule can be transferred to the aromatic part. This energy will be localized in the aromatic group, the decomposition of which gives low yields of hydrogen and scavengeable radicals. The excitation-energy and/or charge transfer can take place within one molecule, i.e. intramolecularly, or between the aliphatic group of one molecule and the aromatic group of another, i.e. intermolecularly.

These two types of transfer are in line with theoretical calculations [244–246]. Ionization and excitation energy transfer from an alkyl group to an aromatic one is an exothermic process, as indicated by both the ionization potentials and the lowest excitation levels of the corresponding groups.

In order to determine the extent of energy transfer and the relative efficiencies of the inter- and intra-molecular processes, the yields of H_2, polymers, and radicals have been measured in the radiolysis of various alkylbenzenes and biphenyls. These were compared with the yields obtained in the radiolysis of the corresponding alkane–aromatic hydrocarbon mixture, i.e. a correlation was looked for between the yields of ArR and those of the equimolar mixture of ArH + RH (Ar denotes an aryl radical).

To estimate the yield of energy transfer, the equation [204]:

$$G(H_2)_{obs} = \varepsilon_{phenyl}\ g_{phenyl}\ (H_2) + \varepsilon_{alkyl}\ g_{alkyl}\ (H_2)$$

was used to calculate the value of $g_{alkyl}(H_2)$, which denotes the amount of H_2 formed from the alkyl group on the absorption of 100 eV of radiation energy assuming that the absorption of energy is in proportional to the electron fractions (ε). A value of 0.038, i.e. $G(H_2)$ of benzene, was taken for $g_{phenyl}(H_2)$.

Jones et al. [204] found that, although the yield of H_2 increases with increasing n-alkyl chain-length, the H_2 formed from the alkyl group, $g_{alkyl}(H_2)$ is the same (about 0.5) for various compounds, in contrast to the value $G(H_2) \approx 5$, which is characteristic of aliphatic hydrocarbons. The $g_{alkyl}(H_2) \approx 0.5$ is produced in reactions that are too fast for energy transfer to compete with them.

In the radiolysis of n-butane–benzene mixtures, a value of ~ 0.5 was also found for $g_{alkyl}(H_2)$. This result seems to verify the suggestion that the chemical bond between the aromatic and the aliphatic part of the molecule does not affect the extent of the protective effect, i.e. intramolecular energy transfer is negligible compared with intermolecular energy transfer [204].

Zeman and Heusinger reported [247] that there is little difference between the $g_{alkyl}(H_2)$ values of alkylbenzenes containing in the side-chain not more than four atoms and those of the corresponding n-alkane–benzene mixtures; both values are about 0.5. With a further increase in the length of the side-chain, however, the value of $g_{alkyl}(H_2)$ significantly increases (Fig. 6.6). This implies that the 'protective' effect of the aromatic ring is approximately constant up to a carbon atom number four, and is reduced by a further increase in chain length. On the other hand, in the case of alkylbenzenes with side-chains of more than four carbon atoms the value of $g_{alkyl}(H_2)$ is considerably lower than that of the corresponding alkane–benzene mixture (Fig. 6.6). This indicates that in alkylbenzenes with longer side-chains than that of butylbenzene and in butylbiphenyls, intramolecular energy transfer is an important process, in addition to intermolecular energy transfer. The contribution of intramolecular energy transfer to the 'protective' effect increases with the length of the side-chain, and is a maximum at about heptadecylbenzene [134, 247].

The $G(H_2)$ values of alkylaromatic compounds containing two separate phenyl groups, and those of various alkylbiphenyls, demonstrate the dependence of energy transfer on the length of the side-chain and the occurrence of intramolecular energy transfer [178, 179, 241].

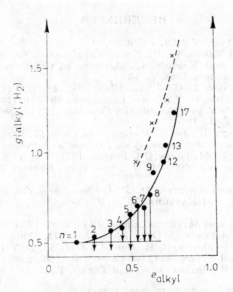

Fig. 6.6. g_{alkyl} (H$_2$)-values as a function of electron fraction of the alkane or the alkyl group (radiation: ^{60}Co-γ)
$\times$ (1 : 1) Alkane–benzene mixtures [134]; $\bullet$ alkylbenzenes [134]; $\downarrow$ alkyl-benzene [204]

However, the distribution of products labelled with tritium, formed in the radiolysis of toluene, ethylbenzene and n-butylbenzene with low-energy ^{3}H$^+$ ions, showed the excitation energy transfer to be independent of the length of the side-chain [248].

From the radical yields obtained with iodine acceptor Bagdasaryan et al. [242] concluded that in the radiolysis of octylbenzene intramolecular is considerably more important than intermolecular energy transfer.

The importance of the intramolecular process is demonstrated also by the influence of the relative positions of the phenyl and alkyl groups on the value of G(H$_2$), observed in the radiolysis of alkylbenzene isomers. G(H$_2$) is lowest, i.e. the 'protective' effect is greatest, if the phenyl group is situated in the middle of the alkyl chain [134]. Similar results were obtained in studies of the crosslinking of dodecylnaphthalene: it was found that the effect of the ring is greatest if it is in the middle, and smallest if it is at the end of the paraffin chain [249].

In the radiolysis of phenylcyclohexane, 1-methyl-4-phenylcyclohexane and 2-cyclohexyl-1-phenylethane, lower values of G(H$_2$) were obtained than with the corresponding alkane–benzene mixtures [208]. This result again shows the higher efficiency of intramolecular energy transfer.

From the above discussion, the diversity of opinions expressed by different authors is clear. Further studies, theoretical as well as experimental, are needed to clarify the relative roles of inter- and intra-molecular energy transfer in the radiolysis of alkylaromatic compounds.

REFERENCES

1. BOYD, A. W., *J. Nucl. Mat.*, **9**, 1 (1963)
2. BURNS, W. G. and MARSH, W. R., *Trans. Faraday Soc.*, **65**, 1827 (1969)
3. WILZBACH, K. E. and KAPLAN, L., *Advances in Chemistry*, Series 82., Amer. Chem. Soc., Washington, 1968, p. 134
4. WEXLER, S. and POBO, L. G., *J. Am. Chem. Soc.*, **91**, 7233 (1969)
5. WEXLER, S. and CLOW, R. P., *J. Am. Chem. Soc.*, **90**, 3940 (1968)
6. *The Current Status of Liquid Scintillation Counting* (Ed. BRANSONE, E. D.), Grune et Stratton, New York, 1970
7. KLEIN, G. and VOLTZ, R., *Int. J. Radiat. Phys. Chem.*, **7**, 155 (1975)
8. VOLTZ, R., *International Discussion on Progress and Problems in Contemporary Radiation Chemistry* (Ed. TEPLY, J.), Prague, 1971, Vol. 1, p. 139
9. BAXENDALE, J. H. and FITI, M., *J. Chem. Soc., Faraday Trans. II.*, **68**, 218 (1972)
10. ZARNEGAR, B. M. and WHITTEN, D. G., *J. Phys. Chem.*, **76**, 198 (1972)
11. BAXENDALE, J. H. and RASBURN, E. J., *J. Chem. Soc. Faraday Trans. I.*, **69**, 771 (1973)
12. BAXENDALE, J. H. and MAYER, J., *Chem. Phys. Lett.*, **17**, 458 (1972)
13. SCHMIDT, W. F. and ALLEN, A. O., *J. Phys. Chem.*, **72**, 3730 (1968)
14. SHINSAKA, K. and FREEMAN, G. R., *Can J. Chem.*, **52**, 3495 (1974)
15. SCHMIDT, W. F. and ALLEN, A. O., *J. Chem. Phys.*, **52**, 2345 (1970)
16. SATO, S., YUGETA, R., SHINSAKA, K. and TERAO, T., *Bull. Chem. Soc. Jpn.*, **39**, 156 (1966)
17. HUMMEL, A., *J. Chem. Phys.*, **48**, 3268 (1968)
18. RZAD, S. J., SCHULER, R. H. and HUMMEL, A., *J. Chem. Phys.*, **51**, 1369 (1969)
19. WARMAN, J. M., ASMUS, K. D. and SCHULER, R. H., *J. Phys. Chem.*, **73**, 931 (1969)
20. WARMAN, J. M., ASMUS, K. D. and SCHULER, R. H., *Advances in Chemistry*, Series 82, Am. Chem. Soc., Washington, 1968, p. 25
21. HENTZ, R. R. and SHERMAN, W. V., *J. Phys. Chem.*, **73**, 2676 (1969)
22. HIROKAMI, S., SHISHIDO, S. and SATO, S., *Bull. Chem. Soc. Jpn.*, **44**, 1511 (1971)
22A. DIXON, R. S., SARGENT, F. P., LOPATA, V. J., GARDY, E. M. and BROCKLE-HURST, B., *Can. J. Chem.*, **55**, 2093 (1977)
23. BECK, G. and THOMAS, J. K., *J. Phys. Chem.*, **76**, 3856 (1972)
24. BURNS, W. G. and BARKER, R., *Progress in Reaction Kinetics* (Ed. PORTER, G.), Pergamon Press, Oxford, 1965, Vol. 3, Chapter 7
25. BURNS, W. G. and JONES, J. D., *Trans. Faraday Soc.*, **60**, 2022 (1964)
26. BOYD, A. W. and CONNOR, H. W. J., *Trans. Am. Nucl. Soc.*, **8**, 421 (1965)
27. SCHULER, R. H., *Trans. Faraday Soc.*, **61**, 100 (1965)
28. BURNS, W. G. and BARKER, R., *Aspects of Hydrocarbon Radiolysis* (Eds GÄUMANN, T. and HOIGNÉ, J.), Academic Press, London, 1968, p. 40
29. YANG, J. Y., STRONG J. D. and BURR, G. J., *J. Phys. Chem.*, **69**, 1157 (1965)
30. PLATZMAN, R., *Vortex.*, **23**, 8 (1962)
31. MAKAROV, V. I. and POLAK, L. S., *Khim. Vys. Energ.*, **4**, 3 (1970)
32. VOLTZ, R., *Radiat. Res. Rev.*, **1**, 301 (1968)
33. HORROCKS, D. L., *J. Chem. Phys.*, **52**, 1566 (1970)
34. PLATZMAN, R., Radiation Research (Ed. SILINI, G.), Elsevier, Amsterdam, 1967, p. 20.
35. COOPER, R. and THOMAS, J. K., *J. Chem. Phys.*, **48**, 5097 (1968)
36. SATO, S., HOSOYA, K., SHISHIDO, S. and HIROKAMI, S., *Bull. Chem. Soc. Jpn.*, **45**, 2308 (1972)
37. FUCHS, C., HEISEL, F. and VOLTZ, R., *J. Phys. Chem.*, **76**, 3867 (1972)
38. MERKEL, P. B. and HAMILL, W. H., *J. Chem.Phys.*, **54**, 1695 (1971)
39. LAWSON, C. W., HIRAYAMA, F. and LIPSKY, S., *J. Chem. Phys.*, **51**, 1590 (1969)
39A. CUNDALL, R. B., ROBINSON, D. A. and PEREIRE, L. C., *Advances in Photochemistry* (Eds PITTS, J. N., HAMMOND, G. S. and GOLLNICK, K.), Wiley, New York, 1977, p. 147
40. KRONGAUZ, V. A., *Int. J. Radiat. Phys. Chem.*, **1**, 465 (1969)
41. HAMMOND, G. S., CALDWELL, R. A., KING, J. M. and KRISTINSSON, H., *Photo chem. Photobiol.*, **7**, 695 (1968)

42. CUNDALL, R. B., *Energetics and Mechanisms in Radiation Biology* (Ed. PHILLIPS, G. O.), Academic Press, London 1968. p. 227
43. GOLUB, M. A. and STEPHENS, C. L., *J. Phys. Chem.*, **70**, 3576 (1966)
44. CUNDALL, R. B. and TIPPETT, W., *Advances in Chemistry.* Series 82. Am. Chem. Soc., Washington, 1968, p. 387
45. CUNDALL, R. B. and TIPPETT, W., *Trans. Faraday Soc.*, **66**, 350 (1970)
46. HENTZ, R. R. and ALTMILLER, H. G., *J. Phys. Chem.*, **74**, 2646 (1970)
47. HENTZ, R. R. and PERKEY, L. M., *J. Phys. Chem.*, **74**, 3047 (1970)
48. EHRL, A., *Atomkernenergie*, **14**, 137 (1969); **14**, 435 (1969)
49. HENTZ, R. R., PETERSON, D. B., SRIVASTAVA, S. B., BARZYNSKI, H. F. and BURTON, M., *J. Phys. Chem.*, **70**, 2362 (1966)
50. FISCHER, E., LEHMANN, H. P. and STEIN, G., *J. Chem. Phys.*, **45**, 3905 (1966)
51. SAITO, N. and SATO, S., *Bull. Chem. Soc. Jpn.*, **42**, 2228 (1969)
52. WEST, M. L. and NICHOLS, L. L., *J. Phys. Chem.*, **74**, 2404 (1970)
53. HORROCKS, A. R., *Can. J. Chem.*, **48**, 1000 (1970)
54. PHILLIPS, D. H. and SCHUG, J. C., *J. Chem. Phys.*, **50**, 3297 (1969)
55. THOMAS, J. K., *J. Chem. Phys.*, **51**, 770 (1969)
56. SKARSTAD, P., MA, R. and LIPSKY, S., *Mol. Cryst.*, **4**, 3 (1968)
57. GANGWER, T. and THOMAS, J. K., *Radiat. Res.*, **54**, 192 (1973)
58. MAGEE, J. L., *Comparative Effects of Radiation* (Ed. BURTON, M.), Wiley, New York, 1960, p. 130
59. BAXENDALE, J. H. and WARDMAN, P., *Trans. Faraday Soc.*, **67**, 2997 (1971)
60. OSTER, G. K. and KALLMAN, H., *J. Chem. Phys.*, **64**, 28 (1972)
61. CUNDALL, R. B. and ROBINSON, D. A., *J. Chim. Soc. Faraday Trans. II.*, **68**, 1133 (1972)
62. THOMAS, J. K. and MANI, I., *J. Chem. Phys.*, **51**, 1834 (1969)
63. MIYAZAKI, T. and YAMAMOTO, M., *Radiat. Phys. Chem.*, **10**, 247 (1977)
64. RICHARDS, J. T. and THOMAS, J. K., *Chem. Phys. Lett.*, **5**, 527 (1970)
65. BENSASSON, R. V., RICHARDS, J. T. and THOMAS, J. K., *Chem. Phys. Lett.*, **9**, 13 (1971)
66. CUNDALL, R. B., EVANS, G. B., GRIFFITHS, P. A. and KEENE, J. P., *J. Phys. Chem.*, **72**, 3871 (1968)
67. LAND, E. J. and SWALLOW, A. J., *Trans. Faraday Soc.*, **64**, 1247 (1968)
67A. ROBINSON, E. A. and SALMON, G. A., *J. Phys. Chem.*, **82**, 382 (1978)
68. DAINTON, F. S., PENG, C. T. and SALMON, G. A., *J. Phys. Chem.*, **72**, 3801 (1968)
69. DAINTON, F. S., ROBINSON, E. A. and SALMON, G. A., *J. Phys. Chem.*, **76**, 3897 (1972)
70. DAINTON, F. S., MORROW, T. and SALMON, G. A., *Proc. Roy. Soc. A.*, **328**, 457 (1972), **328**, 481 (1972)
71. NISHIKAWA, M. and SAUER, M. C., *J. Chem. Phys.*, **51**, 1 (1969)
71A. UENO, T., KOUCHI, N., TAKAO, S. and HATANO, Y., *J. Phys. Chem.*, **82**, 2373 (1978)
72. OGAWA, T., TSUJI, M., TOYODA, M. and ISHIBASHI, N., *Bull. Chem. Soc. Jpn.*, **46**, 2637 (1973)
73. SHIDA, T., *J. Phys. Chem.*, **74**, 3055 (1970)
74. BURNS, W. G., *Trans. Faraday Soc.*, **58**, 961 (1962)
75. BURNS, W. G. and MARSH, W. R., *Trans. Faraday Soc.*, **64**, 2375 (1968)
76. GÄUMANN, T., *Helv. Chim. Acta*, **46**, 2873 (1963)
77. HENTZ, R. R. and RZAD, S. J., *J. Phys. Chem.*, **72**, 1027 (1968)
78. DYNE, P. J. and JENKINSON, W. M., *Can. J. Chem.*, **39**, 2163 (1961)
79. GORDON, A. S. and BURTON, M., *Disc. Faraday Soc.*, **12**, 88 (1952)
80. SHATROV, V. D., TCHKHEIDZE, I. I., SHAMSHEV, V. N. and BUBEN, N. Ya., *Khim. Vys., Energ.*, **2**, 413 (1968)
81. SAUER, M. C. and WARD, B., *J. Phys. Chem.*, **71**, 3971 (1967)
82. AVRAMENKO, L. I., BUBEN, N. Ya., KOLESNIKOVA, R. V., TOLKACHEV, V. A. and TCHKHEIDZE, I. I., *Izv. Akad. Nauk. SSSR, Ser. Khim.*, 2079 (1962)
82A. HOYERMANN, K., PREUSS, A. W., WAGNER, H. GG., *Ber. Bunsenges. Phys. Chem.*, **79**, 156 (1975)
83. PHUNG, P. V. and BURTON, M., *Radiat. Res.*, **7**, 199 (1957)
84. OHNISHI, S., TANEI, T. and NITTA, I., *J. Chem. Phys.*, **37**, 2402 (1962)

85. TCHKHEIDZE, I. I., MOLIN, Yu. N., BUBEN, N. YA. and VOEVODSKII, V. V., *Dokl. Akad. Nauk SSSR*, **130**, 1291 (1960)
86. TOLKACHEV, V. A., MOLIN, YU. N., TCHKHEIDZE, I. I., BUBEN, N. YA. and VOEVODSKII, V. V., *Dokl. Akad. Nauk SSSR*, **141**, 911 (1961)
87. NAGAI, S., OHNISHI, S. and NITTA, I., *Bull. Chem. Soc. Jpn.*, **44**, 1230 (1971)
88. FESSENDEN, R. W. and SCHULER, R. H., *J. Chem. Phys.*, **38**, 773 (1963), **39**, 2147 (1963)
89. SHIDA, T. and HANAZAKI, I., *Bull. Chem. Soc. Jpn.*, **43**, 646 (1970)
90. SHIDA, T. and HANAZAKI, I., *J. Phys. Chem.*, **74**, 213 (1970)
91. OLSEN, K. J., *Trans. Faraday Soc.*, **67**, 2050 (1971)
91A. SHENG, S. J., *J. Phys. Chem.*, **82**, 442 (1978)
92. KNUTTI, R. and BÜHLER, R. E., *J. Chem. Phys.*, **7**, 229 (1975)
92A. MATSUYAMA, T. and YAMAOKA, H., *J. Chem. Phys.*, **68**, 331 (1978)
92B. MATSUYAMA, T. and YAMAOKA, H., *J. Chem. Phys. Lett.*, **57**, 269 (1978)
93. HARDWICK, T. J., *J. Phys. Chem.*, **66**, 117, 1611 (1962)
94. VASILEV, G. K. and TCHKHEIDZE, I. I., *Kinet. Katalis.*, **5**, 802 (1964)
95. GÄUMANN, T. and SCHULER, R. H., *J. Phys. Chem.*, **65**, 703 (1961)
96. CHERNIAK, E. A., COLLINSON, E. and DAINTON, F. S., *Trans. Faraday Soc.*, **60**, 1408 (1964)
97. BURNS, W. G., *VI. Rassegna Internazionale Elettronica e Nucleare*, Roma, 1959, p. 99
98. GURSKY, M. N. and TSOI, A. N., *Khim. Vys. Energ.*, **3**, 187 (1969)
99. ZIMMERLI, B. and GÄUMANN, T., *Helv. Chim. Acta.*, **52**, 764 (1969)
100. GÄUMANN, T., *Helv. Chim. Acta.*, **44**, 1337 (1961)
101. GORDON, S., VAN DYKEN, A. R. and DOUMANI, T. F., *J. Phys. Chem.*, **62**, 20 (1958)
102. WEBER, E. N., FORSYTH, P. F. and SCHULER, R. H., *Radiat. Res.*, **3**, 68 (1955)
102A. SARGENT, F. P. and GARDY, E. M., *J. Chem. Phys.*, **67**, 1793 (1977)
102B. ZUBAREV, V. E., BELEVSKII, V. N. and ZARAZILOV, A. L., *Vestn. Mosk. Univ. Ser. Khim.*, **19**, 456 (1978)
103. DORFMAN, L. M., TAUB, I. A. and BÜHLER, R. E., *J. Chem. Phys.*, **36**, 3051 (1962)
104. PATRICK, W. N. and BURTON, M., *J. Am. Chem. Soc.*, **76**, 2626 (1954)
105. BURNS, W. G. and REED, C. R. V., *Trans. Faraday Soc.*, **59**, 101 (1963)
106. HENTZ, R. R. and RZAD, S. J., *J. Phys. Chem.*, **71**, 4096 (1967)
107. FIELD, F. H., HAMLET, P. and LIBBY, W. F., *J. Am. Chem. Soc.*, **89**, 6035 (1967)
108. HENGLEIN, A., *Z. Naturforsch.*, **17a**, 37 (1962)
109. GIARDINI-GUIDONI, A. and ZOCCHI, F., *Trans. Faraday Soc.*, **64**, 2342 (1968)
110. LIFSHITZ, C. and REUBEN, B. G., *Israel J. Chem.*, **7**, 149 (1969)
111. LIFSHITZ, C. and REUBEN, B. G., *J. Chem. Phys.*, **50**, 951 (1969)
112. VIRIN, L. I., SAFIN, YU. A. and DZHAGATSPANYAN, R. V., *Khim. Vys. Energ.*, **1**, 417 (1967)
112A. ROSENSTOCK, H. M. and McCULLOH, K. E., *Int. J. Mass. Spectrom. Ion Phys.*, **25**, 327 (1977)
113. FIELD, F. H., HAMLET, P. and LIBBY, W. F., *J. Am. Chem. Soc.*, **91**, 2839 (1969)
114. EDLUND, O., KINELL, P. O., LUND, A. and SHIMIZU, A., *J. Chem. Phys.*, **46**, 3679 (1967)
115. EKSTROM, A., *J. Phys. Chem.*, **74**, 1705 (1970)
116. BIRKS, J. B., *Nature*, **214**, 1187 (1967)
117. BADGER, B., BROCKLEHURST, B. and RUSSEL, R. D., *Chem. Phys. Lett.*, **1**, 122 (1967)
118. KROH, J. and HANKIEWICZ, E., *Int. J. Radiat. Phys. Chem.*, **1**, 451 (1969)
119. KROH, J. and HANKIEWICZ, E., *Chem. Phys. Lett.*, **1**, 542 (1968)
120. KROH, J. and BURZYNSKA, E., *Radiochem. Radioanal. Lett.*, **4**, 119 (1970)
121. FISCHER, E., FISCHER, G. and STEIN, G., *Chem. Phys. Lett.*, **2**, 405 (1968)
122. BIRKS, J. B., *Chem. Phys. Lett.*, **1**, 625 (1968)
122A. BECK, G., RICHARD, J. T. and THOMAS, J. K., *Chem. Phys. Lett.*, **40**, 300 (1976)
122B. BECK, G. and THOMAS, J. K., *J. Chem. Soc. Faraday Trans. I.*, **72**, 2610 (1976)
123. CHRISTOPHOROU, L. G., ABU-ZEID, M. E. and CARTER, J. G., *J. Chem. Phys.*, **49**, 3775 (1968)

124. CARTER, J. G., CHRISTOPHOROU, L. G. and ABU-ZEID, M. E., *J. Chem. Phys.*
     47, 3879 (1967)
125. IVANOVA, T. V., MOKEEVA, G. A. and SVESHNIKOV, B. YA., *Opt. Spektr.*, **12**,
     586 (1962)
126. BENSASSON, R. V., RICHARDS, J. T., GANGWER, T. and THOMAS, J. K., *Chem.
     Phys. Lett.*, **14**, 430 (1972)
127. RICHARDS, J. T. and THOMAS, J. K., *J. Chem. Phys.*, **55**, 3636 (1971)
128. GANGWER, T. E. and THOMAS, J. K., *Int. J. Radiat. Phys. Chem.*, **7**, 305 (1975)
129. PLATZNER, I. and THOMAS, J. K., *Int. J. Radiat. Phys. Chem.*, **7**, 573 (1975)
130. JOHNSON, P. M. and ALBRECHT, A. C., *J. Chem. Phys.*, **48**, 851 (1968)
131. BATEKHA, I. G., ALFIMOV, M. V. and SHEKK, YU. B., *Khim. Vys. Energ.*, **3**, 282
     (1969)
132. HENTZ, R. R., *J. Phys. Chem.*, **66**, 1622 (1962)
133. HOIGNÉ, J., BURNS, W. G., MARSH, W. R. and GÄUMANN, T., *Helv. Chim. Acta.*,
     47, 247 (1964)
134. ZEMAN, A. and HEUSINGER, H., *Radiochim. Acta*, **8**, 149 (1967)
134A. DUVALL, J. J. and JENSEN, H. B., *Radiat. Res.*, **70**, 248 (1977)
134B. KOLLÁR, J., FÖLDIÁK, G. and WOJNÁROVITS, L., *Radiochem. Radioanal. Lett.*,
      31, 147 (1977)
134C. FÖLDIÁK, G., *Radiat. Phys. Chem.*, **11**, 267 (1978)
135. RODER, M., *Acta Chim. Acad. Sci. Hung.*, **73**, 443 (1972)
136. VERDIN, D., *J. Phys. Chem.*, **67**, 1263 (1963)
137. INGALLS, R. B., *J. Phys. Chem.*, **65**, 1605 (1961)
138. INGALLS, R. B., SPIEGLER, P. and NORMAN, A., *J. Chem. Phys.*, **41**, 837 (1964)
139. WEISS, J. and COLLINS, C. H., *Radiat. Res.*, **28**, 1 (1966)
140. WEISS, J. and COLLINS, C. H., *Radiat. Res.*, **33**, 274 (1968)
     137 (1965)
141. WEISS, J. and RAO, H. M., *Radiat. Res.*, **32**, 309 (1967)
142. BUBEN, N. YA. and TCHKHEIDZE, I. I., *Zh. Khim. Obshch. Mend.*, **11**, 228 (1966)
143. HOIGNÉ, J. and GÄUMANN, T., *Helv. Chim. Acta*, **46**, 365 (1963)
144. BARSON, C. A. and BEVINGTON, J. C., *Trans. Faraday Soc.*, **55**, 1266 (1959)
145. YAMAMOTO, Y., TAKAMUKU, S. and SAKURAI, H., *J. Phys. Chem.*, **74**, 3325 (1970)
146. YAMAMOTO, Y., TAKAMUKU, S. and SAKURAI, H., *J. Am. Chem. Soc.*, **91**, 7192
     (1969)
147. YAMAMOTO, Y., TAKAMUKU, S. and SAKURAI, H., *J. Am. Chem. Soc.*, **94**, 661 (1972)
148. YAMAMOTO, Y., TAKAMUKU, S. and SAKURAI, H., *Bull. Chem. Soc. Jpn.*, **44**,
     574 (1971)
149. HOIGNÉ, J. and GÄUMANN, T., *Helv. Chim. Acta*, **47**, 260 (1964)
150. HOIGNÉ, J. and GÄUMANN, T., *Helv. Chim. Acta*, **44**, 2141 (1961)
151. VIRIN, L. I., SAFIN, YU. A., DZHAGATSPANYAN, R. V. and MOTSAREV, G. V.,
     Khim. Vys. Energ., **3**, 23 (1969)
152. VIRIN, L. I., SAFIN, YU. A. and DZHAGATSPANYAN, R. V., *Khim. Vys. Energ.*,
     5, 379 (1971)
153. SIEGEL, A. S., *J. Am. Chem. Soc.*, **92**, 5277 (1970)
154. GRUBB, H. M. and MEYERSON, S., in: *Mass Spectrometry of Organic Ions* (Ed.
     MCLAFFERTY, F. W.), Academic Press, New York, 1963, Chapter 10
154A. DUNBAR, R. C., *J. Am. Chem. Soc.*, **97**, 1382 (1975)
154B. JACKSON, J. A., LIAS, S. G. and AUSLOOS, P., *J. Am. Chem. Soc.*, **99**, 7515 (1977)
154C. SHEN, J., DUNBAR, R. C., OLAH, G. A., *J. Am. Chem. Soc.*, **96**, 6227 (1974)
154D. MCCRERY, D. A. and FREISER, B. S., *J. Am. Chem. Soc.*, **100**, 2902 (1978)
154E. HOURIET, R., ELWOOD, T. A. and FUTRELL, J. H., *J. Am. Chem. Soc.*, **100**,
      2320 (1978)
155. KOMATSU, T., LUND, A. and KINELL, P. O., *J. Phys. Chem.*, **76**, 1721 (1972)
155A. YAMAMOTO, Y., TAKAMUKU, S. and SAKURAI, H., *J. Am. Chem. Soc.*, **100**,
      2474 (1978)
156. YAMANOTO, Y., TAKAMUKU, S. and SAKURAI, H., *Bull. Chem. Soc. Jpn.*, **44**,
     2104 (1971)
157. TAKAMUKU, S., SAGI, N., NAGAOKA, K. and SAKURAI, H., *J. Am. Chem. Soc.*,
     94, 6217 (1972)
158. WEISS, J., *Radiat. Res.*, **45**, 252 (1971)

30

158A. KOLLÁR, J. and FÖLDIÁK, G. in *Proc. 4th Tihany Symp. Rad. Chem.* (Eds HEDVIG, P. and SCHILLER, R.), Akadémiai Kiadó, Budapest 1977. p. 133
158B. CHAPIRO, A., JENDRYCHOWSKA-BONAMOUR, A. M. and LELIEVRE, G., *Faraday Disc. Chem. Soc.*, **63**. 134 (1977)
159. CHAPIRO, A., *Radiation Chemistry of Polymeric Systems*, Wiley, New York, 1962
160. CHARLESBY, A., *Atomic Radiation and Polymers*, Pergamon Press, London, 1960
161. CHAPIRO, A., CORDIER, P., HAYASHI, V. K., MITA, I. and SERBAN-DANON, J., *J. Chem. Phys.*, **56**, 447 (1959)
162. KRONGAUZ, V. A. and BAGDASAPYAN, Kh. S., *Zhur. Fiz. Khim.*, **32**, 1863 (1958)
163. PREVOT-BERNAS, A., CHAPIRO, A., COUSIN, C., LANDLER, Y. and MAGAT, M., *Disc. Faraday Soc.*, **12**, 98 (1952)
164. WILLIAMS, F., *J. Macromol. Sci. A*, **6**, 919 (1972)
165. SHENKER, A. P., YAKOVLEVA, M. K., KRISHTALNII, E. V. and ABKIN, A. D., *Dokl. Akad. Nauk SSSR*, **124**, 632 (1969)
166. OKAMURA, S., HIGASHIMURA, T. and FUTAMI, S., *Isotopes and Radiation (Japan)*, **1**, 216 (1958)
167. CHAPIRO, A. and STANNETT, V., *J. Chem. Phys.*, **56**, 830 (1959)
168. CHAPIRO, A., *Makromolekulare Chemie*, **175**, 1181 (1974)
168A. MEHNERT, R., HELMSTREIT, W., BÖS, J. and BREDE, O., *Radiochem. Radioanal. Lett.*, **30**, 389 (1977)
168B. TABATA, Y., *J. Polymer Sci., Symposium* **56**, 409 (1976)
168C. EGUSA, S., TABATA, J., ARAI, S., KIRA, A. and IMAMURA, M., *J. Polymer Sci.*, **16**, 729 (1978)
169. THEARD, L. M., PETERSON, F. C. and HOLROYD, R. A., *J. Chem. Phys.* **51**, 4126 (1969)
170. SCARBOROUGH, J. M. and BURR, J. G., *J. Chem. Phys.*, **37**, 1890 (1962)
171. BURR, J. G., SCARBOROUGH, J. M., STRONG, J. D., AKAWIE, R. I. and MEYER, R. A., *Nucl. Sci. Energ.*, **11**, 218 (1961)
172. BURR, J. G. and SCARBOROUGH, J. M., *J. Phys. Chem.*, **64**, 1367 (1960)
173. BARTONICEK, B., JANOVSKY, I. and BEDNAR, J., *Int. J. Radiat. Phys. Chem.*, **7**, 431 (1975)
174. SWEENEY, M. A., HALL, K. L. and BOLT, R. O., *J. Phys. Chem.*, **71**, 1564 (1967)
175. CAMPBELL, D., SYMONS, M. C. R. and VERMA, G. S. P., *J. Chem. Soc., A*, 2480 (1969)
176. VOEVODSKII, V. V. and MOLIN, YU. N., *Radiat. Res.*, **17**, 366 (1962)
177. HUTCHINSON, W. M., HUDSON, P. S. and DOSS, R. C., *J. Am. Chem. Soc.*, **85**, 3358 (1963)
178. RODER, M., KULES, I. and KISS, I., *Magy. Kém. Foly.*, **72**, 443 (1966)
179. KULES, I. and ERŐNÉ-GÉCS, M., *Acta Chim. Acad. Sci. Hung.*, **58**, 389 (1968)
180. KULES, I., OPAUSZKY, I., KÓSA-SOMOGYI, I. and SCHILLER, R., *KFKI Közl.*, **13**, 137 (1965)
180A. TZIKANOV, V. A., ALEKSENKO, YU. N., TETYUKOV, V. D., KUPRIENKO, V. A., KOBZAR, I. G., KHRAMCHENKOB, V. A., MEXCHERYAKOV, M. P. and ZINCVIEV, V. I., *Nucl. Tech.*, **38**, 187 (1978)
181. BURNS, W. C., WILD, W. and WILLIAMS, T. F. ,*Proc. 2nd UN Intern. Conf. Peaceful Uses of Atomic Energy, UN Geneva*, 1958, Vol. 29, p. 266
182. BOYD, A. W., CONNOR, H. W. J. and MILLER, O. A., *Report AECL-2589* (1966)
183. SCARBOROUGH, J. M. and INGALLS, R. B., *J. Phys. Chem.*, **71**, 486 (1967)
184. KISS, I. and RODER, M., *Magy. Kém. Foly.*, **74**, 336 (1968)
185. LAVROVSKIJ, K. P., PROTSIDIM, P. S. and TITOV, V. B., *Dokl. Akad. Nauk SSSR*, **191**, 1066 (1970)
186. MALAN, O. G., GÜSTEN, H. and SCHULTE-FROHLINDE, D., *J. Phys. Chem.*, **72**, 1457 (1968)
187. HOLROYD, R. A., THEARD, L. M. and PETERSON, F. C., *J. Phys. Chem.*, **74**, 1895 (1970)
188. CHANDROSS, E. A. and DEMPSTER, C. J., *J. Am. Chem. Soc.*, **92**, 704 (1970)
189. HOLROYD, R. A. and CAPELLOS, C., *J. Phys. Chem.*, **76**, 2485 (1972)
190. BROCKLEHURST, B. and RUSSEL, R. D., *Trans. Faraday Soc.*, **65**, 2159 (1969)
191. SAUER, M. C. and MULAC, W. A., *Int. J. Radiat. Phys. Chem.*, **6**, 55 (1974)
192. SAUER, M. C. and MULAC, W. A., *J. Phys. Chem.*, **7͡**, 22 (1974)

193. OKUBO, T., ITOH, N. and SUITA, T., *Mol. Cryst. Liq. Cryst.*, **6**, 227 (1969)
193A. ITOH, N. and OKUBO, T., *Mol. Cryst. Liq. Cryst.*, **17**, 303 (1972)
194. AKASAKA, Y., MURAKAMI, K., MASUDA, K. and NAMBA, S., *Mol. Cryst. Liq. Cryst.*
     15, 37 (1971)
194A. NAKAGAWA, K. and ITOH, N., *Chem. Phys.*, **16**, 461 (1976)
195. AKASAKA, Y., MASUDA, K. and NAMBA, S., *J. Phys. Soc. Jpn.*, **30**, 1686 (1971)
195A. NAKAGAWA, K. and ITOH, N., *Chem. Phys. Lett.*, **47**, 367 (1977)
196. CHONG, T. and ITOH, N., *Mol. Cryst. Liq. Cryst.*, **11**, 315 (1970)
196.A. CHONG, T., NAKAGAWA, K. and ITOH, N., *Chem. Phys. Lett.*, **55**, 107 (1978)
197. BÖHME, U. R. and WOLF, H. C., *Chem. Phys. Lett.*, **17**, 582 (1972)
198. ANDREEV, O. M., SMIRNOV, V. A., SAZHNIKOV, V. A. and ALFIMOV, M. V., *Khim.
     Vys. Energ.*, **9**, 512 (1975)
199. KINELL, P. O., LUND, A. and SHIMIZU, A., *J. Phys. Chem.*, **73**, 4175 (1969)
200. STACHOWICZ, W., KECKI, Z. and MINC, S., *Nukleonika*, **12**, 1135 (1967)
200A. KLEIN, G., *Mol. Cryst. Liqu. Cryst.*, **47**, 39 (1978)
201. POLAK, L. S., CHERNYAK, N. YA., SHAKHRAII, V. A. and SHCHERBAKOVA, A. S.,
     Neftekhim., **1**, 695 (1961)
202. MAKAROV, V. I., POLAK, L. S., CHERNYAK, N. YA. and SHCHERBAKOVA, A. S.,
     Neftekhim., **6**, 58 (1966)
203. KIKUCHI, K., SATO, S. and SHIDA, S., *Nippon Kagaku Zasshi.*, **84**, 561 (1963)
204. JONES, K. H., VAN DUSEN, W. and THEARD, L. M., *Radiat. Res.*, **23**, 128 (1964)
205. MANION, J. P. and BURTON, M., *J. Phys. Chem.*, **56**, 560 (1952)
206. FORRESTAL, L. J. and HAMILL, W. H., *J. Am. Chem. Soc.*, **83**, 1535 (1961)
207. MERKLIN, J. F. and LIPSKY, S., *Biological Effects of Ionizing Radiation at the
     Molecular Level*, IAEA, Vienna, 1962, p. 73
208. MERKLIN, J.F. and LIPSKY, S., *J. Phys. Chem.*, **68**, 3297 (1964)
209. FREEMAN, G. R., *J. Chem. Phys.*, **33**, 71 (1960)
210. BURTON, M. and PATRICK, W. N., *J. Phys. Chem.*, **58**, 421 (1954)
211. STONE, J. A. and DYNE, P. J., *Radiat. Res.*, **17**, 353 (1962)
212. BURR, J. G. and GOODSPEED, F. C., *J. Chem. Phys.*, **40**, 1433 (1964)
213. YANG, J. Y., SCOTT, B. and BURR, J. G., *J. Phys. Chem.*, **68**, 2014 (1964)
214. SAUER, M. C. and MANI, I., *J. Phys. Chem.*, **72**, 3856 (1968)
215. AVDONINA, E. N., *Khim. Vys. Energ.*, **4**, 226 (1970)
216. AVDONINA, E. N., *J. Inorg. Nucl. Chem.*, **33**, 3663 (1971)
217. DYNE, P. J., *Can. J. Chem.*, **43**, 1080 (1965)
218. HORIKIRI, S. and SAIGUSA, T., *Bull. Inst. Chem. Res. Kyoto Univ.*, **43**, 45 (1965)
218A. HIRAYAMA, F. and LIPSKY, S., *J. Chem. Phys.*, **51**, 3616 (1969)
218B. WADA, T. and HATANO, Y., *J. Phys. Chem.*, **81**, 1057 (1977)
219. FREEMAN, G. R., *J. Chem. Phys.*, **33**, 957 (1960)
220. TOMA, S. Z. and HAMILL, W. H., *J. Am. Chem. Soc.*, **86**, 1478 (1964)
221. KENNEDY, M. G. and STONE, J. A., *Can. J. Chem.*, **51**, 149 (1973)
222. DYNE, P. J. and DENHARTOG, J., *Nature*, **202**, 1105 (1964)
223. KROH, J. and KAROLCZAK, S., *Nature*, **201**, 66 (1964)
224. DYNE, P. J. and DENHARTOG, J., *Can. J. Chem.*, **44**, 461 (1966)
225. RAMARADHYA, J. M. and FREEMAN, G. R., *Can. J. Chem.*, **39**, 1769 (1961)
226. BLACKFORD, J. and DYNE, P. J., *Can. J. Chem.*, **42**, 1165 (1964)
227. THEARD, L. M., *J. Phys. Chem.*, **69**, 3292 (1965)
228. JONES, K. H., *J. Phys. Chem.*, **71**, 709 (1967)
229. HORIKIRI, S., SAIGUSA, T. and MORITA, K., *Bull. Inst. Chem. Res. Kyoto Univ.*,
     45, 69 (1967)
230. SHERMAN, W. V. *J. Chem. Soc.*, 5402 (1965)
231. SHERMAN, W. V., *Nature*, **210**, 1285 (1966)
232. RODER, M., *Acta Chim. Acad. Sci. Hung.*, **86**, 211 (1975)
232A. TANAKA, T., MIYAZAKI, T. and KURI, Z., *Int. J. Radiat. Phys. Chem.*, **8**, 645
      (1976)
232B. MIYAZAKI, T., TANAKA. T. and KURI, Z., *Int. J. Radiat. Phys. Chem.*, **7**, 627
      (1975)
232C. MIYAZAKI, T., *Int. J. Radiat. Phys. Chem.*, **8**, 57 (1976)
233. WAKEFORD, B. R. and FREEMAN, G. R., *J. Phys. Chem.*, **68**, 2992 (1964)
234. HOIGNÉ, J. and GÄUM NN, T., *Helv. Chim. Acta*, **47**, 590 (1964)

30*

235. RODER, M., in *Proc. 3rd Tihany Symp. Radiat. Chem.* (Eds DOBÓ, J. and HEDVIG, P.), Budapest, 1972. Vol. 1, p. 323
236. CHAPIRO, A., *J. Chem. Phys.*, **47**, 747, 764 (1950)
237. PROKSCH, E., MAYER, F. and KRENMAYR, P., *Studienges. Atomenergie*, Ber. No. 2291, Ch-135/74, 1974
238. TERENIN, A. N., *Acta Physicochim.*, **18**, 210 (1943)
239. MOLIN, YU. N., TCHKHEIDZE, I. I., KAPLAN, E. P., BUBEN, N. YA. and VOE-VODSKI, V. V., *Kinet. Katal.*, **3**, 674 (1962)
240. PROKSCH, E., *Atomkernenergie*, **13**, 294 (1968)
241. KULES, I. and GÉCS, M., *Proc. 2nd Tihany Symp. Radiat. Chem.* (Eds DOBÓ, J. and HEDVIG, P.), Akadémiai Kiadó, Budapest 1967. p. 319
242. BAGDASARYAN, KH. S., IZRAILEVICH, N. S. and KRONGAUZ, V. A., *Dokl. Akad. Nauk SSSR*, **141**, 887 (1961)
243. BURTON, M., GORDON, S., and HENTZ, R. J., *J. Chim. Phys.*, **48**, 190 (1951)
244. GURNEE, E. F. and MAGEE, J. L. *J. Chem. Phys.*, **26**, 1237 (1957)
245. FUNABASHI, K. and MAGEE, J. L., *J. Chem. Phys.*, **45**, 1851 (1966)
246. McCURBIN, W. L. and BURNEY, I. D. C., *J. Chem. Phys.*, **43**, 983 (1965)
247. ZEMAN, A. and HEUSINGER, H., *J. Phys. Chem.*, **70**, 3374 (1966)
248. KUKLIN, YU. S. and FIRSOVA, L. P., *Vestn. Mosk. Univ. Khim.* No. **1**, 117 (1967)
249. ALEXANDER, P. and CHARLESBY, A., *Nature*, **173**, 578 (1954)
250. HOFER, H. and HEUSINGER, H., *Z. Phys. Chem.*, **69**, 47 (1970)
251. KISS, I., PINTÉR, K. and RODER, M., *Proc. 2nd Tihany Symp. Radiat. Chem.* (Eds DOBÓ, J. and HEDVIG, P.), Akadémiai Kiadó, Budapest, 1967, p. 391
252. HENTZ, R. R. and BURTON, M., *J. Am. Chem. Soc.*, **73**, 532 (1951)
253. BADGER, B. and BROCKLEHURST, B., *Nature*, **219**, 263 (1968)
254. BURNS, W. G., CUNDALL, R. B., GRIEFFITHS, P. A. and MARSH, W. R., *Trans. Faraday Soc.*, **64**, 129 (1968)
255. BIRKS, J. B., CHRISTOPHOROU, L. G. and HUEBNER, R. H., *Nature*, **217**, 809 (1968)
256. WALKER, D. C. and WALLACE, S. C., *Chem. Phys. Lett.*, **6**, 111 (1970)
257. BURTON, M., *Disc. Faraday Soc.*, **36**, 7 (1963)
258. BLAIR, J. M., SMITH, D. B. and PENGILLY, B. W., *J. Chem. Soc.*, **3174** (1959)
259. WALLING, C., *Free Radicals in Solutions*, Wiley, New York, 1957, p. 474
260. HENTZ, R. R. and BURTON, M., *J. Am. Chem. Soc.*, **73**, 532 (1951)
261. BOYD, A. W. and TOMLINSON, M., *Can. J. Chem.* **46**, 3129 (1968)
262. THOMAS, J. K., JOHNSON, K., KLIPPERT, T. and LOWERS, R., *J. Chem. Phys.*, **48**, 1608 (1968)
263. SAUER, M. C., *Int. J. Radiat. Phys. Chem.* **8**, 33 (1976)

7. PAST RESULTS AND FUTURE DEVELOPMENTS

Research in radiation chemistry has produced an enormous amount of experimental data during the past two decades. Radiolysis studies of hydrocarbons have achieved an outstanding position in spite of an apparent decline experienced in recent years.

Classical radiation chemistry of the 50's and the 60's investigated the relations between molecular structure and radiolysis through the analysis of the end-products formed upon irradiation. The formation of products was explained using experiences from studies on thermal cracking, mass spectrometry and photochemistry (mainly mercury-sensitized photolysis) of the given compound. At the early stages of radiation chemistry, chemists attempted to elucidate the mechanism of radiolysis mainly by using scavengers and studying mixtures of hydrocarbons.

Recent advances have been made mostly by means of sophisticated experimental techniques, based mainly on pulsed beams of radiation. It is hoped that many questions originating still in traditional radiation chemistry may now be answered by the use of these new methods. The persistence of interest seems guaranteed by the fact that many outstanding scientists, well reputed for their contributions to traditional radiation chemistry, are active today in these newly conquered fields of radiation chemistry.

The intermediates formed by pulses of radiation are now investigated by observing their optical spectra, ESR signals, or conductivity. The following of the increase or decay of some species is a direct method for obtaining kinetic data. Up-to-date studies on radiation chemistry are carried out using pulsed radiations along with stationary sources, and many laboratories are equipped with vacuum-UV photolysis and laser apparatuses, too. These methods enable radiation chemists to differentiate between processes having ionic or excited intermediates. Ionic reactions are studied by the numerous new methods of mass spectrometry including e.g. ion-cyclotron resonance spectrometry.

The development of this high degree of sophistication in recent experimental techniques have given, however, birth to a new problem. The subjects of studies often seem to be determined by the tools being at hand, although the reverse would, sometimes, be more fruitful. Admittedly, this kind of distortion is frequently encountered in other sciences, as well.

The mosaic-like picture of the radiation chemistry in the 70's is mainly due to this tendency. There are certain subjects on which great efforts have been concentrated while others are seemingly and dissapprovably abandoned. For example, the radiolysis of methane and cyclohexane is far better understood, than that of bridged-ring cycloalkanes, cycloalkenes and alkylaromatic compounds.

When looking into this book, it may appear that radiation chemistry is a large collection of numerical data with very few interrelations, and sometimes with quite a lot of contradictions. Although this may seem to be a more or less common feature of encyclopedic books, it is to be noted that a comprehensive theory of this field is actually lacking, and it is questionable whether such a theory can be developed in the near future. A general and firm theory should describe radiation chemical processes on the basis of the primary physical and physicochemical facts to explain product formation, thus permitting to make predictions, not only qualitative, but also quantitative ones. From a survey of the literature it is apparent that still many more data are required for the development of a reliable theory; an inspection of the recent researches made in radiation chemistry shows great promise that this work is on its way.

The authors wished to provide useful information to experts working in radiation chemistry and related fields. The whole material of hydrocarbon radiation chemistry certainly could not be squeezed in a book; the aim of the work has been to give the reader a relatively complete and comprehensive overview of the field, many times by means of numerical data. A wide variety of phenomena discovered by radiation chemistry have been discussed which are worth studying in more detail, even if further work and more advanced techniques are probably needed to get a deeper understanding of the relations between the structure and reactivity of hydrocarbons.

When evaluating the achievements of radiation chemistry, it should be mentioned in the first place that the application of high-energy radiation rendered possible to produce great amounts of charged species and to study their transport and decay processes under such PVT conditions (e.g. in condensed phases) which had not been possible by other methods of physical chemistry. Earlier, the energy needed for the production of ions had to be provided by the heat of solvation, thus hindering investigations in nonpolar solvents. The studying of the behaviour of electrons in dense fluids started with the application of ionizing radiation, and has become a popular subject to this date, developing in connection with solid-state physics. The contribution of radiation chemistry to our better understanding of free radical reactions was given by the possibility of studying reaction of this kind both in the gas- and liquid phases, and at much lower temperatures than those used in thermal initiation. A number of radical reactions (e.g. chain reactions) have become known in detail, which could be hardly or not at all observed in thermal or photochemical experiments, owing to the much greater selectivity of these latter methods.

A criticism frequently encountered is that the primary energy source in radiation chemistry is enormously more powerful than the energy required for chemical processes (for instance, decomposition or isomerization), i.e. one cracks nuts with a steam hammer. Comparative studies on the radiolysis of hydrocarbons detailed here indicate that 'surprisingly', the nuts yet crack the proper way, i.e. despite energy differences of many orders of magnitude between nuclear and chemical reactions, the latter are selective processes controlled essentially by the structure of the compound irradiated.

In most molecules there are easily breakable bonds, the decomposition of which proceeds at much higher rates than decompositions at other parts of the molecule. These easily breakable linkages often coincide with the weak points found in thermal decomposition, however, there are quite a few notable exceptions, such as some bonds particularly loosened in hydrocarbon molecular ions, and others in certain neutral electronically excited states. The weak points in alkanes are the branchings, in alkenes the bonds in allylic position to the π-bonds. There is a competitive effect: the easily breakable places by their own enhanced rates of decomposition 'protect' the other parts of the same molecule from decomposition, or an other molecule in a mixture of hydrocarbons.

The energy transferred by irradiation induces first those chemical changes that correspond to the specific features of the irradiated molecules: e.g. the decomposition of strained cyclic hydrocarbons with small rings commences with the destruction of the ring, whereas in large saturated rings, where van der Waals repulsion between hydrogen atoms is significant, the main reaction is dehydrogenation. This observation points to possible use of radiation chemical research in stereochemistry.

It is now more or less certain that predictions of the early 60's about the direct industrial utilization of radiation chemical reactions of hydrocarbons have proved to be false; not only radiation-thermal cracking or isomerization, but also preparative radiation chemical halogenation or sulphochlorination (which are outside the scope of this book) have failed to come into use on a commercial scale.

At the same time, radiation chemistry may have a limited practical importance in the laboratory-scale production of hydrocarbon samples, e.g. reference substances for gas chromatography can be prepared by the radiolysis of appropriately selected, readily available compounds in a more economical way than by multistep organic syntheses.

Definite industrial results have been achieved in the field of polymers: radiation cross-linking of polymers is at present (apart from the radiation sterilization of medical products) the radiation technology of the greatest commercial value.

From a theoretical point of view, radiation chemistry provides a powerful tool to get a better understanding of various conventional chemical processes or the structures of some hydrocarbons on the basis of the correlations between chemical structure and radiation chemical processes. These discoveries may help in the selection or development of radiation-resistant materials, which have become more and more important with the advance of nuclear technology. In an indirect way, the results from hydrocarbon radiation chemistry may be of assistance in the perfection of biological radiation protecting systems. Extensive studies carried out recently on the transport of charge carriers in hydrocarbon fluids and solids are also of primary interest for a better understanding of electric insulation.

G. F.

SUBJECT INDEX